EARTH, TIME, AND LIFE

ABOUT THE COVER IMAGE

If you will take this book and open it out flat, the cover image will be fully displayed. The image was acquired by the Landsat 4 satellite in 1984 as it orbited the Earth at 438 miles (705 kilometers) in a sun-synchronous, north-to-south, near-polar orbit. The satellite crosses the equator at approximately 9:45 a.m. Eastern Standard time and provides repeat coverage over any point on the Earth every 16 days at midlatitudes.

The area covered by this image is approximately 13,000 square miles (31,450 square kilometers); the ground scale is 1:500,000. The TM scanner allows any item wider than 33 yards (30 meters) to be observed in seven spectral bands in the visible, near-infrared, shortwave-infrared, and thermal regions. The mirrors on the Landsat 4 satellite reflect energy from the Earth onto 100 silicon detectors which convert the sensed-scene energy into a digital data stream; at the NASA Goddard Space Flight Center in Greenbelt, Maryland this data stream is converted into one of 255 relative reflectance values in each of seven spectral ranges for each 30- by 30-meter picture element. Color-composite imagery as seen on the cover is created by assigning red, green, and blue colors to each picture element for each of three spectral bands.

This color-composite image is produced by superimposing data from bands 1, 4, and 5. The end result is something approaching natural colors, though the filtration causes the shades of blue in the image to be a function of soil moisture; in this winter scene the intense blue on the upper parts of some mountain ranges is snow.

This Landsat scene is 115 by 106 miles (185 by 170 kilometers) in size. In the southwest (left lower) corner is Searles Lake, a whitish-to-bluish (as imaged) alkali lake in a basin west of the Panamint Range, which wears a brilliant aquamarine blue (as imaged) snowcap. Running from northwest to southeast across the left center of the image is Death Valley, a downdropped block whose surface contains the lowest elevation in North America. The central mountainous mass paralleling Death Valley is the Amargosa Range. To the north (top) are some green circular areas of center pivot irrigation; to the east is Amargosa, California, a patch of rectangular green admist the whitish background of the Amargosa Desert. Still farther northeast are the Spring Mountains, their snow cover rendered a vivid blue-green by filtration. Still farther northeast, on the upper right-hand margin, some polyhedrons of white with rectangular green splotches mark Las Vegas, Nevada.

The scenery is this part of the southwestern United States is dominated by uplifted mountain ranges cut by downdropped desert basins, creating the Basin and Range area (see Chapter Twenty-One), It is an area of faulting (see Chapter Seven), volcanism (see Chapter Three), and extension of the continental crust. Its desert landforms (see Chapter Thirteen) are marked by mass wasting (see Chapter Eleven), stream erosion (see Chapter Twelve) and deposition. It can support few people because groundwater is limited (see Chapter Fifteen); it is lonely, stark, wild country, (Image courtesy of Pat Chavez, U.S. Geological Survey Field Center and Flagstaff Image Processing Facility, Flagstaff, Arizona.)

EARTH, TIME, AND LIFE:

AN INTRODUCTION TO PHYSICAL AND HISTORICAL GEOLOGY

SECOND EDITION

CHARLES W. BARNES
Northern Arizona University

John Wiley & Sons

New York • Chichester • Brisbane • Toronto • Singapore

Cover photo by *Flagstaff* Field Center, U.S. Geological Survey
Text and cover design by Michael Jung

Library of Congress Cataloging in Publication Data:

Barnes, Charles W., 1934–
Earth, Time, and Life. 2nd Edition

Includes bibliographies and index.
1. Geology. I. Title

QE28.B24 1988   550   88-73
ISBN 0-471-82598-0

Printed in the United States of America

10 9 8 7 6 5 4 3 2 1

To Lewis Cline, Bob Dott, George Huffman,
Rolland Reid, Stan Tyler, and their colleagues,
with respect and affection

Preface

This book introduces the fundamental concepts of physical and historical geology to beginning college students who have no background in college-level sciences, including geology. For most students, the course in which they use this book will be their only exposure to geology. Therefore this text discusses the whole fabric of geology with minimal use of technical language.

The treatment of the text is based on a major recommendation from a committee created in 1940 by the Geological Society of America to offer suggestions for the improvement of geologic education. The committee's central mandate remains valid today— to offer "only those inferences . . . for which the essential observational data and the logical steps leading to the inference have also been presented." Or as Charles T. Spradling put it, "Knowledge consists in understanding the evidence that establishes the fact, not in the belief that it is a fact."

This book frequently deals with the history of geologic ideas, hoping to show students **how** scientific ideas came to be conceived, were challenged, changed, and then tentatively accepted. The vital connective tissue in understanding geology (and any science) is its history.

Much harm is done in presenting science as eternal truth freshly fallen from the skies and needing to be memorized; the science in this book is an intensely human activity shaped and limited by the world around it. Geology is the result of high human adventure and intense human curiosity, full of fruitful error and self-correcting over time. This science reflects our tremendous human potential to play with ideas and symbols and to use our imagination and our intellect to deny our insignificance.

The book is designed to be used in a variety of formats to serve diverse introductory courses. Each chapter is relatively freestanding, with frequent references to material in other sections. An extensive summary reviews its contents. All boldfaced words in the chapter are collected and listed as key words. A series of thought questions, many of which have no single, obvious answer, follow the key words. These words are defined in the glossary. The suggested readings, which have been brought up to date through 1987, emphasize material easily available even in smaller libraries. Appendices on mineral and rock identification, chemical principles in geology, topographic and geologic maps, measurement conversion, and classification of life-forms are provided at the end of the book. The geologic time scale used is simplified from that of the Decade of North American Geology.

The illustrations have all been conceived exclusively for this book. More than half are new in this edition, and substantial changes were made on many others. They run the gamut from computer-enhanced images of planetary bodies to images taken from the last four centuries of geologic publication.

These images display the historical sweep of our widening curiosity about the earth.

The careful development of a second edition requires attention to feedback from those who were kind enough to use and comment on the first edition. In response to this feedback, the book has been reorganized, creating six additional chapters which either are wholly new or came from parts of the first edition. The entire book has been rewritten; little of the first edition remains. Chapter organization has been much improved, and the book has been updated throughout. New topics are suspect terranes, a review of K − T extinction and its controversial causes, and ultra-SEM crystallography; coverage of the origins of acid precipitation has been expanded. A modernized outline of plate tectonics includes pulsation tectonics and deep-mantle slab penetration theories, updated sections on extensional terranes, deep-earth tomography, new data on earth heat flow, and many other controversial theories that are in developmental stages.

The effort at wholesale revision has been made richer by the assistance of many kind people. I am particularly indebted to Marjorie Dalecheck for her assistance in summer work in the photo archives of the U.S. Geological Survey and to Marcia Goodman for her able support in two summers' work at the History of Science Collections in the University of Oklahoma Libraries. Discussions there with Ken Taylor, Duane Roller, and Dave Kitts were particularly fruitful. Others who lent direct support include Gil Hill, Patrick Petroleum; Archie Hood, Shell Development Company; D. A. Brown, Australian National University; Patricia Hill, geophysicist with the USGS; Bob Houston, University of Wyoming; Michael Collier, geologist–photographer–M.D., Grand Junction, Colorado; Bob Dott, the late Lewis Cline and the late Stan Tyler, all of the University of Wisconsin; Mike Carr, Gerry Sharber, and Baerbel Lucchitta of the Astrogeology Branch of the USGS; Gordon Swann and Ivo Lucchitta of the Flagstaff Field Center, USGS; Ed Wolfe of the USGS Hawaiian Volcano Observatory; George Ulrich of the USGS National Center, Reston, Virginia; Rolland R. Reid, University of Idaho and Homestake Mining Company; the late Elso S. Barghoorn, Harvard University; Trevor Ford, University of Leicester; E. S. Leskowitz at the United Nations; and my colleagues at my home university have all made recognizable contributions. These kind souls are not reesponsible for misstatements or obfuscations; those remain the responsibility of the author.

All who have written for publication know how vital the skills of editors are. My debt to Professor Rolfe Erickson, geologic colleague at Sonoma State University, is incalculable; his pungent, trenchant technical editing much improved the manuscript. The editorial skills of Nancy Terry and Priscilla Todd tremendously sharpened and improved my exposition; I marvel at their skill and professionalism. The late Don Deneck, Bob McConnin, and Clifford Mills served as successive managing editors and added their highly professional expertise in bringing all this together; their artisitc, editorial, and production colleagues at Wiley are tremendously talented.

Finally, no effort of this magnitude would have been possible without the forbearance and enthusiasm of my family. We have shared jointly in both the exhaustion and exhilaration as the preparation of this book enveloped our lives.

CHARLES W. BARNES
Flagstaff, Arizona
December, 1987

Note to Students

Open before you is the result of the combined efforts and talents of several hundred people—editors, artists, compositors, printers, geologists of many kinds, the author, photographers, illustrators, technical editors, archivists, and pilots. This book reflects a tiny fraction of the knowledge gleaned through centuries of scientific thought. A few hints about the way the book is organized and an outline of its underlying point of view will make your use of it more fruitful.

This is a book about ideas. The ideas are, of course, the fundamental concepts that establish the intellectual base for the science of geology (from the Greek *geo*, meaning earth, and *logos*, meaning study of). Today the word *geology* has taken a much larger meaning, as geologists study a wide range of planetary bodies in addition to Earth.

It is also a book about *how* scientific ideas are formulated, how they are challenged, how they are changed, and how they are accepted. Some view science as magic, and a few others view some of its results as both wrong and immoral. This book describes a science that springs quite naturally from human curiosity and scientific knowledge that is ultimately tentative.

Ultimate truth is elusive, as new understandings emerge from challenging current conceptions. Each discovered error is fruitful, fomenting still more study until periods of broad agreement occur; these plateaus are broken by a new insight, a new challenge, new pieces of data, and the process continues. Over a period of time, scientific knowledge is ultimately self-correcting.

The knowledge in this text is broken into five parts: Introduction: Earth Materials and Processes, The Restless Earth, Understanding Landscapes, Earth Resources, and Geologic Time and a Brief Earth History. The whole is a story briefly told, because useful books must hold to reasonable lengths. No geologist, including the author, will find all their favorite subjects here. The book is constructed so that each teacher can develop a special way of sharing the earth with you.

If you should want to read further, each chapter provides up-to-date references that should be available in your college or university library. With these you can extend your knowledge well beyond what is available here. Studying these references and their bibliographies will carry you as far as you wish to go.

If after working with this book you have comments, questions, or suggestions, please write to me in care of the Department of Geology, Northern Arizona University, Flagstaff, AZ 86011, USA. I would enjoy hearing from you, and I will certainly respond.

C.W.B.

Contents

CONTENTS   XV ◀

PART

1

Introduction: Earth Materials and Processes

As this whimsical cartoon suggests, early humans had to study the earth to use it successfully; they were the first geologists. (© 1984 by Sidney Harris - American Scientist Magazine, reprinted by permission.)

CHAPTER *one*

Introduction

Science is built up with facts, as a house is with stones.
But a collection of facts is no more a science than a heap of stones is a house.

JULES HENRI POINCARÉ (1854–1912)

We live in a world transformed by science and the technologies that are its offspring. To say that ours is the age of science is simply to state the obvious.

What is less obvious is where science comes from, what fundamental assumptions underlie scientific knowledge, and how that scientific knowledge is both gained and tested. Explaining how science works, outlining the origins of science, reviewing the interrelationships among the sciences, and providing an introduction to the science called geology are the goals of this chapter.

WHAT IS SCIENCE?

Science is an expression of human curiosity—the desire to know. Science is the pursuit of knowledge for the sake of sheer curiosity. It serves no practical purpose to ask the overtly simple questions—why is the sky blue or why are large animals so rare? Yet there have always been those who ask such apparently useless questions just because they want to know. It is that search for seemingly useless knowledge that has so transformed our world.

Science seeks and organizes observable information into testable and predictive explanations of the relation between effect and cause. A mature science has developed a framework of theories, that organizes current knowledge into explanations and predictions having wide application. Science insists that our world can be understood.

Science is one expression of our *humanity*. Other expressions include our desire to create beauty, to love and be loved, to belong, to defend what we cherish, and to seek a purpose and meaning in life.

Science is also a body of knowledge about the physical world. That body of knowledge is formed from the analysis of sensed data. That analysis results from the application of logical reasoning as limited by our past experience, our intuition, and our current understanding.

THE REQUIREMENTS OF SCIENCE

Science requires *observable information*. Observation provides the facts of science, but not the explanations. Gaining information is an early and funda-

mental activity in science, but unrelated phenomena yield no meaning. As the quotation that introduced this chapter suggests . . . a collection of facts is no more a science than a heap of stones is a house. Information must first be *organized* or classified if it is to be recast into explanations.

Science insists on *organized* information. Classification is the "halfway house" in any science, accurately reflecting our current grasp of the relationships we see. Classification limits all that follows, for our search for causal relationships is limited by how we first organize what we're attempting to explain. The history of scientific effort is full of breakthroughs that were made when old facts were viewed from the perspective of a new organizing principle. As an example, consider Nicolaus Steno who classified objects found within rocks in a wholly new way, and thereby provided the first explanation of the origin of fossils (see the section Establishment of the Field).

Science requires that explanations be *testable*. If a proposed interpretation cannot be repeatedly and rigorously tested, that solution is not science. This principle illustrates why science so utterly rejects so-called "creation science." That set of explanations seeks to interpret by appealing to supernatural, one-time events—an accounting that cannot be tested for either its truth or its falsity. It is also the requirement for testability of propositions that gives science its "unfinished look." Science is a process of constant seeking, not a set of finished truths.

Science claims *predictability*. One of the surest tests of a proposed explanation is its reliability in predicting what *will* occur. The science of weather forecasting places its predictions on the line each day; its accuracy is a measure of our understanding of the physics of the atmosphere. As explanations move closer to truth, they become increasingly powerful predictive tools. Claims of predictability allow us not only to explain the present but also to say what has happened in the past (to *postdict*). Postdiction, the ability to infer events that happened in the past, is one form of prediction common to all sciences, greatly expanding their explanatory power.

THE PURPOSES OF SCIENCE

The central purpose of science is to *explain*. The certainty attached to our explanations is described by the terminology employed.

1. A **hypothesis** is a tentative and speculative explanation advanced for testing. A hypothesis is often expressed in expanded form, with numerous qualifiers, and is commonly modified or completely discarded.

2. A **theory** is a preliminary explanation now well tested and generally agreed upon as an accurate statement of current understanding in the field. It is subject to modification as new tests or new facts arise.

3. A scientific **law** or **principle** embodies the highest level of confidence, based on having survived, and having been modified by, numerous episodes of rigorous testing. A law or principle provides broad explanatory and predictive powers and is commonly expressed in compact, abstract (quantitative) forms. A principle is believed to represent truth, still subject to further modifications by experience.

A hypothesis is modified into a theory and a theory into a principle over a period of time. The vast majority of hypotheses and theories are discarded, with only a few evolving into explanations of much greater power. *Forming and testing explanations of natural phenomena are the fundamental activities of science.* Science requires that concepts developed with our mind fit the data observed with our senses.

Science searches for *causal* relationships. Science seeks repetitive relationships in nature. It explains those relationships in terms of cause and effect, process and product. In that context we discover a principle central to all scientific thought—the *assumption of rationality.*

▼

THE ORIGINS OF SCIENCE

▲

Scientific thought is founded on an assumption so basic that few people question it, or for that matter even think about it, yet the assumption is not only unproved, but unprovable. *Science springs from the underlying assumption that the universe is orderly, and that for every effect there is a cause, for every cause an effect.* This assumption is based on our experience, which repeatedly confirms that this assumption is not in error.

Science would insist that our universe is a system operating according to fundamental, knowable principles. Our universe is not capricious. We cannot imagine a science without this implicit idea—that the universe is rational, *and therefore can be understood.*

If, on the other hand, our universe were irrational—that is, if things just happened for no particular reason—our task of understanding is made hopeless. We then live in a universe so inscrutable, so full of endless possibilities, that all we could do is to bow before the whims of nature and live in a world devoid of intelligible meaning.

In a world where absolutely anything is possible at anytime, how could we plan? How could we defend against the unknown? How could we find our place in a universe where nothing made sense, and destruction could just as easily follow from one plan of action as another?

PRESCIENTIFIC CULTURES

In both ancient and modern cultures that have no science, the answers are the same. To explain and to defend ourselves, we create myths and we practice magic. Myths help us to explain the unexplainable, and magic may seem to help us to control the uncontrollable. In each of these, and in many other ways, we seek to explain and control our world—basic needs if our lives are to have some security and meaning.

The problem with myth and with magic, with its incantations and rituals, was that it didn't always seem to work—we might say its predictive power was near zero. When people discovered what *did* control something as crucial as planting the year's food crops—careful storage of seeds, careful treatment of the soil—magic was slowly replaced by a primitive agricultural technology, a forerunner to science.

Technology put to use what was known, whereas religion attempted to protect against disaster by asking the unknown(s) to work on our side. Earliest technologies and religion were both intensely practical, dealing with the issues of survival. Only much later, as agriculture flourished and civilizations prospered, was there time for leisure and contemplation. Scientists appeared beside the technologists, and theologians alongside the tribal priests.

The distinction between science and technology remains today. Science attempts to *explain* natural phenomena, while technology attempts to *control* or modify natural phenomena. In order to control nature, one must first understand it. But how to understand it?

THE GREEK CONTRIBUTION BEFORE ARISTOTLE

Science, as distinct from technology, began in the Greek (Achaian) culture in Ionia, in what is now westernmost Turkey 800 years before the birth of Christ. These early Greeks assumed, based on religious grounds, that nature "played fair." They saw the earth as comprehensible, and the forces of nature as understandable. The path to that understanding lay in applying pure logic to fundamental assumptions about the earth.

Understanding, to the Greeks, began with a statement of undoubted truth, obvious and accepted by all, termed an **axiom.** From each axiom, a set of *inevitable* consequences logically followed. This method of reasoning, called **deduction**, will be discussed further. Deduction is a wholly intellectual activity, little tied to the observable world. Early Greek (pre-Aristotelian) science was an *intellectual* activity, not an *observational* one.

If derived consequences "made sense," *in intellectual terms*, the explanation or hypothesis was true. Listed are a series of hypotheses from Thales of Miletus (640–546 B.C.); watch the ideas grow in power as the logic unfolds:

1. All things are full of gods. (This was the *axiom.*)
2. Therefore, all inanimate things are immortal, like the gods.
3. Therefore, all things persist, though they may change from one form to another.
4. Therefore, there is one substance, from which everything is made.
5. Therefore, there is a principle that governs all things.

Heraclitus (c. 540 B.C.–480 B.C.) made one critical addition to Thales's fourth hypothesis—that there is one substance, of which all is made. Heraclitus suggested that not only is there only one substance, which we today would call **matter**, but also that matter is forever in the process of change. As he phrased it, *Nothing endures but change.* Thus, we should study not only matter but also *changes* in matter. The assumption that all of nature is formed from a unified substance, which is constantly in a process of orderly change, is at the heart of modern science.

Consider another example of how the Ionian Greeks saw science as solely an intellectual activity, one in which observation, before Aristotle, played a limited role. Question: what was the shape of the universe?

To the Ionians, the sphere was the most beautiful shape of all, hence the universe *must* be spherical. Since our earth, for the Greeks, was at the center of the universe, it too must be spherical. Yet, everyday observation told the early Greeks that the earth was *flat.* Clearly, then, observation was not to be trusted.

Observation, however, was *not* a part of earliest Greek science. That investigative tool was to come largely from Aristotle (384–322 B.C.), a careful and meticulous observer, whose insistence on observation *combined* with logic was to found *modern* science.

Modern science still agrees with Thales that all of nature is formed from one substance. Modern science agrees with Heraclitus that that one substance is forever in the process of change. But modern science, unlike Greek science, is highly *inductive*, using repeated observation as its major source of inspiration and proof.

SCIENCE AFTER ARISTOTLE

Early *assumptions* about the sphericity of the earth later guided the search for *observational* evidence to "prove" that the earth was spherical. Watch Aristotle shift early Greek science by *adding* observation as a check on the beauty and harmony of an idea. How could observation "prove" the earth was spherical?

Aristotle's first proof was that the shadow the earth casts on the moon during a lunar eclipse is always curved. If the earth were any shape but spherical, then at some orientation of the earth, the shadow wouldn't be curved. Aristotle also noted that stars rise sooner for people in the east than in the west, and travelers heading south see previously invisible stars appearing above the southern horizon. Both observations would be impossible if the earth is flat; instead the surface of the earth must be curved from north to south and from east to west. Only the surface of a sphere is curved in all directions, hence the earth is spherical.

Observation provided the *expected* answer, but it didn't raise the initial question. Rather, the initial question arose from philosophical concepts of what was beautiful, harmonious, and simple.

Today, scientists propose explanations that are tested against the observable world and make observations which refine their conceptual models. The most elegant conceptual models are those that find harmony, reveal beauty, and possess simplicity; our debt to the Greeks is enormous.

THE LOGIC OF SCIENCE

Another of our debts to the Greeks comes from their study of the principles of reasoning, termed *logic*. A review of two of these principles may illustrate the extent of that debt.

Scientists analyze the observable world using two fundamental and interrelated methods of argument. These methods are *inductive* and *deductive* reasoning.

Inductive Reasoning

Inductive reasoning draws general conclusions from particular instances. Inductive reasoning starts with observations and derives general explanations of the observations from them.

As an example, if we observe that every morning the sun comes up, after some length of time, we may infer that the sun will rise tomorrow morning—that is, we can now predict, based on experience. Reasoning from our experience with a few days, we now reason that *all* days begin with sunrise. Let us now try an *inductive leap,* in which we move from observation of detail to its general explanation. If we now want to explain *all* sunrises, we may propose that the earth rotates on its axis *or* that the earth is still and the sun revolves about us; *either* leap satisfies the observations we have made.

We have explained what we see; there may well be other cases, not yet experienced by us, that would knock our first explanation into a cocked hat. Hence, explanations reached by induction may enjoy a high level of probability, but they are *not* certainty—other cases, other experiments, other experiences, all may change our explanation. No generalization stands unless it repeatedly passes the tests of still more observations and experiments.

Induction commonly leans on *empirical* evidence, that is, evidence gained from observation and experience. Induction is the basic process by which we learn, *and is the fundamental basis of modern science.* Information from the natural world comes to us as impressions from our senses, as the model scientist meticulously gathers more and more data, without leaping to premature inductive explanations.

But induction has its limits. The natural world is so full of data that no one can hope to make sense of it through experience and experiment alone. To bring the experimental and empirical phase, temporarily, to a close, scientists infer hypotheses and test the consequences of those hypotheses by deduction as well as by continued critical observation.

Deductive Reasoning

Deductive reasoning begins with a demonstrable or accepted general principle as a basic premise, and draws a conclusion that *must* follow, by necessity, from the initial premise. Deduction commonly proceeds from a general principle or axiom to a particular conclusion. As we have seen, it was the *only* method of reasoning employed by early Greek natural philosophers. Once all the self-evident axioms had all their logical implications worked out, philosophic knowledge of the natural world seemed to be complete. There was apparently nothing more to be learned until 2000 years after Aristotle, when experiment and induction gave science its renaissance.

As often stated, induction reasons from the particular instance to the general explanation, while deductive reasoning proceeds from the general explanation to the particular instance. Conclusions reached by deduction are as certain as the initial premise, while conclusions reached by induction are commonly regarded as only probable.

No quantity of inductive testing renders a generalization absolutely valid. The inductive method cannot make generalizations about what cannot be observed. Induction, consequently, has its limits.

To deduce is to infer logical consequences from an initial premise, *known to be true.* If the Law of Gravity (two objects experience a force or acceleration between them that is proportional to their mass and inversely proportional to the distance between them) is taken as true, then the loose apple and the earth accelerate toward one another, and not in any other direction. The inference that apples fall into the earth is an absolutely *inescapable* conclusion based on our knowledge that objects are attracted to one another. Deduction is an analytic process.

Deduction, too, has its limits. Sir Francis Bacon in a classic argument for empiricism as the basis for science put it well . . . *the subtlety of nature is vastly superior to that of argument.* Induction is the creative and synthetic side of science, while deduction is the logical and analytic side. Working together, these two methods have provided a series of checks and balances on each other, and a body of scientific knowledge whose application has revolutionized our world.

Scientists create theories not only to draw conclusions but also to organize and limit what is to be studied. Theories act like lenses, focusing study on topics germane to their investigation. If theory states that the gravitational force (G) between two objects (m_1 and m_2) is proportional to the product of their mass and inversely proportional to the square of the distance (*d*) between them, then we can quickly limit our study of the forces experienced by free-falling objects to those cases predicted by the theory.

WHAT IS A THEORY?

Our word, theory, comes from the Greek word *theoria,* which shares the root *thea* with our word, theater. In a theater we are observers and spectators, not directly involved in the action. Thus, to propose a theory is to contemplate from a distance, to observe, to be detached from nature—at least this was the view of pre-twentieth-century scientists.

Today, scientists increasingly see themselves as a part of nature, deeply affected by it in a cause and effect relation. Scientific knowledge comes increasingly from within the world it so profoundly affects. It also arises because of the explosive growth in the advancement of scientific knowledge.

▼

BOX 1–1

THE GROWTH OF SCIENCE

Science is the dominant human adventure that sets off our century from all earlier times. Yet, previous centuries had scientists, and they made fundamental discoveries. Theirs were the shoulders on which later giants stood.

But few people realize how rapid, in human terms, has been the growth of science, or rather, the sciences, for they now are many and varied. Thales of Miletus (ca. 600 B.C.) was the earliest western scientist in any classical sense; he was the first to ask the question: Of what is the universe made?

Twenty-two centuries later, in the time of Galileo Galilei (1564–1642), the first person to examine the skies with a telescope, or Nicolaus Steno (1638–1686), the first person to decipher the history within stratified rocks, the world's practicing scientists could have all found seats together in a large lecture hall. Science in the seventeenth century was a private enterprise—individuals labored alone, their work little known outside of a very small circle of colleagues.

Science as a major human enterprise really began late in the eighteenth century, but even then, science was the province of a small, highly educated elite. Provoked by philosophic, political, scientific, and industrial revolutions, science became institutionalized in the late eighteenth century, and began an explosive period of growth that continues, unabated, today. From 1750 to 1950, the doubling time, that period of time required for the number of scientists to double, was only 15 years.

Yet, even by the end of the nineteenth century, most of the world's scientists could still know one another. Now, that is, quite simply, a physical impossibility (Figure 1–1).

There are more geologists at an international meeting today than there were scientists of all persuasions in the entire world only 80 years ago. About 5 percent of all people who have ever lived are alive today, but nearly 90 percent of all scientists who have ever lived are now living.

▲

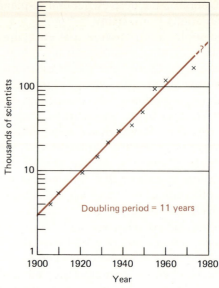

FIGURE 1–1

Rate of increase in number of American scientists since 1900. (Source, American Men and Women of Science, first edition published in 1906, fourteenth edition published in 1979. Permission of the R. R. Bowker Company, a division of Reed Holdings, Ltd., copyright 1979 Xerox Corporation.)

THE ROLES OF THEORIES

A theory defines truth, where truth is simply an explanation of current knowledge. Defined in this way, truth changes in time, as new ideas and new phenomena cause the theory to be readjusted to them. Thus theories are never ultimate or final truth, but rather are increasingly closer approximations of that ultimate truth. Empirical or observational statements—life forms change in time—provide absolute certainty without explanation. Theoretical statements—life forms evolve owing to natural selection—provide explanation without absolute certainty.

A theory *organizes* what is known. A theory relates information in logical and rational ways, defining the boundaries of what is known from that yet unknown. Theories are structured ideas.

A theory *limits possibilities*. Darwin's theory of evolution by natural selection requires that the environment limits survivability of less than ideally adapted individuals. Consequently, we don't swim in crocodile-infested tropical rivers.

A theory must be *able to be proved false*. German astronomer Johannes Kepler (1571–1630) proposed an early theory of planetary orbits that had each planet moving in a perfectly circular orbit—a concept that was beautiful, harmonious, and simple. But fellow Danish astronomer Tycho Brahe's (1546–1601) precise observations showed Kepler's original theory needed a subtle revision. Brahe's contribution to astronomy was to collect enormous quantities of high-precision data on the positions of heavenly bodies without any thought as to what the data might mean, a classic example of pure observation. Only when enough of Brahe's observational data were at hand could Kepler, by induction, falsify his own first theory and recognize that Brahe's data were, instead, consistent with elliptical orbits for the planets moving about the sun.

Often a theory predicts what observations are needed *to confirm or deny the theory.* Kepler's laws of planetary motion state that planets sweep across equal areas of their elliptical orbit in equal lengths of time (Figure 1–2). If that is so, then their orbital speed must vary from point to point—a prediction that *can* be checked. Making the necessary observations of orbital speed brilliantly confirmed the theory that suggested the observation.

Theories lead, in time, to *more general explanations* of varied phenomena. Each idea connects with others, eventually to form a more general theory or principle. Only after Kepler had formulated his laws of planetary motion could Newton formulate the still more generalized principles of gravitational attraction already described $G = \dfrac{m^1 \times m^2}{d^2}$ that explained *why* planets orbit another body in elliptical orbits. As Newton said, . . . *if I see farther than others, it is only because I have stood on the shoulders of giants.*

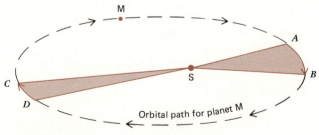

FIGURE 1–2

Sketch of a planet M in elliptical orbit about the sun S at one of the foci of the ellipse. The area of *ABS* and *CDS* are equal. Since the planet *M* sweeps out equal areas in equal time, its orbital speed must be greater when traveling from *A* to *B* than from *C* to *D*. The greater orbital velocity on the arc *AB* is caused by the acceleration produced by *M* being closer to the sun in that part of the ellipse; Newton recognized that the force experienced by planet *M* was an inverse function of the distance squared between the sun and *M*. It was that relation between force and distance that *caused* the orbit to be elliptical.

Theories are often *intuitive,* running counter to "common sense." One of the "giants" before Newton was Galileo Galilei, seventeenth century astronomer, whose advocacy of a heliocentric (sun-centered) universe brought him before the Inquisition. The Inquisition's threats of torture forced him to recant, and he spent the last eight years of his life under house arrest. In the *Two New Sciences,* Galileo analyzes the change in velocity of a freely falling body. There are two ways of measuring the observable acceleration:

1. Change in velocity for each successive foot of fall, or,

2. Change in velocity for each successive second of fall.

But which method of measurement is better? Galileo made an intuitive guess and chose the second method, believing *intuitively,* that a free-falling object is uniformly accelerated in time, but not in distance. On the basis of this unproved hypothesis, he selected a model that would yield the simplest and most elegant description of free fall, and then proceeded to test his hypothesis by deducing consequences that could be tested by experiment. By rolling balls down an inclined plane, for there was no way in Galileo's time to accurately time the rate of motion of a body in vertical free fall, he was able to show that the rate of fall of balls was *independent* of their weight. Galileo also demonstrated that each ball moved at a constantly accelerating velocity. Each second of fall brought an increase in velocity of 32 feet per second; Galileo had measured the acceleration caused by the attractive force between object and earth.

In this single incidence, Galileo displayed many facets of the scientific enterprise: the interplay of inductive and deductive reasoning, the creation of a hypothesis from our human imagination, a reckoning with an element of choice in how to proceed, the use of experiment and observation to check a consequence of deduction; and a refinement toward elegance and simplicity.

This was science at its best—intuitive, human, full of chance and uncertainty, each new discovery damning old truths to uneasy exits. Aristotle had taken it as a self-evident axiom that the speed of an object's fall was proportional to its weight. As Galileo's balls moved down the plane at a speed independent of their weight, Galileo destroyed Aristotelian physics in a single experiment.

The science that this book is about is that kind of science, intuitive, human, full of chance and uncertainty. The natural world uncovered by scientific effort is a world of strangeness and ambiguity; our human sense of perplexity at all this is one affirmation that science is working. But how are sciences born? How does a science develop?

DEVELOPMENT OF A SCIENCE

Looking backward at the history of any science, there are four interconnected stages in the development of a science:

1. Establishment of the field.
2. Classification of objects within the field.
3. Generalization to hypotheses.
4. Abstraction of the central ideas.

ESTABLISHMENT OF THE FIELD

Alfred North Whitehead (1861–1947), distinguished English mathematician and philosopher once suggested that it *takes an unusual mind to analyze the obvious.* The first stage in the development of any science is just that—the recognition that a field of inquiry exists where one had not existed before, perhaps because the answers all seemed obvious.

What could seem more obvious than a rock? Yet, when unusual minds starting asking questions about fossils *in* a rock, the science of geology edged away from myth and magic. This intriguing story is a classic example of how sciences develop, and where they come from.

The hero of our story is Nicolaus Steno, a Danish scientist of the seventeenth century, whose published scientific work was largely in anatomy (the glands of Steno, in our mouth, are named after him). In 1667, Steno dissected the head of a modern shark and compared modern sharks' teeth to curious objects, termed *glossipetrae* or "tongue stones," (Figure 1–3) which were found by the thousands *inside* local stratified rocks. Steno proposed that the *glossipetrae* or tongue stones were, in fact, sharks' teeth, because they were precisely identical in shape, internal form, and composition to modern sharks' teeth. Detailed similarity in form implied, in a rational earth, similarity in origin. We may call this the **principle of similarity.**

Since the prevailing view, which Steno did not question, was that the world had been formed in six days, a very few thousand years earlier, "tongue stones" were a puzzle. How did sharks' teeth, by the thousands, get *inside* rock?

FIGURE 1–3
Steno dissected the head of a modern shark, and showed that its teeth were identical to objects called "tongue stones," which had been found solidified inside sedimentary rocks. On that basis, he disputed the ancient belief that "tongue stones" had somehow *formed within* the rock as the rock formed. (From N. Steno, 1667, *Elementorum Myologiae*, courtesy of the History of Science Collections, University of Oklahoma Libraries.)

CLASSIFICATION OF OBJECTS

As we have previously suggested, classifications, other than alphabetic, are not neutral; they both reflect and direct our search for knowledge. The way we order objects clearly expresses the way we think are related.

Steno *chose*—there's the element of human choice again—to classify the *glossipetrae* as *solids within solids.* Solids enclosed within other solids became a *class* of objects whose origins could be studied. The problem of tongue stones had been posed in an entirely novel way—an unusual mind had begun to analyze the obvious, the vital prerequisite of great science.

Steno proposed that the tongue stones must have been in solid form earlier than the rocks that enclosed them, *because the teeth impressed their form upon the surrounding sediment.* Using logic, Steno had proposed a **principle of molding**: an object that impresses its shape upon another must be older than the surrounding material that was molded to its surface. If we find tire track impressions on a soft road, the road was there first; passage of the tire came later.

GENERALIZATION OF HYPOTHESES

Had Steno stopped short at recognizing that tongue stones were sharks' teeth, that the sharks' teeth were older than the surrounding sediment, and that solids within solids was a classification of the natural world worthy of study, his story would be a minor footnote in the history of geology. Ideas gain their power when they can be generalized from a few members of the class to all members—the inductive leap.

Assuming similarity in form proves similarity in origin, Steno reasoned that stratified rocks were formed as the deposits of rivers, lakes, and oceans, and the sharks' teeth within them had come from animals once swimming in those waters. Therefore—here comes the inductive leap—sedimentary rocks were not formed *with* the earth, but later, and incorporate the teeth and bones of the animals who lived during the time of their formation.

Since marine fossils are often found high in the mountains and distant from the sea, these fossils, which were solids before the sedimentary rock enclosing them, must have formed in a sea, which has now disappeared. Thus, the earth has kept on shifting its surface, for seas and land have changed places, and mountains have formed where there once was ocean floor.

Steno had used the principle of molding and the principle of similarity not only to classify objects, the second step in science, but also to reconstruct a part of the history of the earth's surface, an enormous extension of the power of two simple principles. Inference grew from inference, and a logical system of great explanatory power had been created.

ABSTRACTION OF THE IDEA

The power and usefulness of important ideas can be further extended by abstracting the central principle, the kernel of the entire idea. Many of these abstrac-

tions are most powerfully and conveniently expressed in mathematical terms as $F = ma$ (Force required is the product of mass and acceleration) or $E = mc^2$. (Energy equals the product of mass and the square of the speed of light), whereas others find their power in more qualitative statements. Steno's contribution, a recognition of the meaning of fossils, is in that latter category and falls out of still another extension of his principle of similarity.

Past processes, in or on the earth, cannot be observed. Only the products of the process remain. If we want to infer the ancient processes that formed an object, like tongue stones, we must find our key in the object itself. The most certain guide is detailed similarity with modern objects, like sharks' teeth, formed by processes we can directly observe. In this way, knowledge of the present allows us both to infer the past, to postdict, and infer the future, to predict.

Because Steno understood how a solid body could be enclosed within a solid through a process of nature—by recognizing tongue stones as sharks' teeth and the enclosing rock as formed from sediments—he was able to *order* geologic events in time. Steno had, by analyzing the obvious, established some fundamentals of the science of geology.

THE SCIENCE CALLED GEOLOGY

Steno's story conveys some of the concerns of the science called geology. Geology is the science that studies:

1. The solid earth, including origins of rocks, minerals, mountains, continents, ocean basins, and the forces that change them,

2. The history of life on earth, including the relation between organisms and the environment they modified,

3. The history of the earth through time, including methods of reconstructing both the past sequence and dates of events,

4. The earth for commodities of value, including gemstones, ground water, construction materials, energy resources, chemicals, and ores,

5. The earth for hazards, including volcanic eruptions, landslides, and areas of potential earthquakes and,

6. Other solid astronomic bodies in our solar system, including moons, asteroids, and other planets besides earth.

The field of investigation claimed by geology overlaps into astronomy, biology, chemistry, and physics. Geology is an integrative science, using its own techniques and premises and those of all of the other sciences to study the earth and other planetary bodies. Geology is a heavily interdependent science, borrowing from its allied disciplines for many of its tools (Table 1–1).

The interdependent character of geology gives it one of its special characteristics—it thoroughly integrates all of the other sciences in the search for knowledge about the earth. Each of the allied sciences gives rise to a wide range of specialties within geology.

GEOLOGIC SPECIALTIES

Geologists who work primarily with principles from physics include *geophysicists*, geologists concerned with the origin and variation of the earth's magnetic

TABLE 1.1
Scale of Dependence of the Sciences

Mathematics: Stands alone, validity depends on definitions of factors.
  Physics: Dependent solely on mathematics.
    Chemistry: Dependent on physics and mathematics.
      Biology: Dependent on physics, chemistry, and math.
        Geology: Dependent on all of the above.

field and heat flow, variations in the attraction of the earth's gravitational field, the origin and prediction of earthquakes, and the use of manmade small explosions to accurately depict the orientation and probable composition of underground formations (Figure 1–4). *Structural geologists*, are concerned with theories of the conditions and mechanisms of rock deformation, strength of materials, and the history of the forces that produce folded, sheared. and ruptured rocks.

Geologists who work primarily with chemical

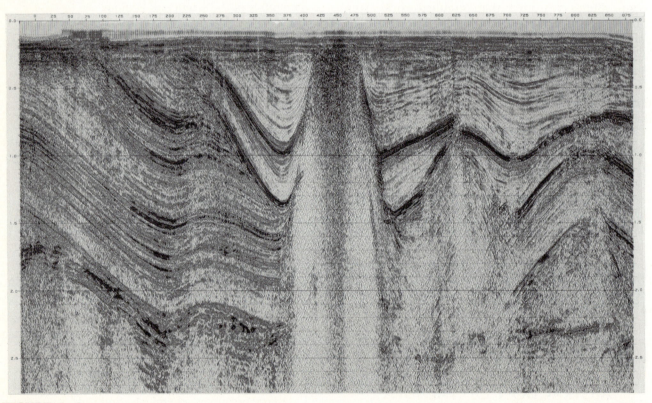

FIGURE 1–4
This is an example of an image of a small part of the upper layers of the earth. The image is obtained by sending vibrations into the earth by explosions, and recording the profiles created as the energy bounces off the layering of the rocks beneath. (Profile courtesy of Seiscom Delta, Inc. Used with permission.)

principles include *petrologists*, who study the origin of rocks, including the chemical environments, as well as pressure and temperature conditions that lead both to the formation of rocks and their modification. *Mineralogists* explore the internal structure of minerals, the chemical composition of minerals, and the environments where minerals form; they establish methods of mineral identification.

Biological principles form the basis for another broad range of specialties within geology, including *paleontologists*, geologists concerned with interpreting the history of life on earth through time and with identifying and classifying the vast parade of life types that have preceded us. Paleontologists study ancient life forms which ranged in size and complexity from dinosaurs to bacteria. *Paleoecologists* study the interpretation of ancient environments, as reflected both by the kinds of rocks and the kinds of life that formed in them; their products are often speculative maps of ancient geographies.

This list is hardly exhaustive, for there are several hundred specialties and subspecialties recognized within the earth sciences. The list also recognizes only one aspect of geology—the concerns directed toward acquiring knowledge for knowledge's own sake. Geology has another equally important side—the applied.

Applied geologists include *economic geologists*, mineralogists, petrologists, structural geologists, and geophysicists who search the earth for economically valuable ore deposits, using highly specialized geophysical and chemical techniques to find their hidden targets. *Engineering geologists* specialize in finding safe places to construct things such as dams, factories, or residences. *Petroleum geologists* use multiple techniques to search for economic deposits of oil or natural gas. *Coal geologists* search for an allied energy resource. *Hydrologists* specialize in the search for usable underground water, perhaps the earth's greatest single resource. In construction areas they may reverse their goals and help rid the area of excess ground water.

Like all sciences, geology has both a curiosity-directed aspect, academic research, and a need-directed branch, applied research, with both avenues deeply intertwined. At its center, geology seeks, simply, to know.

BRIEF HISTORY

Applied geology began when humans first recognized that rocks could be used as tools and weapons; the earliest stone tools were chipped about 750,000 years ago. That first use of stone tools (Figure 1–5)

> **BOX 1–2**
>
> ### KNOWLEDGE, POWER, CERTAINTY, AND MYSTERY
>
> The deepest attraction of science is in its power to understand the mystery it reveals and its way of exploring the world. Even with all its capabilities, science cannot yield unchanging truth, but offers instead another more subtle form of power.
>
> As English philosopher Sir Francis Bacon (1561–1626) wrote *Nam et ipsa scientia potestas est, Knowledge, itself, is power.*
>
> Notice the Latin word translated as "knowledge"—the word is *scientia*. Science, then, is knowledge. In Latin *sciens* is the present participle of the verb to know, which means to separate one thing from another, or to discern.
>
> Bacon was not making the claim that knowledge gave one power over the world. Rather, Bacon argued that full knowledge, really knowing something, causes one to think in certain ways, and not in others. *Knowledge, then, yields power because knowledge promotes right thinking.* Thinking rightly is power in its most wonderful sense—power for a lifetime of discovery.
>
> Does science bring certainty? Absolutely not.
>
> Science is founded on uncertainty. Each time we learn something new, comfortable old ideas must be discarded. We are always working from error toward less error; scientific fact is ultimately tentative. Science finds mystery where others report certainty; quoting Einstein . . . the mysterious. . . *is the source of all true art and science.*
>
> Karl Pearson (1857–1932), distinguished statistician of nearly a century ago, phrased it well: *Does science leave no mystery? On the contrary, it proclaims mystery where others profess knowledge . . . There is mystery enough here, only let us clearly distinguish it from ignorance . . . . The one is impenetrable, the other we are daily subduing.*

---

began what archeologists term the Paleolithic Age; that time also marks one of the first human attempts to classify experienced phenomena. The classification must have been a simple one; some stones were easily chipped to sharp edges, while others proved unsuitable. Some rocks were useful; others were not.

Early humans were geologists of a sort (see cartoon opening this chapter) as they learned which rocks in their vicinity made the best tools. Today the finest surgeon's scalpels are still made from *obsidian*, a volcanic glass first chipped to a sharp edge

FIGURE 1–5

Modern projectile points. Early people also worked stone to a sharp edge. The best stone for this purpose is both hard, dense, and homogeneous. (From J. W. Dawson, 1880, *Fossil Men and Their Modern Representatives*, Fig. 25, courtesy of History of Science Collections, University of Oklahoma Libraries.)

hundreds of thousands of years ago by an inquisitive ancestor. There is still no sharper edge.

The Neolithic Age, beginning about 10,000 B.C., added agriculture, quarrying, ceramics, and mining to the list of ways in which early humans exploited the earth. Some pottery dates back 9000 years, while earliest mining appeared 7000 years ago. Quarrying became a highly advanced art. In 2700 B.C., ancient Britons quarried massive stone slabs weighing up to 50 tons from what is now southwestern England, and moved them 25 miles (40 km). On what is now the Salisbury Plain, these slabs were erected, over a period of a thousand years, to form Stonehenge (Figure 1–6), an ancient astronomical observatory, laid out with high standards of precision. Think of the skills required to quarry rock slabs of such size and weight, move them that far, and erect them in precise order over a millenium (a thousand years); estimates of the labor required approach 18 million man-hours.

Early humans had learned to successfully exploit the earth to meet their basic needs for tools, weapons, pottery, ornaments, cosmetics, and a way to predict the seasons. By trial and error they determined what techniques and materials worked and what did not. They may have wondered why some stones could be worked to an edge, and others could not, but there was no method to allow them to unravel the riddle. The earth had been possessed, and exploited, but it had not been understood.

Thales of Miletus (640–546 B.C.), in writing about magnetism, provided the first attempt to understand the earth. He believed that there is one substance from which all was made (we've mentioned his ideas before in this chapter); that single substance was water. In about 570 B.C., Anaximander of Miletus, an associate of Thales, recognized fossils, and correctly interpreted their origin.

Around 1200, Chu Hsi wrote *In high mountains I have seen shells. . . embedded in rocks. The rocks must have been earthy materials in days of old, and the shells must have lived in water. The low places are now elevated high, and the soft material turned into hard stone*, an idea presented four centuries before Steno rediscovered the same principles. In 1556,

FIGURE 1–6

These stone monoliths, which weigh many tons, still stand upright on the Salisbury plain of England, testifying to the skill and imagination of humanity. (Photograph courtesy of the British Tourist Authority.)

FIGURE 1–7

A typical underground mine, as recorded in *De Re Metallica*, by Georgius Agricola, printed in 1556. Those underground worked by the light of tallow lanterns—dim and dangerous work. (Courtesy of History of Science Collections, University of Oklahoma Libraries.)

Georgius Agricola posthumously published *DeRe Metallica*, a classic description of mining technology (Figure 1–7). In *De natura fossilium* published in 1546, Agricola provided an elegant description of minerals. The Latin adjective *fossilis*, meaning dug up, is the root for the word fossil.

In 1543 Nicolaus Copernicus (1473–1543) published a theory that placed the motionless sun at the center of perfectly circular orbits for the planets and called for the earth to rotate on its axis and orbit the sun. His theory, with its perfect circles was elegant and came much closer to providing predictive power for the location of planets than all of the elaborate sun-centered schemes of the preceding 14 centuries; it remained for Johannes Kepler to place the sun *at* the foci of slightly elliptical planetary orbits in 1609. Observation of planetary positions had finally produced a theory that had excellent predictive power; Greek astronomy with its perfect circles and spheres had been destroyed. Twenty-three years later, Galileo Galilei faced the Inquisition for insisting that Coper-

nican (and Keplerian) hypotheses were more than hypotheses—they were fact.

In 1665, the first scientific journal, *Journal des Savants* was published; the communication of scientific information was becoming easier. In the same year Robert Hooke (1635–1703) identified fossil plants (Figure 1–8), and later described them as *monuments of nature. . . though difficult, it would not be impossible to raise a chronology out of them.* In 1669, Nicolaus Steno (1638–1686) published his analysis of solids within solids, on interfacial angles in crystals (see Chapter Two), on the origin of sedimentary rocks and fossils, and the elementary principles of sequence in stratified rocks.

In 1788, **James Hutton** (1726–1797) published a new theory of the earth, which supported the plutonist view; suggested that the earth was unimaginably ancient; and proposed one version of what is termed **uniformitarianism.** Uniformitarianism is an assumption that the earth's processes have produced similar products throughout all of time. Hutton's published work was plagued by a terribly wordy, disorganized writing style; outside of a small circle of Scottish friends, his revolutionary ideas were

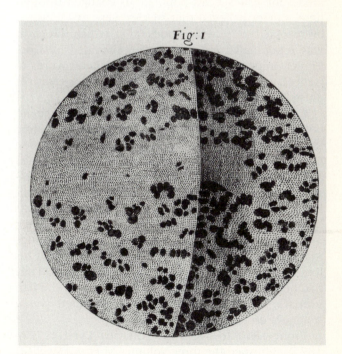

FIGURE 1–8

In 1665 Robert Hooke used a new instrument—the microscope—to compare charcoal to petrified wood, and thereby prove the organic origin of petrified wood—an issue that was much in doubt at the time. (From Robert Hooke, 1665, *Micrographia*, Plate X, Fig. 1, courtesy of the History of Science Collections, University of Oklahoma Libraries.)

all but ignored. In 1802 his friend John Playfair published a summary of Hutton's work, recast into some of the most elegant scientific prose ever written. Hutton's ideas began to have an enormous impact, essentially establishing the theoretical basis for the whole of geology for the next century.

William Smith (1769–1839), English civil engineer, published the first geologic map of England in 1815. Smith had recognized that similar layers of rock had the same assemblage of fossil forms in whatever part of England he found them; therefore, unique groups of fossils were common to any one sedimentary layer. For the first time an individual layer could be recognized as the same age as another layer hundreds of miles away by studying the assemblages of fossils they both contained. The power of geology to reconstruct ancient geographies was greatly enhanced, and another question was obvious: *how* did fossil life change through time intervals recorded in the layers of rock?

Baron Georges Leopold Chretien Frederic Dagobert Cuvier (1769–1832) in 1812 established both vertebrate paleontology and comparative anatomy as sciences, and simultaneously offered a catastrophic model for *how* species populations changed through time. Cuvier believed that species were extinguished by repeated worldwide floods, after which the earth was repopulated by new species provided by God. Cuvier, imprisoned by a belief in the literal words of Genesis, was firmly both a neptunist and a catastrophist.

Catastrophism was a popular model to explain both the extinction and change of life forms in time as well as the origin of landscapes. Catastrophic forces and events, like the flood endured by Noah (as described in Genesis, chapters 7 and 8), were necessary if one believed the Holy Bible was a literal calendar of earth's history. In 1823 Rev. William Buckland published one of many books that explained most geologic phenomena by the universal deluge that took place at the time of Noah; followers of his line of reasoning were known as **diluvialists.** They had been arguing for a century or more with the **fluvialists**, who believed rivers had deposited young sediment and cut their own valleys.

From 1830 to 1833, **Sir Charles Lyell** (1797–1875) published the three volumes of *Principles of Geology*, probably the most influential textbook in the history of geology. In it he proposed a new version of uniformitarian theory (see Hutton's Uniformity of Nature) as an attack on both catastrophism and diluvialism, and a plutonic theory as an attack on the dying theory of neptunism. He suggested, like Hutton, that the earth was very ancient, giving it a far

different chronology from that laid down by Moses. More than any other person, Lyell attempted to set the young science of geology free from the impossible constraints imposed by biblical literalism. Steno, Hutton, and Lyell had in turn established the theoretical foundations for the science of geology by the early 19th century.

The founding of the Geological Society of London in 1807 marks the time when a thousand years of scattered observations about the earth from the previous thousand years were being collected into a science, with its own unique theoretical structure. In 1879, the U. S. Geological Survey was established; the science of geology had been called into subsidized public service.

In 1907, B. B. Boltwood (1870–1927), using radioactive methods, calculated the ages of certain minerals from their uranium/lead ratios; his results ranged up to 2200 million years. In 1912, Alfred Wegener (1880–1930) first proposed the theory of continental drift (see Chap. 10). Continental drift was a theory that the continents had moved slowly in relation to each other, and had a different grouping in the geologic past. In 1962, Harry H. Hess (1906–1969) put together the theory of continental drift, the theory of convection currents, and his work on oceanic ridges to propose what has become known as **plate tectonics** (see Chap. 10). Hess's hypothesis was so speculative that he first called it *an essay in geopoetry.* Plate tectonics, a theory which holds that the earth's outer layer is broken into a dozen or more rigid plates in constant slow relative motion, has since become the dominant theoretical foundation for the whole of modern geologic inquiry, integrating all aspects of geology into a unified theory of the earth.

On July 20, 1969 Apollo XI astronauts Neil Armstrong and "Buzz" Aldrin landed on earth's moon, and collected samples of soil and rocks. Our organized and directed curiosity had brought us from picking up earth rocks for tools to picking up moon rocks for the benefit of all mankind. The human adventure that is science continues.

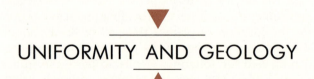

UNIFORMITY AND GEOLOGY

Geologists are sometimes asked when did geology begin? Does an individual science have a birthday?

A science is established when its central theory allows practitioners to reason from examined to unexamined cases. A new science often begins by shattering old ideas. When did these events occur for geology?

GEOLOGY AS AN ESTABLISHED SCIENCE

It would be tempting to pick 1669 as the inception of geology. In that year Nicolaus Steno published his study of solids within solids, the interfacial angles of crystals, and the origin of sedimentary rocks and the fossils within them. In the same study he recognized that mountains and sea floor had exchanged places, implying an earth much older than commonly believed at the time, and he established several fundamental principles that allow the sequence of events recorded in layers of sedimentary rocks to be read. Such an incredible list of "firsts" certainly establishes Steno's contributions as central to the development of geology.

Yet, 1669 cannot be marked as the beginning of geology for two crucial reasons:

1. Steno's ideas were a series of vitally important principles in search of still wider application. Steno never "did anything" with his ideas; he never tried to cast them into a coherent theory which explained the workings of the earth.
2. Steno's ideas changed nothing. Their author never pursued them, and for more than a century, no one saw their revolutionary implications.

The year 1788 marks as sure a birthday as geology will ever have. In that year, James Hutton, Scottish physician, agricultural expert, and amateur geologist, published a comprehensive theory of *why*—and therefore *how*—the earth worked. From our point of view, two centuries later, several aspects of Hutton's theory were seriously in error, but his *Theory of the Earth* revolutionized human understanding of the earth and provided the foundations of modern geology. To understand how revolutionary Hutton's theory was, we need first to understand the major concept his theory overturned.

HOW OLD IS THE EARTH?

The past is not open to direct inspection, so either we must infer conclusions about the past from available data (a form of postdiction), or we must depend on the testimony of others. Lacking a theory that would allow inferences about the past, people based their views on the testimony of others, and the only testimony available was biblical.

The Mosaic chronologies and Genesis were a generally accepted account of a recently created, unchanging world. Seventeenth-century people, from humblest peasant to the most distinguished scientist, accepted biblical testimony as the *only* source of truth. The role of science was simply to *discover* the laws of nature that had long ago been ordained by the Creator.

By various biblical accountings, Creation had occurred a few thousand years before the birth of Christ; some scholars even set the day and hour of creation. The whole course of history, in a biblical view that lasted through much of the eighteenth century, comprised six cosmic "days" of a thousand years (one millenium) each. When Christ was born, four of these days since creation had already elapsed. Another 1.7 "days" had passed before the lifetime of eighteenth-century people. Therefore, only a few hundred years remained until 6000 years since Creation would be completed. After 6000 years would be the promised second coming of Christ and the Last Judgment, to be followed by the promised seventh millennium, completing the seven biblical "days" of the earth's total history (Figure 1–9).

It was clear to all, and explicitly stated in the Bible, that the earth had been created by a loving Creator for the express purpose of benefitting mankind. The earth had been created much as it is now. The only significant event that had altered it at all was the Noachian flood (Figure 1–9).

James Hutton's major premise was that the Creator had created the earth for the comfort of humankind when he laid the theoretical foundations of geology in 1788. Ironically, in arguing from that premise, he destroyed the Bible as a geology textbook, just as Galileo had destroyed it as an astronomy textbook.

Galileo had said it well: *The Bible teaches us how to go to heaven, not how the heavens go.* Now Hutton, arguing primarily from theological and philosophic principles, would also insist that the Bible also does not tell us "how the earth goes." In the process, he not only shattered the Bible as geologic textbook, but he also attacked diluvialism, neptunism, catastrophism, any sense that the earth had a beginning or an end, and any conception that we and the earth had been created together. It was from that kind of intellectual wreckage that modern geology was born.

HUTTON'S "THEORY OF THE EARTH"

James Hutton (1726–1797) was born in Edinburgh, Scotland, attended school there, and continued his

FIGURE 1–9

The Christian concept of earth history is depicted in this frontispiece from a popular seventeenth-century text. Christ's left foot rests on a formless dark earth that, clockwise, takes form and light, and is then flooded (notice the ark). The modern earth is at the bottom center is followed by the promised conflagration, the New Earth, and the Consummation, respectively. (From Thomas Burnet, 1680, (1702 ed.), courtesy of the History of Science Collections, University of Oklahoma Libraries.)

collegiate eduction in Europe. He took a degree in medicine, but as a rather wealthy man, he chose instead to run a highly progressive, profitable farm near Edinburgh.

Hutton was drawn to the study of soil formation, an obvious concern for a farmer. He realized, after intensive observation and study, that rocks *gradually* change into soil, which is *slowly* washed into the sea. The same processes that form the soil also wash it away, a cruel paradox for both the farmer and the philosopher that was Hutton. Since God would not be cruel to his chosen creatures, there *must* exist a restoring force.

Hutton did not stumble across an informative exposure of rock nor was he the first to pursue metic-

ulous fieldwork. Rather, he *deduced* the necessity of a restoring force from a deeply felt conviction that the earth was created by God for the sole benefit of his chosen creatures. Having deduced that there *must* be such a force, he then set out to find both the theoretical and observational evidence to support his deduction.

The restoring force he proposed was the earth's internal heat. He suggested that the earth was one vast ''heat-engine,'' a natural conclusion for someone whose friends included James Watt, the inventor of the modern steam engine which fueled the Industrial Revolution. It was heat that lifted the earth's surface back up and formed igneous rocks like granite.

Many of his most important observations in the field were made to confirm, *not to decide*, his deductions about how the earth worked. When he published the last draft of his work in 1795, he confessed *the only granite I had ever seen when I wrote my Theory of the Earth was . . . (at a street corner) . . . and no more.*

The Perpetual Machine

To Hutton, the earth was a perfect machine, created by God, fueled by heat, set to run forever for the benefit of mankind.

Hutton could see that the earth's surface was always changing, but that change had little sense of direction. Instead, there was only a cycling of the materials of the earth from one form to another then to another and then back to the first. He could, in his own words . . . *find no vestige of a beginning, (and) no prospect of an end.*

In 1795, Hutton published his expanded reconstruction, entitled *Theory of the Earth*. That work is a comprehensive, deeply theoretical attempt to explain the earth's history, its present, and its future in terms of an efficient machine set in motion for the benefit of humankind.

Hutton's theoretical framework for the fledgling science of geology rested on two unique contributions:

1. Hutton recognized that the *earth must be ancient* beyond human comprehension. His evidence was both observational (see Figure 1–10) and deductive, since heat as a restoring force meant that mountain and sea must endlessly occupy the same area throughout time. The recognition of earth's antiquity, as uplift and decay followed one another in cycle after cycle, was Hutton's contribution, both to geology, and to humankind.

FIGURE 1–10

The scene is at Siccar Point in England, an area described by Hutton as displaying an angular unconformity. The layered rock beneath is tilted up at a sharp angle to horizontal, beveled, then covered by the top nearly horizontal layer. An angular unconformity is observational proof of a theory that the earth has repeatedly experienced cyclic uplift and erosion. The immensity of past geologic time is a deductive inference that follows from the observational proof. (Photo courtesy of Stanley S. Beus; used by permission.)

2. Hutton championed *internal heat* as the power that caused rocks to expand and lifted sea floor up into lofty mountains. This force was the previously unrecognized counterbalance to the familiar forces of rock decay and erosion which steadily lowered the continent's surface. The existence of an uplifting force, countering the forces that wore down the earth, prevented total destruction of the land, and meant that earth history was a cycle of decay and renewal. If the earth history was an endless cycle of decay and renovation, the earth was both indestructible and immortal.

As evidence for the earth's antiquity, Hutton proposed not only a new theory of uplift but also a key piece of observational evidence. Hutton deciphered the past events recorded by an **angular unconformity** (Figure 1–10), an exposure of one set of rock layers laying across an eroded surface cut across steeply dipping layers.

Hutton recognized that to form an angular unconformity, old layers of loose sediment had been:

1. Consolidated on the sea floor into solid rock,

2. Then further consolidated and fused by internal heat,

3. Then broken and tilted by the uplifting forces,

4. Then uplifted above sea level, where,

5. Erosion cut across the tilted layers.

One cycle of consolidation, fusion, tilting, uplift, and erosion was complete. But for an angular unconformity to occur, a *second* cycle must occur.

6. The eroded area was lowered down to below sea level, and covered again by sea water,

7. Loose sediment was deposited as a horizontal layer across the eroded edges of the old, tilted layers,

8. The sediment was consolidated on the sea floor into solid rock,

9. Then further consolidated and fused by internal heat,

10. Then uplifted into the air again, where,

11. Erosion has exposed it to our view.

In Hutton's view, an angular unconformity represented evidence that the rocks of one complete cycle had been placed atop the rocks of still an older cycle of deposition, consolidation, uplift, and erosion. An angular unconformity was the *observational proof* both of Hutton's theory of earth cycles and the immensity of time. How much time had it taken to make an angular unconformity? No one could know, but angular unconformities suggested ancient seas, ancient mountains, ancient plains, and again, ancient seas, following one another in endless marching order.

As observational evidence for his concept of heat as the great engine that drove earth uplift, Hutton established the igneous (fire-formed) nature of two common rocks, granite and basalt. The neptunists had argued that these rocks were crystalline sedimentary rocks formed from precipitation from a great sea (the Noachian flood) which had once covered the earth. Had rocks as abundant as granite and basalt been sedimentary like all of the other common rocks, then the earth must once have been totally buried under an ocean and had only recently dried out.

The plutonists, including Hutton, argued that granite and basalt had formed from internal heat; therefore they would record the earth's store of power to lift mountains. If basalt and granite are sedimentary, an ocean had recently left a static earth devoid of power to restore its elevated land. If basalt and granite are igneous, the earth had a power plant, and its history was cyclic. Which was it?

In the hills surrounding Edinburgh, Hutton found his expected answer. A small area of basalt was surrounded by broken, shattered limestone that had been strongly discolored and fused to great hardness. Only a hot fluid could both inject and fuse surrounding rocks. Basalt was of igneous origin; similar exposures elsewhere showed granite had also once been molten. Following another approach, Hutton argued that basalt found intermixed with sedimentary rocks on the sea floor also could not have been of sedimentary origin. Basalt is insoluble in water, so how could basaltic components have gotten into sea water in the first place? There *must* be one single cause for all of the layers on the sea floor, since the layers of sedimentary rock and basalt were so intermixed. Heat both formed basalt as an igneous rock and helped consolidate loose sediment into hard, stratified sedimentary rock. The issues were resolved in Hutton's mind.

Hutton's Uniformity of Nature

The concept of uniformity is the central principle by which geology and all other sciences interpret the past and predict the future. Geologists create an an-

cient dynamic scene from investigation of static objects (rocks and fossils). To do this geologists must reason by analogy, a form of reasoning based on the assumption that if two things are alike in some respects, then they may be alike in other respects. In reasoning by analogy, geologists *assume* that the relation between product and process in the past is the same as that observed today. But how is that crucial assumption justified?

Today we justify the assumption of uniformity purely on the grounds of experience. Our accumulated experience tells us that laws that describe natural processes do not change through time; consequently we assume the principle's validity.

The **principle of uniformity** insists that natural processes operating today are similar to those operating in the past. If this principle were shown to be false, then geologists (and others) could not interpret

FIGURE 1–11

Sir Charles Lyell, British geologist and exponent of a strictly cyclic uniformitarian earth history. Lyell's model insisted that rates of change had always been identical to those observed today, and that each cycle led the earth back to the identical point at which the cycle began. The end result was slow change, always leading nowhere. (Courtesy of the History of Science Collections, University of Oklahoma Libraries.)

the past, since we cannot directly observe past processes, only ancient products. Without the assumption of uniformity that past becomes unknowable.

Uniformity and Knowledge

Considering the principle of uniformity immediately raises those most fundamental and daunting philosophical questions "How do we know?" and "What is knowledge anyway?" Uniformity of nature *does not logically follow from an axiom known to be true.* Thus the principle of uniformity is an *inductive* principle. As with all inductions, we can only speak of the principle of uniformity as a principle with which we have some degree of confidence; it cannot be shown by logical principles to be an undoubted truth.

The issue of the validity of uniformity was an important philosophical concern in Hutton's time. To quote David Hume, Scottish philosopher and friend of Hutton *If there be any suspicion that the course of nature may change and that the past may be no rule for the future, all experience becomes useless and can give rise to no inference or conclusion.* The principle of uniformity withstood the attack of David Hume and other eighteenth-century philosophers as all began to accept uniformity as the most probable way the earth works. The next assault on the principle came from its nineteenth-century champion, British geologist Sir Charles Lyell (Figure 1–11), who damaged the usefulness of uniformity while attempting to defend it.

LYELL AND UNIFORMITY

Sir Charles Lyell (1797–1875) was born the year James Hutton died, and became one of the foremost British geologists of the nineteenth century. He took Hutton's **uniformitarianism**—the word had been coined by William Whewell (1794–1866), one of Lyell's many critics—another step, but it was a limiting, not a broadening step.

Lyell recognized that the *catastrophists* represented a rearguard action against science itself. Geology, which was still struggling to become a science, could never proceed if miraculous intervention was permitted as an agent of past geological change. If there were *no* limits to what was possible, there would be no method by which to decipher the past or forecast the future. In short, if supernatural intervention is a process by which the earth's surface was changed, there can be no science of geology.

The catastrophists had no choice but to argue that violent events had shaped the earth, for many believed that they inhabited an earth only a few thousand years old. Obviously the gradual processes that modify the earth's surface during a human lifetime lack the power to raise a mountain, lower the land into the sea, or form a deep chasm. *If the earth was very young, how could these landscapes have been formed?*

To the catastrophists, the answer was obvious. Violent past earthquakes had hurled mountains into the sky, land into the sea, and had split solid rock, forming chasms that were then occupied by rivers. The neptunists argued that floods once covered the entire earth, depositing first all of the crystalline rocks, then the "young-looking" layered rocks, and finally the loose sediment that tops them. Since then, all had been calm, the infrequent small earthquake only a reminder of the Creator's power.

To lay this pseudoscience to rest, Lyell proposed a sweeping extension to Hutton's original concept of uniformity of process. Lyell proposed that

1. Only *currently* acting geologic processes operated in the past,

2. The rates or intensities of these processes have been *constant* throughout time,

3. Only slow, *gradual* cyclic processes have modified the earth,

4. The earth's processes have always been *exactly* as they are now.

Lyell had taken Hutton's concept of uniformity, which asserted that earth processes had been similar throughout all time, and changed it to a theory of gradualism that proposed uniform, gradual cyclic change modifying an unchanging earth.

Lyell was compelled to postulate this new and different concept of uniformity to combat the prevailing notions of catastrophism. To that extent, Lyell and colleagues were successful. The Bible slowly lost its paralyzing grip on earth history, and catastrophism became a discredited theory for understanding the earth's past or future. But unwittingly, in defending a uniformitarian geology against the catastrophists, Lyell had created *two different* theories of uniformity.

1. A theory asserting that natural laws were invariant in space and time. This is uniformity in the approximate sense of Hutton. This theory remains today a central rule guiding development and choice of hypotheses in all sciences. It is an old idea, going back to Herodotus, and very clearly formulated a century before Hutton by Newton in his *Principia* as the second of four rules of reasoning . . . *to the*

same natural effects we must . . . assign the same causes. Hutton's great contribution was not to *originate* uniformitarianism, but to *apply* it to form a comprehensive theory of how the earth worked. Hutton's law of uniformity is a statement about how to think. It is a method for reasoning from examined to unexamined case.

2. A theory asserting that both ancient and modern causes are *identical,* that current processes acting at extremely slow rates, which have been *constant* through all time, have maintained the earth *exactly* as it is today. This alternate, strict theory of uniformity is uniquely Lyell's. His uniformitarianism was not a prescription for how to think, as was Hutton's, but was instead a *testable* theory about the actual history of the earth.

Briefly, in the century and a half since Lyell, geologists have learned that Lyell's strict uniformitarianism was simply wrong. Our earth has evolved from a past dramatically different from today, and continues to evolve toward a future that will be still different. Its history has been punctuated with catastrophic events, like meteorite impacts, that have greatly modified the earth. Gradual change has also been at work. The earth's total history has direction, for through time its processes have differed predictably both in character and in rate. Lyell's vision of the earth as an eternal, steady-state planet, its lands and seas changing places in an endless waltz, fell in time to a more modern version of uniformity.

Today geologists employ uniformity in the general sense originally described by Hutton. The principle of uniformity as commonly used today states that in reasoning from examined case to unexamined case, scientists must assume that:

1. Supernatural intervention has not occurred, and
2. The relation between cause and effect has remained constant in time and space.

Our modern version of uniformity is little more than a restatement of the ancient Greek tradition that the earth will play fair with us. Albert Einstein expressed this same supposition when he suggested that *God may be subtle, but He is not malicious.* Without the simplifying assumption of uniformity to guide the construction and testing of hypotheses, there is no predictable past or predictable future available to any science.

Although uniformity of nature is nothing more than an induction from our experience, it is all that we have. That our search for the past and the future should be guided at its most fundamental level by two interrelated, unprovable, but probable inferences—nature is both orderly and uniform in time and space—is one of the ironies of science, which has so transformed our world.

SUMMARY

1. Science seeks an understanding of nature by forming and testing hypotheses and theories, eventually proposing natural laws or principles of great explanatory power. Relationships among organized data are explained in terms of cause and effect, assuming natural laws describe a rational universe.

2. Science is the dominant human enterprise of our time; perhaps 90 per cent of all the scientists who have ever lived are now alive.

3. Theories organize what is known, limit what is acceptable, can be falsified, and have predictive and postdictive power.

4. Geology is an interdependent science that seeks an understanding of:

▶ the processes that form and modify the earth and other planets, and
▶ the earth's history, its resources, and its hazards.

5. James Hutton in 1788 provided the theoretical foundation for geology by recognizing that the earth's internal heat provided a restorative uplifting force that counterbalanced the forces of decay and erosion. His recognition of multiple cycles of uplift and decay led him to propose that the earth must be of immense age. He theorized that the relation between process and product was unchanging through all time and space, a concept that has been called uniformitarianism or the principle of uniformity.

6. The principle of uniformity is an inductive principle, a probable inference. This principle allows geologists to infer the past, interpret the present, and predict the future for planet earth.

KEY WORDS

angular unconformity	deductive reasoning
axiom	diluvialist
catastrophism	empirical

fluvialist

geology

geosyncline

Hutton, James

hypothesis

inductive reasoning

isostasy

law

Lyell, Sir Charles

matter

molding, principle of

neptunist

plate tectonics

plutonist

postdict

predict

principle

science

similarity, principle of

Steno, Nicolaus

technology

theory

uniformitarianism or uniformity, principle of

EXERCISES

1. Imagine that you lived during Paleolithic times in the midst of a glacial advance. What evidence could be sought to show that the climate in the past was different than it was now?

2. What evidence could have been used by people before Hutton to prove that the earth was very young?

3. Mineral deposits were first classified by their usefulness. Now geologists classify them by their origin. What shift in view does this imply?

4. What are the values of understanding more of the earth's processes? Is this knowledge power, in the sense of Sir Francis Bacon?

5. What kind of evidence would one seek to show that the principle of uniformity was in error?

SUGGESTED READINGS

LAUDAN, LARRY, 1984, *Science and Values*, Berkeley, CA, U. of California Press, 149 pp.
▶ *How scientists argue, and how they reach consensus.*

LEVESON, DAVID, 1971, *A Sense of the Earth*, Garden City, N.Y., Doubleday, 176 pp.
▶ *A thoughtful statement of our kinship and discovery of our earth. An incredibly moving book.*

GOLDSTEIN, MARTIN and GOLDSTEIN, Inge, 1984, *The Experience of Science, An Interdisciplinary Approach*, New York, Plenum Press, 284 pp.
▶ *An insightful study of how scientists do science. Easy reading.*

WHITEHEAD, ALFRED NORTH, 1953, *Science and the Modern World*, Toronto, Ont., Collier-Macmillan Canada, 212 pp.
▶ *A classic review of the origin and nature of science.*

These crystals formed on our moon. They line a small cavity in one of the rocks returned to earth by the Apollo 14 mission. The largest crystal is apatite, a calcium phosphate. Other crystals include plagioclase, a calcium sodium aluminum silicate, and pyroxene, a magnesium - iron silicate. The largest apatite crystal is about 50 micrometers long; the image is formed by a scanning electron microscope. The crystals grew from a hot vapor as the rock was cooling. Since these crystals come from a rock within the Imbrium Basin of the moon, they must be about 3.9 billion years old. (Image courtesy of NASA.)

CHAPTER *two*

Minerals

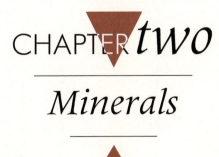

Opinion says hot and cold, but the reality is atoms and empty space.

DEMOCRITUS (c. 460–400 B.C.)

Material objects are of two kinds, atoms and compounds of atoms.

LUCRETIUS (99–55 B.C.)

As the quotations that begin this chapter suggest, reality begins with things unseen, with atoms and empty space. In this chapter we will trace the relation between things unseen—atoms, elements, "compounds of atoms," molecules—and things seen, which we term minerals. We will briefly trace the history of our developing understanding of minerals, describe the testable physical properties of minerals, and examine the relation between minerals and the aggregates of minerals that we call rocks.

The word *mineral* is among the most adaptable of nouns. To druggists, minerals are often combined with vitamins. To rock hounds, minerals are attractive and may be hard enough to polish. To prospectors, minerals have potential dollar value. Early people referred to *anything* dug from the earth as a mineral. Geologists would regard few of these items as minerals; they insist on a much stricter definition. A **mineral** is a *naturally occurring, inorganic, crystalline solid, with a chemical composition and physical properties that vary within narrow limits.* Defined in this way, minerals are naturally formed *matter,* a concept first described in Chapter 1. What then is matter?

COMPOSITION OF MATTER

Matter is that which occupies space and can be perceived. Our modern understanding is that matter is made of combinations of one or more elements. **Elements** are substances composed of only one kind of **atom**, the smallest unit of an element that retains the chemical and physical properties of the element. There are over a hundred chemical elements known, of which 90 occur naturally. The remaining dozen or more elements have been formed only in nuclear reactors and "atom-smashers"; these elements are all radioactive, rapidly decay into other elements, and are not found in nature. In order to understand minerals we must first ask of what atoms are made.

THE STRUCTURE OF ATOMS

The details within an atom cannot be directly observed, though modern instruments allow separation ("atom smashing") of atoms into their many component parts. The model presented here is a simple and useful guide helpful in understanding atoms.

An atom consists of **electrons**, particles having a negative electric charge, that moves within a spherical area whose diameter is approximately one *angstrom,* a unit of length equal to 10^{-10} meter (0.00000000001 m) (Appendix E gives measurement conversions). **Protons**, particles having a positive

TABLE 2–1
Atomic Number and Electron Distribution for Some Common Elements

Element	Symbol	Atomic Number	Number of Electrons in Each Successive Shell Shell Number					
			1	2	3	4	5	6
Hydrogen	H	1	1					
Lithium	Li	3	2	1				
Carbon	C	6	2	4				
Oxygen	O	8	2	6				
Neon	Ne	10	2	8				
Sodium	Na	11	2	8	1			
Magnesium	Mg	12	2	8	2			
Aluminum	Al	13	2	8	3			
Silicon	Si	14	2	8	4			
Chlorine	Cl	17	2	8	7			
Potassium	K	19	2	8	8	1		
Calcium	Ca	20	2	8	8	2		
Iron	Fe	26	2	8	14	2		
Cesium	Cs	55	2	8	18	18	8	1
Gold	Au	79	2	8	18	18	18	1

cyclotron – atom smasher

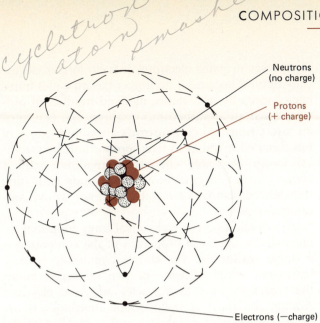

Neutrons
(no charge)

Protons
(+ charge)

Electrons (−charge)

FIGURE 2–1

The oxygen atom, schematically depicted here, consists of a nucleus of eight protons, each having a single positive charge, and eight neutrons, which have no charge. Surrounding the nucleus are eight electrons, each having a single negative charge. Two of the electrons orbit around the nucleus in circular shells close to the nucleus while six of the electrons orbit about the nucleus in circular orbits that define a more distant shell. Since there are eight protons, the atomic number of oxygen is 8.

electric charge equal in magnitude to the charge on an electron, and **neutrons**, particles without an electric charge, are confined to a nucleus at the center of the electron cloud in a region having a diameter of approximately 10^{-12} meter (0.0000000000001 m) (Figure 2–1). The proton and neutron are about 1850 times heavier than an electron, and provide essentially all the mass of the atom.

Most of an atom is empty space, within which the location of any one electron at any one time cannot be pinpointed. A more realistic picture of an atom would show electrons orbiting about the atomic nucleus in clouds that form concentric successively larger shells. The structure of these more advanced models is detailed within the next few pages.

The electrons of an electrically neutral atom—one not combined with another atom—provide a total negative charge equal to the positive charge of the nucleus. The element hydrogen, for example, in its neutral state has a single electron moving about a single proton; the negative charge of its electron is exactly neutralized by the positive charge of its proton. Hydrogen is the simplest atom.

The number of protons in the nucleus is the **atomic number** of the element. The atomic number ranges from one, for the element hydrogen, to 103, for the element lawrencium. Finite quantities of elements are composed of electrically neutral atoms having the same atomic number.

As more and more protons are added, each balancing electron is added in a highly systematic way. The innermost shell can hold only two electrons. Each subsequent shell is filled only when the shell closer to the nucleus is first filled to its capacity. The outermost shell, however, never contains more than eight electrons (Table 2–1). As an example, neon has two electrons in its innermost shell, and a total of eight in its next shell; its atomic number is 10. For sodium, atomic number 11, both of the first two shells are filled to capacity; its eleventh electron must go into the third shell out from the nucleus (Figure 2–2).

Chemists place all the known elements into a **periodic table of elements** (Appendix G), which arranges elements by the number of electrons in their outermost shell. The periodic table of elements has this pattern because it is the number of these outermost electrons that determines many of the chemical

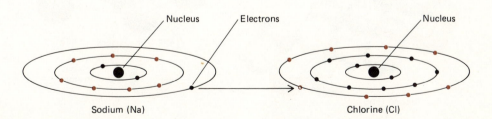

Nucleus   Electrons   Nucleus

Sodium (Na)   Chlorine (Cl)

FIGURE 2–2

In this example, two stable elements are shown. A sodium atom has a total of 11 electrons, one of which is in its outermost shell. A chlorine atom has a total of 17 electrons, seven of which are in its outermost shell. Since having eight electrons in the outermost shell is the most stable condition for any atom, when atoms of sodium and chlorine are brought into contact, sodium gives up its outermost electron easily; now its remaining ring has eight electrons. Chlorine eagerly accepts the donated electron; now its outermost ring also has eight electrons.

properties observed in each element. As an example, similar chemical properties are observed for all elements that have but one electron in their outermost shell, so that lithium, sodium, potassium, rubidium, and cesium are all placed in a single column within the periodic table. Why should the outermost electrons be so important in determining the characteristics of an element?

ATOMIC BONDING

When two or more atoms react with one another, they form, in Lucretius' words, compounds of atoms. The forces that bond or hold atoms together into compounds are electrical; like charges repel one another while unlike charges attract. Each individual atom seeks to fill its outermost shell to eight electrons, the most electrically stable position. For example, neon, with 8 electrons in its outermost shell, is very stable (Table 2–1); consequently it is extremely difficult to combine neon with other elements. The majority of atoms have less than eight electrons in their outermost shell (Table 2–1), and are therefore somewhat unstable or chemically reactive.

In order to provide eight electrons in its outer shell, an atom can either lose, gain, or share electrons with other atoms so that every atom involved in the bonding process gains the "magic eight" electrons in its outermost shell (Figure 2–2). When an atom loses or gains electrons, it is no longer electrically neutral because it no longer has an equal number of electrons to balance its fixed number of protons. Such an atom becomes reactive, and is termed an **ion**. **Chemical compounds** are combinations of ions bonded together in fixed proportions by electrical forces. A stable chemical compound exhibits *overall electrical neutrality*, as compounds exhibiting unbalanced positive or negative charges would continue to react.

The smallest part of a chemical compound that retains all the chemical properties of the compound is a single **molecule**. The simplest molecule is ionic hydrogen, which consists of two hydrogen atoms that react to form two nuclei and one electron. Among the most complex single molecules is the protein ribonuclease, which consists of 1876 nuclei and 7396 electrons.

Bonding among atoms occurs when the unbalanced positive charge on one atom is satisfied by an unbalanced negative charge on an adjacent atom; bonding usually involves electron sharing between adjacent atoms. There are four methods of electron sharing: Van der Waals bonding, ionic bonding, covalent bonding, and metallic bonding. These four methods merge with one another; more than one method of bonding may occur within the same mineral.

In bonded atoms the distance of closest approach between the centers of adjacent atomic nuclei in the

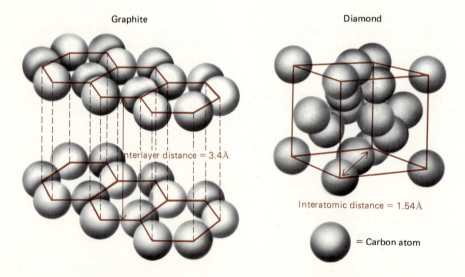

Graphite   Diamond

Interlayer distance = 3.4Å

Interatomic distance = 1.54Å

= Carbon atom

FIGURE 2–3

Both graphite and diamond are composed wholly of carbon atoms. A graphite molecule consists of carbon - carbon bonds in hexagonal arrays that form a sheet of strongly bound carbon atoms. The sheets of carbon atoms are very weakly bound together by Van der Waals bonds with an interlayer distance of 3.40 angstroms. A diamond molecule consists of a huge three-dimensional molecule, only part of which is shown here. The carbon - carbon bonds are largely covalent, and interatomic distance is 1.54 angstroms in a structure that is cubic. Diamond is about 1.6 heavier than graphite, reflecting the closer uniform spacing of carbon atoms in diamond.

molecule is known as the **atomic radius** (Figure 2–3) of the atom in that particular molecule. The size or amount of space occupied by each atom—its atomic radius—and its unbalanced electrical charge determine the shape and strength of the molecule that results when atoms are bound together. For a molecule to be stable, it must be both

1. Composed of atoms whose positive and negative charges are exactly balanced by the unlike charges of adjacent atoms, and
2. Composed of atoms whose differing size allows them to fit into the available space, much as bricks of a unlike size can only be fitted together in a limited number of patterns to make a stable wall.

Van der Waals Bonding

Van der Waals bonding, the weakest of bonds, occurs when neutral atoms are brought together to form a molecule. The atomic radius for these atoms is the distance at which the mutual repulsion of the outer electrons of each atom is exactly balanced by

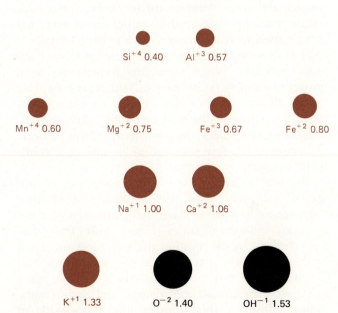

FIGURE 2–4

The spheres represent the overall shape and relative size of 11 common ions. The radii given are those for the ion in ionic crystals; for ions like Si^{4+}, which form largely covalent bonds, the radius given is an approximation. Elements that lose electrons in ionic bonding, thus gaining a positive overall charge, are shown in color. Elements that gain electrons in ionic bonding, thereby gaining an overall negative charge, are shown in black.

the attraction of the nucleus of each atom for the electrons of the other. A common example is the mineral graphite (Figure 2–3). In this example, the bonding between adjacent sheets of carbon atoms is Van der Waals bonding. The distance between adjacent sheets is twice the effective Van der Waals atomic radius for carbon, so the bonding between sheets is extremely weak. The sheets may be easily separated—a process we perform every time we write with a pencil.

Ionic Bonding

Bonding by electron transfer is a common form of bonding within minerals. In **ionic bonding**, one or more of the outer electrons are transferred from one ion to the other (Figure 2–2). Each ion becomes stable by donating to or accepting electrons from the other; the outer shells of both ions wind up with eight electrons. The ionic radius is the distance of closest approach between two ions of opposite charge; the internuclear distance is the sum of the individual radii (Figure 2–4).

Ionic bonding is much stronger than Van der Waals bonding and accounts for the moderate strength and hardness of many common minerals. A common example of a mineral that exhibits ionic bonding is the mineral halite (pronounced hay-lite) or common rock salt, which combines sodium and chlorine (Figure 2–2). Sodium, with one electron in its outer shell, loses that electron to a chlorine atom with seven electrons in its outer shell. The result is a stable outer shell of eight electrons for both ions. As a result of the loss and gain of electrons, sodium and chlorine acquire overall positive and negative charges, respectively, and become ions bound in ionic (sometimes termed electrovalent) bonding. The reaction is explosive; poisonous chlorine and poisonous sodium instantaneously combine to form white crystals of ordinary table salt (Figure 2–5).

Covalent Bonding

Covalent bonding occurs when *some* of the outer electrons of each atom are *shared* rather than transferred as they are in ionic bonding. Covalent bonding is the strongest of all chemical bonds. The atomic radius of each atom is half the distance between the two nuclei, which are tightly bonded together by electrons in the region between the two atoms. The shared electrons between the two atoms cannot be distinguished from one another. Since no electrons are gained or lost, no ions are formed.

An example of covalent bonding is the formation

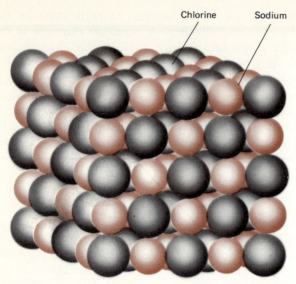

FIGURE 2–5

The sodium chloride molecule is formed by wholly ionic bonding with the transfer of a single electron from sodium to chlorine (see Figure 2–2). The molecule consists of alternating sodium and chlorine atoms that fit together to form a perfect cube. On each of the six "faces" of the cubic molecule, each chlorine atom is surrounded by four sodium atoms and each sodium atom is surrounded by four chlorine atoms.

of a molecule of hydrogen (Figure 2–6). In this simplest of all molecules, the electrons are shared between two atoms of hydrogen, forming molecular hydrogen. The result is that each of the two nuclei has a completed inner shell with two electrons, the most stable configuration. This innermost shell is the only shell filled by just two electrons (see Table 2–1). The mineral diamond exhibits covalent bonding, which explains its great hardness.

Metallic Bonding

Metallic bonding is exhibited by metals. In this form of bonding some electrons are all wholly shared among atoms and move with ease from one atom to another. The overall structure consists of positive nuclei through which free electrons can drift and act as a general "cement." Since electrons are free to move from atom to atom, metals make excellent conductors of electricity.

The chemical bonds in minerals are generally a blend of bond types, consisting of varying degrees of covalent electron sharing and ionic bonding by electron transfer. Van der Waals and metallic bondings are somewhat less common among the two thousand or so known minerals.

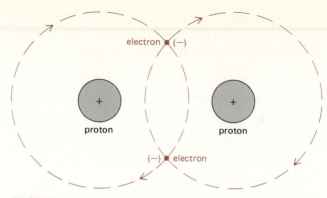

FIGURE 2–6

Molecular hydrogen is formed when two hydrogen atoms bond together in covalent bonding, totally sharing the two electrons between them and intermittently providing each of the two protons a stable inner shell of two electrons. The hydrogen - hydrogen bond is an extremely strong one.

Atoms and Empty Space

We have been describing atoms and empty space, bonded into three-dimensional atomic networks termed molecules. Modern **mineralogy**, the science within geology that studies minerals, asserts that these networks *repeat themselves three-dimensionally in space to form larger frameworks composed of many interlocked molecules.* When these tiny frameworks have grown to visible size, we may call them **crystals**. This theory, like all theories (see Chapter 1), conveys explanation without absolute certainty. Can observation serve as a check on our theory? How can one *see* something a few hundred-millionths of a centimeter across?

The reality within these frameworks can be observed only with extremely sophisticated electron microscopes capable of magnifying atomic structures by many millions of times (Figure 2–7). These images confirm by observation the structures predicted by theory, namely that atoms as they are formed into crystals should sort themselves into positions appropriate to their unbalanced charge and their atomic radius within a three-dimensional framework of other atoms.

For the framework to endure, the structure must be strong. The atoms must fit together in a pattern that can be indefinitely repeated in all directions without collapsing. In the same way that one would build a stone house with stones of many different kinds, size, and shape, each atom within the structure must not only "fit" in terms of its atomic radius but it also must contribute the unbalanced charge necessary to strongly bond its neighbors into a structure having overall electrical neutrality.

The ordering and sorting process that creates stable frameworks of atoms from disorganized matter is termed **crystallization.** With a microscope, we may see crystals form out of solution and thereby may infer the ordering process that forms them. By observing with extremely high-powered electron microscopes, we may see the internal networks of atoms that result from that ordering (Figure 2–7). The agreement between observation at many levels of magnification and atomic theory is a classic example of how ideas are proposed, tested, and then changed until the explanation, a theory, accurately predicts and reflects the certainties observed.

A BRIEF HISTORY OF CRYSTALLOGRAPHY

The study of crystals is termed *crystallography*, and it is among the oldest of sciences. Our term crystal comes from the Greek word *krustallos*, which means ice. The only explanation one could offer 2500 years ago was that glittering crystals were an odd form of ice, frozen so hard that it could not melt. Dozens of minerals were *named*, a step in classification that sug-

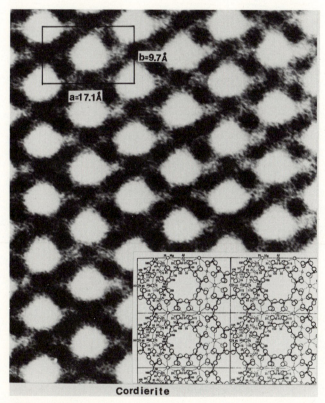

Cordierite

FIGURE 2–7

A high-resolution transmission electron microscope image of the mineral cordierite showing the unit shell. Each of the dark areas represents an individual atom, with the white space representing space between the atoms. The total magnification is 17.7 million times. (From Buseck and Iijima, *American Mineralogist*, vol. 59, pp. 1–22, 1974, copyright by the Mineralogical Society of America. Used with permission.)

FIGURE 2–8

A typical alphabetical list of "minerals," including rocks (*Basaltes*), fossils (*Belemnites*), and minerals (*Amethistus*). Basalt (*Basaltes*) is a common volcanic rock, a belemnite (*Belemnites*) is a common fossil of an ancient squid-like animal, and amethyst (*Amethistus*) is a purple-colored form of quartz used as a semiprecious gem mineral. (List is taken from A. Kircheri, 1665, courtesy of History of Science Collections, University of Oklahoma Libraries.)

gests knowledgeable persons could *separate* one mineral from another, based entirely on appearance. Aristotle (384–322 B.C.) had classified minerals, metals, rocks, and fossils in a single group, a concept that was to confuse mineral classification for nearly 2000 years.

Alphabetical lists of minerals were compiled by the Middle Ages (Figure 2–8). Objects are classified alphabetically, as in a dictionary or phone book, when we know nothing about them or see *no relation* between them. Minerals were seen simply as curious objects that had many uses. Powdered colorful minerals were used as cosmetics, metals smelted from minerals as jewelry, weapons, and simple machines, while many minerals were used as medicines or as talismans, charms thought to possess magic power.

It was Danish physician Niels Stens, whose name was commonly Latinized to Nicolaus Steno (1638–1686) who saw what everyone else had seen, but thought what no one else had thought (see Chapter

One). Steno noted that the angle between like faces of quartz crystals from many localities was *identical* (Figure 2–9). Steno's critical observation was later generalized as the **principle of constancy of interfacial angles**: crystals of a specific mineral have fixed characteristic angles at which the faces, however distorted the faces may be, always meet. This principle, implicit in Steno's work but never stated by him, is the first law of crystallography.

Steno had observed that crystals of a single mineral type were remarkably rigid and repetitive, each in an observable way *exactly* alike. Science has been described as a search for repetition; how could the repetitive relation first observed by Steno be explained? The answer was to wait for more than a century of further work.

In 1783 Romé de Lisle, a French mineralogist, published *Cristallographie*, a book outlining the results of his extensive measurements of interfacial angles. It was de Lisle who first asserted that certain

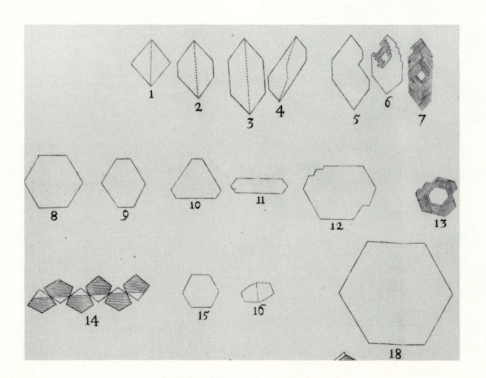

FIGURE 2–9

Plate IX from *Prodromus* by N. Steno, printed in 1669. In these sketches of the outline of the base of hexagonal quartz crystals, notice that for sketches 2 to 13 the angle between the faces remains constant although the length of each face is different in each sketch. Sketch 14 illustrates that Steno must have been able to accurately measure the interfacial angles of hematite; what is presented is a paper model that when cut out and folded along the lines indicated will form a perfect three-dimensional model of the mineral. While Steno never stated the *principle of constant interfacial angles*, he clearly recognized it and was able to accurately measure the necessary angles. (Courtesy of History of Science Collections, University of Oklahoma Libraries.)

interfacial angles are specific to a single mineral species, thereby establishing the principle of constant interfacial angles. He went further and proposed that the shape of the crystal was the same as that of its *integer molecules*, tiny imaginary blocks which were stacked together in space to make the larger crystal. de Lisle thought that these *integer molecules* were the same shape as the crystal or the plane-sided fragments produced when the crystal was cleaved.

A famous countryman of de Lisle's, Abbé René Just Haüy (pronounced a-yoo-eé) (1743–1822), building on the concepts of both Steno and de Lisle, suggested that crystals were composed of minute *"molecules constituantes."* We would call these unit cells today. The **unit cell** is the shape, which when repeated in space, will create the entire crystal. A unit cell exhibits all of the chemical and physical properties of the mineral.

Crystal faces were *produced* when large numbers of these "molecules constituantes" were stacked in rigid three-dimensional arrays, a concept illustrated in one of Haüy's original figures (Figure 2–10). What resulted was a simple geometric shape with constant angles between like faces. Haüy was able to demonstrate that the orientation of crystal faces in space could be predicted as simple ratios of the axes of the unit cell. Haüy also established that crystals *changed* shape *slightly* as a function of small changes in their chemical composition. These minute changes in shape are the result of atoms of different atomic radii slightly warping the unit cells into measurably different shape.

Haüy thought that the variety of crystal shapes exhibited by some minerals resulted from different stacking sequences of the unit cells. From the shapes produced when minerals cleave, Haüy deduced that there were only six possible shapes for the unit cell. Since Haüy proposed that crystal faces were produced by stacking the unit cell, the constant angles between the faces had to reflect both the stacking sequence (Figure 2–10) and the shape of the unit cell. The repetitive forms of crystals and cleavage fragments had suggested the idea of the unit cell to Haüy, but it was an idea lacking observational proof. He was also, of course, unable to *see* the atoms that formed the unit cell. It remained for twentieth-cen-

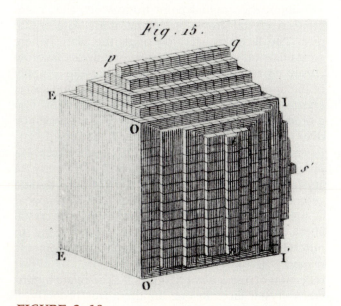

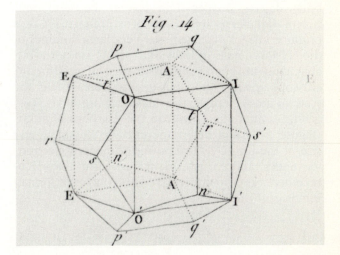

FIGURE 2–10

These two figures are from R. J. Haüy, 1803, *Traite Elementaire de Physique*, Pl. 2. They illustrate his concept of how symmetrical crystals are built up by the symmetrical addition of "integer molecules." Notice that the complex real crystal form shown at the bottom of the figure can be produced by stacking tiny squares in the fashion shown in the top diagram. The "slope" of the faces depends on how the cubic "integer molecules," now termed unit cells, are stacked. In this case, every second row of blocks is repetitively stepped down one step; the slope is 2:1. By

1803 Haüy could relate the measured angles between the faces to the shape of the unit cell. If the slope of a surface can only be created by stepping down perfect squares in simple integer ratios (1:1, 2:1, 3:1, 4:1, etc.), then only certain angles between surfaces are possible. The interfacial angles became the key to describing both the shape of the unit cell and its stacking sequence. (Courtesy of History of Science Collections, University of Oklahoma Libraries.)

tury science to finally pry within crystal frameworks.

The discovery of X-rays in 1895 by Wilhelm Konrad Roentgen (1845–1923) so completely overturned physics that some have called it a second revolution, comparable in impact to Galileo's nearly three centuries before (see Chapter One). A brilliant German physicist, Max T. F. von Laue (1879–1960) thought that a beam of X-rays passed through a crystal might be diffracted by the layers of atoms within the crystal, much as daylight is diffracted into a rainbow of colors when it passes through finely spaced screen wire. An experiment in 1912 confirmed his hunch and gave crystallography its workhorse, an instrument known as an *X-ray diffractometer.*

Using X-rays of known wavelength, the diffractometer has made it possible to observe and study the atomic structure of crystals by passing a beam of energy through the crystal and observing how the internal crystal framework bends and scatters the X-ray beam; the resulting pattern produced when the diffracted X-ray pattern encounters film is unique to each mineral. Among the early crystals studied by X-ray diffraction was halite; data from X-ray diffraction are plotted to yield Figure 2–5.

Although both optical microscopes and X-ray diffractometers are powerful tools, they still cannot "see" individual atoms. Optical microscopes cannot bring objects smaller than several thousand angstroms in diameter into sharp focus. Diffracted X-rays produce images averaged over very large volumes of the crystal; details of individual atoms are not available. Only in recent years has the use of electron microscopy made it possible to bring areas near the scale of atomic radii into relatively sharp focus. Two centuries after Haüy, we can "see" groups of atoms (Figure 2–7) and determine directly the shape and size of the unit cell. The power of modern electronics, optics, and engineering has revealed a world of repetitive order first forecast by the power of human minds centuries before.

PHYSICAL PROPERTIES OF MINERALS

We have learned that a mineral is a naturally occurring, inorganic, crystalline solid, with a chemical composition and physical properties that vary within narrow limits. We have also learned that minerals are composed of atoms locked together by electrical forces into rigid, repetitive frameworks whose shape is expressed in a crystal form, *if the growing crystal had free space in which to grow.* That crystal form is among the many physical properties of a mineral; a **physical property** is a distinctive unchanging characteristic that can be observed or tested. Each mineral has its own unique set of physical properties different from those obtained by observing and testing *any* of the other 2500 or so known minerals. In the same way that no one else has exactly your combination of hair color, fingerprints, facial features, skeletal structure, and dental record, each mineral has its own set of unique properties that identify it (see Appendix A). A summary of some of the most easily observable and testable physical properties follows.

CRYSTAL FORM AND SYMMETRY

Many people think of crystals as rare objects seen in museums or jewelry stores. Actually most inorganic solids are made of crystals, but because most crystals growing in nature usually have to compete for space, perfectly formed crystals are not common.

The external crystal form of a mineral expresses the *systematic* arrangement of crystal faces (Figure 2–11) about the original point from which the crystal grew. A crystal grows by adding atoms to a crystal face. The faces of a crystal mark the plane where crystallization stopped when conditions favoring crystal growth changed. If there is an endless supply of the appropriate atoms, time for growth, and space, crystals grow to enormous size, an angstrom or so at a time. Single crystals 20 yards (18 m) long weighing over 40 tons (380,000 kg) have been described.

The *systematic* arrangement of crystal faces expresses the mineral's **symmetry**, the property of crystals whereby crystal faces always bear a rigid geometrical relation to one another and so reappear at precise angular intervals on rotation of a crystal about a point, a line, or reflection across a plane. In observing crystals, we discover that simple geometric *symmetry operations*—rotation about a line or mirror reflection across a plane (Figure 2-12)—relate a mineral's crystal faces to one another. This property of crystals, first recognized by Haüy, is one easily tested. Pick up an ordinary wooden pencil and rotate it slowly about an axis that is parallel to its length. Every 60 degrees of rotation, a new "face" appears that is identical to the one that preceded it and identical to the one that follows with a *further* rotation of 60 degrees. This is an example of symmetry about an *axis* (Figure 2–12), and this property is termed axial symmetry. The symmetry operation necessary to locate the orientation of the next "face" is rotation of 60 degrees about an axis.

The human body exhibits symmetry about a single *plane* (Figure 2–12) because the right side of the body is nearly the mirror image of the left side. This is termed mirror plane symmetry.

A third type of symmetry results when faces in a series are equidistant from a central point; this is point symmetry. An unsharpened pencil exhibits slightly imperfect point symmetry, axial symmetry about one long axis and three short axes, and planar symmetry about four mirror planes. Can you find all these kinds of symmetry?

Study of the kinds of symmetry that crystals exhibit shows that all crystal forms may be placed into one of six crystal systems. These six crystal systems may be subdivided into a total of 32 classes. There are only 32 possible combinations of symmetry operations that will form crystal faces in three-dimensional space. Each of the six systems has a unique set of symmetry features. By examining the kind of sym-

metry displayed by a crystal, noting its crystal forms, and placing it in one of the six crystal systems, we have begun to discover the physical properties of the mineral.

LUSTER

The physical property termed **luster** describes the way the surface of the mineral reflects light. Minerals that reflect light like shiny metals have *metallic* luster, whereas those that reflect light like fine china have a *vitreous* luster. Other terms used to describe reflectivity include *waxy*, *resinous*, *earthy* or *dull*, *greasy*, *glassy*, *pearly*, and *silky*. Although these terms lack precision, since we might choose different terms to describe the same crystal surface, they do indicate in a general way how the mineral reflects light. For some minerals their luster is a particularly distinctive physical property.

FIGURE 2–11
These glittering crystals of quartz reflect not only light but also the simple mathematics of crystal growth. The covalent bonds between silicon and oxygen form an internal network of molecules whose repetition in three dimensions combines to make the crystal form displayed.

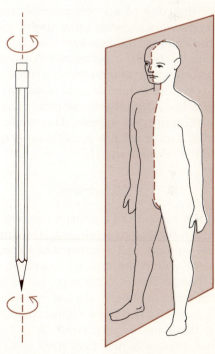

Symmetry about an axis   Mirror plane symmetry

FIGURE 2–12
Examples of two different kinds of symmetry. The pencil is symmetrical about an *axis* parallel to its length. Rotation of the pencil 60 degrees about its axis is the symmetry operation necessary to display, in turn, each of six identical "faces." The human body is approximately symmetrical about a vertical *plane* that bisects the eyes; the colored part of the human figure is the mirror image of the other half.

COLOR AND STREAK

The color of a mineral depends on how the atoms within the mineral selectively absorb certain wavelengths of white light. Mineral **color** is described in familiar terms: deep red, light purple, and so on. Color is among the least useful of physical properties in identifying minerals because their color often depends on impurities hidden within the atomic framework. The mineral quartz, as an example, may exhibit a wide range of colors from clear to milky to yellow to deep brown to rose to lavender, depending on which elements are incorporated in trace amounts. Another cause of mineral color is damage to the atomic framework; radiation damage to quartz yields "smoky" quartz with its characteristic golden brown color. The damage may be reversed and the smoky color cleared by baking the crystal for several hours in an oven. Other minerals, for example some minerals containing copper, exhibit only a single color—vivid blue or green.

Since the color of a larger mineral specimen can be deceptive, a somewhat more useful test is to observe the color of the *powdered* mineral on a white background. The test is performed by vigorously rubbing the unknown mineral across a piece of hard unglazed porcelain termed a *streak plate*. The powdered mineral rubbed onto the streak plate exhibits a color termed the mineral's **streak.** The streak color of a mineral is a more reliable physical property than the color of the whole mineral.

Hardness

Testing the hardness of a mineral involves testing the mineral's resistance to scratching. **Hardness** reflects both the strength and symmetry of electrical attractive forces within the crystal; scratching the crystal causes bonds between adjacent atoms to be torn apart. In order to perform the test one mineral of unknown hardness is rubbed against another object of known hardness, or vice versa. A mineral can be scratched by one harder than itself and can in turn scratch one softer than itself. In order to report the hardness value obtained in numerical terms, a scale of relative hardness of common objects and minerals was established in 1822 by Friedrich Mohs (1773–1839), a German mineralogist (Table 2–2).

The **Mohs' hardness scale** arranges 10 common minerals in order of *relative* hardness: talc, the softest mineral, is designated 1, and diamond, the hardest mineral, is designated 10. The numbers stand for strictly relative differences in hardness; there is, for example, about four times the difference in actual

TABLE 2–2
*Mohs' Hardness Values
for Common Materials*

Material	Mohs' Hardness
Talc	1
Gypsum	2
Fingernail, Halite	2.5
Copper penny, Calcite	3
Fluorite	4
Apatite	5
Ordinary glass or nail	5.5
Pocket knife or Feldspar	6
Streak plate or Pyrite	6–6.5
Quartz	7
Topaz	8
Corundum	9
Diamond	10

hardness between the minerals corundum and diamond as there is between the minerals topaz and corundum.

DENSITY AND SPECIFIC GRAVITY

The **density** of a mineral is its weight per unit of volume, expressed as grams per cubic centimeter. Because the density of a mineral is a number that varies with the units of measurement, for convenience the relative weight of minerals is more commonly expressed as their specific gravity. The **specific gravity** is the ratio of the weight of the substance compared to the weight of an equal volume of water.

Specific gravity is determined by weighing the dry mineral, then adding the mineral to a glass brimful of water, and then weighing the displaced water (Figure 2–13). We owe this method of determination to Archimedes (c. 287–212 B.C.), perhaps one of the greatest scientists of ancient times. As he stepped into a brimming bathtub it occurred to him in a flash that the volume of water slopped out onto the floor equaled the volume of his body, and that the weight of the expelled water equaled the difference in weight of his now partially buoyant body. Excited by his discovery, he supposedly dashed naked into the streets of his home town shouting *"Eureka! Eureka!"* which translated from the Greek means "I've got it! I've got it!"

The specific gravity of common minerals ranges between 2 and 3, with a total range of observed specific gravity from 0.92 for ice to 19.3 for gold (Figure

2–13). The specific gravity of a mineral reflects both the relative *atomic weight* of the atoms within the mineral and the *packing density* or closeness to which the atoms can be packed.

As an example of the effect of the *atomic weight* of atoms within a mineral, consider the mineral aragonite, which is *calcium* carbonate, having a specific gravity of near 3. Identical in internal structure, the mineral cerussite (Figure 2–14), which is *lead* carbonate, has a specific gravity of over 6. The effect of atomic *packing*, on the other hand, is displayed in two minerals of *identical* chemical composition; both diamond and graphite are pure carbon. In graphite, as described earlier, the bonding is weak; atoms are widely separated and the specific gravity is just over 2. In diamond, as described earlier, the bonding is covalent and strong; atoms are closely packed, and the specific gravity is 3.5, nearly twice that of graphite.

FRACTURE

If a mineral is struck hard enough to cause it to break, we may test the *orientation* of the electrical forces that hold it together. If the breakage creates a random and irregular surface, the mineral is said to **fracture**, and we learn that the substance is equally strong in all directions, lacking internal planes of easy failure. Fracture types include *uneven*, the most common, *hackly*, a jagged break typical of the mineral copper, and *conchoidal*, shaped like the inside of a conch or clamshell (Figure 2–15). Conchoidal fractures form in substances that are relatively uniform and homogeneous in all directions, like window glass, flint, and obsidian, a naturally formed volcanic glass. Working flint and obsidian by repeating many

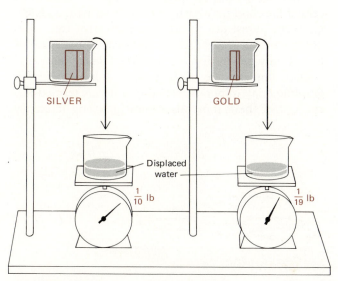

FIGURE 2–13

Sketch of a completed experiment. Bars of gold and silver, each weighing one pound (one-half kilogram) were lowered into beakers brimming with water. All displaced water has been trapped in beakers beneath. The ratio of the weight of the metal to the weight of the displaced water is the *specific gravity* of the metal. *Density* is mass (weight) per volume; in the metric system, density is numerically equivalent to specific gravity. Thus the specific gravity of silver is 10, and its density is 10 grams per cubic centimeter (approximately 625 pounds per cubic foot). The specific gravity of gold is approximately 19. Gold is nearly twice as dense as silver; as shown in the sketch for equal weights, a bar of gold occupies approximately half the space of silver.

FIGURE 2–14

Crystals of cerrusite, a lead carbonate that is twice as heavy as similar crystals of calcium carbonate. The crystals, shaped like match sticks, radiate slightly from a common center of growth. (Photograph courtesy of the U. S. Geological Survey.)

FIGURE 2–15
Natural volcanic glass fractures readily into a *conchoidal* fracture pattern, typically of many homogeneous substances. (Photograph courtesy of the U. S. Geological Survey.)

conchoidal fractures until they merge into a sharp edge was a basic tool-making skill many thousands of years ago. The skill is still practiced today as a hobby by "rock knappers."

CLEAVAGE

If a mineral is struck or wedged, it may split cleanly along closely spaced flat surfaces. The mineral is said to exhibit the physical property termed **cleavage**, and

we know the mineral is not equally strongly bound in all directions. When struck, the mineral readily splits or *cleaves* where the atomic bonds are weakest, and flat planes of ready breakage, called *cleavage planes* result. Cleavage planes are an obvious expression of the ordered internal structure of crystals.

Cleavage was first noted by Steno, and again a century later in 1781 when Haüy accidentally dropped a borrowed specimen of the mineral calcite. To his embarassment, the mineral cleaved into regularly shaped fragments, bounded by three cleavage planes that met at constant angles. No matter how small the fragments, they still displayed the same cleavage planes. To explain what he had seen, Haüy assumed that the mineral calcite was composed of innumerable microscopically small rhombohedra, the "integral molecules" previously described, packed together in an orderly manner to create the shape of a calcite crystal. His accident was a remarkable example of how mishap was turned to good use by someone who thought carefully about what he saw.

Cleavage planes are always parallel to a *potential* crystal face and may easily be confused with a crystal face. The surest test to separate these two categories of flat surfaces is destructive; cleavage planes are produced only by striking the mineral a sharp blow.

The simplest type of cleavage is that exhibited by graphite, clay minerals, and the minerals termed micas. All of these minerals exhibit a strong tendency

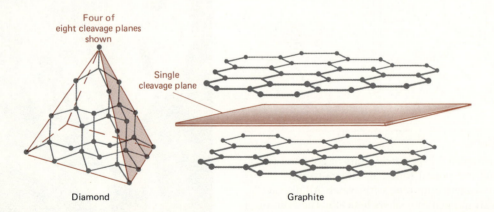

Four of eight cleavage planes shown

Single cleavage plane

Diamond

Graphite

FIGURE 2–16
Diamond, whose cleavage is difficult, and graphite, whose cleavage is extremely easy, split along cleavage planes whose orientation is controlled by their molecular structure. In diamond, carbon atoms are strongly linked in all directions into a complex molecule having overall cubic symmetry. Because of the complexity of the structure, only two of the eight planes of cleavage in diamond are shown in color. The total cleavage pattern for diamond has the shape of an octahedron. In graphite,

carbon atoms are strongly linked together into sheets of atoms in nearly hexagonal arrays, but the sheets are themselves very weakly bonded (see Figure Figure 2–3) to one another. A single plane of easy cleavage, shown in color, allows repetitive separation of each layer from its neighbor. We take advantage of this property by using graphite in pencils. In powdered form, slippery (easily cleaved) graphite is a superb lubricant.

to cleave along single parallel planes, much as a stack of steeply tilted typing paper "cleaves" when sheet after sheet of paper slides off the top of the stack. Powdered graphite, often used as a lubricant, is slippery because one microscopically thin sheet of graphite slips easily over another. Writing with a "lead" pencil leaves trillions of tiny graphite flakes on the paper; the "lead" is a mixture of graphite and clay.

More complex cleavage types include cleavage along two, three, four, eight, and even twelve planes within the same mineral. The ease of cleavage may range from easy, as in graphite, to difficult, as in diamond (Figure 2–16). Some minerals exhibit fracture and no cleavage, as in quartz, while others exhibit only cleavage or cleavage in some direction(s) and fracture in others. However described, cleavage and fracture are physical properties important in mineral identification.

MINERAL CLASSIFICATION

When minerals were first named, as already noted they were classified simply in alphabetic order. As more and more minerals were recognized, such classifications became long and unwieldy. The first attempt to systematize minerals into groups of like kind came from the great Swedish botanist Carl von Linne (1707–1778), better known by his Latinized name Carolus Linnaeus. Linnaeus gave biology its binomial, or *Linnaean taxonomic* system, in which every type of living thing is first given a generic name, designating the group to which it belongs, and then a specific species name. As an example humans are classified as *Homo sapiens*; we are genera *Homo*, species *sapiens*.

THE LINNAEAN CLASSIFICATION

Linnaeus proposed a similar binomial classification for minerals, using Latin names (Figure 2–17). Abraham Gottlob Werner (1750–1817), champion of neptunism (see Chapter One) and an inspiring professor of mineralogy, suggested a similar hybrid Linnaean-chemical classification in 1774. Friedrich Mohs, who gave us our mineral hardness scale, proposed still another Linnaean-type system in 1820. Each of these systems became ever more complex, for the characteristics that defined each of the mineral groups varied from classification to classification. Since minerals are inorganic objects, they lack the

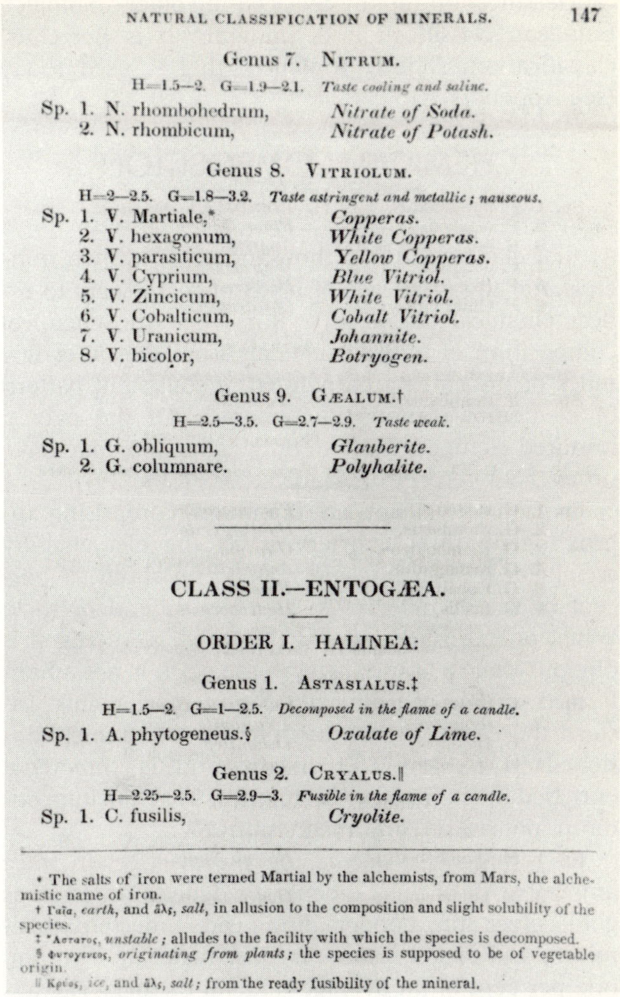

NATURAL CLASSIFICATION OF MINERALS.   147

Genus 7. NITRUM.

H=1.5–2.   G=1.9–2.1.   *Taste cooling and saline.*

Sp. 1. N. rhombohedrum,   *Nitrate of Soda.*
    2. N. rhombicum,   *Nitrate of Potash.*

Genus 8. VITRIOLUM.

H=2–2.5.   G=1.8–3.2.   *Taste astringent and metallic; nauseous.*

Sp. 1. V. Martiale,*   *Copperas.*
    2. V. hexagonum,   *White Copperas.*
    3. V. parasiticum,   *Yellow Copperas.*
    4. V. Cyprium,   *Blue Vitriol.*
    5. V. Zincicum,   *White Vitriol.*
    6. V. Cobalticum,   *Cobalt Vitriol.*
    7. V. Uranicum,   *Johannite.*
    8. V. bicolor,   *Botryogen.*

Genus 9. GÆALUM.†

H=2.5–3.5.   G=2.7–2.9.   *Taste weak.*

Sp. 1. G. obliquum,   *Glauberite.*
    2. G. columnare.   *Polyhalite.*

CLASS II.—ENTOGÆA.

ORDER I. HALINEA:

Genus 1. ASTASIALUS.‡

H=1.5–2.   G=1–2.5.   *Decomposed in the flame of a candle.*

Sp. 1. A. phytogeneus.§   *Oxalate of Lime.*

Genus 2. CRYALUS.‖

H=2.25–2.5.   G=2.9–3.   *Fusible in the flame of a candle.*

Sp. 1. C. fusilis,   *Cryolite.*

---

* The salts of iron were termed Martial by the alchemists, from Mars, the alchemistic name of iron.
† Γαῖα, *earth*, and ἅλς, *salt*, in allusion to the composition and slight solubility of the species.
‡ *Ἀστατος, unstable*; alludes to the facility with which the species is decomposed.
§ φυτογενεος, *originating from plants*; the species is supposed to be of vegetable origin.
‖ Κρύος, *ice*, and ἅλς, *salt*; from the ready fusibility of the mineral.

FIGURE 2–17

This page, taken from James Dwight Dana's first edition of *A System of Mineralogy*, published in 1837, illustrates the *biologic* classification (see Appendix F) of minerals. Thirteen years later, this scheme was abandoned in favor of grouping minerals by chemical subdivisions (see Appendix A), the scheme still used today. (Courtesy of History of Science Collections, University of Oklahoma Libraries.)

sexual parts that were the original bases of Linnaean classification. What was needed was a system by which mineral species could be classified into *logical* groups having related properties.

In 1841 the Swedish chemist Jons Jakob Berzelius (1779–1848), published the first classification of minerals based solely on their *chemical composition.* So great was his authority as a chemist—it was he who proposed the internationally adopted symbols and notation for the chemical elements—that his chemical classification rapidly replaced the older "biologic" Linnaean classification of minerals. It is Berzelius' classification, slightly modified, that is used today (see Appendix A).

CHEMICAL COMPOSITION OF THE EARTH

At first glance, the two thousand or so known minerals and the hundreds of kinds of rocks seem to reflect chemical complexity. But when the chemical composition of typical materials from the outer few miles of the earth are analyzed, a consistent pattern of elemental composition emerges. Of the over a hundred elements known, only eight elements make up over 99 percent of all materials (Table 2-3).

Oxygen and silicon are dominant, comprising approximately three-quarters of all the elements by weight or by percent, and making up about 95 percent of the volume in an average mineral or rock. When one considers that there are 90 elements, it is obvious that the outer layer of the earth is not a haphazard sample of uniformly abundant elements. Instead the earth that we can sample is extremely enriched in oxygen and silicon, and is somewhat enriched in six metals—aluminum, iron, calcium, sodium, potassium, and magnesium.

Everything else occurs, on the average, in trace amounts. All of the elements we depend on, such as gold, silver, tin, copper, zinc, lead, tungsten, chromium, and nickel, *totalled together* make up less than one per cent of an average mineral or rock. The average bit of the earth is overwhelmingly silicon and oxygen, plus a short list of six elements. To assemble the earth, we need but eight ingredients. All else is spice.

THE SILICATE MINERALS

The most common minerals are **silicates**. Each of the minerals of this group is made up of some combination of the metals iron, magnesium, calcium, sodium, and potassium, each possessing a positive charge, combined with negatively charged combinations of silicon + oxygen or silicon + aluminum + oxygen. Silicate minerals are various combinations of these eight elements, but they all are chemically inert or only moderately reactive, have a moderate to low specific gravity, have moderately high melting points, and are fairly hard.

Silicon combines with oxygen to form a silicon–oxygen tetrahedron (Figure 2–18). The unit achieves its stability by placing the small (atomic radius = 0.4 angstrom) silicon ion in the center of four large (atomic radius = 1.4 angstroms) closely packed oxygen ions, the nucleii of which define the four corners of a tetrahedron. The silicon-to-oxygen ratio is 1:4 for an individual tetrahedron. The silicon–oxygen tetrahedron has a total of four more electrons than protons because silicon contributes a charge of +4, whereas each of four oxygen ions contributes a charge of −2 for a total negative charge of −8, providing an unbalanced charge of −4. The excess of electrons in a silicon tetrahedron requires one or more of the positively charged six common metals to form a molecule having overall electrical neutrality. As an example, two atoms of iron or magnesium (each ion has an unbalanced charge of +2) bonding with a single silicon–oxygen tetrahedron will provide an electrically neutral stable molecule of the mineral

TABLE 2–3
Approximate Composition of Earth's Outer Layer

Atomic Symbol	Element	Percent by Weight	Percent by Number of Atoms	Percent by Volume
O	Oxygen	46.60	62.17	94.05
Si	Silicon	27.72	21.51	0.88
Al	Aluminum	8.13	6.44	0.48
Fe	Iron	5.00	2.11	0.48
Ca	Calcium	3.63	2.20	1.09
Na	Sodium	2.83	2.20	1.01
K	Potassium	2.59	1.15	1.49
Mg	Magnesium	2.09	2.05	0.32
TOTAL		98.59	99.83	99.80

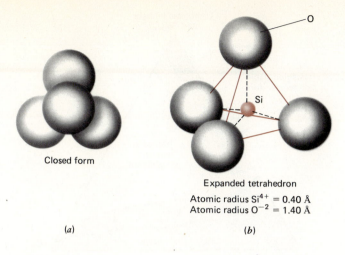

Closed form

Expanded tetrahedron
Atomic radius Si^{4+} = 0.40 Å
Atomic radius O^{-2} = 1.40 Å

(a)   (b)

FIGURE 2–18

Silicon and oxygen combine with covalent bonds to form a structure having the approximate shape of a tetrahedron. It is this shape that provides a stable complex ion having an overall charge of 7–84. Tetrahedra are linked together in space by sharing "corner" oxygen ions and combine with common metals to make silicate minerals (see Figure Figure 2–19).

olivine, composed of independent tetrahedra linked by iron and/or magnesium.

In addition to metals bonding individual tetrahedrons together, other large molecules may be created by sharing oxygen ions between adjacent tetrahedra. If two of the four silicon ions are shared, the silicon to oxygen ratio changes from 1:4, as in the mineral group olivine, to 1:3, as in the mineral group *pyroxene* (Figure 2–19), and the tetrahedra form a single connected chain. If five of the eight silicon ions in two tetrahedra are shared, the silicon to oxygen ratio becomes 1:2.75, as in the mineral group *amphibole*, and the tetrahedra link to form a double chain. If three of the four silicon ions in each tetrahedra are shared, the silicon to oxygen ratio becomes 1:2.5, as in the mineral groups *mica* and *clay*, and the tetrahedra interconnect to form a continuous sheet.

If all the silicon ions in each tetrahedra are shared, there is no more unbalanced charge, the silicon-to-oxygen ratio is 1:2, and a mineral like *quartz* is formed. The tetrahedra now form a complex three-dimensional network. Complex network structure is the most common of all the silicate structures and includes not only the mineral quartz but also the

Mineral		Idealized Formula	Cleavage	Silicate Structure	
Olivine		$(Mg_1Fe)_2SiO_4$	None	Single tetrahedron	
Pyroxene group		$(Mg,Fe)SiO_3$	Two planes at right angles	Single chain	
Amphibole group		$(Ca_2Mg_5)Si_8O_{22}(OH)_2$	Two planes at 56° and 124°	Double chains	
Micas	Biotite	$K(Mg,Fe)_3Si_3O_{10}(OH)_2$	One plane	Sheets	
	Muscovite	$KAl_3Si_3O_{10}(OH)_2$			
Feld-spars	Plagioclase	$(NaAlSi_3O_8 + CaAl_2Si_2O_8)$	Two planes at 90°	Three-dimensional networks	
	Orthoclase	$KAlSi_3O_8$			
Quartz		SiO_2	None		

FIGURE 2–19

Chart of common silicate minerals. Notice that minerals with sheet structures, like the micas, clays, and graphite, exhibit cleavage on one lane, between the sheets.

mineral group *feldspar*, the most abundant mineral group on earth. Feldspars are formed when potassium, or sodium, and/or calcium join with aluminum to *substitute* for up to half of the silicon ions in the total framework.

IONIC SUBSTITUTION

When different atoms are of similar atomic radius and unbalanced ionic charge, they may substitute for each other within a molecular framework. Since minerals form in the natural world, their environment of formation is seldom chemically "pure." Most of the silicate minerals display ionic substitution as "stray" ions get incorporated in the mineral's framework during crystallization.

A good example of simple ionic substitution is the mineral group olivine. As Figure 2–3 points out, magnesium (Mg^{2+}) and iron (Fe^{2+}) are of similar ionic radius and charge. The olivine structure of single tetrahedra $(SiO_4)^4$ accepts *either* iron or magnesium with almost equal ease. A typical olivine contains both, and so olivine is a magnesium iron silicate having a formula of $(Mg,Fe)_2SiO_4$. Pure iron or pure magnesium olivines are known, but are uncommon.

Feldspars and other minerals display *coupled substitution*; the several ions act cooperatively to balance the charge and fill the available space. Figure 2–20 is a *triangular diagram* on which the chemical composition of the three common feldspars is plotted. Each angle (apex) represents the composition of pure feldspar, while the line joining the angles represents the continuous change in composition between any two feldspars, and the interior represents all possible combinations. Each apex represents the composition of one feldspar in its pure state; the side opposite the apex represents zero percent of that feldspar. Any chemical composition can be shown on the diagram since the sum of all ions must be 100 percent.

In feldspars, aluminum, having a charge of $+3$ combines with potassium, having a charge of $+1$, to form an ion pair with a total charge of $+4$. This coupled ion pair is able to replace *one-fourth* of the silicon ions in each tetrahedron. The result is orthoclase feldspar (Figure 2–20). The substitution of potassium and aluminum, together, for silicon, is an example of coupled substitution.

Another example from the feldspars is two aluminum ions $(Al^{3+})_2$ coupled with calcium (Ca^{2+}) to replace *half* of the silicon in calcic plagioclase feldspar (Figure 2–20). Notice that the triangular diagram suggests that no feldspars exist having a composition intermediate between orthoclase and calcic

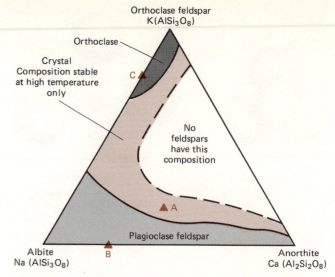

FIGURE 2–20

A triangular diagram of the compositional variation of common feldspars. Each apex of the triangle represents a feldspar whose composition is 100 percent the material listed. The abundance of that component decreases steadily in all directions from the apex, reaching zero at the other two apices and anywhere on the line perpendicular the apex. A feldspar having the composition of point "A" would represent a mixture of 20 percent orthoclase, 40 percent albite, and 40 percent anorthite, and would be stable only when formed at high temperatures and abruptly cooled. A feldspar having the composition of point "B" would be a mixture of 30 percent albite and 70 percent anorthite, while "C" would represent 15 percent albite and 85 percent orthoclase in composition. Minerals "B" and "C" would form by slow cooling.

plagioclase, that there are breaks in the compositions intermediate between orthoclase and plagioclase, and that plagioclase feldspars exhibit a complete smooth transition in composition between sodic and calcic plagioclase feldspars. Chapter Three will describe the implications of this chemical information in terms of the formation of feldspars during crystallization of molten material to form rocks.

ROCKS AND ROCK-FORMING MINERALS

Rocks are natural aggregates of minerals, that are either crystallized into a mass of intergrown crystals or formed from fragments cemented together in water by natural cements. Depending on the environment where they are formed (Figure 2–21), rocks may be broadly classified into one of three types.

1. **Sedimentary rock** is composed of mixtures of minerals, rock fragments, volcanic debris, organic

materials, and often fossils, cemented together by natural cements. Sedimentary rocks are the most common rocks exposed on the earth's surface. Almost all sedimentary rocks are layered or stratified and form in water or fall through the air as **sediment**, loose grains of mineral, rocky, and organic material, which may then slowly be compacted and cemented together to make sedimentary rock.

2. **Igneous rock** is composed of minerals crystallized together into intergrown masses of crystals. A few igneous rocks are made of naturally formed glass that is noncrystalline. Igneous rocks form when earth materials are melted deep within the earth; the minerals crystallize from the molten material as it cools to their freezing temperature, much as ice crystallizes from water on a cold winter day.

3. **Metamorphic rock** is composed of minerals crystallized together into intergrown masses of crystals. These rocks are formed at moderate to high pressures and temperatures within the earth in a wide range of environments. The rocks undergoing metamorphism remain solid during their alteration, but are reduced to a more compact, highly crystalline form. Many metamorphic rocks include platy and needlelike minerals strongly aligned by high pressures.

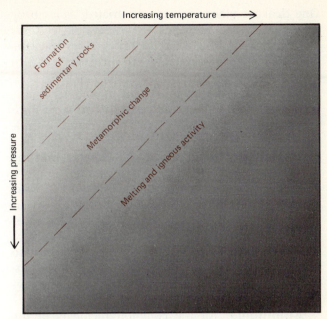

FIGURE 2–21

Environments of rock formation. The chemical and physical processes that form sedimentary rock operate at or near the earth's surface. As temperature and pressure increase, preexisting rocks experience metamorphic changes while still solid. As temperature increases still further, melting may occur and igneous activity begins.

Even with the apparent variety in types of rocks, only eleven minerals dominate their composition. Nine of these minerals are silicates: quartz, orthoclase feldspar, plagioclase feldspar, muscovite mica, biotite mica, amphibole, pyroxene, olivine, and clay. The other two minerals are *carbonates*, chemical compounds that are composed of carbon joined with oxygen, with common metals completing the bond. The common carbonate minerals are calcite and dolomite. All the other two thousand or so known minerals are uncommon or rare. In the same way that eight elements form about 99 percent of all rocks, eleven minerals are the components of almost all rocks. Because these eleven minerals are so common in rocks, they are termed the **rock-forming minerals**. A brief description of their properties and appearance follows (see also Appendix A).

Quartz. The most common single mineral on earth, quartz is abundant in all three major rock types. Lacking cleavage, quartz in many rocks looks like small fragments of broken glass. Harder than any other common mineral, quartz readily scratches steel, glass, and almost all other minerals. Its color in rocks ranges widely, but is commonly clear or cloudy or milky white.

Orthoclase Feldspar. Found generally in igneous and metamorphic rocks, and much less commonly in sedimentary rocks, orthoclase is distinguished by a combination of its salmon-pink to pale-gray color, blocky stubby crystals with two cleavage planes at right angles to one another, hardness just slightly less than that of quartz, and the lack of striations on any of its cleavage planes. It is most commonly found in lighter-colored rocks associated with quartz.

Plagioclase Feldspar. This mineral group has two end members (Figure 2–20), a pure sodic plagioclase termed *albite* and a pure calcic plagioclase termed *anorthite*. The composition of almost all plagioclase feldspar crystals ranges between these two, and may also include small amounts of potassium (Figure 2–20). The plagioclase feldspars demonstrate a phenomena known as **solid solution**, where the composition of a crystal may vary across a limited range of possible compositions. The plagioclase group is more common in igneous and metamorphic rocks than in sedimentary rocks.

Plagioclase crystals are hard and have two well-developed cleavage planes that meet at right angles.

One cleavage plane may appear striated, as if cut by a series of fine parallel lines. The more sodic feldspars come in varying shades of light to medium gray, the color changing to dark gray with increasing calcium content.

Mica. More common in igneous and metamorphic rocks than sedimentary rocks, micas are a family of minerals sharing many characteristics. All micas display well-developed platy cleavage, so that individual flaky layers can be peeled away with the point of a needle. Hardness is moderate, luster is shiny, and the flakes cleaved away are quite flexible. Changes in chemical composition among the micas cause striking changes in overall color. Two micas are most common, a clear to silvery-colored mica termed *muscovite* and a dark reddish brown to bronze-colored mica called *biotite*. Other less common colors are dark green, golden yellow, and lavender.

Amphibole. All the common members of the amphibole group of minerals are very dark green to black and commonly occur as shiny rodlike to needlelike crystals that display two shiny, bright planes of cleavage oriented at 54° and 126° to each other, giving a distinctly rhombic appearance to broken crystals. Amphiboles are a large family of minerals characterized by complex chemistry. They occur much more commonly in igneous and metamorphic rocks than in sedimentary rocks. If the cleavage cannot be easily distinguished, amphibole minerals can easily be confused with minerals of the pyroxene family.

Pyroxene. Rather uncommon in sedimentary rocks, members of the pyroxene family are readily found in both igneous and metamorphic rocks. Most of the common pyroxenes are dark green, have a rather dull luster, and form stubby crystals of moderate hardness that may display two planes of cleavage that are mutually perpendicular. If the cleavage cannot be recognized, pyroxenes and amphiboles may be difficult to distinguish.

Olivine. Essentially unknown in sedimentary rocks and less common in metamorphic rocks, olivine is a common mineral in many dark-colored igneous rocks. Crystals are of moderate hardness, lack easily detectable cleavage, and are various shades of dark yellowish green to brownish green. Calcic plagioclase feldspars and pyroxenes are commonly associated minerals; olivine does not normally occur in rocks containing quartz. Olivine looks like greenish quartz, but is softer.

Clay. This large family of minerals may occur in any of the three major rock types *as a mineral formed by the alteration of other minerals*, but the only occurrence of clay as a primary mineral is in sedimentary rocks. In these rocks, it is the most abundant mineral group. Distinguished by very low hardness and specific gravity, most clays will lightly stick to a moistened tongue and may smell faintly musty on a freshly broken surface. Very high powered microscopes are needed to see individual clay crystals, which always exhibit excellent platy cleavage, like the micas.

Calcite and Dolomite. These two minerals are abundant in sedimentary rocks, slightly less common in metamorphic rocks, and, except for one extremely unusual group, are unknown in igneous rocks. Both calcite and dolomite are commonly cream-colored, though many other colors are known. Crystals are of moderately low hardness and exhibit extremely well-developed cleavage on three planes meeting one another at 60° and 120°. Calcite ($CaCO_3$) is calcium carbonate, while dolomite, $Ca,Mg(CO_3)_2$, results when some of the calcium is replaced by magnesium, forming a calcium-magnesium carbonate by ionic substitution of magnesium for calcium. Calcite will react to dilute hydrochloric acid by vigorous effervescence (bubbling), whereas dolomite must be first powdered before adding the dilute acid, and its effervescence is somewhat more subdued. This is the best way to tell them apart.

SUMMARY

1. Protons, neutrons, and electrons combine to make atoms; elements are composed of atoms all of the same kind, having the same number of protons. Combinations of different elements form stable three-dimensional rigid frameworks of atoms, which in turn form molecules of chemical compounds. The stability of these frameworks comes from some combination of Van der Waals, ionic, covalent, and metallic bonding. Molecules are repetitively stacked in space to yield crystals.

2. The growth of understanding about the origin and architecture of minerals depended on the combination of inquisitive minds and increasingly powerful instrumentation. It still does.

3. Crystallization occurs when atoms of the correct charge and size bond together to make electrically neutral stable atomic frameworks. Crystals grow in size by adding layer after layer of atoms. Atoms whose size is almost right and/or combinations of atoms whose *total* charge is correct can also be included by coupled substitution or solid

solution mixing. Crystals form if there is a enough time, if the temperature is below the freezing point for the crystal, and if a supply of the appropriate atoms is present. Lacking enough space, crystals intergrow and do not develop their potential crystal faces.

4. A mineral is a naturally occurring crystalline chemical compound. Its chemical composition and the resultant physical properties may vary within narrow limits because both simple and coupled ionic substitutions are common. All minerals of the same name exhibit constant interfacial angles between like faces and constant intercleavage angles between like cleavage planes, except for slight variations due to compositional variation by solid solution.

5. Crystal faces are the preserved planes along which crystal growth takes place. All crystal faces form symmetrically with respect to one or more points, planes, or axes of symmetry.

6. The most easily tested physical properties of minerals include crystal form, cleavage, fracture, luster, color and streak, hardness, and specific gravity. Testing for these and other physical properties yields a combination of data that helps identify minerals.

7. Minerals are classified into groups in terms of their overall chemistry. Silicate minerals make up the great majority of the outer part of our earth, and are composed almost exclusively of only eight elements.

8. Natural aggregates of minerals form igneous, metamorphic, and sedimentary rocks. Nine silicate mineral groups and two carbonate minerals make up essentially all the minerals in common rocks.

9. Atoms are the letters, minerals the words, rocks the sentences, and the earth is the story. The apparent complexity at each level of observation obscures the simplicity of components and explanatory principles.

KEY WORDS

atom	calcite
atomic number	chemical compound
atomic radius	clay
cleavage	molecule
color	neutrons
covalent bond	olivine
crystal form	orthoclase feldspar
crystallization	periodic table of elements
density	physical property
dolomite	plagioclase feldspar
electron	principle of constancy of interfacial angles
element	
fracture	proton
hardness	pyroxene
igneous rock	quartz
ion	rock-forming mineral
ionic bond	rock
luster	sediment
matter	sedimentary rock
metallic bond	silicates
metamorphic rock	solid solution
mica	specific gravity
mineral	streak
mineralogy	symmetry
Mohs' hardness scale	unit cell
	Van der Waals bond

EXERCISES

1. What relation, if any, would you expect between a mineral's hardness and its specific gravity? Why?

2. Why is quartz the most common single mineral?

3. What symmetry of floor tiles can you observe? What is the symmetry of a maple leaf? A sphere?

4. Scientists observe that when two atoms have equal charge, the smaller one has greater bonding energy. Given this information, should iron-rich or magnesium-rich olivine crystallize first at very high temperatures?

5. Gemstones are minerals that are hard, beautiful, and rare. Why must gemstones be hard?

SUGGESTED READINGS

ELWELL, D., 1979, *Man-Made Gemstones*, New York, Halsted Press John Wiley & Sons, 257 pp.
▶ *Fascinating review of a secretive world; beautiful color plates.*

EWING, R. C., February 1978, "The Elegant Symmetry of Crystals" in *Natural History*, pp. 65–71.
▶ *Stunning article and photography*

RICKWOOD, P. C., 1981, "The Largest Crystals" in *American Mineralogist*, pp. 885–907.
▶ *A single 80-ton beryl crystal and other wonders described.*

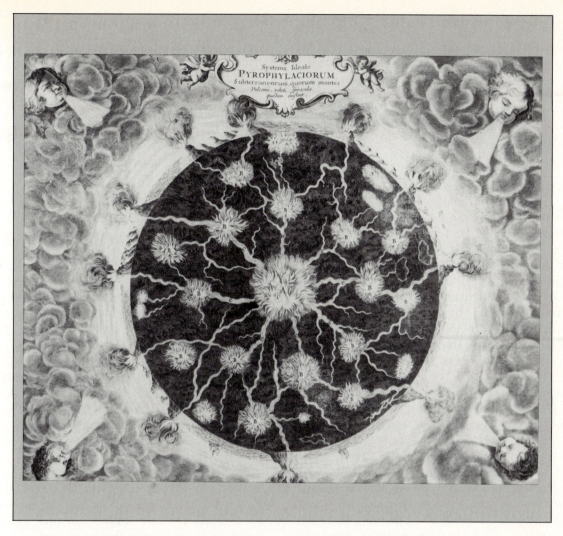

This seventeenth century engraving accurately represents the knowledge of the time. Great fires mysteriously raged within the earth; volcanoes were their "chimneys." (Courtesy of History of Science Collections, University of Oklahoma Libraries.)

CHAPTER *three*

Igneous Rocks

It is difficult to comprehend the violence of this period. Twenty-three known volcanoes operated in Colorado, some of them much larger than Vesuvius. . . . They deposited an incredible amount of new rock, more than fourteen thousand cubic miles. . . . They glowed through the nights, illuminating in ghostly flashes the mountains and plains they were creating.

JAMES MITCHENER, *CENTENNIAL*

James Mitchener, well-known American novelist, is describing the distant past, for there have been no volcanoes operating in Colorado within the brief span of human history. Rather he was describing a scene from many tens of millions of years ago; Mitchener inferred the past from the evidence left behind. The roots of old volcanoes form a large part of the southern Colorado landscape. What now is mute and slumbering must once have "illuminated in ghostly flashes the mountains and plains they were creating."

Active volcanoes create volcanic mountains and plains. They also vividly remind us of the pent-up heat energy beneath our feet. **Igneous rocks** of many kinds are formed from the cooling and crystallization of formerly molten matter. Our earth, apparently so stable and unchanging, retains the power within (Figure 3–1) to reshape its surface without warning. Where once there was sunny street and children's laughter, there may be glowing lava, ash, and destruction. It was American philosopher Will Durant who observed that . . . *civilization exists by geologic consent.*

To understand the source of that power, we first turn to Yellowstone National Park, one of the great places on earth where the power within can be safely studied. As Figure 3–1 suggests, if we could we harness the energy of about 400 Yellowstones, we would provide the total energy needs of the United States. Compared to volcanic areas elsewhere in our solar system, Yellowstone is a pipsqueak. It does, however, provide a window on a dynamic earth.

THE YELLOWSTONE AREA

Yellowstone National Park, with over 10,000 hot springs, geysers, "paint pots," and related phenomena, displays the interrelation between two geologic processes—**plutonism** and **volcanism**. Plutonism refers to a variety of processes that form liquid material within the earth and allow it to cool back to a solid while still within the earth. The liquid material is called **magma**, which is commonly a combination of solids (mineral crystals), liquids, and gases. Volcanism refers to a variety of processes that occur when magma under pressure reaches the earth's surface and erupts as **lava**.

The eruption of Mount St. Helens in southern Washington on May 18, 1980, ejected less than one-eighth of a cubic mile (0.5 cubic kilometer) of vapor–solid–liquid mixture under pressures more than a hundred times atmospheric pressure. The material expanded at supersonic velocities and devastated an area of nearly 200 square miles (500 square kilome-

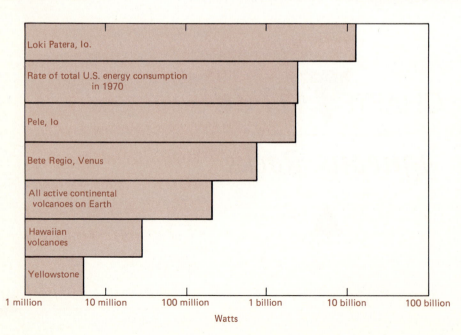

FIGURE 3–1

This graph describes the power output from various sources within our solar system. The Yellowstone area releases about 5 percent of the heat flow from the western United States as a point of comparison. The power released by individual volcanoes on Io, one of the satellites of Jupiter, equals or exceeds the rate of total U.S. energy consumption. If we could harness the output of the volcano Loki Patera on Io, it would meet the power needs of every human on earth.

ters) north of the volcano. The eruption left a crater 1.2 miles (2 kilometers) in diameter. The energy expended has been estimated at 25 to 35 megatons, with ash blasted more than 15 miles (24 kilometers) into the sky. For the millions of people affected, that eruption was a spectacular event.

How then to describe the most recent eruption in the Yellowstone area? Approximately 600,000 years ago, an eruption blasted 250 cubic miles (1000 cubic kilometers) of earth into the atmosphere! Layers of the resulting ash can be recognized over most of the western half of the United States. A **caldera**, the term for a large volcanic depression, about 44 miles (70 kilometers) across was created. Removing two thousand times as much material as the Mount St. Helens eruption (Figure 3–2), the eruption of the Yellowstone caldera must rank as one of the greatest recent catastrophes on earth.

In the hundreds of thousands of years since this event, vegetation has camouflaged the traces of the eruption so that few visitors to Yellowstone have any inkling of what once occurred. Other eruptions of similar kind include one that created the Long Valley caldera, near Mammoth Lakes, California, 700,000 years ago, and another that created the Valles caldera, near Santa Fe, New Mexico, approximately a million years ago.

The mechanism of caldera formation is well known. The eruption of huge volumes of magma from a magma chamber a few miles beneath the earth's surface causes the foundering and partial collapse of the chamber roof. The roof settles downward, leaving a depression at the surface, whose diameter may range from less than a mile to nearly 60 miles (100 kilometers). The depth of the depression ranges up to one mile (0.6 km), so that the overall shape of a caldera is saucerlike.

Yellowstone is a *resurgent* caldera, one where part of the floor has been slowly pushed back upward, probably from the intrusion of fresh magma from below into the magma chamber that created the caldera in the first place. The floor of the caldera has probably slowly risen many thousands of feet and is the roof of a magma chamber (Figure 3–3) beneath it. A resurgent caldera is, then, a depression with a central uplift.

In contrast, a **volcano** is a mound of solid material *added* to the surface of the earth by the extrusion of liquids and solids. It is a high area with a roughly central depression. The depression may be a central crater, commonly formed by explosive eruption of materials. A volcano may also have a caldera, created by collapse of a summit area after rapid eruption of liquids, solids, and gas.

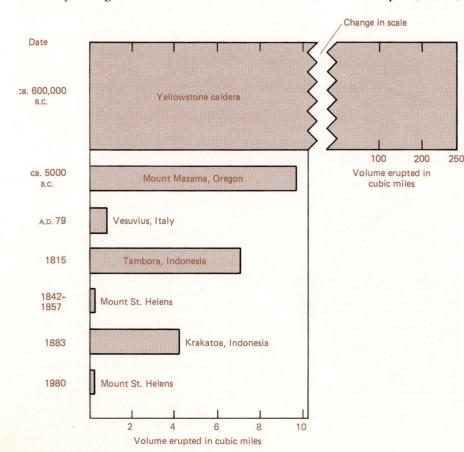

FIGURE 3–2

This chart compiled from data from the U. S. Geological Survey, suggests how trivial the volcanic events witnessed by humans have been compared to the eruption of the Yellowstone caldera. The eruption of Mount Mazama left Crater Lake occupying the caldera that resulted.

In the Yellowstone area, evidence from the passage of earthquake energy gives us a unique glimpse of what a magma chamber may look like. The passage of energy from a distant earthquake is slowed as the energy has to travel through liquid, rather than solid. By mapping the delays experienced by earthquake energy waves passing through the Yellowstone area, geologists have recently been able to infer that there is a partially molten body under Yellowstone, probably consisting of abundant randomly oriented cracks filled with magma. This zone of partially molten rock begins about three to six miles (5 to 10 kilometers) beneath the surface and is approximately pear-shaped, slanting downward toward the southwest. Approximately 30 miles (50 kilometers) wide at the middle, the body extends downward at least 90 miles (150 kilometers), and may extend downward easily twice that far (Figure 3–3).

Yellowstone is a particularly clear example of plutonism and volcanism because the processes are still in operation. It appears that the magma may have its source from the **asthenosphere**, a partially liquid layer about 100 miles (160 kilometers) down, and that it travels upward through the **lithosphere**, the outermost generally solid rocky shell of the earth. While traveling upward through the lithosphere, the magma expands, mixes with and melts other rock, goes through a variety of processes of chemical dif-

ferentiation, and may either cool in place to slowly form **plutonic rock** or erupt through the earth's surface to form **volcanic rock**. Thus both plutonic and volcanic rocks are igneous; they have formed from magma. What then is magma?

MAGMA CHARACTERISTICS

The word magma comes from the Greek root meaning "kneaded mixture." Magma is a mixture of three components:

1. Molten material, commonly of silicate composition;
2. Dissolved gases; and
3. Crystalline silicate and oxide minerals.

MOLTEN COMPONENTS

Eight elements are dominant among molten and crystalline material. These elements and their chemical symbols are oxygen (O), silicon (Si), aluminum (Al), calcium (Ca), iron (Fe), magnesium (Mg), so-

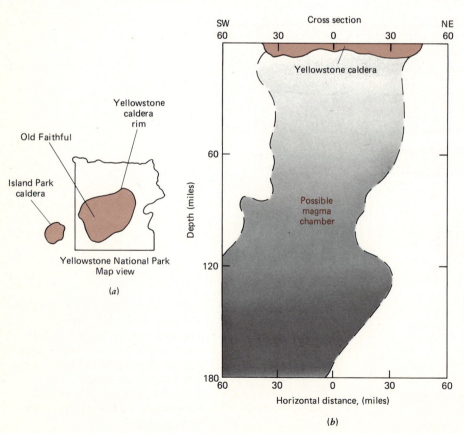

(a)

(b)

FIGURE 3–3

(A) Yellowstone caldera is depicted within the boundaries of Yellowstone National Park. Adjacent Island Park caldera is also shown. (B) A highly generalized diagram taken from distant earthquake data recorded by the U.S. Geological Survey. An area of semiliquid rock appears to extend from about 3 miles (5 kilometers) to perhaps as much as 180 miles (270 kilometers) down, with the percentage of liquid increasing upward.

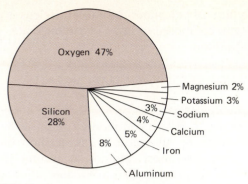

FIGURE 3–4

This pie diagram displays the abundance of elements by weight within the outer part of the earth. These data result from many thousands of chemical analyses of rocks and demonstrate that oxygen and silicon are overwhelmingly abundant in most rocks.

dium (Na), and potassium (K) (Figure 3–4). On chemical analysis, silicon dioxide (SiO_2) is the dominant component, ranging from 45 to 75 percent of the magma. Aluminum averages 15 percent; all other components are in amounts of 5 percent or less.

Since silica (SiO_2) is so dominant, its abundance largely controls the physical properties of magma. The higher the silica content of the magma, the higher the magma's **viscosity** or resistance to flow. Chemical analysis of many thousands of igneous rocks suggests that three types of magma, which grade into one another, are most common.

1. Low-silica magma has temperatures in the range of 900 to 1300°C and exhibits low viscosity—it flows readily. Low-silica lava forms spectacular lava rivers and commonly forms extensive, thin lava flows. Rocks formed from low-silica magma are generally dark in color and are heavy, with specific gravities between 3.0 to 3.3. The most common rock type produced is **basalt** (Figure 3–5), a volcanic rock containing 45 to 50 percent silica, and composed of the minerals calcic plagioclase (see Figure 2–20) and pyroxene, usually plus olivine (see Chapter Two); the much less common plutonic equivalent is **gabbro**.

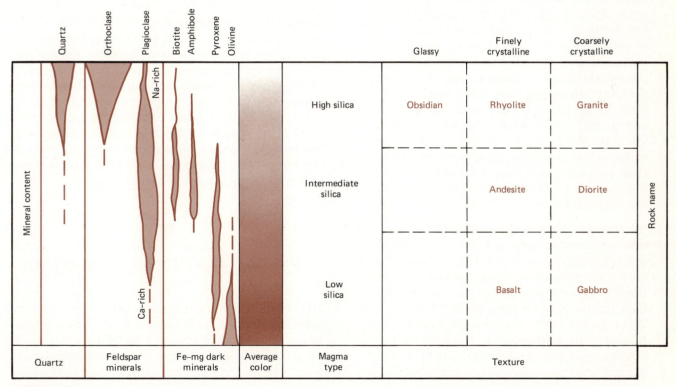

FIGURE 3–5

In this diagram those rocks composed largely of olivine and/or pyroxene are at the bottom with the composition gradationally changing upward, typical of an intrusion zoned from bottom to top by fractional crystallization. To the left are bars of varying width that express the changing mineral composition from bottom to top. To the right are the rock names that express the combinations of mineral content and mineral texture shown. As an example, a basalt is a finely crystalline igneous rock of low-silica magma type; having a dark overall color; composed of olivine, pyroxene, and calcium-rich plagioclase. A porphyritic basalt is shown in Figure 3–8.

Basaltic lava forms extensive flows and volcanic cones with gentle profiles. If on eruption there is rapid escape of dissolved gases, the eruptions may become more vigorous, with incandescent fountaining of glowing lava, and blasting of fragmental and porous material into the air. Entrapped gas may be released as the lava cools, leading to a porous texture on the tops of many lava flows. The Hawaiian volcanoes are formed from low-silica lava.

2. High-silica magma has temperatures in the range of 6 to 800°C and is far more viscous—it flows less readily. Because of the high silica content, the silica tetrahedra in high-silica lava start to link even before crystallization is complete, and these highly interlocked networks impede flow. The lavas are pasty and extrude as thick, bulbous flows or domes. Rocks formed from high-silica magma are light-colored and generally light in weight, having specific gravities between 2.4 to 2.7. The common rock type produced is **granite** (Figure 3–5), a plutonic rock containing 70 to 75 percent silica, and composed of orthoclase (see Figure 2–20) and quartz, usually plus some micas and/or amphiboles (see Chapter Two). The less common volcanic equivalent of granite is **rhyolite**. Rhyolitic lava eruptions are commonly violent because the pasty lava explosively releases the dissolved gases within. Incandescent clouds of ash and gas may travel many miles from the source of eruption and settle to form vast sheets of welded ash. *Obsidian*, a structure-less glass (see Figure 2–15) of generally rhyolitic composition, may form when high-silica lava chills so quickly that there is no time for crystals to form. The lavas in the Yellowstone area are of the high-silica type.

3. Intermediate-silica magma exhibits properties between those of high- and low-silica magmas and may form by an intermixing of these two within the lithosphere. Rocks formed from these magmas are of intermediate color and weight; their specific gravities range from 2.7 to 3.0. When cooled on the earth's surface these magmas form **andesite** (see Figure 3–8), a volcanic rock commonly containing 55 to 65 percent silica and composed of amphiboles and moderately sodic plagioclase (see Figure 2–20) crystals. The plutonic equivalent is **dacite**. Intermediate in viscosity, these magmas form both moderately stiff lava flows and numerous explosive eruptions of ash, a combination that forms the symmetric conical shape we think of

as typical of volcanoes. Mount St. Helens is a volcano composed of lavas of intermediate composition.

GASEOUS COMPONENTS

The **volatiles** or gaseous components are dominated by water vapor, which commonly comprises 85 percent or more of all volcanic gas. Carbon dioxide (CO^2) is the next most common at around 10 percent, while compounds of nitrogen and sulfur constitute a distant third. Other gases present in small quantities include compounds of hydrogen with chlorine, fluorine, bromine, and iodine. Volatiles total 1 to 5 percent by weight; generally they increase the fluidity of lava, lower its melting point, and raise the probability of explosive eruption. Because of their greater viscosity, high-silica magmas tend to trap most of their dissolved gases; it is the rapid expansion of these high-temperature high-pressured gases as magma nears the surface that causes extremely violent eruptions.

CRYSTALLINE COMPONENTS

The crystalline components are those minerals whose melting points are above the prevailing temperature. About nine minerals compose the over 700 varieties of named igneous rocks. They were first described in Chapter Two among the rock-forming minerals; their names and identifying properties are also given in Table 3–1. As the magma continues to cool, crystals grow at the expense of the fluid, and an igneous rock composed of masses of intergrown crystals eventually forms (see Figure 3–5). It is to the processes and sequence of crystallization that we now turn.

▼

CRYSTALLIZATION OF MAGMA

▲

The crystallization of magma is unlike our ordinary experience with liquids such as water. The crystallization of water (see Figure 3–6) into crystals of ice on a cold window occurs almost instantaneously when the temperature and pressure are right. Boiling, the conversion of liquid to gas, also occurs quickly at a fixed temperature and pressure. But water is a homogeneous material of a single chemical composition; its observed behavior (see Figure 3–6) involves three phases—solid, liquid, or gas—of a *single* substance.

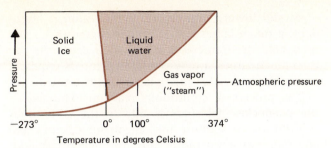

FIGURE 3–6

This diagram displays the temperatures and pressures at which ordinary water converts from solid to liquid to gaseous form. Notice that at temperatures above 374° C , water can *only* exist as a gas, regardless of pressure applied. All water within magma exists only as a dense vapor, and is highly reactive.

Magma is a complex three-phase (solid, liquid, and gas) material made up of anywhere from 10 to 40 or more elements, though dominated by the abundant three (see Figure 3–4). Each of the dozen or more minerals that form within magma as it cools has its own unique chemical composition and therefore its own crystallization or freezing point. Consequently, *each mineral forms while other minerals also crystallize in overlapping sequences of temperature ranges*—from those with high melting and freezing points down to those with low melting and freezing points. As an additional complication, the chemical composition and internal structure of many minerals *changes* as temperature declines; most minerals *react* with the magma as it cools. The complex sequence of events in a cooling magma was finally worked out by an American chemist in the 1920s.

THE BOWEN REACTION SERIES

Norman L. Bowen was an American geologist and chemist whose attention, early in this century, was drawn to the problem of crystal formation and reaction in cooling magma. In the course of his studies, which lasted nearly 50 years, Bowen applied both the principles of physical chemistry and high-temperature laboratory experimentation. He was able to recognize two distinct but overlapping schemes of magma–crystal reaction:

1. Continuous reaction, and
2. Discontinuous reaction.

Operating simultaneously within cooling magma, these two crystallization processes are termed the **Bowen reaction series**.

Continuous Reaction.

The plagioclase feldspars (see Figure 2–20, Table 3–1, and Figure 3–7) form a group of minerals whose composition exhibits gradational variability. From *anorthite*, the pure calcium end member to *albite*, the

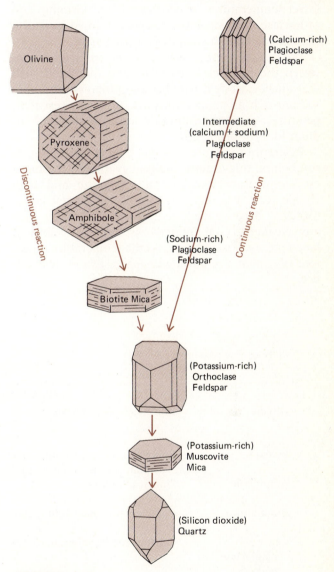

FIGURE 3–7

Although crystals are drawn here to illustrate their appearance, perfect crystals like these rarely form within magma, since the growing crystals compete for space in a fluid of diminishing volume. The more common shape and arrangement of crystals within igneous rocks is shown in Figures 3–8 and 3–22. This diagram displays the sequence of crystallization within magma as described by Bowen and includes both the continuous reaction series and the discontinuous reaction series. Quartz is the last mineral to form, while olivine and calcic plagioclase are the first.

sodic-rich pure sodium end member, any intermediate composition is possible.

Bowen observed that as magma cooled, the first plagioclase feldspar to form (at the highest temperature) was always rich in calcium and aluminum. As temperature dropped, the early formed crystals became more sodic in overall composition. At still lower temperatures, plagioclase ceased forming, and orthoclase, the potassium feldspar, formed instead if the magma contained sufficient potassium.

What Bowen saw was a *continuous reaction* between the plagioclase feldspars and the liquid such that plagioclase feldspars became progressively more sodium-rich as temperature fell. At the lowest temperatures, potassium abruptly took the place of sodium to form orthoclase.

Reference to Figure 2–4 will help explain the reason behind this orderly chain of events. At the highest temperatures, an ion of small size and high charge (Ca^{+2}) is required to hold the crystal lattice together, because high temperatures instantly melt any weaker crystal frameworks formed by ions of larger size or lesser charge. As the temperature drops, ions of similar size but lower charge (Na^{+1}) will suffice.

At still lower temperature, if potassium is abundant, a much larger ion of similar charge (K^{+1}) can now hold the framework together. Because the ion is so much larger, a new crystal structure is required, and orthoclase *rather than* plagioclase is formed.

The process of continuous reaction exhibited by the plagioclase feldspars is due to the continuous substitution of sodium and silicon for calcium and aluminum as temperature falls. The plagioclase feldspars form a **continuous reaction series**, a series of minerals whose composition varies with temperature by continuous reaction between the crystallizing mineral and the liquid around it.

Discontinuous Reaction

Bowen observed that the darker minerals such as olivine, the pyroxenes, the amphiboles, and biotite mica (see Figure 3–6) exhibited a totally different crystallization sequence. Each of these four mineral groups has its own separate crystal lattice, so they *cannot* show gradational change among themselves. Like the plagioclase feldspars, however, each successive mineral formed at lower temperatures requires more silicon and incorporates more ions of larger size and smaller ionic charge.

TABLE 3–1
Igneous Rock-Forming Minerals

Name	Chemical Composition	Appearance
Dark-Colored Minerals		
Olivine	Iron-magnesium silicate	Bottle green to yellowish greenish brown, glassy, stubby, fractured crystals or granules
Pyroxene	Complex iron, magnesium, calcium, sodium silicate, or aluminosilicate	Very dark green to black, dull, stubby crystals or granules; right-angle cleavage
Amphibole	Similar to pyroxene	Very dark green to black, shiny, prismatic needles
Biotite mica	Complex iron-potassium aluminosilicate	Greenish brown to reddish, bronze, or chocolate brown shiny flakes
Calcic plagioclase feldspar (anorthite)	Calcium-sodium aluminosilicate	Dark gray to dark greenish gray rectangular to lathlike crystals, minute stractions on some cleavage faces, occasionally irridescent
Light-Colored Minerals		
Sodic plagioclase feldspar (albite)	Sodium-calcium aluminosilicate	Light gray to cream-colored rectangular to lathlike crystals, minute striations on some cleavage faces
Orthoclase feldspar	Potassium-sodium aluminosilicate	Salmon pink to creamy white rectangular crystals
Muscovite mica	Potassium aluminosilicate	Clear, sparkling flakes
Quartz	Silicon dioxide	Clear to cloudy white glassy fragments; conchoidal fracture, no cleavage; look like glass fragments in the rock

As magma slowly cools, olivine commonly crystallizes first. As temperature drops, pyroxene abruptly begins to form *at the expense of olivine*. If enough silica and time are available, the olivine may be totally converted to pyroxene, and not a trace of olivine will be left. As the temperature continues to drop, the pyroxene reacts with the melt to form amphibole, and all traces of pyroxene may be destroyed. At still lower temperatures, the amphibole reacts with the magma to form biotite mica.

Such a series of "jumps," in which one mineral group reacts to form a totally different mineral group is described as *discontinuous reaction*, and the minerals olivine, pyroxene, amphibole, and biotite form a **discontinuous reaction series**. The series is distinguished by the dark-colored minerals reacting within a cooling magma by either partial or total conversion to one another as temperature falls.

FRACTIONATION OF MAGMA

A study of the textures of igneous rocks had suggested long before Bowen that there was a sequence of crystallization of minerals from magma. *Slow cooling allows time for large crystals to form; rapid cooling results in tiny crystals or a noncrystalline glass*. If the magma at first cools slowly at depth then moves upward as a partial melt and rapidly cools, the result is an igneous rock having two distinct grain sizes; such a rock is said to be **porphyritic**. Many volcanic rocks are porphyritic and so neatly preserve the evidence of sequence of crystallization. The early formed minerals are large **phenocrysts** while the later minerals form a finely crystalline *groundmass* (see Figure 3–8).

Many geologists recognized that phenocrysts tended to be olivine and/or pyroxene plus dark-colored calcic plagioclase feldspars. The overall chemical composition of these early minerals is quite different than the bulk composition of the magma from which they crystallized. The early formed minerals are much richer in iron, magnesium, calcium, and aluminum and much lower in silica than their parent magma; they are also somewhat denser.

Charles Darwin (1809–1882), much better known for his contributions to the theory of organic evolution, was among those who realized that the sinking of the early formed denser minerals to the bottom of the magma body would *change the chemical composition of the remaining melt by selectively removing heavier elements*. He suggested that such a process might be a cause of variation among igneous rocks.

The segregation of olivine, pyroxene, and calcic plagioclase depletes the magma of iron, magnesium, calcium, and aluminum and leaves the magma relatively enriched in sodium, potassium, and silicon. By selectively removing successive groups of crystals, an originally basaltic magma *fractionates* or splits into *different* magmas, each of which is increasingly enriched in elements that will form the lighter-colored, lighter-weight minerals. A magma, whose original composition was basaltic, could thus produce a *series* of magmas ranging in composition from basalt to rhyolite.

Large slowly cooled masses of basaltic magma may complete one of two crystallization sequences:

1. **Fractional crystallization** occurs when processes fractionate or split the minerals that

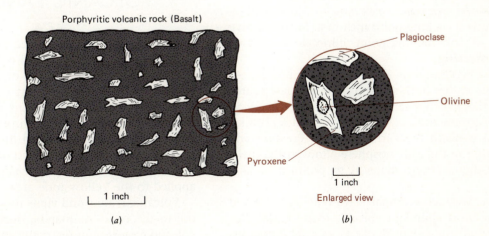

Porphyritic volcanic rock (Basalt)

Plagioclase

Olivine

Pyroxene

1 inch

1 inch

Enlarged view

(a)   (b)

FIGURE 3–8

This is a sketch of a porphyritic (composed of two distinct crystal sizes) volcanic rock. The enlarged view shows an olivine remnant *within* a pyroxene crystal; in this example the discontinuous change of olivine to pyroxene was not completed when the rock became solid. The rock is basalt (see Figure 3–8); the finer-grained material is tiny plagioclase crystals and glass, while the larger crystals are olivine, pyroxene, and calcic plagioclase.

form away from the remaining magma; one such process involving crystal sinking was suggested by Darwin. The lowermost layer of the cooled rock will consist largely of olivine and/or pyroxene plus calcic plagioclase (see Figure 3–5). The composition will grade upward into andesitic rock, composed of amphibole and intermediate plagioclase. Still higher will be rocks of rhyolitic composition; orthoclase and quartz are the "leftovers" when heavier minerals are successively withdrawn from a magma. None of the solid rocks has a composition like that of the original magma; instead each rock type reflects the composition of the magma remaining after one "crop" of minerals has been removed. Yet, the *overall composition of the vertical column of rocks has remained the same.*

2. **Homogeneous crystallization** occurs when turbulence within the cooling magma keeps crystals in constant contact with the liquid. With this kind of turbulent mixing, it is impossible for crystals to separate from the liquid in which they form. Rather the crystals remain in contact with the liquid and react with it through either discontinuous or continuous reaction. With homogenous crystallization, a magma of basaltic composition cools to form basalt. A small piece of the solid rock has the same bulk composition as the original magma.

GENERATION OF MAGMA

In the same way that the first minerals to form have compositions unlike their parent magma, *the first liquid to form when a rock is heated has a different composition than the rock from which it melts.* Minerals having the lowest melting point melt first. Such selective melting of low-melting-point minerals is termed **partial melting.**

Magma generation is a partial melting process. As mentioned earlier in this chapter, the ultimate source of most magmas appears to be the earth's *asthenosphere*, a partially fluid layer about 100 miles (160 kilometers) beneath the earth's surface. Pressure and temperature in the asthenosphere allow partial melting of asthenospheric material to produce basaltic magma.

Basaltic lava is the primary lava type on earth and forms the most abundant volcanic rock—basalt. Basalt forms all the ocean floors as well as numerous volcanoes and flows on both ocean floor and continent.

Continental basaltic volcanoes may have smaller amounts of andesitic and even rhyolitic rocks associated with them. These andesitic and rhyolitic rocks may result either from fractional crystallization of a basaltic parent magma or by partial melting of basaltic rock, with the lower melting point liquids forming a magma of andesitic or rhyolitic composition. Partial melting *always* produces a more silica-rich liquid than its source rock.

Andesitic magma is generated *only* beneath continents where the melting of silica-rich sedimentary rocks mixed into basaltic lava produces a more silica-rich magma by magma contamination and mixing. Other andesitic lavas may be produced by partial melting of basaltic/gabbroic rocks. Rhyolitic lavas are generated only beneath continents. Rhyolites appear to form by combinations of the partial melting of andesitic rocks, fractional crystallization of basaltic lava, and complete melting of sedimentary rocks. Each of these three processes produce high-silica magmas that may cool to form rhyolite or granite.

The only primary magmas are those of basaltic composition. Andesitic and rhyolitic magmas appear to be formed by

1. Fractional crystallization of primary basaltic lava,
2. Contamination of basaltic magmas as they encounter silica-rich sedimentary rocks, and
3. Partial melting of rocks within the lithosphere.

The vast range of igneous rocks—over 700 varieties have been named—result from the complex interplay of these three processes.

VOLCANIC LANDFORMS

A volcano is an upland added to the earth's surface by the eruption of intermixed volcanic gases, liquids, and solids. Near the summit of a volcano is often a **crater** which marks the **vent** through which materials erupt. The crater or vent is connected downward to the magma source through a complex series of cracks, collectively termed the volcanic *conduit* or *pipe*. Particularly large depressions, formed by explosion and collapse, are called *calderas*, a term we have applied to the Yellowstone area.

Volcanic craters and vents may not always be central to the cone. As magma pushes its way into the volcano, cracking of the crater may lead to eruptions from the base of the volcano or anywhere on its flanks. A volcano may erupt from numerous locations at once.

The possible combinations of magma chemistry, volatile content, and temperature lead to a wide range of volcanic landforms. Most volcanoes, however, can be assigned to one of four types:

1. Cinder cones,
2. Shield volcanoes,
3. Composite cones or stratovolcanoes, and
4. Lava domes

CINDER CONES

A **cinder cone** is built predominantly of cooled fragments of basaltic lava blasted into the air under high gas pressure. Rock fragments of volcanic origin are termed **pyroclastics** (*pyro* means fire, while *clastic* means fragments). Accumulations of consolidated pyroclastic fragments form rocks that may be classified as *either* volcanic or sedimentary rocks.

The average size of a fragment within a cinder cone is that of a walnut, generally between one-tenth of an inch (0.2 centimeters) and an inch (2.5 centimeters). Such fragments are generally termed *cinders*. Finer pyroclastic material is called **ash**, and coarser material is variously described as **blocks** and **bombs** (see Figure 3–9).

As the cinders fall back from the air, they form a steep-sided conical hill (see Figure 3–10) composed of unconsolidated cinders. Most cinder cones are less than a thousand feet (300 meters) high and occur in groups either on the flanks of major volcanoes or within relatively older basaltic volcanic areas. The span of their eruptive lifetime is often only a few years. In general, the larger the cinder cone, the greater the distance to the next cone, which suggests that each cone "drains" an underground area proportional to its size.

The unconsolidated loose cinders piled up on the earth's surface are easily washed and blown away. If the loose material is slowly removed, the central pipe or conduit, filled with hard source rock may remain for a time. Such a pipe forms a **volcanic neck**; it is the exposed "plumbing" after the cone has been washed away. A volcanic neck gives us some sense of what the "inside" of a volcano must have looked like as it was forming.

SHIELD VOLCANOES

When highly fluid basaltic lava is extruded, a slightly domed broad upland, termed a **shield volcano**, is formed. Shield volcanoes have a surface slope of only a few degrees. Shield volcanoes may represent discontinuous eruptions over a period of millions of years.

The island of Hawaii at the southeastern corner of the Hawaiian Island chain is the largest and youngest of the islands. Hawaii is composed of four interfingering shield volcanoes whose eruption stretches back more than a million years. Their aggregate diameter exceeds 90 miles (150 kilometers). The summit of the largest shield volcano, known as Mauna Loa (see Figure 3–11) stands 13,680 feet (4200 meters) above sea level and a total of 30,000 feet (9200 meters) above the sea floor. Mauna Loa thus rises higher above the sea floor than Mount Everest does above sea level. The total volume of extruded lava required to form the island of Hawaii exceeds 10,000 cubic miles (40,000 cubic kilometers). Given its age and volume, the average eruption rate has been a cubic mile of basalt per century.

Mauna Loa is the largest active volcano on earth, though it pales by comparison with shield volcanoes on some of the other planets (see Figure 3–12). Its

FIGURE 3–9
A volcanic bomb, caused when a blob of ejected lava was streamlined and frozen during the passage of lava through the air. The bomb is about 2 feet (37/85 meter) long. (Photograph by H. T. Stearns, courtesy of the U.S. Geological Survey.)

FIGURE 3–10
The rounded conical form of this cinder cone northeast of Flagstaff, Arizona, is fairly typical. The cone is little more than a pile of cinders which resulted from eruption and free fall of cinders on to the earth's surface. (Photograph by G. K. Gilbert, courtesy of the Geological Survey.)

neighboring cones are Kohala, Hualalai, and Mauna Kea. Yawning on the eastern slope of Mauna Loa is the Kilauea (pronounced keel-ah-way'-ah) caldera, which is a part of the Hawaii Volcanoes National Park and adjacent to the U. S. Geological Survey's Volcanological Observatory. The Kilauea area is dotted with instruments that deftly record the activity beneath it and make Kilauea and its enormous parent, Mauna Loa, the most thoroughly studied and best understood volcanoes on earth.

Volcanic activity in the Kilauea caldera is nearly continuous and usually amounts to gentle eruptions of steam and smoke. Larger eruptions of liquid lava occur, and still more rarely, fountains of basaltic lava may blast into the air to heights of 600 feet (180 meters), a height greater than the Washington Monument. U. S. Geological Survey geologists have studied these numerous forms of eruption "close-in," and there have been a few very severe burns when curiosity outran luck. The lava has been repeatedly sampled for chemistry, mineral content, volatile makeup, and temperature. Huge outpourings of lava have created lava lakes that have taken decades to cool. Sampling of the cooling lava at many stages has confirmed and refined the experimental work of Bowen done 70 years before.

FIGURE 3–11

The summit of Mauna Loa, Hawaii, from the air. Mauna Loa is the largest shield volcano on earth, though it is dwarfed by some planetary neighbors (Figure 3–12). The summit of Mauna Loa, here above the clouds, stands 13,680 feet (4200 meters) above sea level and a total of 30,000 feet (9200 meters) above the sea floor.

Mark Twain visited one of the lava lakes a century ago and recorded in *Roughing It* his own irreverent description of what it was like to watch the lava lakes of the Kilauea area (see Figure 3–13) at night.

. . . here was a vast, perpendicular walled cellar, nine hundred feet deep in some places, thirteen hundred in others, level-floored, and ten miles in circumference! . . . The greater part of the vast floor . . . was as black as ink . . . but over a mile square of it was ringed and streaked with a thousand branching streams of liquid and gorgeously brilliant fire! . . . Imagine it—imagine a coalblack sky shivered into a tangled network of angry fire! . . . The smell of sulphur is strong, but not unpleasant to a sinner.

Earthquakes associated with the eruption of basaltic lava suggest that the magma source for Kilauea is at least 35 miles (60 kilometers) beneath it. As the magma forces its way upward, it accumulates in a shallow chamber beneath the summit of Kilauea, causing the summit of the crater to inflate, swell, and tilt as the pressure rises. Cracking of the stretched, tilted surface is signaled by swarms of small earthquakes, sometimes as many as 1000 per day. A rapid increase in earthquake intensity and number indicates an impending eruption. During the eruption the summit deflates and tilts back. After the eruption, the volume of lava extruded equals the amount of summit subsidence.

When deep earthquakes are studied, many of them reveal bursts of **harmonic tremor**, a peculiar quivering motion of cracks carrying highly pressured magma. This zone of deep earthquakes is somewhat southeast of Kilauea and is actually under the ocean floor. About 30 miles (50 kilometers) southeast of Kilauea is Loihi, an underwater volcano discovered in the 1950s. It appears that the zone of deep earthquakes records the location of the magma source that feeds both Kilauea and Loihi. If eruption continues at current rates, Loihi will build itself above sea level in a few tens of thousands of years and be the youngest of the Hawaiian islands. Loihi might grow to the size of Mauna Loa in half a million years or so. Building Hawaiian shield volcanoes is very much an ongoing process.

COMPOSITE CONES OR STRATOVOLCANOES

Among the most beautiful of all mountains are the volcanoes variously termed **composite cones** or **stratovolcanoes**. Formed from the intermittent

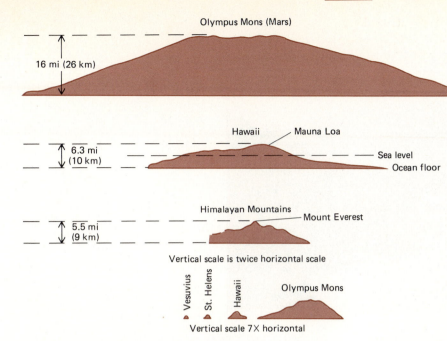

FIGURE 3–12

Comparative size of various mountains and volcanoes on earth with Olympus Mons, a volcano on Mars that would totally cover the State of New Mexico if placed on earth.

of generally andesitic lava, composite cones are formed of roughly alternating layers of pyroclastics and lava (see Figure 3–14).

Among well-known composite cones are Mount Taal in the Philippines, Mount Fuji in Japan, Mount Garibaldi in British Columbia, Mount Hood and Mount Jefferson in Oregon, Mount Baker, Mount Rainier, Mount Adams, and Mount St. Helens in Washington. Crater Lake, in Oregon, is a "beheaded" composite cone named Mount Mazama,

FIGURE 3–13

Halemaumau crater at night in a 1961 eruption. Large slabs of cooler black lava float on incandescent liquid. As you can see, Mark Twain described it well. (Photograph by D. H. Richter, courtesy of the Geological Survey.)

which lost its summit in a series of huge explosions about 7000 years ago. The remaining volcano collapsed to form a caldera which later filled with rain water to form Crater Lake.

Mount St. Helens was a symmetrical volcanic cone located in southwestern Washington, about 50 miles northeast of Portland, Oregon. It is one of the 15 or so Cascade Range volcanoes and domes (see Figure 3–15). Most of what can be seen has formed within the last 10,000 years, but the volcanic center may be at least 40,000 years old. Within the last 4500 years Mount St. Helens has had a high frequency of eruptions. The catastrophic eruption in May of 1980 occurred after some two months of events, including numerous small to moderate earthquakes, small eruptions and explosions with ash and steam clouds, small avalanches, and the formation of several small craters. As time wore on, harmonic tremors were recorded, which was evidence for motion of pressured magma under the mountain. Two weeks before the May 18th blast, the north slope of the mountain began to bulge with an average displacement of five feet (1.6 meters) per day; parts of the upper north flank had moved a total of nearly 300 feet (90 meters) prior the blast.

On May 18, 1980, at 8:32 A.M., a moderately large earthquake accompanied the largest landslide ever recorded in historic time as half a cubic mile (2.3 cubic kilometers) of the north flank slid downwards, suddenly releasing pressure on the pent-up magma that had risen into the summit. A devastating lateral blast was followed almost immediately by a vertical blast as an ashy column of pyroclastic debris rose more than 12 miles (20 kilometers) above the vol-

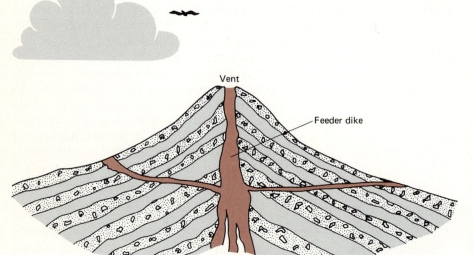

Vent

Feeder dike

? ?

FIGURE 3–14

(A) Mount Mayon in the
Philippines displays the classic
shape of a stratovolcano.
(Photograph by H. Gannett,
courtesy of the U.S. Geological
Survey.) Sketch (B) illustrating
what lies within Mount Mayon.
The stratovolcano is composed of
alternating layers of lava flows (in
color) and fragmental ash (in
black) and is fed through a central
vent with a pipe-like feeder dike
beneath. Smaller dikes may feed
small flank eruptions on the sides
of the cone.

cano. The explosion was heard 200 miles (320 kilometers) away. A *pyroclastic flow*, a volcanic eruption
consisting of glowing hot particles of ash suspended
in turbulent, expanding gas, swept to the north and
northwest up to 17 miles (27 kilometers), burying the
area with up to 180 feet (55 meters) of mud and ash.

Mount St. Helens erupted explosively five more
times during 1980, and in so doing began to rebuild
its elevation by creating a central dome in the crater
formed by the explosions. Since 1981, the central
dome has been enlarged by the addition of an average of 35 million cubic feet or 1 million cubic meters
per month of lava. At this rate, it will take about two
centuries for Mount St. Helens to regain its former
size.

That scenario, however, is highly unlikely. It is
more likely that future moderate to violent eruptions
will blow away all or part of the growing central
dome. Such eruptions may punctuate centuries of
relative quiet. Mount St. Helens is a typical stratovolcano formed by thousands of years of intermittent
cone building and cone destruction. The process can
only continue.

Lavas of the Mount St. Helens dome are largely
dacite, a volcanic rock intermediate in composition
between andesite and rhyolite (see Figure 3–5).
Rhyolitic lava contains a much higher percentage of
silica than Hawaiian basalt and is about a million
times more viscous. Such viscous lavas form lava
domes, the fourth type of volcanic cone.

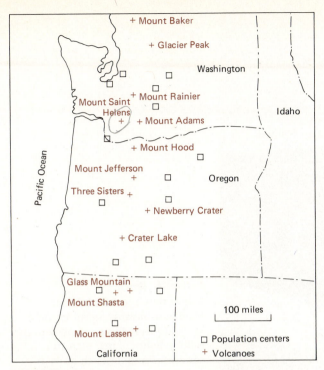

FIGURE 3–15
Mount St. Helens is one of a dozen volcanoes within the Pacific Northwest that have erupted within the last 10,000 years. These volcanoes pose a hazard to those living nearby.

LAVA DOMES

Lava domes, sometimes called plug domes or simply volcanic domes, are formed when extremely stiff (highly viscous) lava is extruded. Lava domes are usually fairly small as they are too viscous to flow any distance from their source. Most lava domes grow by expansion from within, so their internal structure is a series of pasty flow lines.

As the dome grows, its outer surface may shatter and form spiny knobs over the magma source. Many domes occur within the craters or on the flanks of composite cones, while others may build from flat ground within a volcanic field.

Examples of lava domes include Lassen Peak and Mono Dome in California and Mont Pelée in Martinique, West Indies. The eruption of Mont Pelée in 1902 accompanied the growth of a dome and buried the coastal town of St. Pierre about four miles (6.5 kilometers) away under an incandescent *pyroclastic flow* of white-hot ash and gas. Such a flow, called a **nuée ardêntes** (translated from the French as "fiery ash cloud," and pronounced new-á ahr-dahnt') may move at speeds in excess of 60 miles (100 kilome-

ters) an hour as a glowing avalanche, burning and flattening everything in its path within tens of miles from the vent. The 1902 eruption suffocated and burned the 28,000 citizens of St. Pierre in a few minutes on May 8. Only two people survived, one because he was a prisoner in a poorly ventilated dungeon.

▼

OTHER VOLCANIC FEATURES

▲

Volcanic activity produces a wide variety of landforms other than constructional landforms like volcanic mountains. Other easily visible results of volcanism include:

1. Lava flows,
2. Lava tubes,
3. Lava plateaus, and
4. Dikes and sills.

LAVA FLOWS

Not all lava erupts with explosive violence or forms a mountain. If the lava is a thin fluid, any gas may easily escape and the lava spills quietly out of multiple vents to form a **lava flow**. Many lava flows are tonguelike in map view and often build out from the base or anywhere on the flank or top of a basaltic volcano (see Figure 3–16).

Rivers of moving lava may flow at speeds from 3 to 30 miles (5 to 50 kilometers) per hour, depending on the slope. Cooling lava may form rubbly blocks ranging in size from shoe boxes to houses; the resulting rough texture on the surface of the flow is called by its Polynesian name **aa** (pronounced ah-ah). The flow shown in Figure 3–16 is an aa flow, which typically is extremely difficult and even dangerous to walk across. Eruptions of more fluid basalt form wrinkled, ropy surfaces on flow tops; such a texture is termed **pahoehoe** (pronounced pa-hoy'-hoy) (see Figure 3–17).

LAVA TUBES

Sometimes the rivers of flowing basaltic lava crust over on top, while the lava beneath continues to flow. As the eruption slows the lava beneath may drain away, leaving empty caves known as **lava tubes**. These tubes range in diameter from a few feet to several tens of feet and can sometimes be followed underground for thousands of feet.

FIGURE 3–16
A lava flow on the Uinkaret Plateau in northwestern Arizona. The dark flow fills in a valley in the rolling, light-colored plateau surface. (From a sketch by W.H. Holmes, courtesy of the U. S. Geological Survey.)

FIGURE 3–17
This lava flow from Kilauea volcano, Hawaii exhibits the "ropy" texture on its surface that is termed pahoehoe. (Photograph by R. T. Holcomb, courtesy of the U. S. Geological Survey.)

LAVA PLATEAUS

When quiet eruptions of extremely fluid lava recur over tens of thousands of years, the flows may stack one on top of another and cover huge areas. An area buried under an extensive cover of multiple lava flows is a **lava plateau**. The Columbia River Plateau of Washington and Oregon is an example of this type of volcanic activity (see Figure 3–18).

The Columbia River Plateau encompasses an area of 100,000 square miles (260,000 square kilometers). Flow on flow is stacked to total heights that can reach more than a thousand feet (300 meters). Individual lava flows on the Columbia River Plateau can be traced for distances in excess of 100 miles (160 kilometers) and often bury old soils formed on the surface of the underlying flow (see Figure 3–19). To cover such an immense distance and to spread so uniformly, the lava must have been as fluid as water. The total volume of lava erupted over several million years is 35,000 cubic miles (140,000 cubic kilometers), a volume equivalent to a single lava flow 750 feet (230 meters) thick covering the entire state of Texas.

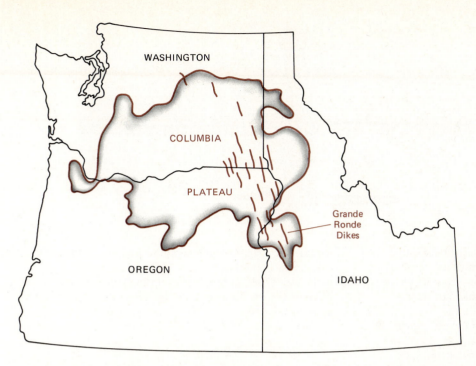

FIGURE 3–18

A sketch map of the location of the Columbia Plateau, one of the world's larger lava plateaus. Lavas were fed to the surface through numerous systems of dikes, of which only one set are shown here. Most of the feeder dikes cannot be located, as they are buried under layer after layer of younger lava.

DIKES AND SILLS

Volcanoes are fed magma through an intricate series of underground cracks. After cooling, the former cracks remain filled with hard igneous rock. Where those rock-filled cracks cross the layering of preexisting rock, the igneous rock is termed a **dike** (see Figure 3–20). Where the pressured magma splits a layered rock and intrudes between the layers, the resulting tabular rock layer is termed a **sill** (see Figure 3–20).

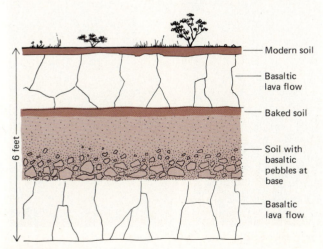

FIGURE 3–19

Sketch from field notes of an area near Walla Walla, Washington. An overlying basaltic lava flow buries and bakes an underlying soil zone. Within the unbaked soil, leaf fossils are visible.

Dikes form when magma solidifies in the cracks in the overlying rock through which magma fed upwards to a volcanic flow or cone. Many dikes are termed *feeder dikes* where it is reasonable to connect the dike with a known eruptive center. The Grande Ronde dike system served as feeder dikes to the Columbia Plateau. These dikes extend over distances of 120 miles (200 kilometers) and have a total combined width of 30 miles (50 kilometers); their intrusion must have caused considerable extension in the rocks they split. After lengthy periods of erosion the overlying volcano can be eroded away; leaving behind a **volcanic neck** (see Figure 3–20), formed when the feeder dikes are left exposed to view (see Figure 3–21).

Since dikes and sills originally form underground, the textures of the resulting rock often reflect slower cooling rates than those experienced by volcanic rock extruded directly onto the surface. As such, dikes and sills connect the volcanic world above with the plutonic world beneath. It is to that plutonic world that we now turn our attention.

▼

PLUTONISM

▲

Plutonic rocks form from the slow underground cooling of magma and therefore are made of crystals large enough to be easily seen. Inspection of plutonic

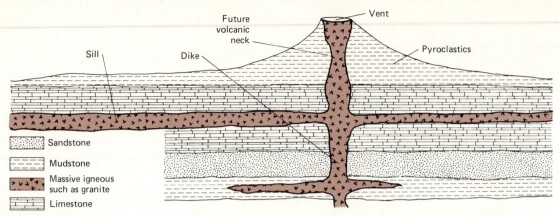

Future
volcanic
neck

Vent

Sill

Dike

Pyroclastics

Sandstone

Mudstone

Massive igneous
such as granite

Limestone

FIGURE 3-20

A schematic cross-section through the earth beneath a volcano. The vent is fed through feeder dikes. Magma flowing upward under pressure not only cuts across the layering of overlying rocks to form a dike but may also wedge layered rocks apart to form a sill that approximately parallels the layering of the older rock.

Notice that a sill, unlike a lava flow (Figure 3–19), will bake rocks both above and below it. If the loose pyroclastic material that makes up the volcanic cone is worn away, the upper part of the feeder dike is exposed as a volcanic neck.

rocks with a simple magnifying glass will reveal the texture shown in Figures 3–22 and 3–23. Minerals have intergrown to form a dense mesh of interlocking crystals.

The word *plutonic* is from Pluto, the Roman god of the underworld. Our word volcanic derives from Vulcan, the Roman god of fire and metalworking. Early people living in the area of a little volcanic island in the Mediterranean Sea off Sicily saw smoke, sparks, and flame occasionally issuing from the island and thought that the island must be the chimney of Vulcan, blacksmith to the Roman gods. They named the island Vulcano. The words plutonic and volcanic have come down to us from these ancient times.

As mentioned, dikes and sills are rock bodies that straddle the classification of objects into plutonic and volcanic associations. Depending on how much overlying rock has been worn away, we may see dikes and sills that fed volcanic hills at the surface or much more deeply seated dikes from magma chambers many miles within the earth (see Figure 3–24).

LACCOLITHS

If a deep-seated magma source pumps magma into a sill, the pressure of the intruding magma may be sufficient to lift the upper layers of the intruded rocks (see Figure 3–24). Magma under pressure arches the overlying rocks and leaves behind a much-enlarged sill, now referred to as a **laccolith**. A laccolith is a

FIGURE 3-21

These dikes exposed near Alamosa, Colorado, are more resistant to weathering and erosion than the adjacent rock. The dikes therefore form vertical walls marching across the Colorado landscape. (Photograph by G. W. Stose, courtesy of the U. S. Geological Survey.)

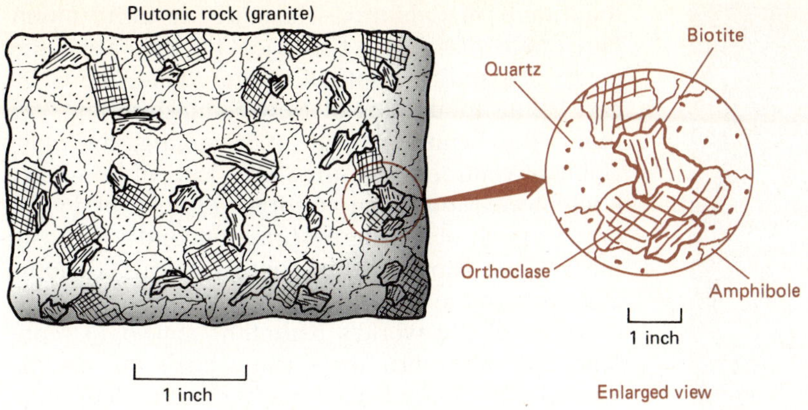

Plutonic rock (granite)

Quartz

Biotite

Orthoclase

Amphibole

1 inch

1 inch

Enlarged view

FIGURE 3–22

A sketch of a specimen of a common igneous rock displaying the coarsely crystalline texture of a plutonic rock. The enlargement of an area of the rock shows the interpenetrating pattern of crystals that forms when crystals grow in competition with one another. The plutonic rock is granite (Figure 3–8).

fairly common intrusive or plutonic body, usually composed of intermediate- to high-silica rocks. When exposed to view after erosion has stripped away part of the overlying rocks, laccoliths are composed of finely crystalline porphyritic rocks.

Laccoliths also provide an interesting lesson in magma hydraulics. The maximum height of a volcanic vent is limited by the pressure required to pump magma to the summit. For eruption to occur, the weight, and thus the height, of the column of magma from the summit to the magma source must be less than the pressure exerted on the magma chamber by the overlying rocks and expanding volatiles within the magma. Magma is able to reach the summit of a volcano because its specific gravity is less than that of the surrounding solid rock; the greater the specific gravity (density) contrast the higher the magma can reach.

The density contrast between the intermediate- to high-silica magmas that form laccoliths and the sedimentary rocks they commonly intrude is small; thus feeder dikes beneath a laccolith cannot connect to a deep source. Moreover, for a laccolith to form, the pressure from the magma below must actually *lift* the total weight of the layers of overlying rock. Since, to quote Johannes Kepler "Nature uses as little as possible of anything," laccoliths form *very* near the earth's surface, thus minimizing the pressure difference required for uplift when the density contrast is low. Laccoliths represent a compromise.

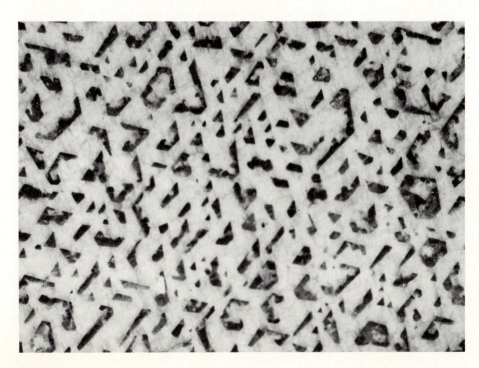

FIGURE 3–23

This plutonic rock's texture is unusual because only two crystals—quartz and orthoclase—crystallized from the magma. In the competition for space as they grew, two crystals formed this texture, reminiscent of hieroglyphic writing. The rock is often termed a graphic granite. (Photograph by W. T. Schaller, courtesy of the U. S. Geological Survey.)

FIGURE 3–24
Light-colored plutonic rock forms a dike that invades darker-colored rocks near the margin of a batholith. (Photograph by J. R. Stacy, courtesy of the U. S. Geological Survey.)

BATHOLITHS

A **batholith** is a mass of coarsely crystalline plutonic rock whose exposed area exceeds 40 square miles (approximately 100 square kilometers). If the magma mass currently beneath the Yellowstone area cools in place (see Figure 3–3) and the forces of erosion slowly strip off the overlying several miles of covering rock, the roof of a batholith will be exposed. The shape of the total batholith may be somewhat pear-shaped based on the models presented in Figure 3–3. Other batholiths in other places may be more sheetlike in form, perhaps representing the depth at which temperatures were high enough to cause partial melting of the rocks.

There are many places within the United States where one can see batholithic masses of plutonic rocks. The largest of these areas is the Sierra Nevada mountains of California—imagine an entire mountain range made up of interlocking crystals! Other areas include the whole of central Idaho, the Green and White mountains of New Hampshire and Vermont, the Laramie Range of Wyoming, the Llano Uplift of central Texas, the Black Hills of South Dakota, Stone Mountain in Georgia, Pike's Peak in Colorado, much of northern Minnesota and Wisconsin, the Wichita Mountains of southwestern Oklahoma, and parts of the Cascade Range in Washington.

Many of the world's batholiths appear to represent nothing more than giant pools of magma, formed and slowly cooled many miles underground. Like laccoliths, most batholiths are formed from intermediate- to high-silica magma. The common rock types range from *quartz diorite*, the intrusive equivalent to the *dacite* lavas of Mount St. Helens, to *granite*—the most common plutonic rock. Depending on the level of erosion, the boundaries of these clearly igneous batholiths may range from those that reflect forceful intrusion into cold wall rocks near the top of the batholith to boundaries with wall rocks deeper within the magma chamber; these margins are highly altered by intense heat near the batholithic base.

A few batholiths may have formed by wholesale transformation, without melting, of in-place sedimentary rock. The processes proposed for transforming sedimentary rock into crystalline "igneous-appearing" rocks are poorly understood and very controversial. One proposed process argues that as sedimentary rocks were heated while deep underground, ions moved into the rock in superheated steam, converting the rock directly from sedimentary rock to a rock composed of intergrown crystals of feldspars and other minerals.

There is little evidence at the margins of some batholiths or in their component rocks to indicate that the marginal rocks were ever fluid; in the conventional sense not all batholiths may be strictly igneous rock bodies. The margins of these quasi-igneous batholiths are often gradational, so that sedimentary rocks pass imperceptibly into metamorphic rocks that grade imperceptibly into crystalline rocks composed of feldspars and quartz.

Many gradations between these two types of batholiths are reflected in a wide range of margin types. The varying nature of batholithic margins was recognized more than a century ago and touched off a lengthy and continuing debate among geologists. The origin of crystalline rocks in some plutonic settings is still controversial. The world miles beneath our feet may still contain some surprises.

SUMMARY

1. Igneous rocks form when magma cools slowly to form coarsely crystalline plutonic rock or lava cools rapidly to form finely crystalline volcanic rock or volcanic glass. Pyroclastic rocks form when the fragments produced by violent eruptions are cemented or welded together; they are hybrid igneous–sedimentary rocks.

2. Magma forms from partial melting in the asthenosphere and is of low-silica basaltic composition. Magmas of all other compositions arise from contamination of primary magma as it melts other rocks, fractionation of primary magma into intermediate- and high-silica residual magmas, and partial-to-complete melting of lithospheric rocks.

3. The shape of a volcanic cone is determined by the silica and volatile content of the lava. These two factors control lava viscosity. Viscous light-colored low-density high-silica lava produces lava domes and short, thick lava flows. With the addition of volatiles, nuée ardêntes and pyroclastic rocks, whose fragments may be welded together, result. Intermediate lavas create composite cones formed from ash produced during dome destruction and lava flows erupted during dome construction. The viscosity of these lavas may be a million times that of low-silica lavas, which produce shield volcanoes, thin, widespread lava flows, lava tubes, and lava plateaus. Low-silica lavas are dark-colored, high-density lavas of low viscosity. Cinder cones are formed when low-silica lavas erupt under high gas pressure; steep-sided cones reflect the angle of repose of angular cinders.

4. Volatiles are largely water vapor and carbon dioxide; their abundant release in the early earth may have given the earth its oceans and primitive atmosphere. Volatiles lower the viscosity of magma.

5. Plutonic rocks form dikes and sills beneath volcanic areas as well as laccoliths and batholiths. Not all batholiths are former magma chambers; a few seem to have formed by intense metamorphism of sedimentary rocks.

KEY WORDS

aa	ash
andesite	asthenosphere
basalt	lava plateau
batholith	lava tube
block	lithosphere
bomb	magma
Bowen reaction series	nuée ardêntes
caldera	pahoehoe
cinder cone	partial melting
composite cone	phenocryst
continuous reaction series	plutonic rock
crater	plutonism
dacite	porphyritic
dike	pyroclastic
discontinuous reaction series	rhyolite
fractional crystallization	shield volcano
gabbro	sill
granite	stratovolcano
harmonic tremor	vent
homogeneous crystallization	viscosity
igneous rock	volatiles
laccolith	volcanic neck
lava	volcanic rock
lava dome	volcanism
lava flow	volcano

EXERCISES

1. Olympus Mons, the largest shield volcano on Mars is 370 miles (600 kilometers) across and 17 miles (27 kilometers) high. It is three times as high as either Mount Everest or Mauna Loa, the largest volcano on earth. If placed on earth, Olympus Mons would totally cover the state of New Mexico. Would the source of magma for Olympus Mons be deeper or shallower than the source for Mauna Loa? Why?

2. Given the data in Question 1, would the height of Olympus Mons tell you anything about the force of gravity on Mars as compared to Earth?

3. The estimated volume of Olympus Mons is 250 times that of the island of Hawaii, which appears to have formed in a million years. IF the average

rate of eruption of Olympus Mons and the Hawaiian Islands were the same—a dubious assumption—how long did it take to make Olympus Mons? Interestingly, your answer will agree with a date derived for Olympus Mons from an entirely different line of reasoning.

4. Individual lava flows on Mars can be traced for distances greater than 120 miles (200 kilometers). Would this suggest a high- or low-eruption rate? Would this suggest high or low viscosity for the lavas?

5. If a magma chamber started feeding magma of lower density into the feeder dike miles beneath the earth's surface, would that raise or lower the probability that a laccolith could form near the surface? Does your answer square with the observation that all laccoliths are composed of relatively low-density igneous rock?

6. If you saw a phenocryst consisting of a core of olivine surrounded by a rim of pyroxene surrounded by an outer shell of amphibole, how would you explain the origin of what you had seen?

7. Why must porphyritic rocks be the product of separate intervals of slow cooling and an interval of rapid cooling?

SUGGESTED READING

BULLARD, FRED M., 1976, *Volcanoes of the Earth*, Austin, University of Texas Press, 441 pp.
► *The lore of volcanoes, superbly done.*

DECKER, ROBERT AND DECKER, BARBARA, 1981, *Volcanoes*, San Francisco, W. H. Freeman & Co., 244 pp.
► *Authoritative review in paperback for students.*

VITALIANO, DOROTHY B., 1973, *Legends of the Earth: Their Geologic Origins*, Bloomington, Indiana University Press, 305 pp.
► *Witty, provocative, and thoroughly researched, this is must reading for anyone who delights in the earth.*

This photograph was taken from the Apollo 8 spacecraft looking south at the large moon crater Goclenius and numerous smaller craters. The rocky surface of our moon was heavily cratered by meteorite impacts billions of years ago. In the absence of an atmosphere, the impact scars remain sharp billions of years later. On the earth, records of meteorite impact are gone after a few hundred million years. (Photograph courtesy of NASA)

CHAPTER *four*

Weathering

What greater folly can there be than to call gems, silver, and gold precious, and earth and dirt vile?

GALILEO (1564–1642)

Rotting wood, rusting pipe, peeling paint, and the potholed street—all these attest to the power of sun, water, and ice to alter everyday things. The value of weathering depends on one's perspective; the same weathering processes that leave the farmer scraping the peeling paint form the soil that provides a livelihood.

Soil is one of our greatest national assets. It is the bridge between the world of food and the world of minerals. With the exception of seafood, everything we eat comes in some way from the soil. Solar energy is expended in making soil, and that energy flows through soil to plants and animals and on to us. An apple tree is nature's way of converting soil to something edible; we, in turn, transform that apple to human flesh. We *are* the earth, converted into another of its endless forms.

This is a chapter about energy. It is not about the electrical energy that binds atoms in orderly arrays (Chapter Two) or the heat energy released by volcanoes (Chapter Three). It is about the energy from the sun, which interacts with water, life, and atmosphere to change solid rock to crumbling soil. Change is everywhere in the natural world.

PERSPECTIVES ON WEATHERING

Weathering is the disintegration of rock into fragments and the decomposition of rock into soft, water-rich minerals and soluble chemical compounds. In **mechanical weathering** rock fragmentation is the dominant process. In **chemical weathering** rock decomposition and chemical change predominate. Although these two types of weathering can be separated in discussion, in the natural word they normally operate together and mutually reinforce one another.

Weathering occurs on or near the earth's surface, at the interface between the *atmosphere*, the gaseous envelope surrounding the earth, the *hydrosphere*, the water that flows on and within the earth, and the *biosphere*, a collective term for all that lives. Taken together, the combined assault of life, water, atmosphere, and sun slowly reduces rock to soil, iron pipe to rust, and wooden beam to rotten timber.

Weathering is the response of rock materials to exposure. The products of rock weathering are known as **regolith**, a collective term for all weathered products. The earth's regolith consists of various mixtures of **soil**, which supports vegetation, broken rock, and chemically altered minerals. Where weathering is intense, the regolith covering solid rock may be many feet deep; where weathering is weak, regolith is thin.

A twentieth-century French chemist, Henri Louis Le Châtelier, proposed a principle that explains many types of response processes like weathering. **Le Châtelier's principle** states that *a system at equilibrium, if disturbed, reacts so as to regain equilibrium under the new conditions.* While this idea might at first glance seem to govern only those reactions in a test tube, the earth's surface is also a chemical laboratory.

It is a laboratory where the addition of solar energy, the energy of living things, changes in temperature, and the addition of water and natural chemicals disturb the old equilibrium expressed in the minerals and textures of exposed rocks. The rocks, a system in equilibrium with the higher pressure and temperature environments in which they originally formed, react to the disturbance created by the water-rich, lower-temperature environment of the earth's surface by weathering. Regolith, a new system in equilibrium with the new conditions, is

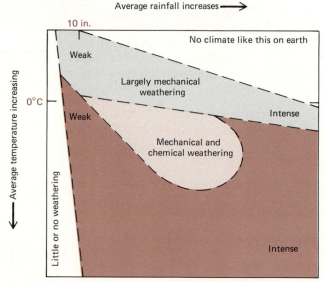

FIGURE 4–1

Weathering types shown as a function of climate. Notice that the intensity and type of weathering is largely controlled by the availability of water. Mechanical and chemical weathering (the colored area) are most effective when temperatures are somewhat above freezing in a moderately rainy environment.

formed. Challenge is followed by response. Rock, a system of minerals, has been disturbed; soil, a system of new minerals, results.

Even the word suggests that weathering is related to weather and climate. Observation of weathering rock suggests that two factors, *temperature and moisture,* control the intensity and type of weathering.

Increase in temperature enhances weathering in two ways. A 10° C increase in temperature *doubles* the *rate* of chemical reactions that form weathered products. The total organic content of soil also *rises exponentially* with temperature, which leads to increased microbial weathering of soil minerals as microorganisms "feed" directly from the soil's mineral fragments.

The absolute necessity for weathering is water. Below some threshold moisture content (Figure 4–1), neither chemical or mechanical weathering occur. Not only is water an ideal solvent, a participant in many chemical reactions, and a necessity for life within the soil, but it also is a devastatingly effective agent in shattering rocks where freeze/thaw cycles are common.

▼

MECHANICAL WEATHERING

▲

In mechanical weathering, a rock is broken into smaller and smaller fragments of otherwise unchanged rock. There are four processes that lead mechanical weathering:

1. Crystal wedging, *(ice)*

2. Root wedging,
3. Erosional unloading, and
4. Rock expansion due to heating and/or water intake.

CRYSTAL WEDGING

The most spectacular and abundant form of crystal wedging is the growth of ice crystals in cracks within rocks. When liquid water changes to ice, it undergoes a 9 percent volume *increase.* The pressure applied by ice crystal growth in an enclosed crack has been measured and reaches 15 tons/in^2 (2115 kg/cm^2). This stress is much greater than the breaking strength of average rock. Although water in rock cracks is seldom completely confined, even a small fraction of the stress produced by growing ice crystals may cause a rock to shatter on a bitterly cold night.

Frost wedging, the shattering of solids by growing crystals of ice, is a familiar form of crystal wedging. It is also a familiar problem on streets and curbs, where potholes wreak havoc with car suspensions in late winter. *Frost heaving,* the irregular expansion of moist soil as ice crystals grow within it, may cause damage to building foundations, driveways, and plant roots and is another variation of frost wedging. In mid-latitude climates, where temperatures pass across the freezing temperature for water many times a year, frost wedging is commonly the dominant form of mechanical weathering.

Ice is not the only crystal species whose growth causes the fragmentation of rocks. In some desert climates, the growth of salt and gypsum crystals within the cracks and pores of a rock will cause effects much like that of ice. *Salt weathering* has been described in many coastal areas, where crystals of salt from seawater help wedge rocks apart. Salt weathering has also been recognized in arid areas as diverse as Saudi Arabia and Antarctica.

Perhaps the best-known example of salt weathering is the continuing decay of the Sphinx, symbol of Egypt. The limestone of which the Sphinx is constructed contains natural salts. The Aswan High Dam has raised the level of underground water in the area of the Sphinx and heightened local humidity. Both processes add water to the stone and the air around it. As water condenses on the Sphinx at night, it dissolves the salt near the surface of the statue. When evaporation occurs during the next day, the salt crystallizes again within the pores of the rock. The growing crystals of salt exert great pressure on the rock surface, causing it to chip. The situation is further aggravated by the high-salt mortar originally used to restore the flaking monument. Workers working on the Sphinx in the morning report that "you can hear the stones popping like potato chips." At the present rate of weathering, the 64-foot-high (20-meter-high) figure will be a mound of limy dust in 5 to 10 centuries.

ROOT WEDGING

Plant roots in search of moisture and nutrients may grow into natural rock fractures. As the roots grow, they also are capable of providing pressures of many tons per square inch and wedging (Figure 4–2) rocks apart. Other forms of organic activity, including animal burrowing, help expose additional broken rock to the surface. While not as obvious as a tree roots wedging boulders into fragments, the activities of animals do much to churn the upper layers of the regolith and bring fresh material to the surface.

FIGURE 4–2

Notice the power of a root, here wedging a granite boulder into fragments. (Photograph by G. K. Gilbert, courtesy of the U. S. Geological Survey.)

FIGURE 4–3

Unloading fractures in granite, Yosemite National Park, California. (Photograph by H. W. Turner, courtesy of the U. S. Geological Survey.)

EROSIONAL UNLOADING

Erosion describes a group of processes that transport weathered material. These processes include failure of regolith under the tug of gravity, and the carving and removal of rock and regolith by running water, wind, and glaciers. While weathering alters rock while it remains in place, erosion removes the newly weathered regolith. Thus these two processes work hand in hand, each assisting the other.

An example of how erosion assists weathering is **erosional unloading**, a weathering process produced when erosion strips away the thousands of tons of rock that overlaid rocks formed at great depths. Ordinary rock is an elastic solid, a fact known to anyone who has ever bounced a rock off a solid surface. Rocks deep within the earth experience the weight of overlying rocks from the time of their formation. As the overlying rock is removed by erosion, the underlying elastic rock expands upward.

Much of the expansion that is caused by erosional unloading takes the form of fracturing, the only method of expansion available to brittle rocks at the surface. These fractures are usually roughly parallel to the earth's surface, with spacing between cracks decreasing upward. The fractures remind one of the layers of an onion gradually being peeled away (Figure 4–3).

Such **sheeting** or spalling, the splitting off of thin surface layers of homogeneous rocks, is common. The explosive failure of rock termed "rock bursts" in underground mines and open rock quarries has a similar origin, made more spectacular by the quite rapid release of overlying pressure. The concept of unloading fractures is an old one, traceable back to a turn-of-the-century study of the origin of the spectacular granite domes in Yosemite National Park (Figure 4–3) by the American geologist G. K. Gilbert. Although not everyone would agree that curving sheet fractures are wholly the result of unloading, the majority of sheeting in massive rocks reflects the expansion of rocks toward the sky.

As water enters sheeting fractures, it reacts with the feldspar minerals, causing them to swell. Such volume increase accentuates the separation or spalling of surface layers. **Exfoliation** describes the collection of processes that cause rocks to spall off into thin, curving layers (Figure 4–4). Exfoliative processes probably include pronounced weathering of corners where fractures meet, expansion caused by erosional unloading, and expansion caused as minerals take on water. Whatever the collection of processes, the results of exfoliation remain the same— thin, slabby sheets of loose rock lying on the rock surface (Figure 4–4). Rock climbers have long ago learned to be wary of exfoliated rock surfaces.

FIGURE 4–4

This boulder is exhibiting exfoliation, the slabbing away of outer shells, due to the addition of water. This hydration caused the minerals near the surface to swell and formed thin fractures parallel the rock surface. (Photograph by J. R. Stacy, courtesy of the U. S. Geological Survey.)

FIGURE 4–5

Granular disintegration of granite near Midlothian, Virginia. Pocket watch indicates scale. (Photograph by J. B. Woodworth, courtesy of the U. S. Geological Survey.)

OTHER CAUSES OF ROCK EXPANSION

In addition to rock expansion caused by to erosional unloading, the surface of rocks expands in the heat of the sun and shrinks in the cold of the night. Attempts to duplicate this process many times in the laboratory have failed to yield any evidence that alternate heating and cooling of rocks directly causes them to fragment.

While it appears that rocks do not undergo "fatigue" failure from repeated cycles of heating and cooling, adding a little water to the surface causes it to begin to discolor, pit, and spall within a relatively short time. As we have noted, water seems to be a universal requirement for weathering of any kind.

Any unwary camper who has built a fire directly on a massive homogeneous rock has learned that the intense heat of a fire causes explosive results as thin fragments of "rock shrapnel" are released. The same process operates at a much larger scale during forest fires, where large areas of rock may be intensely heated. Examination of a recently burned-over area will reveal a surface littered with thin, roughly circular slivers of rock. Before we attempted to control forest fires, lightning made them fairly common, and explosive rock expansion due to heating was a significant form of mechanical weathering.

A third method of rock expansion results from **hydration**, the addition of water to the chemical structure of minerals. Hydration causes minerals to swell in size, producing new minerals of greater volume. The chemical aspects of hydration are discussed in the next section of this chapter.

Because minerals within a rock are densely compacted together, the swelling of minerals, caused by the taking on of water, will shatter and displace nearby minerals. In this way a solid rock may be changed to one composed of loosely connected minerals. The material may still look like rock on casual observation, but any impact completes the process, and leaves only a granular loose aggregate (Figure 4–5). **Granular disintegration**, as this process is called, is an early stage in rock weathering and a bane to rock climbers, because the rock surface becomes loose, crumbly, and covered with mineral and rock fragments ready to roll. Hydration thus contributes to granular disintegration and assists exfoliation, both of which reflect an increase in rock volume as all the fracture space is added to the rock.

The surface area newly exposed by any form of mechanical weathering is increased by approximately the cube root of the number of new particles created; as an example, shattering one rock into eight rock fragments doubles (2 is the cube root of 8) the surface area exposed. Because the ratio of surface area to volume *increases* as particle size *decreases*, continuing disintegration of a rock exposes more and more surface area to chemical attack. Thus mechanical weathering promotes chemical weathering.

CHEMICAL WEATHERING

Chemical weathering changes the *chemical composition* of the rock in predictable ways. As a generalization, chemical weathering *adds* water to the rock and *removes* various metals in solution. Abundant water, higher temperatures, and sloping topography tend to cause higher rates of chemical weathering. Water accelerates the rate of chemical weathering in two distinct fashions:

1. Water moves through the rock acting as a solvent and a mild acid (like vinegar). As long as water can move through the **weathering zone**, the part of the exposed rock affected by chemical weathering processes, weathering can continue. As in any chemical reaction, for the process to continue, the products of the reaction must be continually removed from the site of reaction.

2. Water in quarrying and carrying away the weathered products from the weathering zone, exposes fresh rock beneath to further attack. If the products of weathering reactions are not removed, weathering can only continue until the weathering zone thickness equals the depth of water penetration. At that time, weathering stops, since the soil material exposed to the atmosphere is now in equilibrium with the atmosphere, and the potentially reactive rocks beneath the weathering zone are buried under their own alteration products.

Both of the above processes involving water exemplify Le Châtelier's principle. The development of a weathering zone composed of regolith is the result of a system disturbed and regaining equilibrium under new conditions.

Temperature is a second factor that accelerates weathering. Higher temperatures mean more rapid chemical reactions and greater participation by microorganisms in the decomposition of rock. Higher temperatures have two other effects, both of which

lower the weathering rate. Higher temperatures evaporate water that might otherwise leach through the weathering zone. Warming water also drives out carbon dioxide gas; since it is carbon dioxide dissolved in the water that makes it acidic, warming the water makes it less acid.

Sloping topography also accelerates weathering, as it causes some of the precipitation to run off and erode the surface, constantly exposing more rock to both chemical and mechanical weathering. In contrast, flat low areas may receive abundant moisture, but if drainage is sluggish and the products of chemical reactions tend to remain, further decomposition is inhibited.

Chemical weathering includes four processes, all of which interact as they attack the exposed surfaces of rocks and minerals. The processes can be expressed as a series of generalized chemical equations whose reactants (the original parent materials) and products (the results of the reaction) must reflect the chemical environment and the equilibrium between reactants and products. These processes are

1. Hydrolysis, the solvent action of water,
2. Carbonation, the reaction of carbon dioxide with rock,
3. Oxidation, the reaction of oxygen with rock, and
4. Hydration, the chemical combination of minerals with water.

HYDROLYSIS

Water easily dissociates (ionizes) into its component ions:

EQUATION 4–0

$$H_2O \rightarrow H^+ + OH^-$$

In chemically pure water, the quantity of the hydrogen ions (H^+) is precisely equal to the quantity of the hydroxyl ions (OH^-); the water is neutral, that is, it is neither acidic nor alkaline. *Acid solutions are characterized by excess hydrogen ions; alkaline solutions are characterized by excess hydroxyl ions.*

If an acid is added to water, the relative concentration of H^+ increases and the relative concentration of OH^- decreases, so that the product of multiplying the two concentrations remains equal. The acidity or alkalinity of a solution in water is described by a number on the **pH scale.** The number representing pH ($1 \div \log H^+$) is the *inverse* logarithm of the hydrogen ion concentration in the solution. *Decreasing* pH numbers indicate *increasing* acidity. A solution

TABLE 4–1
pH of Some Common Substances at 25° C

Material	pH
Hydrochloric acid	0.1
Gastric juice in stomach	1.7
Lemon juice	2.0–2.2
Tomato	4.0–4.4
Urine	5.0–7.0
Blood	7.35–7.5
Seawater	7.75–8.25
10% ammonia solution	11.8
2% trisodium phosphate (TSP)	11.95
Lye (sodium hydroxide)	13.73

that is neutral has a pH of 7.0. A solution having a pH of 4.0 is 10 times more acidic than one of pH 5.0 and 1000 times more acidic than a neutral solution whose pH is 7.0. pH falls slightly with rising temperature, for higher temperatures promote dissociation or ionization of water into its ionic components. The pH of pure water at 0 °C is 7.47 (slightly basic or alkaline) while at 100 °C the pH is 6.1 (slightly acid). Table 4–1 lists the pH of several common substances.

Because the hydrogen ion is an extremely chemically active ion, it is able to take the place of metals in aluminosilicate minerals (see Equation 4–1). Increasing acidity implies a higher concentration of hydrogen ions and ever more vigorous removal of metals from minerals. But even neutral water is sufficiently ionized so that **hydrolysis**, the displacement of positively charged ions (commonly metals) by the hydrogen ion in water, can readily occur.

Consider the following example of the weathering of granite, a rock rich in orthoclase feldspar, in a tropical environment. As inspection of Figure 4–1 will suggest, the tropic zones of the world should be zones of intense chemical weathering, dominated as they are by high temperatures and high average rainfall. The reactions are shown in stages, with pure, neutral water the only thing added. Both mineral names and chemical formulas are given. All *soluble* compounds are shown in **bold face**.

EQUATION 4–1

orthoclase + water → illite + **silica** + **potassium hydroxide**

$$3KAlSi_3O_8 + H^+ + OH^- \rightarrow KAl_2(Al,Si_3)O_{10}7(OH)_2 + \mathbf{6SiO_2} + \mathbf{KOH}$$

Equation 4–1 describes the change of orthoclase to a mica like clay, *illite*, with both dissolved silica and potassium hydroxide removed in solution from the weathering zone. Notice that H^+ has displaced *some* of the potassium in orthoclase, forcing the potassium out in solution.

EQUATION 4–2

illite + water → kaolinite + **potassium hydroxide**

$$2KAl_2(Al,Si_3)O_{10}(OH)_2 + 5H^+ + 5OH^- \rightarrow 3Al_2Si_2O_5(OH)_4 + \mathbf{2KOH}$$

Equation 4–2 describes the alteration of illite to *kaolinite*, a clay devoid of potassium, as all potassium has been flushed out of the weathering zone. Kaolinite is a stable mineral in most environments, but in the tropics, even kaolinite is out of equilibrium with the abundant supply of water and high temperature. Under tropical conditions, the following reaction is completed.

EQUATION 4–3

kaolinite + water → gibbsite + **silica in solution**

$$Al_2Si_2O_5(OH)_4 + H^+ + OH^- \rightarrow 2Al(OH)_3 + \mathbf{2SiO_2}$$

Equation 4–3 describes the final hydrolysis of kaolinite into *gibbsite*, a mineral that is pure aluminum hydroxide, one of several minerals in the aluminum ore termed *bauxite*. Even silica has been washed out of the weathering zone; all that is left in equilibrium with intense tropical weathering is the insoluble, inert material gibbsite.

In each of the preceding reactions, hydrolysis is revealed by repetitive reactions. *Hydrogen atoms replace potassium atoms, and both potassium and silica are placed into their soluble forms and leave the weathering zone in solution.* Hydrolysis always involves hydrogen ions replacing metals such as potassium, sodium, calcium, iron, and magnesium, which then leave the weathering zone in solution.

One way to promote even more active hydrolysis is to increase the concentration of hydrogen ions, that is, to make the environment more acidic. If the water is more acidic, more hydrogen ions are present than when the process proceeds with pure, neutral water.

Nature provides many sources of natural acids. The role of both natural and human-caused acid precipitation will be discussed later in this chapter. There are three natural source of acids.

1. Humic acids produced by decaying vegetation,
2. Sulfuric acid produced by oxidation of natural sulfides, and
3. Carbonic acid produced by addition of carbon dioxide to water.

Humic Acids

Humic acids form when rainwater trickles through decaying vegetation. The complex organic acids that result may yield pH values as low as 4 (a moderately strong acid). Leaching (hydrolysis) through a mat of decaying pine needles or oak leaves may be intense and leads to the formation of *podzolic* soils discussed later (Figure 4–6) in this chapter.

Sulfuric Acid

Sulfuric acid is produced by the reaction of water and oxygen with natural sulfide minerals, such as pyrite, as shown in the following equation.

Equation 4–4

pyrite + water + oxygen → iron (ferrous) sulfate + sulfuric acid

$$2FeS_2 + 2H^+ \ 2OH^- + 7O_2 \rightarrow [2FeSO_4 + 2H_2SO_4]$$

Sulfuric acid is a highly ionized acid, placing a very heavy concentration of hydrogen ions into the environment; it is therefore a very strong acid.

Carbonic Acid

Carbonic acid is formed when water reacts with carbon dioxide gas. The resulting acid is weakly ionized and is therefore a very mild acid, found in all carbonated beverages. Because carbonic acid is, however, so abundant, weathering due to carbonic acid is classified as *carbonation*.

CARBONATION

Carbonation is a form of hydrolysis characterized by dissolving material in carbonic acid, a weak but abundant acid. Carbonic acid is formed in the following way.

EQUATION 4–5

water + carbon dioxide gas → carbonic acid

$$H_2O + CO_2 \rightarrow H_2CO_3$$

Much of the carbon dioxide that combines with water in the ground comes from decaying organic matter, releasing carbon dioxide. Because carbon dioxide is a gas heavier than air, it remains near the ground surface. The dissociation (ionization) of carbonic acid proceeds as follows.

EQUATION 4–6

Carbonic acid → Hydrogen ion + Bicarbonate ion

$$H_2CO_3 \rightarrow H^+ + HCO_3-$$

An additional source of carbon dioxide is our atmosphere, which naturally contains approximately 0.03 percent carbon dioxide. As rain falls through the atmosphere, the rain interacts with the carbon dioxide to make mildly acidic rain. Rain, uncontaminated by pollution, has a pH ranging from 4.9 to 6.5. Rain in equilibrium with carbon dioxide has a pH of 5.6, so that almost all rain, even in the absence of pollutants, is mildly acidic, with the dominant acid being carbonic acid. As an example of the effect of carbonation, here is the effect of carbonic acid on orthoclase feldspar.

EQUATION 4–7

Orthoclase feldspar + carbonic acid + water → gibbsite clay + potassium bicarbonate + silica

$$2KAlSi_3O_8 + 2(H^+ + HCO_3\text{-}) + (H^+ + OH^-) \rightarrow Al_2SiO_5(OH)_4 + 2KHCO_3 + 4SiO_2$$

Notice that the effect of carbonation, as opposed to hydrolysis, is to produce soluble metals as bicarbonates (Equation 4–7), rather than as hydroxides (Equations 4–1 and 4–2). The intensity of attack is also increased, as more hydrogen ions are available within the weathering zone. Had the acid employed been sulfuric acid, the metal would have been converted into the soluble sulfate; if the acid were nitric, the resulting metal would have been converted into the soluble nitrate.

The net effect of hydrolysis, carbonation, or dissolution of any material by acid solution is the same:

1. Residual stable clay minerals are left behind,
2. Metals are placed into solution, the compound chemistry of which depends on the kind of hydrogen ion donor employed,
3. Some loss of silica into solution occurs, particularly in intense chemical weathering environments.

If a granite is exposed to weathering, its feldspar crystals will slowly dull, become rotten, and turn to clay, with the calcium, sodium, and potassium that once held them together being replaced with hydrogen ions. The metals, and perhaps some of the silica in solution, will largely leave the weathering zone. Quartz, the other common mineral in granite, is extremely resistant to weathering. As granite weathers, the feldspar turns to clay, a process releasing the quartz crystals once interlocked among the feldspar grains. The quartz enters the weathering zone as sand grains, where the grains remain as part of the soil or regolith until eventually both sand grains and clay minerals are washed to the sea. In the sea, waves wash and sort the grains of quartz and the flakes of clay; the loose sediment eventually settles into layers which may later be formed into sedimentary rock (see Chapter 5).

But what of the darker-colored minerals within granite? Although hydrolysis and carbonation affect them as well, releasing still more clay and soluble silica, another process termed *oxidation* plays an important role in converting solid aluminosilicate minerals to insoluble brightly colored oxides.

OXIDATION

When oxygen from the atmosphere or water combines with minerals, the process is termed **oxidation**. It is a particularly important process in the chemical weathering of the common dark minerals, which normally contain large amounts of iron and/or magnesium. As an example of oxidation, consider the effect of the oxidation of the mineral *fayalite*, an iron-rich olivine.

EQUATION 4–8

$$\text{fayalite} + \text{oxygen} \rightarrow \text{hematite} + \textbf{soluble silica}$$

$$2Fe_2SiO_4 + \ _2 \rightarrow 2Fe_2O_3 + \textbf{2SiO}_2$$

The effect of oxidation has been to take a common igneous mineral, remove a part of its silicon dioxide (silica) content, and leave an insoluble deposit of dark red hematite behind. *Hematite* is an iron (ferric) oxide, and is normally insoluble in water. Hematite, and another iron oxide, *limonite*, are both relatively soft minerals and commonly range in color from dark red to lavender to golden brown. Because the colors are bright and durable, many forms of iron oxides are used as pigments in paint; the traditional deep red barn owes its color to the use of hematite as the primary pigment.

Both hematite and limonite, when they occur in huge quantities, are used as iron ores. More commonly they are in trace amounts in many kinds of sediment and sedimentary rocks, where they form the wide range of colors from tans to maroons to lavenders to buff to golden browns that are so common in sedimentary rocks.

Oxidation generally converts iron minerals to forms that are insoluble. In the absence of oxygen, iron (ferrous) compounds are largely soluble in water. Consequently the amount of oxygen in the water affects the solubility of iron compounds. Oxidation also affects many other metallic ions, such as magnesium, manganese, and uranium. The effects are much the same, as the oxidation of a metal to various forms of its oxides normally results in changes in both solubility and color of the new mineral.

HYDRATION

Pure water can also lead to chemical changes by another process called **hydration**, the *addition* of water to a mineral, which may or may not form a different mineral. The hydration of any mineral causes the old mineral to take on water and increase in volume. As mentioned earlier, the hydration of feldspars may be partly responsible for granular disintegration and exfoliation. The hydration of hematite forms limonite, while the hydration of the mineral anhydrite forms gypsum, as follows:

EQUATION 4–9

$$\text{anhydrite} + \text{water} \rightarrow \text{gypsum}$$

$$CaSO_4 + 2H_2O \rightarrow CaSO_4 \cdot 2H_2O$$

The change from anhydrite to gypsum increases the volume of the new mineral by about 40 percent.

Although such swelling is greater than usual, swelling is a common result of hydration. Many clay minerals swell when they are wetted without forming a new mineral species. When the volume increase is large, such expansive clays present very difficult construction problems. Highways built over such clays may soon look more like roller coasters than highways, and structures built over expansive clays commonly suffer cracked foundations, jammed doors and windows, and severely cracked floors.

Each of the four chemical weathering methods leaves its own distinctive signature in the weathering zone. Hydration leaves swollen expanded minerals that contain water. Oxidation leaves brilliantly colored largely insoluble compounds of iron, manganese, and magnesium. Carbonation leaves behind clays and transfers metals out of the weathering zone as bicarbonates. Other acids produce soluble metallic humates, sulfates, and nitrates, while also leaving behind abundant clay. Hydrolysis similarly forms clay minerals and transports metals away from the weathering zone as soluble hydroxides. Table 4–2 tabulates the products of the weathering of an average igneous rock by a combination of the processes we have discussed.

In summary, chemical weathering produces large amounts of clay minerals, some quartz grains, and small amounts of iron oxides, which primarily act as pigments. These residual products largely remain in the weathering zone. Leaving the weathering zone are small amounts of silica in solution as well as soluble compounds of potassium, sodium, calcium, and magnesium in watery solution. Chemical weathering leaves the slightly acid quartz and clays on the continents and sends the alkaline metals to the ocean; chemical weathering sorts on a grand scale.

TABLE 4–2
Products of Weathering

Original Mineral	Residual Products	Soluble Products
Quartz	Quartz grains	Silica (in some areas)
Feldspars	Clay minerals	Silica (in some areas) K^+, Na^+, and Ca^{2+} compounds
Micas, pyroxenes, and amphiboles	Clay minerals	Same as feldspars
	Hematite	
	Limonite	
Olivine	Hematite	Silica
	Limonite	Mg^{2+} compounds

SOILS

Among the products left behind by chemical weathering, none is more valuable than soil, for soil supports all that we eat. Although the forces of erosion also gnaw away soil, carrying the insoluble minerals toward the sea, the balance that remains is the temporary parking lot for weathering's valuable residues. In time, all soil is washed downslope into the sea where it is slowly reformed into sedimentary rock (see Chapter Five).

Geologists and soil scientists divide the outer part of the earth into two units, which are often transitional. The upper unit consists of mixtures of unconsolidated material, including rock debris; volcanic ash; loose material deposited by glaciers, streams, and wind; and accumulations of plant material. This upper unit has been called both regolith and soil. To a farmer, soil is that which supports plant growth. To the engineer, soil is a mixture of materials suited for foundations or excavation. To the geologist, soil is weathered rock mixed with decaying plant material. Each profession sees soil in a slightly different way.

Beneath soil is ordinarily hard rock, commonly termed **bedrock** to clearly indicate that it is in place, while the overlying soil may have been transported to its location. The boundary between bedrock and the overlying soil ranges from gradational, where the soil has formed in place by the alteration of bedrock, to abrupt, where the soil has been carried in and deposited on top of bedrock.

The shape and properties of a body of soil depend on the effects of climate, life forms, parent rock materials, topography, and time. An extremely young soil, or one transported from somewhere else, as for example soil along a river bank, lacks any observable changes due to weathering. Such immature or transported soils lack the separation into roughly horizon-

tal layers termed soil **horizons** that are common in most more mature soils that have formed in place.

A distinctive soil type can be mapped and is composed of horizons or different layers that roughly parallel the land surface. A vertical section downward through all the horizons of a soil is termed a **soil profile** (Figure 4–6). Classification of **zonal soils**, showing well-developed horizons and soil profiles, depends on the kind and vertical sequence of horizons that developed when the bedrock was converted into soil. In turn, the sequence of horizons depends on the bedrock type and climate in the area.

The soil horizon is a zone of relatively homogeneous chemistry, mineralogy, texture, and appear-

A Horizon
Upper part, very dark brown, rich in humus and clay. Lower part sandy and pale gray. Boundary transitional to horizon beneath.

B Horizon
Yellow-brown, blocky clay. Occasional limestone fragments up to ½ inch. Drainage fair, boundary transitional.

C Horizon
Limestone fragments to 3 inches with interspersed deep orange clay. Boundary transitional.

D Horizon
Mottled, thin-bedded limestone, pale yellow-gray. Some ¼-inch seams of brownish clay near upper boundary.

FIGURE 4–6
A typical soil profile for a podzol. Notice the transitional boundaries between horizons. (Sketch taken from field notes; area is in southern Dane County, near Madison, Wisconsin.)

ance that is separated from the zones above and below it by measurable changes in the soil's properties. Figure 4–6 shows the soil profile exhibited by a well-zoned soil in southeastern Wisconsin. At its base, unaltered bedrock is termed the **D horizon**, while moderately altered bedrock is the **C horizon**. The **B horizon** is the *zone of accumulation*, where soluble and insoluble material from the layers above it accumulate. The surface layers are termed the **A horizon** and represent the *zone of leaching*, where both the insoluble clays and the more soluble metallic compounds may be moved to the B horizon below. The A horizon also may contain large amounts of organic matter, both living and dead.

Zonal soils merge imperceptibly with azonal soils, those soils lacking obvious divisions into layers or horizons. As one example of an azonal soil, recently exposed rock may yield thin soils with only the beginnings of distinguishable zones or horizons. The influence of bedrock is particularly well displayed in these poorly developed or immature soils. Limestone bedrock will yield limy soils, while quartz-rich rocks will yield sandy soils. Other types of azonal soils are transported soils, including the sandy soils in river bottoms and in sand dune fields in arid areas.

As soils gain maturity and well-developed zonation after prolonged exposure to the dominant climate, the influence of the bedrock type is much subdued, though never erased. Given still more time, the development of horizons in the soils becomes ever more striking, and soils formed under the same climate but on unlike bedrock types become quite similar. Consequently mature soils reflect the climate that formed them.

MAJOR SOIL TYPES

There are six major zonal soil groups whose soil profiles reflect their maturity and the climate that formed them (Figure 4–7). This classification of mature, well-zoned soils is an old and traditional one. Many other classifications have been proposed, but this one has the advantage of being simpler than more recent classifications. A still more generalized soil classification breaks all of the world's soils into two types, *pedalfer* and *pedocal* soils.

PEDALFER SOILS

The **pedalfers** are moderately to strongly leached soils composed mostly of iron and aluminum oxides.

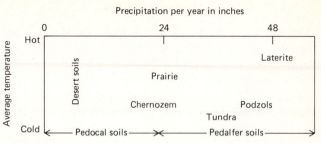

FIGURE 4–7

Relation between climate and dominant zonal soil type. Approximately two-thirds of the world's soils are either desert or laterite soils. Both present formidable handicaps to agriculture.

The word *pedalfer* consists of three syllables, each derived from Latin and having its own meaning; ped = soil, al = aluminum, and fer = iron. Pedalfer soils are characteristically acidic and form in humid climates that receive more than 24 inches (60 centimeters) of rain each year. They include tundra, podzolic, and lateritic soils.

Tundra soils

This group of immature soils develop in subarctic regions where the ground is not permanently frozen at the surface, but is permanently frozen at some depth beneath the surface. The result is an acid, poorly zoned soil which is often waterlogged. Accompanying tundra soils are peat bogs, marshes, and major construction problems in trying to design a stable foundation for buildings in an environment where the upper soil turns to mush every summer.

Podzol soils

Podzolic soils form in cool, rainy climates where trees are abundant; they are characterized by strong zonation, including a bleached, nearly white A horizon underlying a surface mat of partly decayed tree/leaf/needle litter. The soil profile displayed in Figure 4–6 is that of a podzolic soil. The B horizon is characterized by a well-developed central zone of abundant iron oxides and aluminous clays of yellow brown to reddish brown color. Podzols are typically thick, strongly zoned, intensely acid soils, best developed on porous parent rock, and closely tied in origin to the trees that grow in the soil that they modify. The term *podzol* is a Russian word best translated as *ash soil*, a descriptive name for the intensely leached and bleached part of the middle A horizon.

Laterite soils

Laterite soil derives its name from the Latin word *later*, a brick, which aptly describes both the color and properties of this soil when it is dry. Lateritic soils form in the tropics in environments of high rainfall and high temperatures. Lateritic soils have been cut into blocks and used as building materials in the tropics with great success.

Leaching in lateritic soils is so intense that *all* of the soluble silica *and* metals are washed away in solution. What is left behind is the most insoluble of natural materials, iron and aluminum oxides. These are the most intensely leached soils on earth and are therefore poor soils for agriculture. Although not all tropical soils are laterites, where they are common they form the base for an impoverished agricultural economy.

The A horizon is thin and is often composed of concretions of iron and aluminum oxides. The B horizon is thick and composed of concretionary clay, while the underlying C horizon is composed of mottled and bleached zones that grade imperceptibly into solid rock. Lateritic soils tend to be extremely thick soils, reflecting the intensity of the weathering environment. Where the bedrock is rich in feldspars, lateritic soils may consist of nearly pure aluminum oxide; such soils may be termed *bauxite* and mined as aluminum ore.

PEDOCAL SOILS

This large group of soils forms in environments that receive less than 24 inches (60 centimeters) of rain in an average year. They are characterized by far more soluble materials than pedalfer soils. For this reason they are among the richest soils on earth and form the dominant soils for the world's grasslands and prairies. The term **pedocal** translates as follows: pedo = soil, cal = calcium, the most distinctive element in these soils. However, unlike the pedalfers, whose lateritic soils are so intensely leached they lack basic plant nutrients, the pedocal soils formed in extremely dry areas have the opposite problem. Because of the lack of rainfall, desert soils are *too* rich in soluble compounds, and the excess amounts of alkali metals render desert soils difficult to use in the absence of extensive irrigation. Pedocal soils include three soil types; desert soils, chernozem soils, and prairie soils, which are transitional to pedalfer soils.

Prairie Soils

Prairie soils are strongly zoned soils that are transitional (Figure 4–7) between the pedocal and pedalfer soil groups. In North America the soils of large parts of south-central Canada and the central United States are of the prairie type. As the name implies, these are grassland soils. They have a thick, well-developed, dark, organic-rich A horizon that grades slowly downward into a clay-enriched, light brown B horizon. Increased warmth and moisture remove accumulations of calcium carbonate from prairie soils; this lac of calcium carbonate is the major characteristic that sets prairie soils apart from their close associates, the *chernozems*.

Chernozem Soils

Chernozem soils occupy a slightly drier, cooler climatic belt than the prairie soils. The overall soil profile of chernozems is much like that of prairie soils, except that calcium carbonate is always present in the B horizon. The calcium carbonate has been leached from the soil horizons above, mostly as the soluble calcium bicarbonate, but is *redeposited* in the deeper soil layers, instead of being flushed out of the weathering zone as in more humid regions. Leaching in chernozemic soils is less complete than in prairie soils.

Both chernozem and prairie soils are widespread soil types. They form in zones of low to moderate rainfall in temperate climates worldwide. The native vegetation is short grass for chernozem soils and tall grass for the slightly wetter prairie soils. Both are extremely productive soils for farming, as they are only moderately leached, fairly well drained, neither particularly acidic nor alkaline, and contain moderate amounts of organic matter in the A horizon.

The zone of carbonate redeposition in pedocal soils reflects incomplete leaching. The depth to the carbonate-rich layer—often called the **caliche** zone or **hardpan** layer—is directly related to average rainfall (Figure 4–8). In areas still more arid than those in which chernozem soils form, the carbonate layer forms closer to the surface. As aridity increases, calcium carbonate layers are joined by still more soluble materials including gypsum, and eventually even by sodium and potassium salts, among the most soluble materials on earth. The transition from the chernozems to less well-leached soils brings us to the alkali desert soils, where hardpan or caliche layers are near or at the surface in the most intense desert regions.

Desert Soils

This large soil group forms in areas receiving less than 10 to 15 inches (25 to 38 centimeters) of precipitation a year. Little leaching has occurred. The soil is alkaline and contains abundant soluble alkali metals. The effect of parent material is obvious in these extremely immature, poorly zoned soils. The soil consists mostly of mechanically weathered rock fragments. The environment is one of strong oxidation, with little organic matter available. Desert soils are also found in polar regions because both areas desert and polar environs are characterized by a lack of fluid water. Desert soils are too rich in soluble alkalis for most types of plant growth; only salt-tolerant plants survive. Desert soils are naturally "overfertilized."

The absence of chemical weathering, other than strong oxidation, is reflected in a surface phenomena unique to desert regions. What little moisture available may condense on rock surfaces at night, slightly dissolving and oxidizing the iron and manganese compounds in the rock. The next morning brings total evaporation, and the oxidized iron and manganese compounds are brought to the surface as a thin layer of what is called **desert varnish** (Figure 4–9).

Given time, soils reflect climate and range from immature desert soils overenriched in soluble materials to the intensely leached laterites that are depleted of soluble materials. It is the soils of the temperate climates, with moderate rainfall, that feed the world's hungry.

MINERAL STABILITY

Studies of soils and rocks in the weathering environment reveal an interesting pattern of mineral stability. Repeated studies have shown that minerals rich in silicon - oxygen bonds are most resistant to weathering, while minerals rich in iron and magnesium and relatively depleted in silicon - oxygen bonds weather most rapidly. Table 4–3 lists the relative reactivity of minerals in the weathering environment.

When we see this table, we are again reminded of the explanatory power of Le Châtelier's principle, mentioned in the beginning of this chapter. Those minerals formed at the highest temperatures (olivines, pyroxenes, amphiboles, and calcium-rich plagioclase feldspar) are highly reactive in the weathering zone. Their equilibrium is *most* disturbed by being placed in a low-temperature, water-rich environment.

Minerals that form at progressively lower temperatures and that are progressively rich in the strong silicon - oxygen bonds (the micas, sodium-rich plagioclase feldspar, orthoclase feldspar, and quartz) are

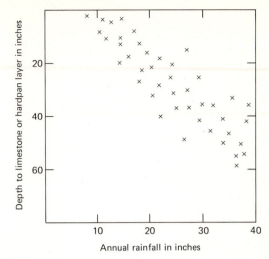

FIGURE 4–8

Relation between average rainfall and depth to top of zone of lime accumulation. Note that in areas where rainfall averages less than 15 inches of rain a year the hardpan is at or near the surface. That makes construction in these areas more difficult. (Adapted from Hans Jenny and Frank Leonard, 1934, *Soil Science*, 38, 363–381, with permission. Copyright © 1934, The Williams & Wilkins Co., Baltimore.)

much less susceptible to chemical weathering. Minerals *formed by* weathering (the clay mineral group and the iron and aluminum oxides) are highly resistant to further weathering. They *are* in equilibrium with their environment, so they experience little further change.

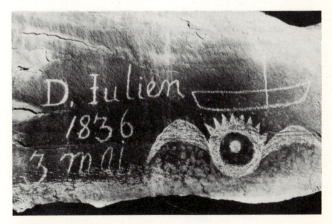

FIGURE 4–9

Desert varnish often served as "blackboard" for rock graffiti of all kinds. Here an early French explorer of the West left his mark near Hell Roaring Canyon, Utah. Nearly 150 years later, the inscription scratched through the old desert varnish remains unchanged in an arid environment. The rate of varnish formation must be extremely slow. (Photograph by K. Sawyer, courtesy of the U. S. Geological Survey.)

TABLE 4–3
Mineral Susceptibility to Chemical Weathering

Most Susceptible	Less Susceptible	Least Susceptible
Olivines		
Pyroxenes		
Amphiboles		
Anorthite (Ca-rich Plag.)		
	Biotite mica	
	Albite (Na-rich plag.)	
	Orthoclase feldspar	
	Muscovite mica	
	Quartz	
	Clay minerals	
		Aluminum and
		Iron Oxides
		(Gibbsite,
		Bauxite,
		Limonite,

As with many scientific principles, the explanatory power extends from a chemistry laboratory to the rainswept realities of exposed rock and soil. Reference to Figure 3–6 should suggest one more connection to you. Note the relation between mineral stability in the weathering zone (Table 4–3) and the position of the mineral in Bowen's reaction series. Given your knowledge of Le Châtelier's principle, would you have predicted that the earliest minerals to form in a magma are the most susceptible to chemical weathering, while the last to form are the least susceptible?

As suggested in Chapter One, science is far more than just a collection of facts; it is an *activity* of finding order among facts. The ordering concepts are creations of our human minds. Science is rationally organized experience. Le Châtelier's principle is an organizing principle that allows us to see weathering as a *response*, not an isolated *fact*. Rocks, at equilibrium with the higher-pressure and higher-temperature environments where they formed within the earth *react* when exposed to the atmosphere. The response is predictable using Le Châtelier's principle. The rocks will react so as to absorb water and increase their volume in the wet, low-pressure environment of the earth's surface. The minerals whose equilibrium is *most* disturbed—the high-temperature minerals—will be most reactive (Table 4–3).

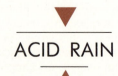

ACID RAIN

During the last several decades, in both Europe and North America, scientific evidence has been accumulating that air pollution resulting from the burning of both petroleum-based and coal fuels may result in deposition of rain, snow, or fog whose acidity is well above normal. Collectively called **acid rain**, regardless of precipitation type, the acid reaction products are often deposited in locations remote from the pollution source as air currents move the suspended polluted air for hundreds or thousands of miles. As the effects of acid rain have even crossed national boundaries, international tensions rise between those nations receiving the damage from acid rain and those nations whose industrial sources have allegedly caused the pollution.

THE CHEMISTRY OF PRECIPITATION

Acids occur naturally in the atmosphere due to the solution of carbon dioxide in water, forming dilute carbonic acid, or the oxidation of naturally occurring compounds of nitrogen and sulfur, forming dilute nitric and sulfuric acid. The natural acidity of rainwater has often been quoted as pH = 5.65, the acidity of water in equilibrium with the atmospheric concentration of CO_2. However, the addition of windblown dust, naturally occurring oxides of sulfur and nitrogen, and other naturally occurring compounds, yields pH values that range from 4.9 to 6.5 for "natural" rain. (Remember that each *decrease* in pH number signals a tenfold *increase* in acidity; rain having a pH of 4.9 is approximately 70 times more acid than rain having a pH of 6.5). Thus "normal" rain, snow, or fog, uncontaminated by human activities has a wide range of acidity.

The Human Addition

Fuel combustion adds a variety of compounds of sulfur and nitrogen to the atmosphere. Historically, our

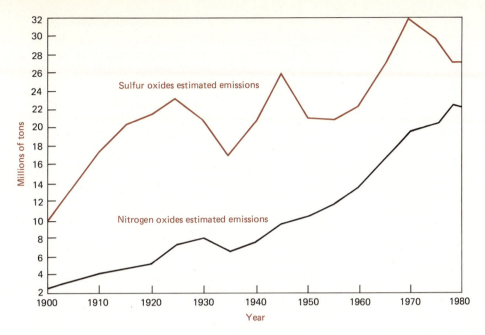

FIGURE 4–10

This graph depicts estimates of historical emissions of various sulfur oxides (the colored line) and nitrogen oxides for the years 1900–1980 in the United States. (Data courtesy of the National Acid Precipitation Assessment Program.)

consumption of fuels has increased every decade, with increasing emission of sulfur and nitrogen oxides of various kinds (Figure 4–10). When these compounds react with sunlight and moisture, the result is visible smog and an increase in the acidity (a lowered pH) of precipitation. More than half of the acidity of precipitation on a worldwide basis may be due to natural sources, but human-caused acidity may dominate in some areas. It has been recently estimated that 90 to 95 percent of precipitation acidity in eastern North America (east of 90° west longitude) may be the result of human activities, primarily the combustion of coal.

Figure 4–11 displays the average value of pH measured in precipitation in the United States and Canada in 1983. The data define a ''bulls-eye'' pattern centered on the Ohio River valley. We know of no natural causes that would account for such a pattern. The area of highest acidity does, however, coincide with an area of heavy industry and urbanization along the Ohio River and eastern seaboard. If we plot the *emissions* of sulfur and nitrogen oxides on the same map (Figure 4–11), the area of extremely acid rain coincides with the area of high emissions. Thus it appears that the sulfur and nitrogen oxides produced by combustion are readily transformed by the moisture in the atmosphere into nitric and sulfuric acid, which then fall with the rain or snow, making that precipitation abnormally acidic. Fog has much the same effect, putting rock and all life forms into direct contact with a mist composed of moderately strong acids.

About two-thirds of the acidity falling in eastern North America is thought to be the result of sulfur

oxides interacting in the atmosphere with moisture, while the remaining third is the result of nitrogen-containing compounds. For the western half of the continent, more than half of the acidity is associated

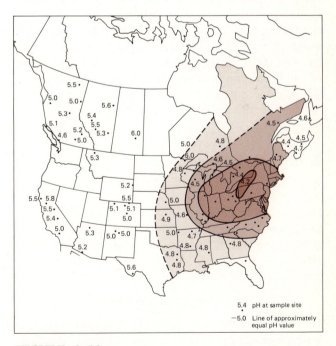

FIGURE 4–11

This map displays the average pH of precipitation over the United States and Canada for the year 1983. Note the extreme acidity of the Ohio River valley area and the alignment of pH values parallel to major storm tracks. The export of acid precipitation to northeastern Canada has strained relations between the two countries. (Data courtesy of the National Acid Precipitation Assessment Program.)

with nitrogen compounds. The two major gases (various sulfur and nitrogen oxides) that react to form acids usually reside in the atmosphere for only a few days before they are returned to earth as precipitation. Thus areas of deposition are commonly along the storm tracks from sources of pollution.

Other areas that have been experiencing the effects of acid rain include Scandinavia, and much of western Europe, particularly the Federal Republic of Germany. The effects have been noted worldwide; even the pristine air of the arctic and antarctic regions now bears the telltale signs of excess acidity. The arctic also preserves a record in the snow of historical acidity; the pH of precipitation found preserved in glaciers is generally above 5, with a 180-year-old snow fall in Greenland yielding pH values ranging from 6.0 to 7.6. Acid precipitation is a recent phenomena.

EFFECTS OF ACID RAIN

Rainfall at Pitlochry, Scotland, on April 10, 1974 yielded a pH reading of 2.4; that is 6 times more acidic than vinegar. Summer rain in the northeastern United States routinely yields pH values of 4.0, a value 15 times the acidity of "normal" rain and a 1000 times the acidity of distilled water. Such rain falling on limestone bedrock is quickly neutralized, much as antacid pills neutralize excess stomach acidity. The neutralization creates accelerated weathering of the limestone bedrock. This is but one example of the damage created by acid rain. The types of damage can be subdivided into three main categories.

1. Effects on lakes and streams,
2. Effects on vegetation, and
3. Effects on rock and other materials.

Each of these effects leads to other more subtle effects, for the earth's surface is a laboratory with tightly and subtly interlocked systems of chemical change.

Lakes and Streams

Lakes and streams surrounded by limestone maintain a near-neutral pH at the expense of increased weathering rates of the limestone. For lakes and streams surrounded by other kinds of rock, their water chemistry reflects the pH of incoming precipitation. A recent comparison of high-elevation surface waters in Washington state with waters in similar geologic settings in New Hampshire reveals an average difference in pH of nearly 2 points. That difference corresponds to nearly a hundredfold increase in acidity in the far Northeast, a difference that can only be explained by human-caused pollution of the air.

The net effect of increasing the acidity of both surface and underground water is to accelerate chemical weathering and to mobilize (dissolve) materials from the weathering zone that would normally remain insoluble. The most dramatic and far-reaching consequence of increased acidity is the mobilization of aluminum from the crystalline rocks of the Northeast as well as from the generally pedalfer-style soils.

The solution of aluminum from aluminosilicate minerals occurs rapidly as pH values drop below 4.0; the aluminum is released primarily as colloidal aluminum hydroxide ($AlOH_3$), the same material found in some well-known liquid antacids. The effect is the same; the aluminum hydroxide neutralizes the excess acidity, but at the expense of accelerated chemical weathering of rock and soil, which places excess aluminum within both ground and surface water.

The effect on life within the water can be devastating, particularly because the increase is so rapid, with little time for life to adjust to the radical change in water chemistry. The most devastating effect is on fish, where the excess aluminum in the water causes deposits of aluminum on the fish's gills, leading to slow death by suffocation. Some of the formerly great fishing streams and lakes in the Northeast are already dead, while others are dying. These simple facts illustrate the complexity and the interdependence of our environment; burning coal to make heat produces acid rain, which leaches excess aluminum from rocks and soils, which then kills fish. Who would have imagined fish suffocating in New Hampshire lakes because we fire up a boiler in Akron?

Vegetation

Atmospheric pollution may be taken up directly by the foliage on plants. Likewise, plant roots draw on the acid water circulating through the soil. Not all pollutants are harmful, however, as the excess nitrogen added to many vegetated areas could potentially function as a nutrient. The nutrient pathways and complex biochemistry involved in acid rain deposition is still not fully understood.

Atmospheric pollution is causing extensive damage, particularly to high-elevation forests in podzolic soils in Scandinavia, the United States, Canada, and western Europe. Forest damage has included dieback of spruce and fir at high elevations, reductions in growth rate, yellowing of foliage, and partial to total defoliation. Nowhere has that damage been better documented than in the Federal Republic of Germany. There nearly 50 percent of the total forested area of Germany shows areas of severe damage; the total affected area is about 19 million acres (3.7 million hectares). Each later survey turns up ever more extensive damage, and surveys in other parts of Europe including Switzerland, Austria, Sweden, Denmark, and the Netherlands are also beginning to reveal the ill effects of acid rain.

In addition to dieback and stunting of growth, vegetation in affected areas show bizarre symptoms of abnormal growth, suggesting that plant hormone imbalances have occurred. Alteration of soil chemistry by acid rain, leading to a decrease in the availability of some nutrients and an increase in the level of aluminum and other toxic materials in the soil, has been suggested as one major cause of these developments. Other suggestions have been that the leaves directly take in excessive nitrogen from the acid rain and have the effect of overfertilizing the plant, leading to an imbalance of nutrients and reduced winter hardiness. Oxidants, such as ozone and waste products from automobile exhaust may also play a role. As ozone concentration, aluminum mobility, and acid deposition all appear to have increased simultaneously in the eastern United States, it may well be that combinations of these factors are the agents causing such devastation of high-elevation forests.

Which raises an interesting question. If acid rain is leading to dieback, stunted growth, and abnormal tree growth, what is it doing to us?

Materials

The list of building materials that has experienced severe weathering seems endless. Galvanized coatings on roofs and gutters are being severely damaged. The copper plating covering the refurbished Statue of Liberty is being thinned and blackened. Brownstone houses in the Northeast are showing signs of accelerated weathering. Limestone and marble statues and monument stones are rapidly being destroyed and defaced, with the attack of sulfuric acid changing the limestone and marble surfaces to gypsum, a hydrated calcium sulfate (see Equation 4–9).

Plutarch called the Parthenon, the temple built atop the Acropolis near Athens 24 centuries ago *a spirit impervious to age . . . dressed in the majesty of centuries.* For the last 24 centuries Plutarch seemed to be correct. Although the Romans turned it into a brothel, Christians into a church, and the Turks into a powder magazine that blew up in 1687, little seemed to faze the Parthenon.

After resisting human onslaughts for 24 centuries, The Parthenon is now crumbling under the emissions of automobile exhausts and of the factory smokestacks of Athens. The combined damage of 24 centuries has been dwarfed by the damage of the last 50 years. The details of sculpture on many faces adorning the temple are gone; the marble of the Parthenon is quite literally washed away in the acid rain. Things were made worse by the iron bars used for reinforcement in earlier attempts at restoration; their corrosion has hastened the decay of one of the world's artistic treasures.

From whatever constellation of causes, the introduction of large volumes of sulfur and nitrogen oxides into our atmosphere causes damage both more obvious and more ominous with each passing day. Not all weathering is natural or benign.

SUMMARY

1. Weathering is the response of materials to a water-rich mildly acidic atmosphere, to daily temperature change, to life activities, and most particularly to water.

2. The results of mechanical weathering are fragmentation of rock into smaller fragments of unchanged rock. Mechanical weathering is accomplished by crystal and root wedging, erosional unloading, and rock expansion due to heating and intake of water.

3. The results of chemical weathering are the decomposition of the original rock into quartz grains, abundant clay minerals, and small amounts of iron oxides. This material plus that formed by mechanical weathering create regolith or soil atop bedrock. Soils are natural bodies of similar character; well-developed mature soils display horizonation or zonation into distinct layers.

4. The soluble products of chemical weathering are small amounts of silica in solution as well as larger amounts of the soluble salts of potassium, sodium, magnesium, and calcium. Under quite restricted conditions, large amounts of silica, iron, and even aluminum compounds may be released in solution.

5. The overall chemical effect of weathering is to leave mildly acidic soils temporarily affixed to the continents and to send alkaline materials in solution to the oceans. In time all material is washed from the land into the sea, where it consolidates in layers to form sedimentary rocks (Chapter Five).

6. The primary control over soil type is climate. In climates dominated by high temperatures and heavy rains, the soils may be so leached that all that is left is the most insoluble iron and aluminum oxides, which form an acid soil. In arid climates the soils retain almost all their soluble materials, so that the soils become "overfertilized" and highly alkaline. The great agricultural soils of the world are those formed at mid-latitudes in areas of moderate rainfall.

7. Acid rain appears to result from the interaction of atmospheric moisture with the products of combustion. The patterns of emission of sulfur and nitrogen oxides in industrial areas approximate patterns of excessively acidic rain. Acid rain leads to mobilization of aluminum in soils, fish kills, destruction of vegetation, and intense chemical weathering of building materials of all kinds.

KEY WORDS

A horizon

acid rain

B horizon

bedrock

C horizon

caliche

carbonation

chemical weathering

D horizon

desert varnish

erosion

erosional unloading

exfoliation

frost wedging

granular disintegration

hardpan

horizon

hydration

hydrolysis

horizon (soil)

Le Châtelier's principle

mechanical weathering

oxidation

pedalfer

pedocal

pH scale

regolith

sheeting

soil

soil profile

weathering

weathering zone

zone of accumulation

zonal soil

EXERCISES

1. Since the moon has no atmosphere, what kinds of weathering could occur on the moon? Which of the three major groups of rock—igneous, sedimentary, and metamorphic—*must* make up the moon's surface?

2. The pH value in your stomach is below 2—so acidic that stomach gastric juices are capable of dissolving meat. If your stomach routinely digests meat, why doesn't it digest itself? (Ask a doctor!)

3. In a pile of weathered rock debris, would mica, quartz, or amphibole be most likely to form nearly spherical grains? Why?

4. Some beach sands in Hawaii are composed almost entirely of olivine, derived from the waves pounding the nearby olivine-rich basalt. In the tropical climate typical of Hawaii olivine *should* be quite unstable. Explain the apparent contradiction.

5. Assuming you would like a tombstone that lasts, what kind of a tombstone should you instruct your heirs to order?

6. As erosive processes remove the uppermost part of a soil, isn't the parent material for a soil really the B horizon? Discuss.

7. Geologists have sometimes called soils "shale factories." Can you suggest the reasons for this view?

SUGGESTED READING

CARROLL, DOROTHY, 1970, *Rock Weathering*, New York, Plenum Press, 203 pp.
▶ *Thorough, readable discussion of the topic*

NATIONAL ACID PRECIPITATION ASSESSMENT PROGRAM, 1988, *Annual Report*, Washington, D.C., EOP Publications, 102 pp.
▶ *Updated each year, this reviews progress on acid rain*

WINKLER, ERJARD M., 1973, *Stone: Properties, Durability in Man's Environment*, New York, Springer-Verlag, 230 pp.
▶ *Excellent review of the weathering of worked stone by a scientist who "has a genuine personal affection for commercial stone."*

In this photo from space, the Mobile River pours tons of sediment into Mobile Bay and the Gulf of Mexico. The swirling patterns in Mobile Bay and the Gulf of Mexico define plumes of sediment on its way to settling in quiet water and slowly becoming sedimentary rock. (Skylab photograph, courtesy of NASA.)

CHAPTER *five*

Sedimentary Rocks

A handful of sand is an anthology of the universe.

DAVID McCORD

Sedimentary rocks originate at or near the surface of the earth. Most are formed when the products of weathering accumulate. Sedimentary rocks may be formed from mineral or rock fragments—termed *sediment*—weathered from older rocks, from direct chemical precipitation of dissolved minerals in lake or seawater, through the accumulation of decayed and modified plant debris, or from accumulation of skeletal debris from either continental or marine life, and from direct secretion of carbonate materials from marine or freshwater plants and animals.

Many sedimentary rocks contain contributions from several of the above sources. Since they form in natural, open surface environments, there are few limitations on what components can be mixed together and modified to make a sedimentary rock. Most sedimentary rocks contain fractions that are mechanical, chemical, and even biologic in origin. In terms of the potential array of constituents, sedimentary rocks are rather complex.

Sedimentary rocks are also the rocks that reveal most about the earth's past, for combinations of certain constituents, textures, mineral compositions, and preserved evidence of past life—termed *fossils*—provide abundant information about past environments of deposition and, therefore, ancient geographies.

The products of life and the products of chemical and mechanical weathering form the materials that are incorporated into sedimentary rock. The word **sediment** comes from the Latin word *sedimentum*, which means settling. Thus sediment is that which settles in air or water. Common kinds of sediment include mineral and rock particles, chemical compounds precipitated from and settling through water, and fragments of dead organisms including leaves, shells, and skeletal material. The common character of any sediment is that it is heavier than air or water and thus settles or moves downward.

When sediment settles into layers and consolidates into hard rock, the result is **sedimentary rock**, a rock composed of sediment. Of the rocks exposed to view on the earth's continents, 66 percent are of sedimentary origin. They are clearly the most commonly seen of all rock types. Yet, in terms of their total volume in the outer 10 miles (16 kilometers) of the earth, sedimentary rocks comprise only 5 percent of the total rock. Those two facts, integrated, suggest that sedimentary rocks are widespread, but thin, a proposal consistent with their origin *on* the earth's surface.

Sedimentary rocks are the only one of the three major categories of rocks that contains fossils. **Fossils** are any preserved evidence of past life, including footprints, bones, casts and molds of hard parts, carbon films, and so forth. Since fossils tell us about ancient geographies in which the animals or plants lived and provide evidence of the rock's relative age, the recognition and understanding of fossils preserved in sedimentary rock are very important.

Sedimentary rocks also have major economic importance, serving as the underground reservoirs for natural gas and oil. Sedimentary rocks are also quarried as the only source for some fertilizers, many construction materials, aluminum and iron ore, and for a wide range of other economic uses. Coal, America's most abundant energy resource, is a sedimentary rock formed in ancient swamps. Coal is strongly compacted and modified plant material; it furnishes an unusual example of a common sedimentary process termed **lithification**, the conversion of loose sediment into solid sedimentary rock.

THE FORMATION OF SEDIMENTARY ROCKS

Four processes operate, often together, to convert or lithify loose sediment into durable sedimentary rock. Those processes are

1. Compaction—loss of volume, including loss of fluids,

2. Precipitation—chemical crystallization from solution,

3. Cementation—binding grains together with a natural glue, and

4. Recrystallization—intergrowth of crystals under pressure.

COMPACTION

As sediment is deposited, most commonly the materials settle through water to the floor of a lake, river, or ocean. The mass of individual fragments may be more than doubled in volume by the water trapped between them. As more and more sediment is deposited on top of the original material, the weight of the overlying sediment will cause large volumes of water

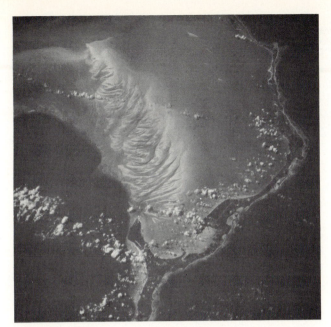

FIGURE 5–1

This is a aerial view of Eleuthera Island in the Bahamas. This photograph was shot from the Space Shuttle Columbia. The swirling patterns in the center delineate areas of calcium carbonate mud deposition over a coralline reef bank; the water has been turned a milky azure blue in color by the precipitation of the white muds. (Photograph courtesy of NASA.)

to be expelled, and the sedimentary fragments settle into a stable arrangement with little space between the fragments. This volume loss is termed **compaction**, and it is the common first step in transforming sediment into rock.

The net effect of compaction is to bring grains into much closer contact with one another, so that the volume decreases and the density increases. Compaction occurs as large volumes of water are expelled.

PRECIPITATION

Many sedimentary rocks are formed primarily from chemical compounds that have been forced by chemical changes to crystallize (to precipitate) in water. Boiling a pan of water dry furnishes an example of this process; the sides of the dry pan usually contain tiny crystals of the minerals once dissolved in the water. Dry lakes in deserts furnish another example, for the glittering crystals that line the lake floor were precipitated from water as it evaporated.

Precipitation is a process by which a chemical compound in solution is changed from its soluble (dissolved in water) to its insoluble crystalline form.

The end result is the formation of mineral crystals which settle to the bottom in the area of deposition. In the ocean the conversion of calcium bicarbonate to crystals of calcium carbonate by both warming temperatures and life processes is an example of precipitation. As previously explained (see Chapter Four), warming of seawater causes dissolved carbon dioxide gas to come out of the water, while many life processes also lower the amount of dissolved carbon dioxide; the loss of carbon dioxide from the water causes the precipitation of calcium carbonate. In some areas of warm water and reefs, the tiny needle-like crystals of calcium carbonate are so abundant that they make the ocean water milky in color (Figure 5–1).

CEMENTATION

The fragments and crystals comprising most sediment must be bound together to make a durable rock, a process termed **cementation**. In the water-rich environments in which almost all sedimentary rocks form, precipitation of materials like silica, iron and manganese oxides, and carbonates of calcium and magnesium form natural cements that bind the grains together. Clay minerals in combination with other chemical materials may also serve as cement. The natural cements that hold sediment together serve the same purpose as mortar for bricks.

Matrix is the finer-grained material filling in the voids between coarse fragments, while **cements** bind the fragments of sediment into solid rock (Figure 5–2). The combination of matrix and cement forms a durable, hard rock. Most cements form from precipitation of dissolved minerals, including quartz, calcite, hematite, and clay. The process of adding cement may take many centuries and is aided by the pressure of overlying sediment. Pressure causes fragments in contact with one another to dissolve at contact points, yielding cementing materials in solution.

RECRYSTALLIZATION

When tiny mineral crystals are placed under pressure accompanied by mild increases in temperature, they may react by **recrystallizing**—that is they may intergrow into a mass of fewer, but larger, crystals. This process, particularly common in many rocks containing carbonate minerals, leads to a compact, tough durable rock. This process may also form cement having the same composition as the detrital fragments. Such a self-cemented rock may become so hard that it breaks across the fragments of which it is composed.

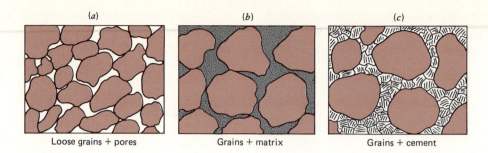

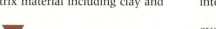

FIGURE 5–2

(A) Loose grains are slightly compacted, and space between the grains is totally open. (B) The pore space is filled with fine-grained matrix material including clay and very fine rock fragments. (C) The grains "float" in a sea of crystalline cement that binds the fragments together into a durable rock.

▼
ENVIRONMENTS OF FORMATION
▲

Chemical sedimentary rocks form wherever the chemistry of the natural environment causes dissolved materials to precipitate from water as minerals. An example is the evaporation of seawater, which forces normally highly soluble minerals to precipitate.

Sediment accumulates in environments where little energy remains. Wherever there is insufficient energy available to keep a particle in motion, it will settle. If there is insufficient energy to move it again, sediment remains, and eventually becomes compacted by the weight of overlying sediment. The addition of cement, created both by pressure solution of its grains and the addition of external cementing materials, slowly changes sediment into a hard, durable *detrital* sedimentary rock composed of layers of sediment.

On earth low-energy environments are generally at the base of some sloping surface where either moving water or moving air slow down and drop the sediment they have been carrying. Sediment is often deposited and then later eroded again. As an example, estimates of the time required for a sand grain to move from Minnesota to the Gulf of Mexico via the Mississippi River approximate 2 million years. During most of that time a sand grain is idle, temporarily deposited in a bayou or sandbar; after a prolonged time, it is washed away again and moved a bit further down river.

CONTINENTAL ENVIRONMENTS

Even when sediment is trapped in an environment of low transportative energy, the trap is temporary on a time scale of millions and millions of years, for everything on a continent is slowly being washed down toward the sea. Areas of deposition on a continent include river systems, dune fields, swamps, and lakes.

FIGURE 5–3

This photograph of a typical lake sediment illustrates the very fine grain size and extremely well-developed stratification that typifies many lake deposits. The average grain size of these particles is less than 17/8256 millimeter. The rock specimen is from an ancient lake bed in Wyoming. (Photo by W. H. Bradley, courtesy of the U. S. Geological Survey.)

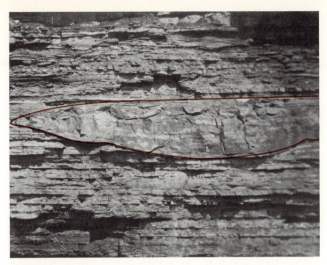

FIGURE 5–4

The colored line outlines an ancient stream channel cut into older sedimentary rocks and covered by younger sedimentary rocks. The scour and fill texture is typical of a stream deposit. (Photograph courtesy of the U. S. Geological Survey.)

FIGURE 5–5

Cross-bedding in modern sediment. Note geologic hammer measuring about one foot (0.3 meters) long furnishes scale. (Photograph by H. E. Malde, courtesy of the U. S. Geological Survey.)

Lakes

On that time scale of millions of years, lakes are transient features. Since all lakes trap sediment, they are doomed to fill up with the sediment being fed to them, which often includes organic debris. Sediment deposited in a lake is often fine-grained and have very conspicuous layering (Figure 5–3). If the lake dries up at some time in its history, distinctive deposits of evaporite minerals like gypsum may occur.

Rivers

Sediment deposited anywhere on the banks of an active river system displays a great range of variability. Sediment is deposited one day, only to be scoured away the next. River-deposited sediment commonly displays channel shaped (Figure 5–4) wedges of sediment that once filled an abandoned stream channel. The shape of the wedge records the cross-sectional shape of the ancient channel.

Dune Fields

Wind that deposits sand in dunes blows from one direction one day and a different direction the next. The result is layers of sand deposited as a mixture of tilted layers that intersect and cut one another (Figure 5–5). On close inspection, each sand grain will be seen to be polished and frosted by the frequent collision with other dry grains in a desert windstorm. Since there is little material available to cement the sand grains in a dune field, such deposits are rarely preserved, but are quite distinctive if they are.

Swamps

Another particularly distinctive continental environment is the swamp. Characterized by very dark colored, very fine-grained sediment and abundant organic matter, swamps are among the lowest-energy environments on earth. Among the common rocks formed in many swamps are various forms of coal, the only rock that burns. Swamps may grade laterally into rivers, with their characteristic river deposits, or into coastal zones, which also have a unique assemblage of sediments.

MARINE ENVIRONMENTS

Lower-energy marine environments include the deep ocean floors, the continental slope and rise (Figure 5–6), and the continental shelf. Unique oceanic environments include reefs of many kinds. Still other environments include a bewildering array of volcanic island environments, lagoons, and tidal estuaries. Only a few of the more common marine environments will be reviewed.

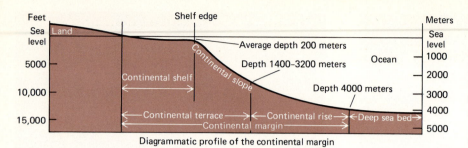

FIGURE 5–6
Diagrammatic profile of the continental margin. The continental shelf totals about 15 percent of exposed area of the continent. (Diagram modified from U. S. Geological Survey.)

The Continental Shelf

Defined as that part of the continent submerged under less than 600 feet (200 meters) of seawater, the **continental shelf** is a zone of sediment deposition on very gently sloping areas immediately adjacent to the continent. Continental shelves, if fully exposed by withdrawal of the sea, would add approximately 15 percent more area to the world's continents. Nearest the continent, the shelf grades upward into the beach environment. Seaward, the shelf grades offshore onto the continental slope (Figure 5–6). The shelf is a region of shallow water and abundant light and life, an ideal environment for the formation of chemical and detrital sedimentary rock.

Sediment deposited on the shelf tends to be washed back and forth many times by wave action. In a general fashion the finer-grained sediment is washed furtherest offshore, and the coarser-grained sediment is deposited nearest the continent. Rivers, glaciers, and wind drain the continents of approximately 20 billion tons of sediment each year. The great majority of this sediment load is deposited on the continental shelf, where it is compacted and cemented into well-layered sedimentary rocks that are extensive in a horizontal direction. In terms of volume, *most sedimentary rock is formed on the continental shelf.*

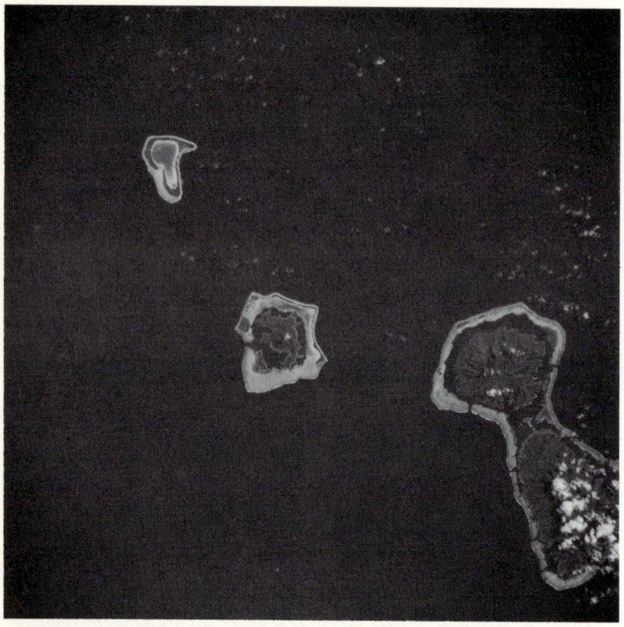

FIGURE 5–7
Aerial view of the Society Islands in French Polynesia; the northernmost island is Bora-Bora. The islands are volcanic and surrounded by a fringing reef, which causes a lagoon. The accumulation of lime muds in the quiet waters of the lagoon is mixed with the weathered products of the volcanic rocks in a tropical environment. Such atolls are formed as colonies of reef animals grow on the flanks of sinking volcanoes. (Photograph courtesy of NASA.)

The Continental Slope and Rise

Sediment from the earth's continents largely reaches this deeper ocean zone by submarine landsliding from the continental shelf margins. The sediment deposited on these steeper slopes is characterized by repetitive abrupt changes in layering upward, each change defining a new deposit from another landslide. Contorted and convoluted bedding and a variety of distinctive scour marks (produced by submarine landslides scouring the sediment surface) also mark sediment deposited in these areas.

Deep Sea Bed

Little sediment reaches the deep ocean floor. Most of what is deposited in this very deep water consists of the finest clays blown from the land and finally deposited on the ocean surface. Such fine clays may take many years to sink to the ocean depths. Another distinctive kind of deep-sea sediment is abundant shells of tiny marine organisms who live far above the deep sea floor, but whose shells slowly fell to the floor after the animal dies.

Atolls

An atoll is a ring of limestone rocks that surround a submerged volcanic island; at its center is a lagoon (Figure 5–7). As a mixed environment, an atoll receives carbonate skeletal and carbonate mud debris that is typical of many reefs plus a wide variety of volcanic rock debris. The marine environment inside the fringing reef is characterized by extremely low energy, though sediment deposition in the lagoon may be vigorously interrupted by tropical storms that send seawater surging through the fringing reefs. These storm deposits can be recognized by abrupt upward changes in layering, from thin layers of lime mud to layers composed of broken fragments of reef and large fragments of volcanic rock.

Estuaries

An estuary is a drowned river, formed when sea level rises against a coastal river (Figure 5–8). The life forms in many estuaries reflect the mixture of seawater and freshwater that blend within them. The

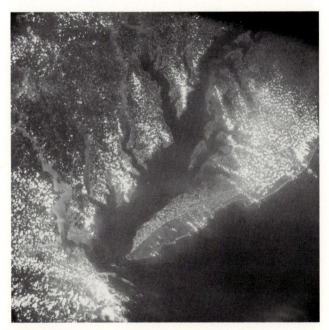

FIGURE 5–8
Aerial view of the Chesapeake Bay region, including portions of Virginia, Maryland, and Delaware as photographed during the Apollo-Soyuz test project. Land is light in color, while the streams and the Atlantic Ocean are dark. The white spots are clouds. All of the streams in the area have been flooded by rising sea level and form estuaries whose pattern is a typical stream pattern. (Photograph courtesy of NASA.)

layering pattern of sediments also reflects the constant ebb and flow of tidal currents, for estuaries are areas dominated by complex currents of different density and velocity.

CHARACTERISTICS OF SEDIMENTARY ROCKS

The arrangement of the material that makes up sedimentary rocks largely depends on the processes that affect loose sediment before it is compacted and cemented into hard rock. Although few processes are totally unique to a single depositional environment, some visual characteristics of sedimentary rock are quite distinctive and may convey a great deal of information about the environment in which the sediment was deposited. We shall describe only four of the many material arrangements unique to sedimentary rock:

1. Bedding or stratification,
2. Cross-bedding,
3. Ripple marks,
4. Mud cracks and raindrop pits.

BEDDING OR STRATIFICATION

The single most distinctive characteristic of sedimentary rocks is its widespread layering (Figure 5–9). This layering, often termed **bedding** or **stratification**, results when the air or water currents that deposit sediment sweep the sediment into thin layers spread over large areas. Stacked one on top of another, the stratification of sedimentary rocks reflects many combinations of subtle changes during deposition. Alterations in layering can reflect changes in current velocity or direction, changes in sediment source, changes in climate in either the weathering zone or the depositional area, changes in sea level, and a host of others.

Layering is displayed as vertical variations in rock color, grain size, and layering type. Each variation in the rock is due to variations in the sediment source, sediment transport mechanisms, and the environment of accumulation and lithification.

Our common experience (Figure 5–9) is that the layering of undisturbed sedimentary rocks in expo-

FIGURE 5–9
View of the Dead Horse Point area, Utah. All the rocks are of sedimentary origin and layering or stratification is very prominent. (Photograph courtesy of the U. S. Geological Survey.)

sures that range in size from hand-sized specimens to canyon walls forms a series of approximately parallel and horizontal lines. Where exposures will allow, it is generally possible to trace an individual layer for some distance in all directions. Most layers of sedimentary rocks are, in three dimensions, a thin sheet extending in all directions.

CROSS-BEDDING OR CROSS-STRATIFICATION

If instead of maintaining parallelism, the layering in sedimentary rock is a series of gentle arcs and curves that intersect and cut off one another, the rock then exhibits **cross-bedding** or *cross-stratification*. The layering sweeps across the horizon in great swirls (Figure 5–10); each layer is stacked at high angles with respect to its neighbor.

Certain variations of cross-bedding are distinctively formed by running water, where rivers are frequently cutting back and forth across a wide channel. Each time the river current sweeps a new channel, it cuts out part of an old deposit; each time a river cuts away at its bank, it stacks a series of layers at an angle across still older layers. Much more commonly, cross-bedding indicates that the rock formed originally within a sand dune, and shifting wind is responsible for this rather striking sedimentary structure (Figure 5–10).

RIPPLE MARKS

Anyone who has ever watched water flow over loose sediment or seen the wind blow sand across a basin has observed ripples; commonly ripples form when slightly turbulent air or water forms an undulating surface of alternate crests and valleys (Figure 5–11). **Ripple marks** may be formed as wind blows across water and sets it into motion; the oscillating motion of the water may scour the sediment surface beneath into a series of fine ripple marks. As river currents flow, the currents may create miniature dunes in the sediment in the channel floor. Sand dunes are themselves a type of very large ripple mark.

By carefully analyzing the shape of ripple marks, it is sometimes possible to determine the direction of current flow that created the ripples. Observation has shown that the more gently sloping part of a ripple mark faces the direction of current flow. This may allow the reconstruction of slopes in an ancient geography long since changed.

MUD CRACKS

Drying mud shrinks as it loses water, and that shrinkage cause the mud surface to crack. Such **mud cracks** (Figure 5–12) are the inevitable result of loss of volume. The shape of the surface bounded by cracks in somewhat random, though five- and six-

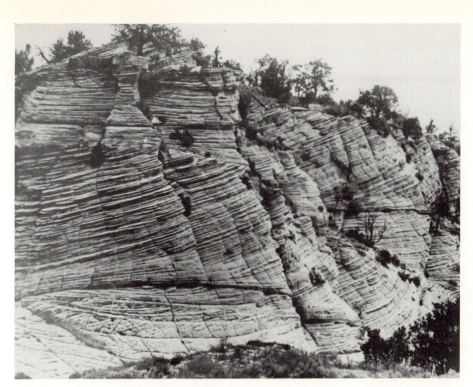

FIGURE 5–10
This is an exposure of sedimentary rock in Zion National Park, Utah. The sandstone displays excellent cross-bedding, one of the lines of evidence that suggests that this rock was once part of an ancient field of sand dunes. (Photograph courtesy of the U. S. Geological Survey.)

FIGURE 5–11
Ripple marks are prominent on the bedding surface of a former beach sandstone exposed in Jefferson County, Colorado. (Photograph by J. R. Stacy, courtesy of the U. S. Geological Survey.)

FIGURE 5–12
Mud cracks in modern mud. Abundant small craters were formed by raindrop hitting the surface when it was still moist and soft. (Photograph by N. K. Huber, courtesy of the U. S. Geological Survey.)

sided polygons are common.

The mud cracks recorded in Figure 5–12 also record another episode in the area; the original muddy surface is pitted by the impact of rain drops that made miniature impact craters. *Raindrop imprints* are clear evidence that the muddy surface was once exposed to air, and rain. The combination of raindrop imprints and mud cracks is proof positive that the sediment was deposited in a continental environment where it dried and cracked.

In the absence of raindrop imprints, mud cracks do not as clearly indicate continental conditions. Mud cracks have been observed to form in mud beneath miles of seawater. Here the mud is often *colloidal*, meaning that individual clay flakes are of such tiny size that they interact with water to form a natural gel, much like a gelatine dessert. Colloids have the unique ability to expel water from themselves, even while they are covered by water. If you have left a gelatine dessert uncovered in your refrigerator you may have noticed that in a few days the gelatine will crack and "weep" water on its surface. This is an everyday example of shrinkage cracks that form in aging colloidal material.

There are dozens of sedimentary structures that yield sometimes unique environmental information about the processes that operated during deposition of the sediment as well as those that modified the sediment after deposition. This brief review of four of them may indicate some of the diversity of information recorded in sedimentary rocks.

CLASSIFICATION OF SEDIMENTARY ROCKS

Sedimentary rocks are classified into two broad overlapping groups:

1. **Clastic** (from the Greek *klastos* meaning broken) or **detrital** sedimentary rocks composed largely of fragments cemented together.
2. **Chemical** sedimentary rocks, composed largely of precipitated crystalline minerals.

Rocks that consist of fragments with small amounts of cement are termed clastic or detrital sedimentary rocks. Rocks that consist primarily of chemically precipitated material, with few fragments, are termed chemical sedimentary rocks. As with all of nature, our attempts to classify the natural world

prove difficult; many rocks have nearly equal amounts of fragments and "cement." They span our artificial classification boundary between detrital and chemical sedimentary rocks, and their classification becomes arbitrary.

Both types of sedimentary rock are most commonly formed in water, and both reflect the environments in which they form. Clastic sediment accumulates in lower elevations where there is little or no current to disturb it. Chemical sedimentary rocks from from precipitation of minerals in both freshwater and saltwater. Chemical sediments collect in environments of low chemical energy.

Once sedimentary rocks are formed, movement of the earth's surface upward will bring them once again into the weathering zone, while movement downward will move them into environments where either metamorphism or melting are probable. *Weathering, metamorphism, or melting apply increased energy* to the sedimentary rock and effect changes.

CLASTIC OR DETRITAL SEDIMENTARY ROCKS

The fragmental material that makes up clastic rocks displays a wide range of composition. Among the more common types are volcanic rock or mineral fragments, quartz grains, granitic rock fragments, fine-grained micaceous muds and dark volcanic rock fragments, calcite (calcium carbonate), dolomite (calcium - magnesium carbonate), and clay. The composition of sedimentary components can be altered in many ways in the lengthy process of formation of a sedimentary rock. Because of this great variety, the composition of many detrital or clastic rocks is a somewhat subordinate criterion in classification.

Clastic sedimentary rocks are classified *on the basis of dominant fragment size* (Table 5–1), with fragmental composition reflected either in modifying adjectives, as in *arkosic* sandstone, or in distinctive names, as in *coquina*.

In terms of abundance, the most common sedimentary rock is shale or mudstone, followed by sandstone, followed by limestone. Limestone is commonly a hybrid rock, composed of both chemical and detrital limy components which makes it particularly troublesome to classify.

Shale or Mudstone

As described in Chapter Four, the chemical weathering of most rocks releases large quantities of clay minerals into the weathering zone. Soil-forming pro-

TABLE 5–1

Classification of Clastic Sedimentary Rocks. Terms in Parentheses are Modifying Adjectives for any Rock Name Shown in Capital Letters.

Dominant Grain Size	Dominant Mineral or Fragment						
		Quartz	Orthoclase	Plagioclase	Calcite	Dolomite	Clay
	Volcanic Rock or Mineral Fragments	(Quartzose)	Granitic Rock Fragments (Arkosic)	Dark Volcanic Rock Fragments (Graywacke)	Shell Material (Calcareous)	Shell Material (Dolomitic)	(Argillaceous)
Boulder (256 mm)	Volcanic breccias and agglomerate ↑	Fragments of this size rarely consolidate and form rock. The only significant exception is *tillite,* consolidated glacial debris.					
Gravel (2mm)	Lapilli	CONGLOMERATE (if fragments are rounded) BRECCIA (if fragments are angular)			↑ COQUINA (shell hash) ↑		
Sand (1/16 mm)	Tuff — Pyroclastics	SANDSTONE			Calcarenite — LIMESTONE	DOLOSTONE	
Silt (1/256 mm)	Ash	SILTSTONE					
Clay (1/256 mm)	Ash ↓	No such rocks			Micrite ↓	↓	CLAYSTONE or SHALE

cesses form still more clays. The abundance of sedimentary rocks containing clay minerals should come as no surprise.

Shale is an extremely fine-grained sedimentary rock composed dominantly of clay minerals. Shales display very well-developed, thin laminations or bedding, a property termed **fissility**. Clay-rich rocks lacking fissility are termed **claystones** or **mudstones**. Most shales and claystones are very soft rocks; their surface feels quite smooth to the touch. Since shales and mudstones tend to form very subdued topography and are commonly poorly exposed, we are rarely aware of their extraordinary abundance.

The average particle size for shales and mudstones is less than 1/256 mm (see Table 5–1). In this size range, the dominant material is clay, although volcanic ash, quartz grains, and carbonate minerals are also somewhat stable in this size range. At this extremely small size, the surface area of these grains is so large compared to the volume of the grain that chemical weathering destroys most minerals before they reach this size. Electrical charges at the surface of each grain become the dominant effect; most shales form by the clumping together (termed *flocculating*) of clay flakes resulting from their electrical attraction to one another.

Sandstones

Formed of clasts and grains between 1/16 mm to 2 mm, **sandstones** and their finer-grained version, the **siltstones** (see Table 5–1), form a series of generally well-stratified rocks that feel distinctly gritty to the touch. Within this size range all minerals that have survived weathering are relatively stable, and the compositional range is wide, resulting in the formation of tuffs and various types of sandstones including quartzose, arkosic, greywacke, calcareous, and dolomitic varieties (see Table 5–1).

Sandstones containing abundant feldspars and dark-colored volcanic rock fragments suggest a source that has been relatively free of chemical weathering, since fragments of olivine, glass, pyroxenes, amphiboles, and dark-colored rock fragments composed of these minerals are highly reactive to chemical weathering (see Table 4–3), with feldspars somewhat less so. Only quartz is essentially unaffected by chemical weathering in most climates, and it is this mineral that dominates in most sandstones. Where the word sandstone is used without qualifying adjectives, the term normally refers to a rock composed dominantly of sand-sized quartz fragments.

Cementing agents run the gamut. Common cements are various mixtures of silica, clays, carbonate minerals, and iron oxides. The cementing materials partially fill the space between grains (see Figure 5–2), and merge imperceptibly with **matrix**, the finer-grained materials that help to fill intergrain space. The matrix materials may, in part, be formed as grains are compacted and fragments are altered by chemical reaction with surrounding fluids. Matrix may also be a part of the material furnished to the site of deposition.

The range of different grain sizes in a clastic sedimentary rock is expressed by the term **sorting**. A well-sorted rock is one in which the range of grain sizes is small, whereas a poorly sorted rock is one in which the range of grain sizes is large. Moderately to poorly sorted rocks may have an adjective added to express the subordinate grain size, as in the term *conglomeratic* sandstone (see Table 5–1).

Rocks whose average grain size exceeds 2 mm are termed **conglomerate** or **breccia** (see Table 5–1). Although less common than other fragmental rocks, these rocks composed primarily of rock fragments, suggest a nearby source for the gravelly and bouldery sediment and rapid transport by energetic currents.

Sandstones composed dominantly of calcareous materials (those formed from calcium or calcium-magnesium carbonate minerals) receive special names, such as *coquina* or *calcarenite* (see Table 5–1). Even more commonly, they are simply called limestones or dolostones, with adjectives attached to indicate their largely clastic origin.

Limestones and Dolostones

The third most abundant sedimentary rock type, limestones and dolostones, are often of mixed clastic and chemical origin. They are listed both in Tables 5–1 and 5–2.

Limestone is a sedimentary rock composed dominantly of the mineral **calcite**, which is pure calcium carbonate ($CaCO_3$). Closely related to limestone is the rock **dolostone**, which is composed of the mineral **dolomite**. Dolomite is pure calcium-magnesium carbonate (($Ca,Mg)CO_3$). Collectively, these two rocks are often known colloquially as "carbonates," since they often form together and form an interesting chemical puzzle.

TABLE 5–2
Classification of Chemical and Organic Sedimentary Rocks. In Order Not to Make This Table too Complex, Minor Occurrences Outside of Common Environments Have Been Ignored.

Environment of Formation	Dominant Mineral or Minerals							
	Calcite	Dolomite	Gypsum	Halite	Organic Matter	Hematite and Magnetite	Silica	Phosphates
Open ocean	Limestone					Banded iron formation	Diatomite	
Restricted ocean		Dolostone			Rock phosphate			Rock phosphate
Evaporating lagoon			Gypsum					
Intense evaporation				Rock salt				
Freshwater lake	Marl and tufa						Diatomite	
Swamp					Coal			
Waterfalls and springs	Travertine							
Caves	Stalac(g)mites and dripstone							
Hot springs	Travertine						Geyserite	

BOX 5–1

"THE DOLOMITE PROBLEM"

Both dolostone and limestone form in the seas today, but in extremely unequal amounts. Dolostone is formed only in shallow water in conditions of intense heat; it is rare among modern rocks. Limestone is forming all over the world's ocean, though it forms more abundantly in shallow, warm tropical water.

Limestone is formed in part because the oceans of the world are nearly saturated to supersaturated in calcium carbonate. A variety of processes, most of them involving the interaction of life, release calcium carbonate as limestone muds and as a bewildering array of skeletal tissue formed by creatures like microscopic zooplankton, corals, oysters, starfish, fish, and marine mammals.

The rarity of modern dolostone and the abundance of modern limestone are all the more puzzling in view of the abundance pattern of calcium versus magnesium salts in the ocean; the modern ocean contains five times as much dissolved magnesium as calcium. Why in an ocean in which magnesium chloride is the second most abundant dissolved compound should so few modern dolostones form by precipitation? Turning to the abundance pattern displayed by older rocks only deepens the mystery.

In older and older rocks, the mineral dolomite and the rock dolostone become progressively more abundant. *Why* should *much more* dolomite have formed on the ancient earth than is forming today? This dilemma has been called "the dolomite problem."

The data on modern versus ancient abundance appear to violate the principle of uniformity, which asserts that the processes that modify the earth are unchanging through all of time. If "the present is the key to the past," why is dolomite so rare at present and why was it so common in the past?

A professor at Memphis State University has recently discovered that superimposed on the general decline of dolostone production as we near modern time are periodic *increases* in dolostone production. These increases coincide in time with periodic worldwide rises in sea level. It appears that periods of greater dolostone formation coincide with periods of high sea level.

Periods of high sea level provide far more extensive, quite shallow continental shelves. It appears that the formation of dolostone is favored by extremely shallow continental shelves, which in turn foster high evaporation rates that selectively enrich the remaining shelf water in magnesium. The magnesium-enriched seawater then chemically interacts with the underlying calcium carbonate soft sediments, converting them into the calcium-magnesium carbonate mineral—dolomite.

The overall trend of decreasing dolomite production through time suggests that extremely shallow marine environments were much more common in the past than now. This scenario is quite consistent with what geologists know about the abundance of past marine life types.

The resolution of the "dolostone problem" turned out to support, rather than challenge, the principle of uniformity in its explanatory power. The unique origin of dolostones points us back to a past when the earth *was* somewhat unlike it is today. It was an earth whose continents were thinner and lower, whose sea floors were higher, and on which extremely shallow seas were quite abundant. This model is discussed in the last few chapters of this book, and its probable driving mechanism, a theory termed *plate tectonics* is discussed in Chapter Ten.

The earth in the distant past was, however, an earth where product was still related to process in a uniform, logical way. That our earth is sufficiently different today that the processes making dolostone are rare is only a testimony to the power of a dynamic, evolving earth. That we can read backward into an earth unlike today is testimony to the power of human minds.

CHEMICAL SEDIMENTARY ROCKS

This large and extremely varied group of rocks are those formed predominantly by chemical changes, including precipitation of minerals from both freshwater and saltwater, evaporation, and by a variety of complex processes altering both marine and continental environments.

Reference to Table 5–2 will suggest the variety of rocks in this category. Many of them have economic value, including gypsum (construction materials), salt (seasoning and chemical industry), rock phosphate (fertilizer and chemical industry), banded iron formation (iron ore), and diatomite (used as an abrasive and a filtering material). Coal provides an increasing amount of our industrial energy needs as we slowly learn how to burn it without excessive pollution.

Notice that limestone and dolostone are listed in both sedimentary rock classification tables. We experience limestone in many freshwater settings including lakes, waterfalls and springs, hot springs, and caves.

Chemical sedimentary rocks are classified on the basis of chemical composition. Their dominant texture is that of interlocking crystals, since most chemical rocks form by precipitation and evaporation in/from water. At the most basic level, the arrangement of their crystals is similar to that of igneous rocks, which also form by crystallization of minerals from a fluid.

SEDIMENT SOURCES AND SEDIMENTARY RECYCLING RATES

As already suggested, two-thirds of the rocks exposed on the continents are of sedimentary origin. The remaining one-third are equally divided into igneous and metamorphic rocks. The exposed igneous rocks of all ages are half volcanic and half plutonic. Almost all the volcanic rocks that remain exposed to view have formed within the last few hundred million years.

Both plutonic and all metamorphic rocks have the inverse relation. The most abundantly exposed plutonic and metamorphic rocks are older than 600 million years. We should not conclude from these data on exposed area for rocks of different age that volcanism is becoming more common while plutonism and metamorphism are becoming less so. It is the amount of exposure area that is related to age, *not* their absolute abundance within the earth's outer shell.

The abundance by age of each exposed rock type is actually a function of *where* the rock forms. Plutonic and metamorphic rocks are both formed tens of miles within the earth. We should expect rocks formed at great depth not to be exposed to view until the miles of overlying rock have been stripped away, and this takes hundreds of millions of years. Only our most distant descendants hundreds of millions of years from now will see the plutonic rocks currently forming miles under Yellowstone (see Chapter Three).

At Yellowstone, it is the extremely young volcanic rocks that are now exposed to view. Since volcanic rocks form *on* the surface, we can *expect* that the youngest volcanic rocks will be the most abundantly exposed. Progressively older and older volcanic rocks have had more and more "opportunities" to be stripped away by weathering and erosion; fewer of them are left to view as time continues.

SEDIMENT SOURCES

Since it is rock exposed on the continents that largely furnishes sediment, we can now predict that on an overall worldwide basis that *one-twelfth of exposed rock is volcanic, one-twelfth is plutonic, one-sixth is metamorphic, and the remaining two thirds is sedimentary*. These proportions suggest the probable mix of sources of raw materials for making sedimentary rock.

These figures are still not quite right, because the majority of surface area with plutonic and metamorphic rocks is topographically low (most of these rocks are buried beneath sedimentary rocks), and not furnishing much sediment. With this revision in mind, it appears that something like 80 percent of the sources for sedimentary rocks are older sedimentary rocks, with volcanic rocks which tend to form higher elevations, furnishing at least another 10 percent. The contributions from plutonic and metamorphic rocks on a worldwide basis are probably less than 10 percent.

SEDIMENTARY ROCK RECYCLING

Study of the relation between rock age and exposure area for sedimentary rocks (Figure 5–13) yields re-

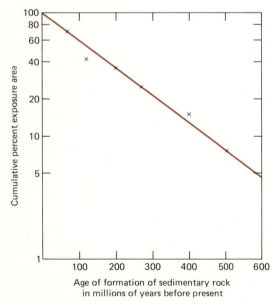

FIGURE 5–13

Relation between age of deposition of the sedimentary rock (on the horizontal axis) and the percentage of exposure area occupied by it (on the vertical axis). Notice that rock 130 million years old or younger comprises half the total exposure area of sedimentary rocks. (Adapted from Harvey Blatt and Robert Jones, 1975, *Geological Society of America Bulletin*, p. 1085. Used with permission.)

sults comparable to those for volcanic rock. The older the sedimentary rock, the less of it is left exposed to view.

Of all rocks from the oldest seven-eighths of all earth history—rocks older than 600 million years—only 5 percent are sedimentary. The other 95 percent of these ancient rocks are wholly of plutonic or metamorphic origin. The oldest sedimentary rocks are almost entirely gone.

The relation between sedimentary rock age and amount of exposed area on the present land surface can be described by a rather precise mathematical statement. The relation is a *lognormal* relation, that is the straight line on Figure 5–13 results because the spacing between the cumulative percentage figures on the vertical axis is proportional to the logarithm of the number. Put more simply, for equal units of time (the horizontal axis), exponentially increasing units of area (the vertical axis) are produced. Plotted on semi-log paper, the result is a straight line.

Inspection of the graph indicates that the doubling rate for area is 130 million years (see Figure 5–13). One-half of all surface area occupied by sedimentary rock has been formed within the last 130 million years. Phrased another way, one-half of all *exposed* sedimentary rock is younger than 130 million years. Of the remaining one-half, half are between 130 and 260 million years old. Of the remaining one-quarter, half are between 260 and 520 million years old, and so forth.

These data suggest a consistent picture. *In 130 million years (about 3 percent of all earth history) half of all exposed sedimentary rocks are recycled into other sedimentary rocks of younger age.* The older the rock, the less its exposed area, as each 130-million-year cycle gnaws away one-half of what was left by the previous 130 million years of weathering and erosion.

Hutton's principle of uniformity (see Chapter One) asserted that earth processes have remained the same through all earth history. It is perhaps surprising to find that the *rate* of weathering and erosion of sedimentary rocks has remained essentially the same for many hundreds of millions of years.

The recycling rate for sedimentary rock is extremely high. If half of all sedimentary rocks are recycled in 3 percent of all earth history, the odds of truly ancient sedimentary rocks surviving become vanishingly small. That is exactly what we see; the first seven-eighths of all earth history is represented by only 5 percent of all sedimentary rocks. Put in reverse, 95 percent of the sedimentary record of 87 percent of earth's history is gone.

Since we have suggested that approximately 80 percent of sedimentary rocks are formed by recycling other sedimentary rocks, a high recycling rate should be expected. Sedimentary rocks are a thin blanket on a restless earth. On the cosmic time scale, they are transient records of low-energy environments.

SUMMARY

1. Sedimentary rocks furnish a record of both marine and continental low-energy environments where erosional processes dump weathered material, termed sediment. The rocks are formed by various combinations of compaction, precipitation, cementation, and recrystallization.

2. The range of components that can form a sedimentary rock is extremely wide, since the environments in which sedimentary rocks form are environments open to random additions from multiple sources. If the rock is dominantly composed of cemented fragments, it is termed a clastic rock and classified on the basis of fragment size. If the rock is dominantly composed of precipitated and recrystallized material, it is termed a chemical sedimentary rock and classified on the basis of its chemical and mineral composition.

3. The greatest *volume* of sedimentary rocks is formed on the continental shelves, though all areas of low gravitational or chemical energy may receive sedimentation.

4. Almost all sedimentary rocks have bedding or stratification. This nearly universal characteristic reflects the formation of sediments by dispersal of materials in a fluid medium, where the particles settle out into horizontal layers. The sedimentary rocks formed on continental shelves are a three-dimensional sheet.

5. Sedimentary rocks form two-thirds of all rocks exposed to view on the continents; igneous and metamorphic rocks together in equal parts make up the remaining third. Sedimentary rocks are rapidly recycled; half of all exposed sedimentary rocks are recycled within each 130 million year period. Approximately 80 percent of the source for new sedimentary rock is the weathering and erosion of both soil and older sedimentary rocks.

6. The fossils in sedimentary rocks, combined with the wide variety of sedimentary structures and compositional variations, provide a record of past environments.

KEY WORDS

bedding	fissility
breccia	fossils
calcite	limestone
cement	lithification
cementation	matrix
chemical sedimentary rock	mud crack
clastic sedimentary rock	mudstone
claystone	precipitation
compaction	recrystallization
conglomerate	ripple marks
continental shelf	sandstone
cross-bedding	sediment
cross-stratification	sedimentary rock
detrital sedimentary rock	shale
dolomite	siltstones
dolostone	sorting
	stratification

EXERCISES

1. Devise a classification scheme for sedimentary rocks based solely on ease of excavation. Who would use such a classification?

2. How would you explain the following statement of fact to a friend? The older the rock is, the more likely it is to be plutonic or metamorphic.

3. If we knew that the earth's supply of heat energy was decreasing through time—it is—would that be consistent with your answer to Question 2?

4. If the same mass of granite weathered in a dry climate, it might produce an arkosic conglomeratic sandstone, while if it weathered in a tropical climate it would produce a claystone. Would the composition of either sedimentary rock clearly reflect the climate in which it formed?

5. Sand and gravel businesses are the cornerstone of the construction industry. If one constructed a graph of sand and gravel pits in use versus their elevation above sea level, what would such a graph look like? Why?

6. If the earth lost all of its internal energy, but the sun still shone, what kind of rocks would be exposed after a few hundred million years? Would there still be continents? Why?

SUGGESTED READINGS

STANLEY, STEVEN M., 1986, *Earth and Life Through Time*, New York, W. H. Freeman and Co., 690 pp.
▶ *Chapters 1–4 provide an up-to-date review; emphasizes environments*

REINECK, H. E., and SINGH, I. B., 1973, *Depositional SedimentaryEnvironments*, New York, Springer-Verlag, 535 pp.
▶ *Elegantly illustrated review of sedimentary environments and rocks*

This is a metamorphic rock, strongly layered in compositional bands that are crumpled by the intense temperature and pressures under which they formed. Cutting across the obvious layering is a very subtle planar feature at a high angle. The white ruler is one foot long.

CHAPTER *six*

Metamorphic Rocks

No single thing abides.
Fragment to fragment clings, and thus they grow
Until we know them and name them.
Then by degrees they change and are no more
The things we know.

LUCRETIUS (56 B.C.)

More than two millennia ago Titus Lucretius Carus, Roman philosopher and poet, described the objects of the materials world in accurate terms—"no single thing abides." This chapter is about rocks that "by degrees" have changed "and are no more the things we know." The process is one of metamorphosis in which solid rock, subjected to environmental changes, is converted to rock of quite another kind. Metamorphic rocks are another of nature's endless experiments with material things. They are rocks transformed from other rocks.

ENVIRONMENTS OF METAMORPHISM

Many of the processes that form sedimentary rocks may be readily observed, studied, and understood. Similarly, the surface environments in which they form can be analyzed in detail. By observation one can study all stages in the transformation of rock to regolith, to eroded material, to sediment, to compacted and cemented sedimentary rock. Laboratories can duplicate aspects of many of the sedimentary environments and processes. Volcanic rocks share many of the same advantage; once lava has erupted, the processes that affect lava's change into volcanic rock may be easily studied in the field and duplicated in the laboratory.

Metamorphism, like plutonism, occurs deep within the outer earth. Its processes cannot be studied by direct observation. Rather, erosion of overlying rock presents us with the product—metamorphic rock—and from that product we must work our way toward understanding the processes that formed it.

Metamorphism, literally means *change in form* (meta = change; morph = form). **Metamorphic processes** are the processes that alter solid rocks under conditions of temperature, pressure, or both that differ from those under which the rock originally formed. By definition, metamorphic processes occur at temperatures lower than the rock's melting point,

but above temperatures typical of sedimentary environments. Metamorphic changes occur in rock while it remains solid.

The domain of the three most common styles of metamorphism is a pressure and/or temperature region intermediate between the areas of sedimentary and igneous activity (Figure 6–1). Other, less common styles of rock reconstruction have also been described as metamorphism. This text will outline a total of five varieties of metamorphism.

1. **Shock metamorphism**—meteoritic impact causes the formation of new, abnormally dense minerals having a shattered texture accompanied by oriented conical structures unique to shock metamorphism. These metamorphic changes occur in rocks immediately adjacent to the impact crater (Figure 6–2). An abrupt episode of extreme pressure and high temperature creates structures which are less common on our earth than on other planets (see Figure 13–2), for our earth is largely protected from meteorite impacts by its atmosphere. Because of the rarity of shock metamorphic products on earth, this type of metamorphism is not further considered in this chapter.

2. **Cataclastic metamorphism**—the formation of highly sheared textures in rocks granulated

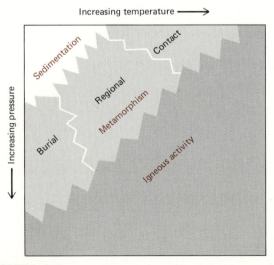

FIGURE 6–1

Summary sketch of the domains of sedimentary, igneous, and metamorphic activity. All boundaries are gradational.

FIGURE 6–2

This photograph of Meteor Crater, near Flagstaff, Arizona, displays the common (see chapter opener photo, Chapter Four) circular scar made by the impact of a large, extremely high velocity meteorite on the earth's surface. A metallic meteorite struck the limestone-covered plain about 50,000 years ago. (Photograph courtesy of the U. S. Geological Survey.)

under high pressure along earthquake faults. The metamorphic changes occur in rocks immediately adjacent to the fault. In this style of metamorphism, minerals are granulated and smeared (Figure 6–3). Commonly, new minerals do not form, and the metamorphism is confined to a

FIGURE 6–3

Cataclastically deformed gneiss, showing characteristic texture. Notice the streaky ''smeared'' appearance of the rock that is exposed in Haywood County, North Carolina. The rule is 6 inches (15 centimeters) long. (Photograph by J. B. Hadley, courtesy of the U. S. Geological Survey.)

textural change. Cataclastic metamorphism is not further considered in this chapter.

3. **Contact metamorphism**—the formation of new minerals stable at high temperature in a rock intruded by magma. The metamorphic changes are confined to the vicinity of the intrusion. An abrupt episode of high temperature followed by slow cooling brings about contact metamorphism. New minerals are commonly formed, and the texture of the rock is changed; many contact metamorphic rocks are very hard and dense.

4. **Burial metamorphism**—the formation of new, abnormally dense minerals in rocks exposed to elevated pressures within the earth. Burial metamorphism occurs over extensive areas and may be observed in many thick sequences of sedimentary rock. It also occurs on ocean floors where rocks are in collision with the continents, an idea discussed more fully in Chapter 10. The resulting rock exhibits textural change and growth of new minerals; the formation of dark blue varieties of amphiboles may be particularly diagnostic of burial metamorphism.

5. **Regional** or **dynamothermal metamorphism**—the formation of new minerals and textures in a large volume of rock exposed both to high temperatures and high pressures deep within the outer earth. This process forms the majority of metamorphic rocks (Figure 6–1). Most regional

metamorphism occurs during the collision of continents and the uplift of major mountain ranges, ideas discussed more fully in the next four chapters. Regional metamorphic rocks are exposed over an average of one-sixth of continental land areas.

METAMORPHIC TEXTURES AND PROCESSES

The **texture** of a rock is its overall appearance in terms of the arrangement of grains that compose the rock. Thus we can say that granitic texture means that the crystals composing the rock are arranged in a manner similar to that of a granite—that is, composed of interlocking crystals of various sizes in a randomly oriented arrangement.

Interlocking crystals in a random three-dimensional arrangement imply growth of crystals from a stationary fluid, where pressures are equal in all directions so that there is no force to align the growing crystals. If we say that a rock exhibits clastic texture, we indicate that the rock is composed of fragments of various sizes that are cemented together; the rock most commonly exhibits a striking layered arrangement of grains termed *bedding* or *stratification*. Bedded

or stratified textures form when grains precipitate or settle out of air or water. As these two examples suggest, each texture—each spatial arrangement of grains—is directly related to the processes that formed the rock.

METAMORPHIC TEXTURES

Since the formation of metamorphic rocks occurs within the earth, it is only their textures and minerals that allow us to infer the processes that once affected the rock. We may check our inferences with laboratory work, though the high pressures and temperatures, as well as the immense length of time required to form metamorphic textures somewhat limits our ability to duplicate them in assemblages of silicate minerals. Much of our understanding of metamorphic textures has come from laboratory experiment with materials softer and more reactive than silicate minerals; much of the research involves concepts and techniques from experimental ceramics and metallurgy. Well-equipped laboratories studying metamorphic textures have used various mixtures of beeswax, iron filings, lead, ductile crystals of various chemicals, and silicone polymers (like "Silly Putty") to work out analogues of processes that occur many miles within the earth.

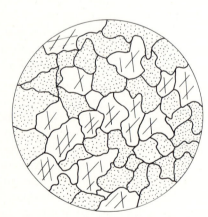

FIGURE 6–4

Sketch of the common texture of a nonfoliated metamorphic rock. The scale is enlarged from natural size to display relations among grains. The rock is a quartz-rich (dotted grains) marble. Notice that the crystals are of approximately equal size, with each crystal interfingering with its neighbor. This texture is the result of crystals growing at the same time and competing for space among themselves. Compare this nonfoliated texture with Figure 6–5.

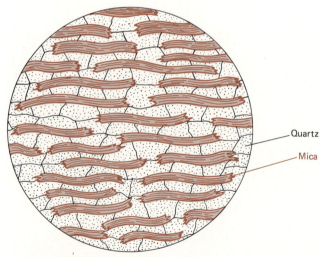

FIGURE 6–5

Sketch of the common texture of a foliated metamorphic rock. The scale is enlarged from natural size to display the relations among grains. The rock is a quartz-rich mica schist. The boundaries of quartz grains are faint and form approximate hexagonal shapes. The mica flakes are sharply defined; their parallelism defines the direction of foliation. The foliation in this case is schistosity. Compare this foliated texture with Figure 6–4.

Textures that are unique to metamorphic rocks fall into one of two categories: *nonfoliated* and *foliated*.

Nonfoliated Texture

Nonfoliated texture consists of a mass of randomly oriented interlocking crystals (Figure 6–4). Rocks with nonfoliated texture break in random fracture patterns across the interlocked crystals.

When viewed with the naked eye, the texture of nonfoliated metamorphic rocks is like that of many igneous rocks and most crystalline chemical sedimentary rocks. All three categories of rocks are formed from mutually interpenetrating crystals, which in igneous and chemical sedimentary rocks grow within a stationary fluid. To the experienced eye, subtle differences in mineralogy and texture generally allow separation of crystalline rocks into igneous, chemical sedimentary, or nonfoliated metamorphic categories.

Foliated Texture

Foliated texture is characterized by interlocking crystals that are strongly aligned approximately parallel to one another. Platy or needlelike minerals emphasize foliated textures in common metamorphic rocks (Figure 6–5) Foliated rocks form wherever abundant platy or needlelike minerals grow in parallel orientation during metamorphism. Foliated texture may be further emphasized by compositional banding of minerals in layers parallel to strongly aligned minerals. In the absence of either compositional banding or strongly aligned platy and/or needlelike minerals, the rock will exhibit nonfoliated texture.

METAMORPHIC PROCESSES

Both nonfoliated and foliated textures are formed by the same basic processes. Indeed, foliated and non-foliated metamorphic rocks are often mixed together within the same rock exposure. The dominant processes during contact, burial, or regional metamorphism are *recrystallization* or *neocrystallization*.

Recrystallization

Recrystallization is a term that describes the growth of larger crystals at the expense of smaller crystals of the *same* mineral. Finely ground material with its abundant surface area is more chemically reactive than the same material in coarser chunks. A familiar example is mechanical weathering that enhances chemical weathering by disintegrating rock into smaller particles. Since sedimentary rocks are largely composed of finely ground material, when sedimentary rocks are deeply buried and exposed to higher temperature and/or higher pressure, chemical reactions begin to occur between adjacent grains *while the rock remains solid*. Thousands of small crystals recrystallize into hundreds of coarser crystals within a solid rock.

The process of crystal growth in a solid is not a familiar one to us. Only the process of annealing metal furnishes a natural analogue. Heating powdered metal causes the small shredded grains to begin interacting (annealing) with one another. Grain size increases as the surface area of smaller grains is consumed. The rate of reaction among the grains decreases as grain size increases. Growing crystal runs into growing crystal, mutual grain boundaries continue to grow, and a new texture at equilibrium with higher temperature is established as the heat energy that initiated the crystal growth dies away.

When limestone, a rock composed of fine-grained calcite, is exposed to temperatures and pressures well out of the range typical of the sedimentary environment, the small calcite crystals in limestone recrystallize to much coarser calcite crystals; the resulting metamorphic rock is a *marble* (Table 6–1). Figure 6–

TABLE 6–1
Classification Chart of Common Metamorphic Rocks*

Nonfoliated Texture		Foliated Texture	
Hornfels	*Any rock type.*	**Slate**	*Mud.*
Quartzite	*From sandstones.*	**Phyllite**	*Varied sources.*
Marble	*From dolostone and limestone.*	**Schist**	*Phyllite or slate.*
Greenstone	*From dark volcanics and pyroclastics.*	**Greenschist**	*Dark volcanics.*
Granite	*From clay, feldspar-rich sediments or*	**Blueschist**	*Greenschist.*
	silica-rich metamorphic rocks.	**Amphibolite**	*Dark volcanics and dolostone.*
Granulite	*Varied sources.*	**Gneiss**	*Schist.*

Source Rock Type is given it *italics*; Rock name is **bold face.

BOX 6–1

RECRYSTALLIZATION AND SURFACE AREA

Much of the energy that drives the process of recrystallization comes from the *loss* in grain surface area when recrystallization occurs. Since grain surface areas are so reactive, the *loss* of grain surface means that the resulting rock has become more stable (less reactive). The magnitude of the energy that drives recrystallization is partially dependent on the difference in surface area between the grains in the original rock and the crystals in the recrystallized rock. The larger the difference, the more energy is available to cause recrystallization.

Assume a rock composed of cubes, each of which is D inches (or centimeters) on a side within a much larger cube of rock that is a inches (or centimeters) on each side. The surface area of each small cube face is D^2 and the total surface area of each small cube is $6D^2$. The total surface area of the large cube is $6a^2$.

Since (from Chapter Four) the surface area created by shattering one grain is increased by the cube root of the *number* of new particles created, the cube root of the ratio of surface areas (or sides) of the larger cube to the smaller will tell us the total number of smaller cubes within the larger cube. The total number of small cubes within the large cube is $(a \div D)^3$.

If we deal with a large cube in which $a = 1$ in, the total grain boundary of all the small grains within the one-inch cube is computed as follows:

EQUATION 6–1
Total Grain Boundary Area ≅ Number of small cubes
in a 1-in² cube × Area of one cube

$$TGBA \cong (1 \div D)^3 \times 6D^2 \cong 6 \div D$$

or

Total Grain Boundary Area ≅ $6 \div D$

But grains in rocks are not ideal cubes. Laboratory measurements have confirmed that the formula given approximately doubles observed values for irregular grains. For irregular grains, then, a more realistic formula is TGBA ≅ $3 \div D$. Let us use this formula to compute the total grain boundary area within a one inch cube of a common rock.

The grain boundary area in a one-inch cube, composed of grains that have an average diameter of 0.004 inch (an extremely fine-grained size typical of a rock like *chert*) is about 750 square inches per cubic inch ($3 \div 0.004$) of rock. Recrystallize this extremely fine-grained rock into a quartzite having a grain size of 0.1 inch, and the grain boundary area is now only 30 square inches per cubic inch of quartz-

ite. Recrystallization has caused the *loss* of 720 square inches (4645 square centimeters) of surface area in *each* cubic inch of rock.

In the same way that mechanical weathering makes a rock *more* chemically reactive by increasing its surface area, metamorphic recrystallization makes the rock far *less* chemically reactive by diminishing its surface area. Part of the energy needed to recrystallize a rock comes from the consumption of reactive crystal surfaces.

4 sketches the texture typical of a marble. Since calcite in not a platy or needlelike mineral in growth habit, marble is a nonfoliated rock.

Neocrystallization

Neocrystallization differs from recrystallization in only one particular. In **recrystallization,** both "old" and "new" minerals are the same; only the texture is changed as grain size increases. In **neocrystallization,** "old" minerals of several different kinds react to form a different mineral(s). Equation 6–2 shows an example of neomineralization.

EQUATION 6–28

calcite + quartz → wollastonite + carbon dioxide

$$CaCO_3 + SiO_2 \rightarrow CaSiO_3 + CO_2$$

Two minerals, calcite and quartz, react with one another at high temperatures to grow a completely new mineral—wollastonite. As carbon dioxide gas continues to escape, the reaction continues until all of the quartz is converted into wollastonite, after which neomineralization stops.

PRESSURE SOLUTION

Both recrystallization and neocrystallization involve a complex group of processes occurring in the solid state. One of these processes can be described as **pressure solution**, in which a mineral dissolves only in areas under high pressure and is redeposited only in zones of low pressure (Figure 6–6).

As the process continues, foliated rocks form when growing platy minerals align themselves roughly *perpendicular to the dominant pressure*. This alignment of platy or needlelike minerals is predictable, because less energy is expended by the minerals growing at 90° to the dominant force. Minerals attempting to grow parallel to the applied force are dis-

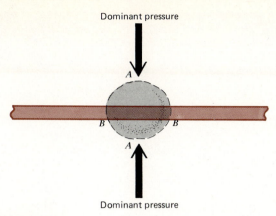

Dominant pressure

Dominant pressure

FIGURE 6–6

Under high pressure a mineral (shown in gray with dashed-line boundaries) dissolves at points A, the areas of highest pressure. The dissolved mineral material is redeposited in the areas of lowest pressure, labelled as B. As the process of pressure solution and redeposition continues, the "old" mineral gradually grows to form the shape shown in color. Repeated many times with naturally platy or needlelike minerals, the result is a foliated rock.

solved as fast they attempt to form; the pressure adds more energy to the mineral lattice than it can accept and remain solid.

Solids dissolving under pressure are not an everyday experience. Ice skaters are unaware that the pressure of their bodies directed on the narrow blade temporarily melts the ice over which it passes. Anyone watching a huge glacier is unaware that it moves by dissolving at its base, which is under many tons of pressure from the overlying ice. Those who have attempted to pack snow into a snowball may realize that if one presses extremely hard, the snowball partially melts and leaks water.

Rocks buried deep within the earth experience the same effect. Minerals with overall blocky shape dissolve under pressure and recrystallize as nonfoliated rock. If the recrystallizing minerals are normally platy, flaky, or needlelike, the resulting rock will be foliated.

▼

ORIGINS OF LAYERING IN METAMORPHIC ROCKS

▲

Unlike sedimentary rocks, in which essentially all layering has a common origin, the settling of materials in a fluid, foliated metamorphic rocks may have

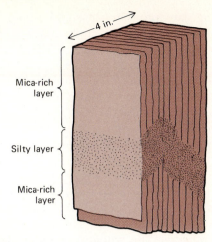

Mica-rich layer

Silty layer

Mica-rich layer

FIGURE 6–7

An example of slaty cleavage. This sketch, taken from field notes, is of a specimen of slightly metamorphosed sedimentary rocks near Ely, Minnesota, known as the Knife Lake Group. Notice that the planes of slaty cleavage, which are vertical in this sketch, offset former horizontal *relict bedding*. The relict bedding must originally have been alternating layers of siltstone and shale. Slaty cleavage is well developed in the mica-rich layers but nearly absent in the silty layers.

four kinds of layering, each of which has its own unique origin. Layering in foliated metamorphic rocks forms from any one or combination of four origins:

1. Slaty cleavage
2. Schistosity
3. Gneissosity or compositional banding
4. Relict bedding

SLATY CLEAVAGE

Slaty cleavage is a structure of controversial origin, but most likely it is formed by the rapid expulsion of water from clay-rich sediment under conditions of near-surface temperature and elevated pressure. The upwelling water rotates the tiny clay flakes until they are parallel to flow lines, much as logs are lined up parallel to current flow in a river. This type of layering is most common in the rock type termed *slate* (see Table 6–1). It is characterized by the extremely strong alignment of very low temperature micaceous clays, such that the whole rock readily splits or cleaves into very thin sheets (Figure 6–7). The mineralogy and texture is that of a highly lithified and altered sedimentary rock.

We take commercial advantage of slaty cleavage

by splitting slate into thin layers and using it as flooring, roofing, and as felt-covered surface in the highest-quality billiard and pool tables. The mineral content of most slates is more typical of a *thoroughly* compacted sedimentary rock than of a conventional metamorphic rock, but by long tradition slates are classified as metamorphic. They illustrate the difficulty in classifying those rocks that form on or near the boundary line between categories.

SCHISTOSITY

Schistosity is a rock texture with striking parallelism of flaky mica and other platy or needlelike amphibole minerals. The whole rock has a sheen, and the texture is totally dominated by the abundance of flaky or needlelike crystals (see Figure 6–5). The mineralogy and texture is that of a totally recrystallized metamorphic rock. Schistosity is the normal rock texture for the rocks *phyllite* and *schist* (see Table 6–1).

GNEISSOSITY

Gneissosity (pronounced nice'-ossitee) is a rock texture characterized by rough parallelism of flaky and needlelike minerals in a rock where feldspar and *compositional banding* dominate. Compositional banding is a crude layering of minerals into bands of unlike composition. Compositional banding may be of several origins and does not necessarily represent relict bedding, which is described next. Most commonly, compositional banding appears to form by processes unique to high temperatures and pressures, when the metamorphic rock is becoming somewhat plastic and highly susceptible to chemical and dynamic changes. It is the common texture of the high-pressure, high-temperature metamorphic rock termed *gneiss* (pronounced as "nice"; see Table 6–1).

Gneiss (Figure 6–8) is often a handsome rock and much used as an ornamental building or monument stone. Commonly the aligned layers of platy minerals and compositional banding parallel to one another. Rarely zones of aligned minerals may crosscut compositional banding in rocks with complicated histories of deformation.

RELICT BEDDING

When metamorphism has not been too intense, the former layering of the old sedimentary or volcanic rock may be preserved as **relict bedding** (see Figure 6–7). The original variation in composition commonly is preserved in relict bedding, but differences

FIGURE 6–8

An example of a gneiss whose origin is probably a hybrid of igneous and metamorphic processes. The gneissosity and the compositional banding are somewhat contorted; the rock clearly flowed while solid. (Photograph by J. B. Hadley, courtesy of the U. S. Geological Survey.)

in color, grain size, and arrangement of grains may also be preserved. Relict bedding need not parallel slaty cleavage, schistosity, or gneissosity, as its origin predates any platy features of metamorphic origin.

CLASSIFICATION OF METAMORPHIC ROCKS

Like other crystalline rocks, metamorphic rocks are classified on the basis of their texture and their mineral composition (Table 6–1). A brief review of the distinguishing features of common metamorphic rocks follows.

NONFOLIATED ROCKS

Nonfoliated metamorphic rocks, which are made up of interlocking, randomly oriented crystals and thus have a crystalline texture, may be formed by regional, burial, or contact metamorphism. A few common nonfoliated metamorphic rocks are *marble, hornfels, quartzite, greenstone, granulite,* and *granite.*

Marble

Composed of interlocking crystals of *calcite*, marble can be formed by both contact and regional metamorphic processes. If the origin is contact metamorphism, the marble will only form adjacent to an igneous intrusion into limestone and grade rapidly into unaltered limestone a short distance away. *Dolomitic marble* forms from the metamorphism of dolostone. Marble is often used as a monument stone and is a premier material for sculpture.

Hornfels

This group of rather nondescript rocks forms solely by contact metamorphism. Hornfels may be of many different compositions and may commonly only be identified with certainty if the specimen is collected adjacent to an igneous intrusion. Hornfels is commonly fine-grained, rich in silica, hard, dense, and fractures in conchoidal (rounded) surfaces.

Quartzite

Like marble, quartzites are very common metamorphic rocks, formed by burial, regional, and contact processes. Since quartzites are metamorphosed sandstones and cherts, they are composed largely of the mineral *quartz*. Quartzites tend to be dense, sugary-textured hard rocks.

Greenstone

An aptly named rock, most greenstones (Table 6–1) are nondescript dark yellow-green rocks containing a variety of generally dark green, dark yellow green, and grayish platy minerals in a random, smeared texture. They form primarily from the regional metamorphism of basalt and similar rocks. They are a particularly common rock type among many ancient rocks.

One type of greenstone, rich in the minerals talc and chlorite, is known as *soapstone* and is sometimes used for laboratory bench tops and heavy wood-burning stoves. *Jade*, a semiprecious stone, is also a variety of greenstone, rich in specific varieties of green amphiboles.

Granulite

Formed by very high temperature, high pressure metamorphism, granulites appear to form in a dry en-

vironment very deep within the earth. The textures are coarsely crystalline, with common minerals including pyroxene, garnet, and plagioclase feldspar. Granulites form from very high-grade regional metamorphism.

Granite

Formed by poorly understood processes at the highest grades of regional metamorphism, some exposures of granite have gradational boundaries into metamorphic rock and appear to have formed by wholesale transformation of sedimentary rocks. The idea of metamorphic formation of granite remains quite controversial among geologists, but it appears that some granites of rather distinctive mutual boundary textures form in this fashion. As one British geologist put it decades ago, "There are granites and there are granites." It requires a careful investigation of mineralogy, intergrain textures, sequence (or absence of sequence) of mineral formation, and boundary relations in the field to establish whether granite has formed by metamorphism or cooling of a fluid.

FOLIATED ROCKS

Formed primarily by regional metamorphism and less commonly by burial metamorphism, foliated rocks are composed of interlocking crystals but are dominated by parallel flakes of micaceous minerals and ÷ or needlelike crystals of amphiboles and other minerals lying in the plane of foliation. Among common foliated rocks are slates, phyllites, schists, gneisses, blueschists, and amphibolites.

Slates

As mentioned earlier, it is not clear that slate has ever been exposed to temperatures typical of many metamorphic rocks, but slate is traditionally regarded as the result of very low-grade metamorphism of claystones and shales. Distinguished by an extremely pervasive slaty cleavage, the slabby texture of slates makes them easy to distinguish.

Phyllites

The mineralogy of phyllites indicates the rock has been largely recrystallized and is dominated by micaceous minerals that are so small as to be barely visible. Most phyllites reflect light with a soft sheen and have moderately developed schistosity. Phyllites form from low to medium levels of regional metamorphism of a variety of rocks.

Schists

Coarser than phyllites, schists display the same schistose texture. In most schists the micaceous minerals are coarse enough for individual flakes to be seen with the naked eye. Schists are commonly the product of medium grade regional metamorphism of phyllites. If the schist is composed of numerous bluish to purplish amphiboles, it is termed a *blueschist*, which may indicate that it formed by burial metamorphism. If a schist is dominated by green micas, it is a *greenschist* (Table 6–1), and results from the metamorphism of certain phyllites, dark volcanics, and pyroclastics.

Gneisses

The abundance of feldspars and the presence of gneissosity distinguishes gneisses from schists. Gneisses form from high grades of regional metamorphism of schists, which causes some micas to be recrystallized into feldspars, forming compositional banding. As pressures and temperatures become extreme, micas disappear, and gneisses may grade into granulites or granites. Characterized by a coarse gneissosity, gneisses display compositional banding in roughly parallel layers. As temperatures near the melting point, the compositional banding may become quite swirled in appearance, reminding one of the patterns made by scum in an eddy of a stream.

Often the only difference between a gneiss and a granite is textural. Gneisses display gneissosity and compositional banding, while granite displays the same mineral content as a gneiss but in a random crystalline texture. As mentioned earlier, it appears that some granites form by the metamorphic transformation of gneisses, whereas others may form by the melting of a gneiss which then slowly cools to form granite. Granite appears to have multiple origins. In 1596 Caeselpinus defined granite as "*a grained rock*", literally, one composed of grains. Although such a wholly descriptive definition is useless today, it reminds us that the origin of some granites remains enigmatic four centuries later.

Amphibolite

With composition consisting of amphiboles and calcic plagioclase feldspars in roughly equal amounts, amphibolites are the common result of medium- to high-grade metamorphism of dolostones, dark-colored volcanic rocks, and the sedimentary rocks formed from weathering of dark volcanic rocks.

▼

WHAT IS A SOLID?

▲

Most of our experience tells us that fluids are materials above their freezing point that flow, and that solids are substances below their freezing point that are rigid and never flowing. Yet there are experiences that challenge these ideas. An asphalt road on a blistering hot summer day retains the tread pattern of tires, and glaciers are frozen rivers of ice that still move majestically down the valleys that hold them at rates of a few yards or meters per year. Vertical glass windows in century-old houses are always thicker at the bottom than at the top because the glass has slowly flowed downward.

Even rock flows. Marble benches in old cemeteries, public buildings, or campuses have sagged and warped under their own weight. Large deposits of salt flow so readily under their own weight that they move upward many miles and may *intrude* into overlying rocks (Figure 6–9), shattering them and shouldering them aside, much as would an igneous intrusive body. Deposits of gypsum make extrusive "volcanoes" in parts of Utah and Iran, complete with

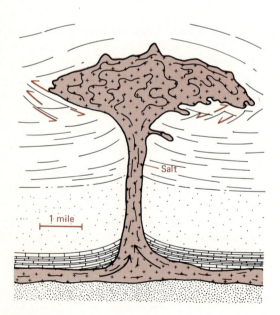

FIGURE 6–9

Schematic cross-section of an area intruded by a salt dome. The bedded salt at depth is of much lower density than the surrounding sedimentary rocks. When placed under the pressure of miles of overlying rock, the salt becomes fluid and moves upward, behaving like any other intrusive mass, except that it is well below its melting point. The layering preserved in the upper part of a salt dome is highly convoluted.

flows from the top of the "crater." All the elements of igneous activity are present in salt domes and gypsum "volcanoes" except one—heat. Gypsum and salt have flowed while they were solid and cold.

What then is a solid? It is something that *does not flow during our experience with it*. We have introduced a new element into our definition; that element is *time*.

Perhaps the best example of the way time affects the behavior of a substance is demonstrated by several types of silicone polymers, including that well known toy material named "Silly Putty." Hit with a hammer, it shatters like glass; it appears to be a brittle solid. Drop it on the floor, and it bounces like rubber; it now appears to be an elastic solid. Leave it on the floor and it flows like molasses; it now acts like a thick, viscous fluid.

We now use terms like brittle solid, elastic solid, or viscous fluid to describe the *behavior* of Silly Putty depending on the *rate* at which we applied force. Rapidly applied force brings brittle behavior while slowly applied force brings fluidity. *Changes in the solidity of material depend on time as much as temperature.*

Our simple experiments with "soft solids" suggests a new way to classify matter; Figure 6–10 presents such a classification. This figure also suggests why metamorphic rocks yield evidence that they have flowed, even though they remained below their melting point. Exposed as they are to high temperatures *near* their melting point as well as high pressure

endured for perhaps millions of years, we should expect them to flow while solid.

METAMORPHISM AND WEATHERING—A STUDY IN CONTRASTS

The central theme of Chapters 2–6 is that the earth is dynamic, always applying energy to change its materials from one form to another. Each product provides evidence of the process that formed it.

Let us compare two processes that change solid rocks into wholly new forms. Metamorphism adds both heat and pressure; weathering operates at surface temperature and pressure (Table 6–2)

THE ROCK CYCLE

If product reflects process, rocks tell stories. The grains of sand in a child's sandbox are the offspring of weathered mountains and at the same time parent to future cliffs of sandstone. Rocks and minerals have both recognizable past and predictable future.

James Hutton, the brilliant Scottish physician–farmer–naturalist introduced in Chapter One, saw this relationship clearly. He recognized that each rock was constantly changing into something else, which would only be changed again. Like Heraclitus (540–475 B.C.) he saw that the earth was always in process, always in the act of becoming something else. In brief, all rocks form from other rocks.

Hutton suggested a concept that is useful in reviewing and understanding many of the ideas presented in Chapters Two through Six. His concept was termed **the rock cycle** and is presented graphically as Figure 6-11.

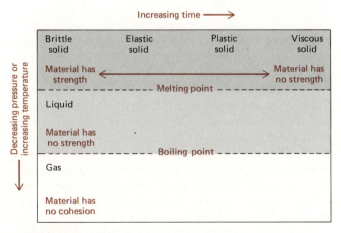

FIGURE 6–10

The behavior of matter. We are accustomed to thinking of material in terms of thermal boundaries (at constant pressure); these boundaries are reflected in the horizontal lines on the graph. The behavior of matter over long periods of time is reflected in the changes from left to right on the graph. No stressed material has any strength over unlimited time.

TABLE 6–2
Comparison of Weathering and Metamorphism

Observed Characteristic	Effect of Weathering	Effect of Metamorphism
Total rock volume	Increases	Decreases
Average rock density	Decreases	Increases
Average grain size	Decreases	Increases
Average metal content	Decreases	Increases
Average water content	Increases	Decreases
Grain surface area	Increases	Decreases
Overall rock texture	Random	Foliated

Hutton recognized that rocks of all kinds when exposed at the surface react by weathering, with weathered products largely eroded and carried downhill to the rivers, lakes, and seas (Fig. 6–11). There the weathered material was slowly cemented and crystallized into sedimentary rock, which through uplift might once again reach the surface. About two-thirds of all rocks exposed at the surface today are of sedimentary origin.

If, however, the sedimentary rock was buried still more deeply, it lost water and was recrystallized into metamorphic rocks under high temperature and pressure (Fig. 6–11). These metamorphic rocks may be slowly uplifted to the surface over many tens of millions of years. About one-sixth of all rocks exposed today are of metamorphic origin and were largely formed hundreds or thousands of millions of years ago.

If sedimentary and metamorphic rocks are buried so deeply as to melt, they form magma. If magma extrudes from the earth's surface, it forms young volcanic rocks, exposed over one-twelfth of the earth's surface. If magma cools slowly deep underground, it forms plutonic rocks (Fig. 6–11). Hundreds of millions of years later, plutonic rocks may also be uplifted and exposed. About one-twelfth of all rocks exposed today are of plutonic origin, having cooled to a solid hundreds or thousands of millions of years ago.

One cycle ends and another begins again whenever rocks are exposed at the surface. Weathering disaggregates and decomposes the ancient crystals and sends "new" material toward the sea once more. The rock, however, goes on changing. As James Hutton phrased it, "the earth has no vestige of a beginning—no prospect of an end."

The concept of a rock cycle is a satisfying summary of the relation between rock types and energy. Two centuries ago James Hutton saw the rock cycle as an expression of the "economy of nature." Everything is reused. Feldspar yields to weathering and releases floods of clay to the rolling sea. Deep within the earth clay is changed to mica and at even higher temperatures to feldspar, which on the surface yields to weathering and returns to clay.

QUO VADIS?

It was on the Appian Way near Rome that Peter was asked a troubling question: "Quo Vadis?" Whither goest thou? Let us ask that same question of the rock cycle. Where does the cycle go?

Cycles have much about them that both fascinates and comforts human nature. They combine the excitement of incessant motion with the comfort of re-

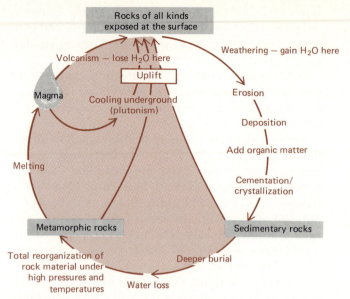

FIGURE 6–11

Schematic diagram of the rock cycle concept. Processes are in color; products are in black. The unshaded portion of the diagram includes all the processes driven by the sun's energy; all these processes involve adding water to material and creating products of low density and high volume in environments of low energy. The shaded portion of the diagram includes all processes that occur within the earth; all these processes involve high-energy environments, loss of water, and heating under pressure. The solar energy received at the earth's surface is more than 5000 times that escaping from the earth's interior, yet it is that tiny trickle of heat escaping beneath our feet that is the probable cause of earthquakes (Chapter Eight) and of the opening and closing over time of the earth's ocean basins (Chapter Ten).

taining the status quo. Renewal always comes from decay—a comforting thought to the only creature that knows it must die.

For many reasons the rock cycle was thought at the end of the eighteenth century to neatly "wrap up" the earth's processes. The concept casts the earth as a machine making endless changes, which to Hutton had *purpose*. In Hutton's words ". . . the whole presents a machine of peculiar construction by which it is adapted to a certain end." That certain end—that purpose—was to replenish the world for humankind by the incessant formation of new soil.

That a cycle which "gave us comfort" also had a purpose made its explanatory power seem even greater. But before we fully accept it, we must remember that the concept of a rock cycle was proposed two centuries ago, when almost nothing was known of the earth's interior. From the point of view of those on the surface, the rock cycle was a closed system, endlessly cycling the materials of the outer

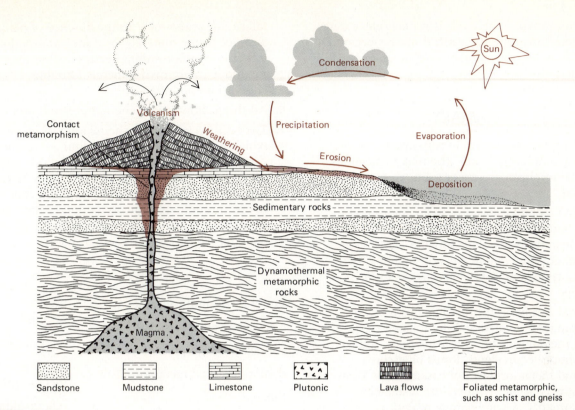

FIGURE 6–12

Summary of the major rock-forming processes. The continents are made up of rock that is quite different in composition from that fed to it by basaltic volcanism taking place beneath the continents.

few miles of the earth. Earth processes were believed to operate the same way through immense periods of time, an idea we have previously termed the *principle of uniformity*.

Our question—"quo vadis?"—was addressed to the earth. The answer from strict uniformitarian principles is—nowhere; the earth was a perpetual motion machine with a purpose, but no on–off switch.

TWO CHALLENGES

We now can challenge the older, steady-state version of the uniformitarian rock cycle on at least two grounds.

1. *Every* geologic process within the earth involves the expenditure of energy from a finite store. To speak of the earth as a perpetual motion machine is nonsense. We must instead assume that the earth is progressing from a distant past with a huge supply of internal energy to a distant future when it will have none. This concept is much more thoroughly discussed in the chapter on geologic time (see Chapter Seven-

teen). We live now somewhere in the middle of that scenario. The earth within is slowly running down.

2. Volcanic products from deep within the earth are basaltic, whereas ile magma originating from the melting of sedimentary rock is more commonly granitic in composition. Igneous rocks that result from the recycling of sedimentary rocks are quite different from those that result from *initial* melting deep within the earth (Figure 6–12).

Continents form from the net transfer of basaltic material from beneath the continents *to* the continents, and that transfer is *one-way*. Once formed, light continental crust is never returned to the deeper regions from which basalt comes. This concept will be much more thoroughly discussed in the chapter on plate tectonics (see Chapter Ten).

In our newer vision, the rock cycle is *not* a closed system going nowhere. Seen in modern terms, the rock cycle is a one-way transfer of material from within the earth to the continents. On the smaller scale, the rock cycle continues to be an excellent model of changes occurring within the outer few tens of miles of the continental surface. On the largest scale imaginable, the earth is *permanently* changing. As Heraclitus said, the earth is always becoming . . . but it is becoming more and more *unlike* the way it once was.

SUMMARY

1. Metamorphic rocks form in response to elevated pressure and temperature. The major categories of metamorphism are contact, burial, and regional.

2. Regional metamorphism of sedimentary rocks forms both foliated and nonfoliated crystalline rocks. Examples are slate, phyllite, schist, gneiss, granite, granulite, quartzite, and marble. Thermal metamorphism of sedimentary rocks forms hornfels.

3. Regional metamorphism of igneous rocks forms a wide range of foliated and nonfoliated rocks. Burial metamorphism of basaltic rock, and its sedimentary equivalents, forms blueschists. Regional metamorphism of basaltic rock forms greenstones in low-temperature water-rich environments. A wide variety of dark-colored phyllites, schists, amphibolites, and granulites form in progressively higher-temperature higher-pressure environments.

4. The dominant processes in all three major categories of metamorphism is recrystallization. Recrystallization is the growth of larger crystals at the expense of smaller crystals. The energy for recrystallization comes from loss of surface area, pressure solution, and increased heat.

5. Foliated textures have strongly aligned platy or needlelike minerals lying in the plane of foliation. Slaty cleavage may arise due to rapid water loss under pressure, strongly orienting their micaceous clay minerals. In phyllites and schists, schistosity results from total recrystallization of the rock with grain growth perpendicular to the direction of greatest pressure. In gneisses, gneissosity forms in the same way as schistosity, along with parallel bands of contrasting mineralogy.

6. If metamorphism of sedimentary rocks or layered igneous rocks is not very intense, former bedding or stratification may be seen as relict bedding. The layering of relict bedding need not parallel any foliation in the rock nor is it necessarily the same as compositional banding displayed in higher grade metamorphic rocks.

7. Nonfoliated textures form through total recrys-

tallization of the rock. In the absence of platy and needlelike minerals, or if they are randomly intergrown, the resulting texture is that of randomly intergrown minerals forming equilibrium boundaries.

8. All things flow, given enough pressure and time.

9. The rock cycle displays the relation of processes which alter material in and within a dynamic earth. The older version of the rock cycle failed to recognize that internal processes may have been more vigorous in the past or that continents, through an immense period of time, have grown in volume at the expenses of the basaltic material which forms the ocean floors.

KEY WORDS

burial metamorphism

cataclastic metamorphism

contact metamorphism

dynamothermal metamorphism

foliated

gneissosity

metamorphic processes

metamorphism

neocrystallization

nonfoliated

pressure solution

recrystallization

regional metamorphism

relict bedding

rock cycle

schistosity

shock metamorphism

slaty cleavage

texture

EXERCISES

1. In some metamorphic rocks, slaty cleavage cuts across relict bedding. How can that be explained?

2. How would you explain the observation that the older an exposed rock, the greater the likelihood that it is plutonic or metamorphic? *Note*: there are two *distinctively different* explanations for this observation.

3. We infer that the ancient earth had far more internal heat energy than it does now. Is that consistent with either of your answers to question 2?

4. In the light of our observation that nothing—over a long enough period of time—is solid, why is our earth spherical? Why are all solid astronomic bodies over a certain size spherical?

5. In areas of repeated deformation, we may observe schistosity of one composition cutting schistosity of another composition. How could that be explained?

6. What do we mean by our expression "as solid as a rock?"

7. What feature would make it easy to separate a schist from a marble?

8. What is the difference between slaty cleavage and schistosity?

SUGGESTED READINGS

BEST, M. G., 1982, *Igneous and Metamorphic Petrology*, New York, W. H. Freeman and Co., 433 p.
► *Up-to-date and straightforward. Assumes mineralogy background*

HARKER, A., 1930, *Metamorphism*, London, Methuen, 214 pp. (Reprinted in 1976, Halstead Press, New York.)
► *More than half a century old but extremely well written with really excellent drawings of metamorphic textures. A classic.*

PART

2

The Restless Earth

The elongate folded mountains of the Sierra Madre Oriental near Nuevo Leon, Mexico, remind us of crumpled carpet. This is a classic example of ridge and valley structure, a series of alternating upfolds and downfolds similar to the structures of the central Appalachian Mountains of North America. (Photograph courtesy of NASA.)

CHAPTER *seven*

Structural Geology

Our human experience with rocks is that they are strong solids. They fail only when attacked with sledge hammers or blasting powder.

Yet naturally deformed rocks tell another story. As the photo that opens this chapter suggests, the naturally occurring layering in rocks may be formed into intricate shapes more reminiscent of the flow patterns produced on the surface of moving fluids. In other exposed rocks a wide range of styles of deformation is evident.

Clearly rocks can be deformed in a wide variety of ways, with the processes leading to failure, which is defined as loss of coherence, strength, or rigidity. Deformation can run the gamut from brittle shattering, typical of window glass, to flowage as a cold solid, also typical of window glass (see Chapter 6). Geologists have long asked questions about how rocks deform and fail. Here are some examples.

1. Under what conditions do rocks fail?
2. What processes occur within rocks as they fail?
3. How many different kinds of failure can we observe?
4. What are the kinds of features that result from failure?

This chapter will answer these questions and raise others. Understanding the processes of rock failure leads to larger and larger questions. What causes earthquakes? How do mountains form? Why doesn't erosion level the continents to sea level? Are continents and seafloors made of the same stuff or are they somehow different? Answers to these questions fill the next several chapters. For now, we turn to simple first questions. What are the forces that shatter and distort "solid rock?"

STRESS AND STRAIN

Geologists who study deformed rocks deal with three fundamental physical properties—*force, stress, and strain.* It will clarify our thinking about rock deformation to describe each of these terms before we look at how these quantities interact to deform rock.

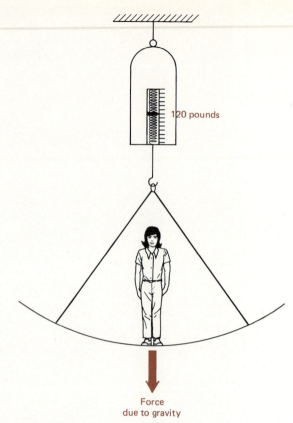

120 pounds

Force due to gravity

FIGURE 7–1
The attraction between the mass of a human being and the mass of the earth beneath causes a force that we measure as weight. The downward force of gravity is balanced by an equal force due to the resistance of the metal spring to extension and twisting.

FORCE

A **force** is a physical agent that causes a change in velocity or shape of matter. Our usual experience with forces encompasses, for example, *weight,* a force created by the mutual attraction of an object with the earth. In order to fully describe a force, *we must specify its magnitude in the direction along which the magnitude of the force is measured.* For weight, we might say that a human weighs (its magnitude is) 120 pounds or 54.4 kilograms as measured in a direction perpendicular to the earth's surface. Other examples of common kinds of force are frictional forces, electrostatic forces, and magnetic forces.

The magnitude of a force that creates motion can be measured by the acceleration in velocity that force causes when applied to a mass; in the format of an equation $F = M \times A$. If a human being weighing 120 pounds free falls from an airplane, the product

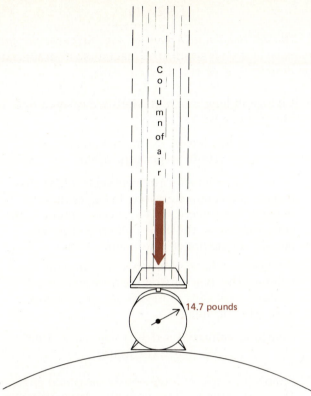

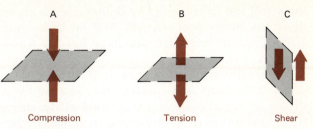

Compression   Tension   Shear

FIGURE 7–3

Stress is measured as the magnitude of force per unit area. When the forces oppose one another in a parallel direction perpendicular to a plane (A), the stress is compressional. If the forces act away from each other in a parallel direction perpendicular to a plane (B), the stress is tensional. If the forces oppose one another in parallel directions parallel to a plane (C), the stress is shear stress.

FIGURE 7–2

If we could "weigh" the column of air above us on a scale whose surface area is 1 square inch (6.45 square centimeters), we would discover that the mass of that column of air exerts a downward force measurable as 14.7 pounds (6.67 kilograms). That air pressure will balance a column of mercury about 30 inches (760 millimeters) tall; such a measurement of prevailing air pressure forms a part of ordinary weather forecasts.

of his or her mass (120 pounds) and the acceleration of their free fall velocity (1 foot ÷ second for each second of free fall) is 120 *poundals*, a unit of force. A more commonly used unit of force is the *dyne*, which is the force required to cause a one-gram mass to accelerate one centimeter per second during each second of acceleration.

If the application of a force does not cause a change in velocity, the magnitude of the force may be measured in other ways. When one steps on a simple scale, the magnitude of downward-directed force, one's weight, is measured by the extent to which a calibrated spring is stretched (Figure 7–1).

STRESS

Rather than dealing with force, geologists usually find it more useful to deal with **stress**, *the magnitude of force per unit of area.* Common units of stress in-

clude pounds per square inch and grams per square centimeter. The best known stress is atmospheric pressure. At sea level on an average day, the mass of the column of air above you applies 14.7 pounds of force to every square inch of material on earth. The figure of 14.7 psi (pounds per square inch) is normal atmospheric pressure (Figure 7–2).

Geologists describe the stress level in rocks in terms of force ÷ area on or along an imaginary plane within the rocks. This is a convenient way to describe the stress magnitude imposed on a rock. Three kinds of stress may be recognized.

1. Normal or **compressional stress**—opposing forces pushing toward each other in parallel directions perpendicular to a plane (Figure 7–3A).
2. Tensile or **tensional stress**—opposing forces pulling away from each other in parallel directions perpendicular to a plane (Figure 7–3B).
3. **Shear stress**—opposing forces acting toward each other in parallel directions *parallel to* a plane (Figure 7–3C).

The application of an imposed force to a cylinder of rock creates all three kinds of stress at the same time (Figure 7–3D). Resisting each of these three kinds of stress is the **strength** of the rock, which is the *magnitude of the stress it can withstand without failure.* If the forces applied create stresses greater than the strength of the rock—its natural resistance to stress—the rock will fail.

As an example, consider applying a force to loose snow. Snow will respond by some compaction to compressional stresses, as anyone who has packed a snowball knows well. Snow has moderate compressional strength. Stack it up in small snowdrifts, and snow will not instantly fail by sliding along a plane; snow has some shear strength. Pull on loose snow and it instantly crumbles; dry snow has NO tensile strength.

Many substances mimic snow in their response to stress. Most rocks and concrete are, however, very strong in compressional stress, moderately strong when exposed to shear stress, and quite weak when exposed to tensile stress. We'll discuss this phenomena again within this chapter.

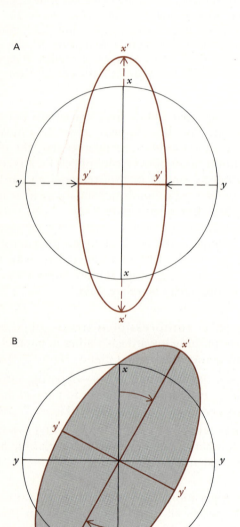

FIGURE 7–4

(A) The line *X - X* is elongated to *X' - X'* while the line *Y - Y* is shortened to *Y' - Y'*. The change in length/original line length describes the percentage of strain. (B) The line *X - X* is not only elongated to *X' - X'* but rotated 30*4o.6. Both the rotation amount and/or the line elongation can be used to describe the extent of strain.

STRAIN

The *deformation undergone by material due to the application of stress is* **strain**. The magnitude of strain can be described in several different ways.

1. Change in length of a previously measured line. The formula is:

   change in line length ÷ original line length × 100% = per cent strain.

   Thus a line originally 100 centimeters long that is now 110 centimeters long indicates that the material has undergone 10 percent strain (+ 10 cm ÷ 100 cm × 100% = 10%) in a direction parallel the deformed line (Figure 7–4A).

2. Change in volume of a previously measured volume. This type of strain is measured in a manner analogous to measuring strain along lines. The formula is:

   change in volume ÷ original volume × 100% = percent strain.

3. Change in angle of a previously measured angle. This type of strain measurement is most common when there has been an overall change in shape of the body. The change of an original 90° angle to a 87° angle would be described as three degrees (3°) of strain in a clockwise or counterclockwise direction (see Figure 7–4B).

Natural materials display a wide range of strain responses to imposed stress. Factors that affect the behavior of materials under stress include their temperature, cohesion (how tightly bound grains are to one another), stress difference between least and greatest imposed stress, and duration of imposed stress difference. Each of these factors is considered in the next part of this chapter, which deals with the types of strain exhibited by stressed materials.

▼

PATTERNS OF STRAIN

▲

A tennis ball at the moment it is struck by the racquet strings flattens against the racquet face, and then almost instantly regains its shape as a spherical object flying down the tennis court. Its strain (change of shape) has been *temporary*. Struck <u>much</u> harder, it explodes. Its strain (change of shape) has been *permanent*. Rocks, like tennis balls, exhibit both temporary and permanent forms of strain.

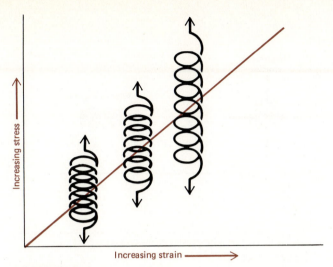

FIGURE 7–5
In the simplest kind of elastic strain, each incremental increase in stress results in proportional increase in strain, which is shown by an extension of the spring. Rubber bands exhibit the same linear relation between stress and strain. Once all stress is removed, all strain disappears, and the spring snaps back to its original shape.

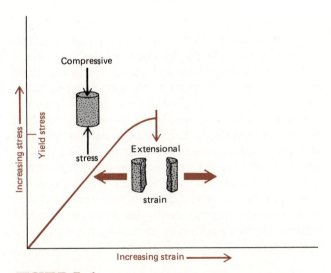

FIGURE 7–6
Brittle strain is characterized by an early increment of elastic strain (the straight colored line) followed by abrupt failure (the colored arrow) of the specimen by extensional strain. In the example shown, a cylinder of rock is placed under compressive stress in a direction parallel its length. The rock responds by elastic strain until its yield stress is reached. At a stress level slightly higher than that, it fails explosively and separates into two pieces on a plane parallel to the direction of applied compressive stress. The same plane is perpendicular to the direction of highest tensional strain, so the rock is literally pulled apart,

TEMPORARY STRAIN

The only common type of temporary or recoverable strain is **elastic strain**, a type of strain that is stored in stressed materials and then disappears when the stress is removed. Pull on a rubber band, and it increases its length—a form of measurable strain due to the tensile stress applied. Let go of the rubber band, and it snaps back to its original length. The energy, *stored* in the extended rubber band, is used up in regaining its original length and shape.

Another characteristic of many forms of elastic strain is that the *magnitude of the strain is directly proportional to the magnitude of the stress* (Figure 7–5). Each increment of stress causes an equal increment of strain. At the conclusion of the experiment the formerly stressed material displays *no permanent strain* (it is still the same shape as when the experiment started).

Rocks display elastic strain when stress magnitudes and duration of stress are both low. Drop a rock on a concrete floor, and the rock bounces off the floor. As we will learn in Chapter Eight, the passage of earthquake energy waves through rock depends on the ability of solid rocks to behave as elastic solids.

PERMANENT STRAIN

Permanent strains are nonrecoverable. The rock mass after permanent strain has suffered a persistent change in length, orientation, or volume. Many permanent strains are preceded by elastic strain, which changes into permanent strain after the strength of the rock has been exceeded. There are three broadly defined types of permanent strain, *each of which is transitional to the other*. These types of permanent strain are

1. **Brittle strain,** a sudden failure along a single plane of high tensile stress, with complete loss of all rock strength (Figure 7–6),

2. **Ductile strain,** moderately abrupt to gradual failure along a series of closely spaced planes of high shear stress, with partial loss of all rock strength (Figure 7–7),

3. **Viscous strain,** gradual failure by fluid flow in a direction parallel to planes of highest compressional stress. The rock retains a measure of strength or resistance to flow; this measure is now termed **viscosity** (Figure 7–8), defined as the ratio of stress to *rate* of strain. The higher the viscosity of a substance, the greater its resistance

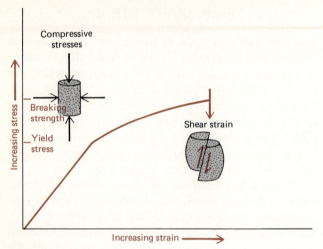

FIGURE 7–7

Ductile strain is characterized by early elastic strain, followed by permanent strain at higher and higher stress levels above the yield stress. Eventually, the rock fails by sliding on a plane that is inclined to the direction of maximum compressive stress; this plane is the plane of maximum shear stress. In the example given, a cylinder of rock under increasing compressional stress first responds by elastic strain, then begins to bulge and change its shape as the yield stress is passed. Failure eventually occurs along an inclined plane of high shear stress, and the two blocks of rock slide past one another. Such failure is called faulting, and the energy release due to faulting would be perceived as an earthquake.

to flow, and the less its rate of strain for a constant stress.

Brittle Strain

After a brief episode of elastic strain, materials exposed suddenly to high stress differentials (the difference between greatest and least imposed stress) may abruptly, even explosively, fail by brittle strain (Figure 7–6). *The stress level at which permanent strains occur* is termed the rock's **yield stress**. The usual result of brittle failure is a shattered rock that has failed along a single plane by extension perpendicular to that plane. Only the area immediately adjacent to the plane of failure exhibits any evidence of strain; most of the rock is uninvolved.

Factors that enhance brittle behavior are cold temperatures, and sudden application of high differential stresses. On an earth where both temperature and overall pressure on a rock increase downward, our experience is that truly brittle behavior only occurs within the uppermost mile or two of the earth.

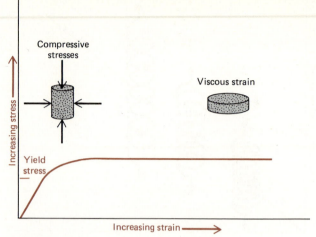

FIGURE 7–8

A rock cylinder exposed to high compressive stresses from several directions and high temperatures over a long period of time may fail by viscous strain. With such failure the cylinder slowly flattens and widens. There is no rupture, and the rock retains much of its strength while continuing to fail by viscous flow. Increasing the stress above the yield stress will increase the *rate* of flow.

Ductile Strain

After a longer episode of elastic strain, materials exposed to moderate stress differentials may begin to fail by slippage parallel to single to multiple planes of high shear stress that are oriented at some angle to the direction of highest imposed compressional stress (Figure 7–7). The usual result of ductile failure is a rock that show failure by slippage along smeared and streaked surfaces. Usually only parts of the total rock are involved; areas between the planes of slippage may or may not give much evidence of strain.

Factors that enhance ductility are warmer temperatures, moderate stress differentials, and moderate stress duration. On our earth, this means that ductile behavior is more common at depths greater than a mile or two.

Viscous Strain

Also incorrectly termed fluid strain, viscous strain occurs when hot solid rocks are exposed to low stress differentials over long periods of time. Under these conditions the rock flows while still maintaining its strength and solidity, an idea first discussed in Chapter Six (see Figure 6–9). The rock flows in a direction perpendicular to the largest imposed compressional stress direction. Every particle of the rock exhibits

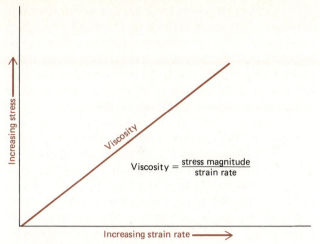

FIGURE 7–9
If a solid is exhibiting a simple type of viscous strain, the rate of flow is directly proportional to the magnitude of the stress. The constant that relates these two variables is *viscosity*. Since viscosity for any one temperature is a constant, increasing the strain rate requires increasing the stress magnitude.

evidence of flowage (Figure 7–8A).

On our earth, viscous flowage occurs deep within the outer earth where temperature and average pressure are both quite high. Viscous flow is common in many foliated metamorphic rocks and may be universal within the inner earth.

For many rocks that have deformed by viscous strain, the *rate of flow is directly proportional to the magnitude of stress*. The ratio that relates stress magnitude to flow rate is **viscosity** (Figure 7–8B). The more viscous the substance, the slower its rate of flow for a given amount of stress. Decreasing the temperature dramatically lowers the viscosity, an observation known to anyone who has tried to pour very cold

motor oil into an engine, or tried to start an engine full of cold oil.

SUMMARY

The boundaries among brittle, ductile, and viscous behavior are gradational, ranging from explosive brittle failure to serene viscous flow. The chief controls are temperature and stress duration. Increasing temperature and stress duration along with declining stress difference will cause a rock to become less brittle and more viscous.

It is to the natural world that we now turn. Although the study of stress and strain and the patterns of strain provide powerful insights into rock behavior, we turn away from the world of theoretical explanation to the world of observational challenge. We will discover that little is more fruitful than good theory.

▼

BRITTLE STRUCTURES

▲

Rocks on the earth's surface exhibit nothing but brittle behavior. Attacked with sledge hammer, dynamite, or even growing ice crystals (see Chapter Five), rocks fail by shattering. They perversely fail to fold, buckle, flow, or fault *during our experience with them*.

Yet exposed folded, buckled, and faulted rocks are common, especially within mountains. Clearly rocks now at the surface have been exposed in the past to conditions within the earth that favored folding, buckling, flowage, or faulting. Within these same rocks is contradictory evidence that indicates highly brittle failure also was common in rock's history. The patterns of strain suggest that rock displayed a wide

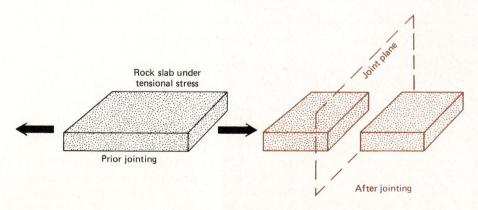

FIGURE 7–10
Many common types of joints result from tensional stress parallel to the layering of the rock. The slab, shown in color, has failed by a true extension fracture, and the plane of failure is termed a joint plane.

range of behavior; we are left with a puzzle to solve—the central activity of science.

JOINTS

The most abundant structures formed by brittle failure in rocks are *joints*. **Joints** are naturally occurring planes of rock fracture with motion commonly perpendicular to the plane of failure. Most joints exhibit evidence that brittle strain occurred along planes of high tensile stress (Figure 7–6 and 7–10), which caused the rocks to be pulled apart along planes perpendicular to layering.

The surface features of a joint plane suggest it is commonly an extremely brittle structure. Since many joint planes are vertical, they must have formed parallel to the downward pull of gravity and reflect pure extension perpendicular to the joint plane. For this reason many joints are termed *extension fractures* or *release joints*. In crystalline igneous and metamorphic rock as well as in layered sedimentary rock joints planes are commonly both vertical and abundant.

Many joint planes occur as a series of parallel planes that repeatedly break the intervening rock. Such groups of parallel or subparallel joint planes are termed **joint sets.** Where these joint planes reach the surface in a semiarid region, the joint planes provide moisture traps; consequently patterns of vegetation may outline the surface expression of the joint sets (Figure 7–11).

It is just as common for joint sets to intersect one another, often in directions that are approximately perpendicular to one another. Joint sets of this type may form quite prominent vertical faces along natural rock exposures. The rocks may appear to be broken into a series of stair steps as each joint plane is prominently expressed (Figure 7–12).

The origin of most joints, especially in layered sedimentary rocks, remains an enigma. Joints are apparently formed early and are commonly affected by other features such as faulting and folding. The common occurrence of joints as planes perpendicular to bedding, and the general observation that the spacing between joints increases in thicker layers both suggest that, in sedimentary rocks, joints release tensile stresses in directions parallel to layering (Figure 7–10). Why there should be early, nearly universal layer-parallel tensile stress is simply unknown. Various theories have been suggested, including tidal stresses and volume expansion on uplift, but no compelling theory has yet emerged.

DUCTILE STRUCTURES

Ductile structures form in response to moderate differential stress and moderate temperatures. Most ductile structures form owing to failure by sliding

FIGURE 7–11
In an aerial view of strongly jointed rocks, rows of dark-colored vegetation outline the joint sets in this area of southern Utah. Only one joint set is observed in this area, but it forms a prominent pattern. (Photograph courtesy of the U. S. Geological Survey.)

FIGURE 7–12

In this area of central Ohio, prominent joint sets are perpendicular to one another, so that the rock naturally breaks into a series of stair steps. The trees on top of the ridge give some sense of scale. (Photograph courtesy of the U. S. Geological Survey.)

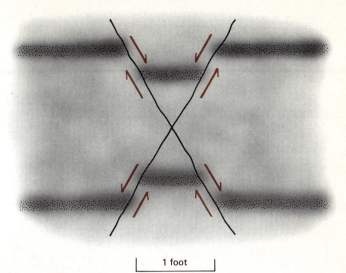

| 1 foot |

FIGURE 7–13

In this view looking directly down onto a flat rock surface, shear joints have offset two dark lines of fragments in the rock along shear joint planes whose surface intersection form a broad "X-shaped" pattern in the rocks. The amount of movement is typically only a few inches or centimeters.

(Figure 7–7) along a plane or planes of high shear stress, a process termed **faulting**. The plane of failure is termed a **fault plane**, and the resulting structure is termed a fault. A **fault** is a planar fracture or series of fractures *along which* sliding motion has taken place. When there is slippage along the fault plane we experience an *earthquake*, a topic discussed in the next chapter.

Among the smallest scale faults are features termed **shear joints**. These misnamed features are actually abundant small faults that intersect each other at acute angles (Figure 7–13). The surfaces of shear joints display smeared, streaked structures that document slip along the fault plane.

FAULT TYPES

Faults are classified on the basis of the orientation of the fault plane and the relative motion of the blocks of rock on each side of the fault plane. Since ductile structures display a complete gradation from rather brittle to nearly viscous characteristics, we should ex-

pect a wide range of strain patterns within the ductile range. The classification used in this book recognizes four major fault types from among the many found in nature.

Normal Faults

Also called *gravity faults*, a **normal fault** has a fault plane that dips generally steeply downward (commonly more than 60°) from the horizontal. Faulting motion is directly down the fault plane (Figure 7–14). Normal faults have received their descriptive name because they are so common.

The block of rock *vertically* beneath the fault plane is termed the **footwall**; that block *vertically* above the fault plane is termed the **hanging wall**. These old terms spring from miners whose underground tunnels might follow a fault plane, so that the hanging wall was "hanging over their head" and the footwall was that block of rock on that they placed their feet.

A normal fault is one in which

1. The hanging wall block has moved downward relative to the footwall block, by slippage down the fault plane.

2. The fault plane dips steeply downward from horizontal. The angle of inclination may vary from 60° to nearly 90°.

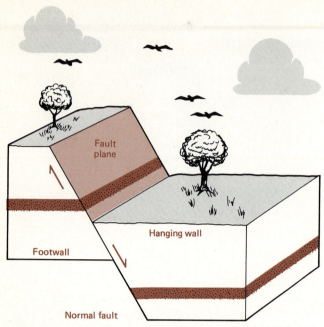

FIGURE 7–14

This block diagram illustrates the pattern of normal faulting. Notice that the rock layer records the amount of offset along the fault plane as the block to the right dropped down.

An exception to this last statement is a special class of normal faults, termed *detachment faults*. This increasingly recognized type of normal fault is characterized by low angles of fault plane dip (often less than 30°). Movement of the hanging block is down a much gentler slope which may have an overall listric (concave, spoon-shaped) surface. Motion along detachment faults is a form of gravity-driven sliding. Detachment faults seem to form most commonly in environments that have undergone active extension and thinning of the crust. We will meet listric detachment faults again in Chapter Twenty-One.

Normal faults form in environments where rocks are placed under horizontal tension while experiencing the downward force of gravity. The combination of stresses produces a rather brittle ductile structure termed the fault plane, and the hanging wall slides downward to form a normal fault. When two normal faults form whose fault planes are inclined toward one another, the resulting downdropped block is termed a **graben**; the high block between grabens is termed a **horst** (Figure 7–15).

Large grabens include the Red Sea, a downdropped area dividing the Arabian Peninsula from Africa, and Death Valley, a downdropped block that forms the area of lowest elevation in North America. Much of the scenery of the southwestern United States owes its origin to grabens, which form desert basins, and horsts, which form high isolated ranges (see Figure 13–8 and 13–10). That part of the United States is often called the Basin and Range—an area of crustal extension where normal faults are extremely abundant.

Reverse Faults

As the name implies, a **reverse fault** is one where the movement of the hanging and footwall blocks is reversed from that along a normal fault—that is, the hanging wall moves upward relative to the footwall (Figure 7–16A). Reverse faults result from strong upward-directed forces that form block uplifts. As reverse faults are generally uncommon, they are not further discussed in this chapter.

Thrust Faults

A **thrust fault** is a reverse fault where the hanging wall is driven upward along a fault plane that angles quite gently from horizontal; the average angle of inclination is only 20°. This fault type is characteristic of strong horizontal compression in an environment of little effective vertical stress.

FIGURE 7–15

Grabens and horsts reflect a specialized pattern of normal faulting, breaking the earth's surface into alternating uplifted ranges and downdropped basins. Since the graben block is vertically above the fault plane, it is the footwall block, and the horst is a hanging wall block. The offset of an individual layer records the amount of offset between the graben and horst blocks.

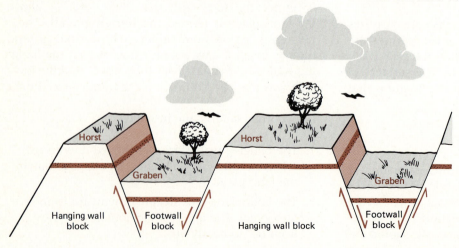

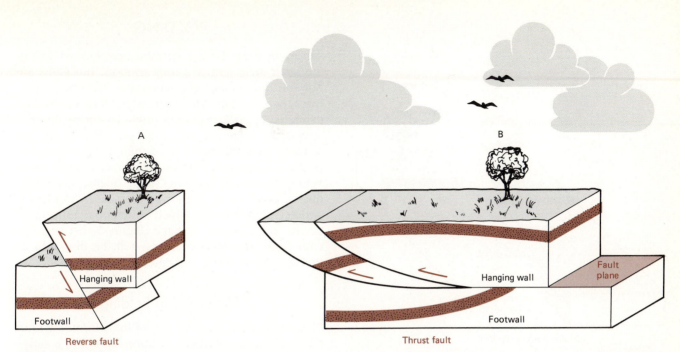

A

B

Hanging wall

Footwall

Reverse fault

Hanging wall

Fault plane

Footwall

Thrust fault

FIGURE 7–16

(A) In a rather rare type of fault, the hanging wall moves up along a steeply inclined plane in a form of deformation known as a reverse fault. (B) Thrust faults are a much more common form of a reverse fault, where the fault plane is very gently inclined. In the example shown here the fault plane breaks up into two individual splinters, each of which is an individual thrust fault.

Located most commonly in mountainous areas, thrust faults commonly are a rather ductile fault, with intense smearing along the fault plane. Thrust faults may form in multiples of intersecting fault planes and stack blocks of rock over one another repetitively (Figure 7–16B). In intensely deformed mountainous zones such as the Alps, thrust faults may extend for very, very large distances and transport huge masses of rocks. In the same way that normal faults may carry the hanging wall blocks several thousand feet downward, thrust faults may carry the hanging wall block thousands of feet upward and many miles horizontally.

Thrust faults occur in the same stress environment that produces many kinds of folding, so that thrust faults and folds are often found together in nature. The processes that allow large blocks of rock to be folded and driven laterally up slopes of a few tens of degrees are not fully understood, but they testify to the power of a dynamic earth to make mountains and strongly deform rocks into relatively ductile strain patterns.

Strike-Slip Faults

A **strike-slip fault** is characterized by a vertical fault plane along which all motion is horizontal. It appears to result from moderate horizontal compression combined with moderate vertical loading due to gravity.

Strike-slip faults are a common fault type and result in horizontal movement of one block of rock past another. The terms hanging wall and footwall are now meaningless, so that different terminology is employed to describe the relative motion of blocks. If, as viewed by an imaginary observer whose feet straddle the fault plane, the right foot is moved rearward by motion along the fault plane, the fault is termed a **right-lateral** strike-slip fault. If the imaginary observer's left foot is moved rearward, it is **left-lateral** strike-slip fault (Figure 7–17). The distinction can also me made by imagining an observer looking from one block to another; if the block *across* the fault plane moves to the right, the fault type is a right-lateral strike-slip fault and so forth.

Perhaps the best-known strike-slip fault in the United States is the San Andreas fault of California. This fault, that angles north- northwestward from the Gulf of California to San Francisco, is a right-lateral strike-slip fault. Sporadic motion along the fault creates some of the most destructive earthquakes experienced anywhere on earth. The San Andreas fault bisects southern California, creating distinctive landscapes along its path (Figure 7–18).

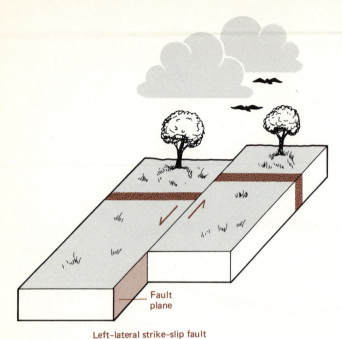

Left-lateral strike-slip fault

FIGURE 7–17

A left-lateral strike-slip fault forms when the motion in the fault plane is wholly horizontal. In the example shown here, the motion is marked by the offset of vertical layers. As is typical of strike-slip faults, the fault plan is vertical.

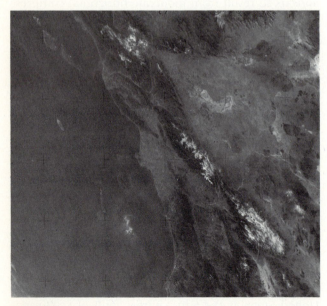

FIGURE 7–18

In this photograph taken from Skylab in 1974, much of southern California near Los Angeles can be seen. The dark-colored areas of coastal California are abruptly separated from the lighter-colored desert areas to the east by two straight lines that mark the surface expression of the San Andreas and Garlock Faults. They intersect at Tehachapi near Gorman, California. (Photograph courtesy of NASA.)

FOLDING

All faults as well as shear joints are formed within the strain field termed *ductile behavior*. But faults are not the only category of naturally deformed structures that are of ductile origin. Some forms of folding exhibited by rock owe their origin to ductile failure.

Broadly there are two processes that distort the shape of an originally horizontal thin rectangular rock layer (Figure 7–19A). The layer may be truly bent or flexed into a curved shape (Figure 7–19B) by ductile processes, or it may be crosscut by numerous parallel planes along which differential movement occurs to create a new shape (Figure 7–19C) by viscous flow. The former process will be discussed in the next section of this chapter.

Flexural-Slip Folding

Flexural-slip folding is typical of folds formed within the uppermost few miles or kilometers of the earth. The layered rock is flexed; the strain due to bending causes the development of a wide variety of joints as well as many other auxiliary structures. As bending continues, layers begin to slide by ductile shear across one another, with each layer sliding over the layer underneath and generally moving toward the crest of the developing fold (Figure 7–20). It is a model of folding often called the "card-deck" model. To demonstrate, pick up a stack of playing or index cards. While holding them tightly, push on the ends, and the center of the card pile arches up to form a fold as each card angles upward on the card beneath it.

More formally this folding style is termed **flexural-slip** folding, as the fold results from a true flexure of the rock layer, which slips on layers above and below it. The layers of rock retain their thickness as measured perpendicular to bedding. Interlayer slip is the dominant process that allows flexuring to continue. The more steeply inclined the layers, the more dominant interlayer slip becomes.

VISCOUS STRAIN

When rock has become so hot and/or is under such long-endured stress that it behaves as a thick fluid, the strain becomes viscous. Other than certain forms of structures unique to high-grade metamorphic rocks, the dominant structure produced by viscous strain is folding, which commonly occurs in certain soft sedimentary rocks as well as in many metamorphic rocks.

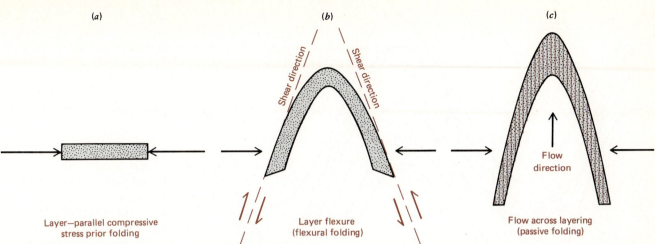

(a)

Layer—parallel compressive
stress prior folding

(b)

Shear direction   Shear direction

Layer flexure
(flexural folding)

(c)

Flow
direction

Flow across layering
(passive folding)

FIGURE 7–19

(A) A slab of rock begins to experience compressional stress parallel its layering. If the rock does not fail by extension fracturing, it may buckle (B) and deform into a folded slab with strong shear stresses parallel the steeply inclined bedding planes. At still higher average stress and temperature levels, the layering may become passive, and its offset (C) may simply mark the viscous patterns in a slowly flowing rock.

PASSIVE FOLDING

When flowage occurs across the preexisting layering of sedimentary or foliated metamorphic rocks, the layering is offset into a new shape that records the differences in velocity of flow *across* the layer. This type of folding is termed a **passive fold.** The preexisting layers in a passive fold are *not* flexed or bent. Instead they are progressively offset by flowage across them, forming the layers into arcuate patterns.

It is the progressively offset old layering that allows us to see the patterns of increasing and decreasing velocity of flow across them. If the viscous flow becomes turbulent, the passive fold forms become complex and then chaotic, but if the viscous flow remains relatively laminar, folded forms of great beauty occur. The occurrence of passive folding can always be recognized because the crests and troughs of folded layers become extremely thickened while the sides of folds become extremely thinned (Figure 7–21).

As can be seen by even this brief discussion, folds attract the eye and have prompted centuries of investigation into the processes that form them. In general, folds seem to form owing to strong horizontal compression that crumples layered rocks into folded and faulted areas. To study folding in more detail, we need to examine basic ideas about the geometry of folds.

FOLD GEOMETRY

An understanding of fold geometry begins with an understanding of fold parts. The two sides of a fold are its **limbs**. The **axial plane** is an imaginary sur-

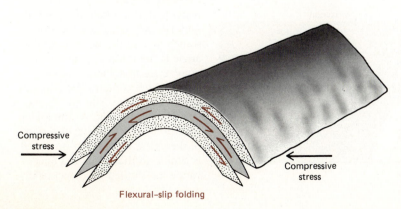

Compressive
stress

Compressive
stress

Flexural-slip folding

FIGURE 7–20

Flexural-slip folding occurs when rock layers truly flex or buckle and begin to slide over one another as the fold continues to grow. Compressive stresses initiate the buckling, and slip between the layers accentuates the slippage, which often streaks and scars the exposed bedding plane surfaces.

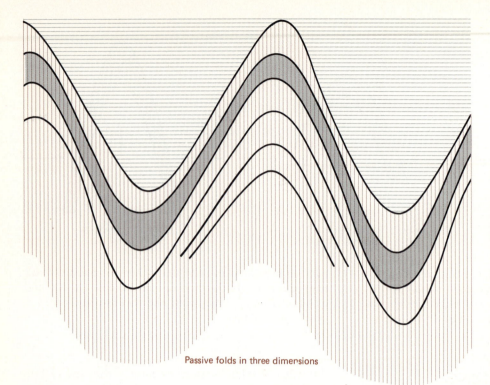

Passive folds in three dimensions

FIGURE 7–21

A passive fold is one in which slip across the layer is the dominant process. The planes of slip may be closely spaced and barely visible as in this sketch, or if they are sufficiently closely spaced, they may be essentially invisible, and the rock will appear to have flowed without leaving any direct evidence of flow behind. The intersection of the pervasive minute planes of slip with the fold limbs may leave delicate lines marking the intersections.

face that symmetrically bisects a fold. The **fold axis** is the line created by the intersection of the axial plane with the surface of the folded layer. The fold axis may be a horizontal line; if the fold axis is not horizontal, the fold is a **plunging fold** (Figure 7–22).

If the layers of rock are deformed into an arch, the fold is termed an **anticline** (Figure 7–23). If the layers are deformed into a trough, the fold is termed a **syncline** (Figure 7–24). As both of these figures suggest, it is rare to find the whole of a fold well preserved and exposed to easy view, but a little imagination allows one to reconstruct the form of the full fold.

If the axial plane of a fold is vertical, the fold is an **upright fold** (Figure 7–25A); if the axial plane is in between vertical and horizontal, it is an **overturned fold** (Figure 7–25B); and if the axial plane is horizontal, it is a **recumbent fold** Figure 7–25C). Recumbent folds particularly display the power subterranean forces not only to crumple the layered rocks

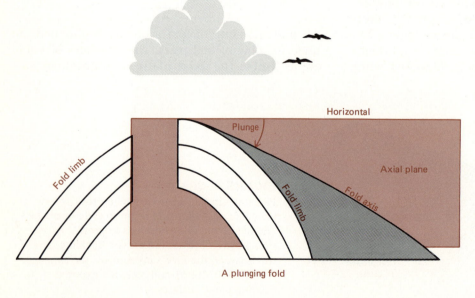

A plunging fold

FIGURE 7–22

If the fold as a whole is tipped, the fold axis will point downward at some angle to the horizontal; that angle is the fold's plunge. The axial plane is an imaginary bisecting plane which contains the fold axis within the plane. The fold limbs are the sides of the fold. As the folded surface intersects the level surface of the earth, the pattern produced is similar to that produced by a canoe half-buried in a level field.

FIGURE 7–23

A river has exposed the upfolded limbs of an anticline. The central part of the fold has been washed out by the river, producing a small cave. (Photograph courtesy of the U. S. Geological Survey.)

but also to turn the folds in the air so that one of the limbs is upside down.

Folds often occur in groups, and these tell us that the region at one time experienced strong compressive stress, which crumpled the layered rocks much as one kicks a rug into a series of folds. If the axis of a fold is traced along the earth's surface, the fold will eventually die out, much as faults do. No area of de-

FIGURE 7–24

A part of a syncline has been exposed by a river, with each of the limbs inclined toward one another. This photograph is nearly a century old; the geologist sports a coat and tie and provides a scale. The layers of rock are crosscut by a highly specialized type of jointing that reflects the intense strain produced by bending thick layers of rock around a tight curve. (Photograph courtesy of the U. S. Geological Survey.)

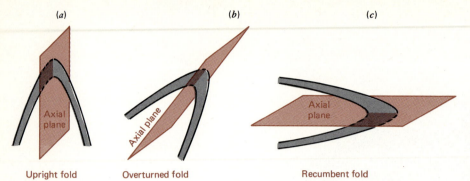

(a) (b) (c)

Upright fold   Overturned fold   Recumbent fold

FIGURE 7–25

The orientation of the fold's axial plane defines three common fold orientations: (A) upright, (B) overturned, and (C) recumbent.

formation is of indefinite size, so that all deformed rocks pass horizontally into undeformed rocks.

If the fold axis is plunging, the folded rocks form canoe-shaped patterns where they intersect the earth's surface. The surface pattern produced by plunging anticlines and synclines is a series of tightly compressed Z-shaped (Figure 7–26) ridges and valleys. This photo from space of part of the Appalachian Mountains reminds us of how descriptive our analogy of kicking a carpet really is. The Z-shaped ridges of the Ridge and Valley region of the Appalachians were produced when Europe and America collided hundreds of millions of years ago (Chapter Nineteen). The flowing patterns of symmetrical ridges and valleys are a record of this awesome collision in the past.

The products of deformation span the range from glass-brittle extension joints to passive folds where

rock flowed as a viscous mass, carrying and distorting the old layering. All of these structures remind us that the rocks beneath our feet are not as solid as they seem. Especially in mountainous areas rocks are riddled with the evidence of past strain.

SUMMARY

1. Forces cause changes in shape or volume of rocks. The force per unit of area is termed stress. Compressional and tensional stress act as parallel forces that are perpendicular to a plane while shear stresses act as parallel but opposing forces along a plane.

2. A rock's strength is a measure of its resistance to stress. Most rocks are strong when exposed to compressive forces, moderately strong when expressed to shear forces, and weak when exposed to tensile stresses.

3. Strain is the rock's response to stress. Elastic strains are perfectly proportional to stress; those strains are fully recoverable when stress is removed. Permanent strains include brittle, ductile, and viscous strains, which are transitional with one another. The yield strength is the stress level at which strains become permanent.

4. Increasing average overall stress, duration of stress, and temperature causes a rock to become progressively more ductile in its response. When rocks become wholly viscous, the rate of flow is directly proportional to stress.

5. Brittle strain in rocks is largely confined to extension joints; strain is limited to motion perpendicular to the joint plane. Ductile strain includes all forms of faulting and flexural-slip folding; with increasing ductility strain becomes progressively more pervasive and may involve much of the rock adjacent to shear planes.

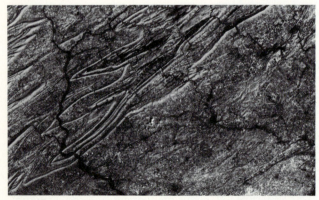

FIGURE 7–26

This is an image taken in 1972 from 568 miles (914 kilometers) in space by a satellite originally called ERTS-1. The Allegheny Mountains, near Harrisburg, Pennsylvania, form a series of flowing curves as plunging anticlines and synclines of the Ridge and Valley part of the Appalachian Range intersect the earth's surface. The Susquehanna, Delaware, and Schuylkill Rivers drain the area to the southeast.

Viscous strain includes passive folding with strain distributed throughout the entire rock.

6. Folds involve the offset of formerly horizontal layers by either true flexuring of the layer or by differential offset by flowage across the layer. Common folds include anticlines and synclines, which form distinctive patterns on the earth's surface.

KEY WORDS

anticline	left-lateral fault
axial plane (of fold)	limb (of fold)
brittle strain	normal fault
compressional stress	overturned fold
ductile strain	passive fold
elastic strain	plunging fold
fault	recumbent fold
faulting	reverse fault
fault plane	right-lateral fault
flexural-slip fold	shear joint
fold axis	shear stress
footwall	strain
force	stress
graben	strength
hanging wall	strike-slip fault
horst	syncline
joint sets	tensional stress
joints	thrust fault
upright fold	viscous strain
viscosity	yield stress

EXERCISES

1. If a series of sedimentary rocks are folded into an anticline and then eroded back down into a flat plain, would the oldest or youngest rocks be exposed in the center of the folded area? Why?

2. It is common in the Appalachian Mountains to discover that the downfolded synclines form the topographic ridges, while the upfolded anticlines form the valleys. How can this be explained?

3. If we added water to a consolidated rock before we deformed it, would you expect that the addition would make the rock more brittle or more viscous? Why?

4. Why are intense folding and thrust faulting commonly associated in time and space?

5. We observe that thicker rock layers in sedimentary rocks have greater horizontal distances between joint planes than jointing formed in thinner layers of the same rock composition. How could this be explained?

6. Hanging wall blocks of thrust faults may move laterally many miles or kilometers. Why don't hanging wall blocks on normal faults move downward similar distances?

SUGGESTED READING

Davis, George A., 1985, *Structural Geology of Rocks and Regions*, New York, John Wiley & Sons, 538 pp.
▶ *There is no finer introduction. Lucid writing and impeccable science.*

The scene in Lisbon, Portugal, during the great earthquake. Ships founder, parts of the city collapse into the sea, steeples topple, and seawater roars into the helpless city. (Courtesy of History of Science Collections, University of Oklahoma Libraries.)

CHAPTER *Eight*

Earthquakes

Nature to be commanded, must be obeyed.
SIR FRANCIS BACON (1561–1626)

Our earth is seldom quiet. Thousands of earthquakes disturb the sensitive quivering pens of delicate recording instruments every single day. Every day somewhere on earth humans feel the earth shudder. The tremor may be momentary and gentle—only a reminder of the forces beneath our feet. More rarely, buildings begin to lurch and sway in tune with the rumbling rock beneath them. Still more rarely, the earth shudders in violent waves, and people die among the remnants of structures that once protected them.

Earthquakes are among our most ancient scourges. Prayers for deliverance from earthquakes are among the oldest written records, because early civilizations in the Mediterranean area had to contend with both volcanic eruptions and a trembling earth. The association of earthquakes with volcanic eruptions led many people to suppose that great fires within the earth were the cause of both volcanism and earthquakes. Indeed, Pompeii (see Chapter Three) had been shaken by a great earthquake in A.D. 63, and then rebuilt, only to await the suffocating ash of the A.D. 79 eruption of Vesuvius.

THE CAUSES OF EARTHQUAKES

When William Shakespeare (1564–1616) wrote ". . oft the teeming earth is with a kind of colic pinch'd and vex'd by the imprisoning of unruly wind within her womb, which for enlargement striving, shakes the old bedlam earth and topples down steeples and moss grown towers," he was registering the common understandings of his time. Water or wind driven into underground caverns as well as fires within those caverns were widely believed to be the cause of earthquakes.

In 1755 a giant earthquake devastated Lisbon, Portugal (see chapter frontispiece), and took the lives of 50,000 people. This tragic event marks a turning point in both philosophy and geology. For philosophy, Voltaire (1694–1778) and others began to ask how a kind Almighty could allow so much suffering to befall so many innocent people. For geology, the Lisbon earthquake was the first one whose physical properties were recorded and studied.

PRE-TWENTIETH CENTURY

Crude measurements of the travel time of the energy released by the Lisbon earthquake showed that earthquake energy moves through the solid earth at a speed approximating that of sound. This could be recognized because the interval between the occurrence of the earthquake in Lisbon and the arrival of the energy waves elsewhere correlated with the distance to the source of the earthquake (Figure 8–1). The difference in time of arrival of the earthquake energy at two towns a known distance apart was proportional to the speed of sound.

It was also recognized that the passage of earthquake energy from place to place indicated that the earth behaves as though it were an **elastic** body— that is, the earth responds to earthquake energy by distorting its original shape and then regaining that shape. The earth vibrates under the impact of the energy released by an earthquake, much as a bell does when it is struck. The energy released by a great earthquake may travel entirely through the earth, ringing the entire earth like a colossal spherical bell. The oscillations of the vibrating earth can be readily felt and heard, and amplified by poorly consolidated or waterlogged soil; they are the basic cause of the collapse of structures during an earthquake.

There were also early attempts to pinpoint the direction of the source of energy. Early scientists real-

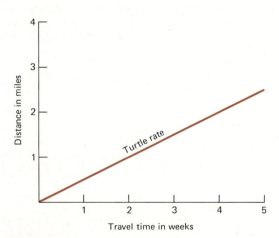

FIGURE 8–1

This whimsical example illustrates an important idea. If a turtle has been lost for four weeks and its travel speed is known 1/2 mile (4/5 kilometer) per week, it must be two miles from its home. Given any two of the three variables (speed of travel, distance, travel time) in the graph, the third variable can be easily found. Such a simple graph is correct as long as the rate is constant.

ized that cooperative recording of the time of arrival of earthquake wave information might yield the exact location of the earthquake. The long road to understanding why earthquakes occur had begun. Great tragedies often mark beginnings, and it was the Lisbon disaster that marked the beginning of serious investigation into the mysteries of earthquake origin.

In the nineteenth century our understanding of how energy travels within the earth was much improved, but the *causes* of earthquakes remained a puzzle. People now recognized that great underground fires or unruly winds in deep caverns were hardly reasonable causes, but the origins of earthquakes remained a puzzle into the twentieth century, when a combination of careful observation and laboratory investigation finally unearthed two dissimilar causes for earthquakes.

In the interim, the *effects* of earthquakes were ever more skillfully recorded and measured. These measurements revealed that earthquakes released energy in several different waveforms, some of which passed entirely through the earth, revealing much about its interior (see Chapter Nine).

Thus came the appreciation that earthquakes bring more than destruction and horror. They also bring a reasonably detailed understanding of the earth's interior, provide an understanding of their causes which may make earthquake prediction possible, and reveal the dynamic processes that modify the earth on the largest scale (Chapter Ten).

TWENTIETH CENTURY

During this century geologists have come to understand in some detail how the energy released by earthquakes is transmitted through the earth, and they have concluded laboratory studies that attempt to mimic the events leading up to an earthquake. We now know that *all earthquakes are caused by the long-term storage of increasing amounts of energy in elastically deforming rocks, followed by the sudden release of this stored energy.*

The storage and release of energy is accomplished in two very different processes:

1. Elastic rebound
2. Dilatancy-diffusion

Elastic Rebound

When an earthquake occurs by motion along a preexisting fault, its probable cause is elastic rebound. A fault (see Chapter Seven) is a generally planar surface along which sliding motion has taken place between the two blocks separated by the fault. Once motion has occurred, the fault plane may serve as a continuing zone of easy slip, allowing continuing sliding along the fault. The fault then exhibits *aseismic creep*. **Aseismic creep** is the phenomenon of essentially continuous motion along a fault, unaccompanied by earthquakes.

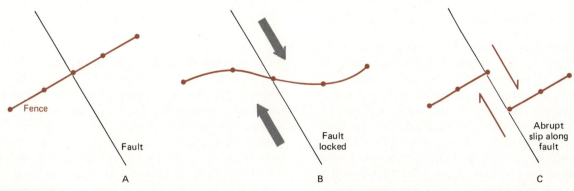

FIGURE 8–2

Aerial view in which the intersection of a fault plane with the earth's surface is shown as a black line. (A) A wooden fence has just been built across the fault. (B) Continuing movement of the earth causes bending of the rocks and the attached fence. The fault plane is locked, and energy is being stored as elastic strain (compare with

Figure 8–10). (C) Stored energy level exceeds friction along the fault; abrupt slip occurs which breaks the deformed fence. The rocks rebound at high speed and move to new positions along the fault. The accumulated strain is released as energetic *seismic waves* (from the Greek *seismos* meaning "earthquake"), which we can both hear and feel as ground motion.

If, however, the fault plane becomes locked, through both the friction produced by the broken rock within the fault plane and the very large stresses acting across it, the fault plane area becomes an extremely strong zone, *resisting further motion.* In this more common situation, the rocks adjacent to the fault zone begin to bend and distort as elastic solids (see Figure 7–7). In the bending process, energy is stored, much as an archer stores energy in a bow by bending it (Figure 8–2).

Eventually the energy stored in the rock exceeds the **breaking strength** of the rock. Like an unlucky archer who bends the bow too far, the old fault explosively fails and blocks of earth shudder past one another at speeds of hundreds of miles per second. The energy, stored as elastic strain in deforming rock over many years, is suddenly released, and the distorted rocks rebound to recover their original shape. **Elastic rebound** is the process by which rocks change their shape as energy is stored, and then regain their original shape as the rock abruptly fails and rebounds or springs back to its original shape (Figure 8–2).

We feel the release of stored energy as an **earthquake**; the vibrations produced by the passage of energy in the form of elastic waves travelling through the earth. The earthquake is the inevitable result of storing energy capable of moving blocks of the earth in relation to each other, rather than steadily releasing it by aseismic creep. Along some very long fault planes, parts of the faulted area steadily creep aseismically, whereas other parts lock up and then sporadically release the stored energy as an earthquake (Figure 8–3). The alternate sticking and then slipping of earth blocks along a fault produce elastic rebound.

FIGURE 8–3

The narrow, linear valley cutting through the low hills in the center of the aerial photograph is the surface expression of the San Andreas Fault in the Carizzo Plain of southern California. The block of rock to the left moves away from the observer at an average rate of a few inches or centimeters per year. When such motion is locked along the fault, strain accumulates in large volumes of strongly deformed rock, setting the stage for disaster. (Photograph by R. E. Wallace, courtesy of the U. S. Geological Survey.)

Dilatancy-Diffusion

Laboratory studies of cylinders of rocks, placed under a wide range of pressures and temperatures, offer the best picture of what may happen deep within the earth. These studies show that the *creation of a new fault also creates earthquakes,* and these earthquakes are preceded by both rock **dilatancy**, volume change, and **diffusion**, the flow of fluid into dilatant rock. These two processes, briefly mentioned here, are described later in this chapter. **Dilatancy-diffusion,** the combination of these two processes together may lead to the creation of a brand new fault plane, accompanied by an earthquake (Figure 8-4).

When about one-half the force that will finally shear the rock has developed, many small cracks begin to appear within the rock, *increasing its volume.* Listening to each tiny, microscopic rupture as the

rock becomes dilatant has been likened to listening to popcorn pop. As each fracture occurs, there is a minute (0.3 percent-2 percent) volume increase. On the earth's surface the hordes of tiny cracks create a measurable swelling or tilting. The result can also be measured as an increase in the resistance of the rock to passage of an electric current through it; the air in the newly formed cracks has high electrical resistance. Still another measurement of this property is a slight decrease (1-2 percent) in the velocity of earthquake waves *from other nearby earthquakes* as they pass through the dilatant rock.

As pressures continue to increase, the minute cracks become larger, and water or other underground fluids begins to *diffuse* (to move by flow

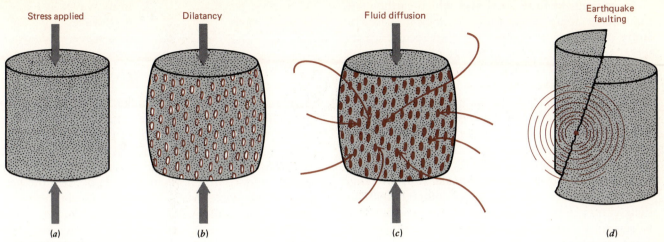

Stress applied   Dilatancy   Fluid diffusion   Earthquake faulting

(a)   *(b)*   *(c)*   *(d)*

FIGURE 8–4

(A) The cylinder of rock is placed under stress. (B) Continued stress causes the cylinder to slightly bulge, and volume is slightly increased as thousands of tiny elliptical cracks form within the rock. (C) Fluid diffuses into the dilatant rock, lubricating the cracks and enlarging pressure across them. (D) Failure of a series of cracks along a plane leads to faulting and creation of seismic waves.

through tiny cracks) into the *dilatant* rock. The continuing influx of water raises the pressure in the now water-saturated cracks. The diffusion of water into the rock can be recognized as a decreased resistance of the rock to the passage of electric current and an increased velocity in the passage of earthquake waves from nearby earthquakes.

The diffusion of fluids into the cracking rock slightly weakens it, as fluid pressures expand and lubricate the cracks. Catastrophic failure occurs as the rock's strength falls to values less than the stress being applied. A fault plane is suddenly created as numerous minute cracks intersect and join to form larger cracks, and the energy stored in the slightly bulging, dilatant rock is abruptly released. The abrupt release of stored energy is felt as rippling waves across the earth; an earthquake has now occurred.

imply that the releases of stored energy will be very large. It simply takes a long time to build up the staggering amounts of energy released by truly large earthquakes.

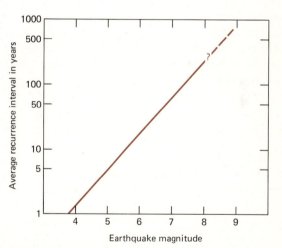

FIGURE 8–5

Since there is a relation between maximum displacement and earthquake magnitude (see Figure 8–11), if we assume a slip rate of 1 inch (2.5 centimeters) per year, we can plot the curve shown for an *average* recurrence interval. The longer the expected slip is forcibly held to zero the greater the displacement and magnitude of the resulting earthquake. It appears that a magnitude 5 earthquake releases the energy stored for 5 years, while a magnitude 7 earthquake releases the energy stored for 70 years. Parts of the San Andreas Fault have not moved in more than 70 years. (After R. E. Wallace, courtesy of the U. S. Geological Survey.)

THE REPETITION OF EARTHQUAKES

Both the elastic distortion of highly strained rock along an old fault zone and the dilatancy-diffusion that precedes the formation of a new fault are processes of energy storage. The longer the energy is stored and the greater the volume of rock that stores the energy, the greater the energy release in the future earthquake. Long intervals between earthquakes

Since faults identify areas over which earthquakes may occur repetitively, they are logical places to study the empirical relation between recurrence interval and the magnitude of energy release. What we see (Figure 8–5) is what we expected. In any one year there are a very few very large quakes, many moderate quakes, and thousands of weak tremors, most of which are below the level of human perception.

A very large earthquake may release in a few seconds the energy stored in the rocks for centuries. The energy released in a great earthquake is in the range of 10^{25} ergs (10,000,000,000,000,000,000,000,000 dyne-centimeters), an amount of energy equivalent to several thousand atomic bombs of the type used in World War II.

MEASURING EARTHQUAKES

The energy released by earthquakes travels around and through the earth as various types of elastic waves. This means that passage of the energy through earth material causes a *temporary* change in shape and/or volume of the material; the rock springs back to its original shape as the energy moves on. The trembling felt as an earthquake occurs is very real, for the earth heaves and shudders as the seismic energy passes. A much simplified classification of earthquake wave types would include:

1. Surface wave:
   A. *L* wave;

2. Body wave:
   A. *P* wave,
   B. *S* wave.

SURFACE WAVES

Surface waves are earthquake waves that are restricted to passage within the earth's surface. Surface waves may vibrate the surface in very complex patterns. Among the simpler types of surface wave is the *L* **wave**, a type of surface wave that vibrates the earth's surface in a horizontal direction. It is the passage of this wave that is responsible for the sickening sway of tall buildings during an earthquake. The *L* wave produces a type of earth motion that is potentially quite damaging to buildings. All earthquakes release energy as surface waves.

L waves are also termed *Love waves*, after the English geophysicist A. E. H. Love (1863–1940) who first recognized them. Their behavior within the earth's surface has made it possible to estimate the thickness of the earth's crust (see Chapter Nine); those data provided the first evidence that the continental crust is considerably thicker than oceanic crust.

BODY WAVES

Body waves are earthquake waves whose characteristics allow travel through part or all of the interior of the earth. Body waves can sometimes be recorded on the side of the earth directly opposite where the

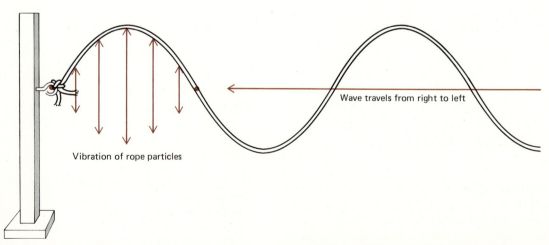

Wave travels from right to left

Vibration of rope particles

FIGURE 8–6

Take a limp piece of line, attach one end to a wall, and set the other end into vertical motion. The waveform travels from right to left. Any *point* on the rope vibrates up and down. The rope does not move horizontally at all. The velocity of the travel of the waveform down the rope is constant. The velocity of vertical vibration of the rope changes constantly from point to point.

earthquake occurred. All earthquakes release energy as body waves as well as surface waves.

Two types of body waves, *P waves* and *S waves*, are of particular importance in understanding how seismic energy travels. The *P* **wave**, also called the *primary wave* (because it is the first earthquake wave felt), is a a *dilational* wave, one that causes volume change. The *P* wave, traveling at the highest velocity of all waves, arrives first, and vibrates particles *parallel to the path of energy travel.* The *P* wave travels through the earth like a sound wave, and moving equally well through rigid solids, liquids, and gases.

The *S* **wave**, also called the *shear wave* or *secondary wave* (because it arrives after the primary wave), is a distortional elastic wave. The passage of energy as a shear wave causes particles in its path to vibrate *at right angles (Figure 8–6) to the direction of energy passage.* The *S* wave travels at a velocity intermediate between the velocities of the *P* and the *L* waves. Like all shear waves, it can travel only through a substance that is a rigid solid.

Body waves passing through the earth may cause the entire earth to vibrate in frequencies ranging from 1 to 3 *cycles per hour.* If we could hear them, the tones would be about sixteen octaves below middle C on the piano. The earth may continue quivering for several weeks after a very large earthquake, producing "music" far too low pitched for human ears to hear. Such "music" can be recorded by special instruments and is evidence that the earth itself is an elastic solid that is capable of resonating like any well-struck bell.

▼

CHARACTERISTICS OF SEISMIC WAVES

▲

A seismic or earthquake wave is difficult to visualize, but tying a long rope to a handy stake will help us imagine what goes on when seismic (earthquake) energy passes (Figure 8–6). The rope will help us understand what is meant by a seismic **wave**, a pattern that travels through a substance while the particles in the substance vibrate back and forth.

If we attach one end of the rope and then shake the unattached end up and down, the result (Figure 8–6) is familiar. A looping pattern or waveform is created and travels down the rope to its attached end. The curving *form* moves *parallel* the length of the rope, while the cord *itself* moves *only up and down,* a seismic wave pattern typical of an *S* wave.

The speed of *forward* motion of the looping pattern or waveform is the **wave velocity**, which re-

mains constant. The rope moves *up and down,* however, at a velocity that is constantly changing. The speed of the up and down motion of the rope is zero at the points along the rope where the vertical motion reverses and increases to a maximum at the crest and trough of each loop. The speed of up and down motion bears no relation to the wave velocity—the speed of forward travel of the waveform.

If we imagine the same process happening within a rock we have a model of the passage of the *S* wave through the rock. The particles are vibrated up and down, while the waveform moves along a straight path from one point to another. For the particles in a rock to vibrate, *they must be connected* so that each particle affects its neighbor. The vibrating substance must also be *elastic,* so that displaced particles will return to their original position as the waveform travels on. The propagation or travel of a seismic wave *requires temporary displacement of particles; the substance must be capable of being temporarily or elastically deformed.*

ELASTIC MODULI

A **modulus** is a number that expresses the degree to which a substance possesses some property. Two moduli, the bulk modulus and the shear modulus, express the extent of deformation of solids and are critical in understanding how rock properties affect the passage of seismic waves.

The **bulk modulus** is a measure of any substance's resistance to change in *volume*—that is, its *compressibility.* The greater the bulk modulus, the more incompressible the material. Concrete has a high bulk modulus and will stand up to very heavy loads without a change in volume. Soft rubber has a low bulk modulus; it is easily compressed.

The **shear modulus** is a measure of resistance to a change in *shape*—the *rigidity* of the material. The greater the shear modulus, the more rigid the material. High-grade steel has a high shear modulus; it is not easily bent.

The greater the value of the two moduli, the more strongly *connected* the parts of the material are and the greater is their resistance to temporary displacement. In thousands of experiments it has been observed that *the more incompressible and rigid a substance, the faster a wave travels through it.* Strongly connected particles in a substance inhibit displacement but speed the waveform along its way.

THE ROLE OF ROCK DENSITY

One other factor that affects the wave velocity is the *density* of the material through which the seismic energy passes. The **density** is the mass per unit of volume, as described in Chapter Two. Density describes the total mass of material in any particular volume. We experience density as the weight or "heft" of material pulled toward the earth by the acceleration of gravity.

The seismic wave experiences density as heavier, more tightly packed material per unit of volume. Therefore larger amounts of dense materials need to be displaced by the passage of the waveform. From thousands of experiments it has been observed that *denser materials slow the wave velocity whereas less dense materials allow more rapid propagation of the waveform.*

SUMMARY OBSERVATIONS

Combining the effects produced both by the various moduli and the effect of rock density, we can state the observed relations in a simple and approximate formula.

EQUATION 8–1

Wave velocity $\cong$ [bulk modulus + shear modulus] ÷ density

This generalized statement tells us that seismic energy travels through the earth as waveforms whose propagation speed depends on *three* measurable criteria. *The wave velocity is directly proportional to the value of the two moduli and inversely proportional to the rock density.*

P WAVE TRAVEL

Equation 8–1 may be used to describe the wave velocity of a P wave traveling through a solid rock. The wave velocity increases with increases in the two moduli and decreases with increasing rock density.

Like solids, liquids and gases have a bulk modulus, but unlike solids, they have *no* shear modulus. By definition neither a liquid nor a gas has rigidity, since either flows into whatever shape of container is available. Neither offers any resistance to shearing forces; *fluids and gases cannot be sheared.*

The wave velocity of the P wave passing through either a liquid or a gas, then, depends solely on the bulk modulus, since the shear modulus is zero. *P waves should travel more slowly through liquids and gases than solids, since the accelerating effect of the shear modulus is now absent.*

P waves can pass through any kind of material. Their speed is, as predicted, diminished in liquids and gases; however, they continue to move as they alternately compress and decompress the elastic material through which they travel.

Sound waves travel in exactly the same way. P

waves are analogous to sound waves generated by a vibrating earth, much like the sound waves generated by a vibrating guitar string. The P wave is the fastest of the waves, traveling through surface rocks at speeds averaging 3 miles (5 kilometers) per second.

S WAVE TRAVEL

The S wave travels or propagates by causing particles to oscillate in a direction *perpendicular* to the direction of travel, like the rope in Figure 8–6. This method of travel causes the substances affected by an S wave to *change their shape, but not their volume.* The effect of S waves on materials is unlike that of P waves, which temporarily change *both* the *volume and shape of solids* and the *volume only* of liquids and gases through which they pass.

If we describe the effect of S it waves on materials, Equation 8–1 is modified to Equation 8–2.

EQUATION 8–2

Wave Velocity $\cong$ Shear Modulus ÷ Density

Notice that the bulk modulus seen in Equation 8–1 is absent from this equation describing shear

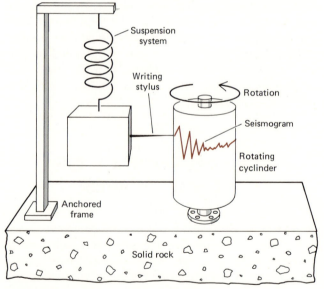

FIGURE 8–7

This sketch of a simplified seismograph reveals its component parts. A rotating cylinder with paper on it is attached firmly to solid rock so that if the earth moves, the cylinder moves with it. Adjacent to the cylinder is a large mass, so carefully suspended that it remains essentially stationary in space during earth movement. Affixed to the stationary mass is one of many types of writing instruments. The motion of the drum relative to the fixed mass is recorded by the writing stylus as a line on the turning drum. The written record is termed a *seismogram,* a record of earth motion. (Sketch modified from U. S. Geological Survey data.)

TABLE 8–1
Variation of Wave Velocity as a Function of Material Properties

Material and Wave Type	Increasing Density	Increasing Resistance to Compression	Increasing Rigitity (Resistance to Shear)	Increasing Fluidity
P Wave in a solid	Slows	Speeds	Speeds	Slows
P wave in a fluid or gas	Slows	Speeds	Rigidity Absent	Slows
S wave in a solid	Slows	No effect	Speeds	Slows or stops
S wave in a fluid or solid	Velocity ≅ 0	Velocity ≅ 0	Velocity ≅ 0	Velocity ≅ 0

waves. This is true because *no compression occurs as shear waves pass*. The *only* temporary change produced by the passage of *S* waves is a change in shape. Because the shear modulus, without the bulk modulus, is the only factor affecting wave velocity, *S* waves travel at a velocity *only about two-thirds that of P waves.*

Because S waves can pass only if they can temporarily change the shape of the materials through which they propagate, they can travel only travel through solids. Only solids are sufficiently rigid that displaced particles can elastically deform and then spring back to their original shape. Since fluids and gases have no fixed shape, displacement of their particles is permanent, which inhibits wave travel.

SUMMARY

The differing properties of *P* waves and *S* waves, namely the relation between wave velocity, wave type, and properties of the materials through which the waves travel, are critical in learning how earthquake waves are recorded. We will also use this same information to interpret the architecture of the earth's interior based on the way it affects body waves passing through it. A summary of the interdependent properties of seismic waves and earth materials is given in Table 8–1.

RECORDING GROUND MOTION

▲

The passage of seismic energy vibrates the earth's surface. These vibrations are *recorded* by a *seismograph,* literally an "earthquake writer." A **seismograph** is an instrument that records (Figure 8–7) the amplitude and arrival time of motion of the earth's surface.

The first seismograph was invented in A. D. 132 by Chang Heng in the earliest attempt to measure the severity and origin of earthquakes. In the intervening 18 centuries, seismographs have been greatly improved in sensitivity and precision, but the principles governing their operation remain the same. One part of the instrument, usually a scribing or writing pen, is suspended in such a way that it is essentially independent of the earth (Figure 8–7) and remains fixed in position in space.

The recording drum is firmly anchored to the earth so that if the earth moves, the drum moves parallel to earth motion. The drum turns at a precisely calibrated rate so that the motion of the drum against the stationary pen provides a record through time of earth motion.

The resulting **seismogram** is a written record (Figure 8–8) of ground motion *in the plane perpendicular to the vertical axis of the pen*. Three seismographs provide a complete seismic observatory, as each seismograph is set up to individually record north-

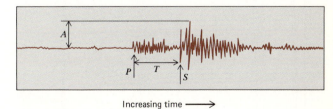

Increasing time ⟶

FIGURE 8–8
In this sample seismogram, the paper strip moved from right to left as the earthquake occurred. The first wave to disturb the earth was the *P* wave known to be the fastest of the seismic waves. The second, slower wave to arrive is the *S* wave. The *amplitude* (A) of the wave is its vertical departure from the centerline; the amplitude shown is the maximum amplitude, necessary to determine the earthquake's *magnitude*. The interval *T* is the *time difference* between the arrival of the two waves and is a function of the distance to the earthquake source.

south, east-west, and vertical components of the total earth motion. The cumulative record from all three seismographs will define ground motion in any direction.

OTHER INSTRUMENTS AND OTHER USES

Additional components of many seismic observatories are *microseismic* apparatus designed to record high-frequency motion (very, very rapid vibrations) from very small earthquakes. These microseismic machines feature extremely high amplification of ground motion and are used most commonly to study weak seismic activity caused by frequent discontinuous motion along faults. Microseismic stations also monitor events that may be useful in predicting much larger earthquakes.

Still other instrumentation may include *long- and ultralong-period seismographs,* used to study the low and very low frequency motion (very, very slow vibrations) produced by very large earthquakes. Another useful device in areas where intense earthquakes may be expected is the *strong-motion seismographs or accelerographs,* which are rather *insensitive* seismographs used to display the extremely violent ground motion of very large earthquakes, that may knock more sensitive conventional seismo-

graphs out of commission.

Seismographs have recently been used to detect unannounced underground nuclear explosions. Powerful underground explosions produce an energy release that is comparable to a small earthquake, and the resultant energy waves travel through the earth in much the same manner. There are, however, minor differences in the "signature" of underground explosions as compared to earthquakes. Seismographs can detect many kinds of underground explosions and separate them from natural events.

Seismologists, geologists who specialize in the study of earthquakes, have become quite sophisticated in the detection and localization of underground explosions. Routine monitoring of the earth has made the underground nuclear test-ban treaty enforceable without the need for on-site inspection.

MEASURING EARTHQUAKES

Earthquakes leave a record in two fundamental ways. The oldest method of measuring earthquakes is to assess the level of damage as well as the events observed during the earthquake. Newer methods measure various components of ground motion. Each of these two separate methods provides information of quite different value.

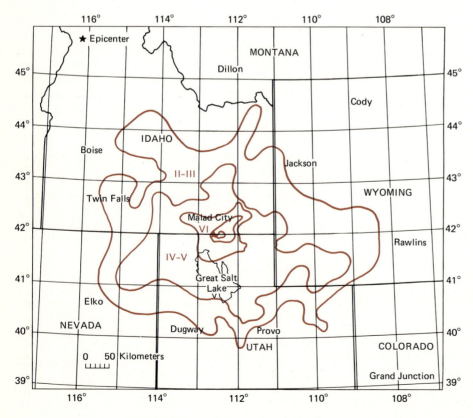

FIGURE 8–9
Modified Mercalli intensities assigned to the March 28, 1975, earthquake (Magnitude = 6.1) at Malad City, Idaho. The isoseismal lines are drawn based on hundreds of returned "Earthquake Report Questionnaires," and include within them all reports of the intensities shown. Intensities reached VII near the epicenter and VIII right at the epicenter (compare with Table 8–2). (Computer-plotted isoseismal map courtesy of the U. S. Geological Survey.)

TABLE 8–2
Modified Mercalli Scale of Earthquake Intensity

I	Not felt except by very few people under special conditions.
II	Felt by a few people, especially on upper floors. Suspended objects may swing
III	Felt noticeably indoors. Standing automobiles may slightly rock.
IV	Felt by many people indoors, by a few outdoors. At night, some are awakened. Dishes, windows, and doors rattle.
V	Felt by nearly everyone. Many are awakened. Some dishes and windows are broken. Unstable objects are overturned.
VI	Felt by everyone. Many people become frightened and run outside. Some heavy furniture moved; some plaster falls.
VII	Most people are in alarm and run outside. Damage is negligible in buildings of good construction, considerable in poorly constructed buildings.
VIII	Damage is slight in specially designed structures, considerable in ordinary buildings, great in poorly built structures. Heavy furniture is overturned.
IX	Damage is considerable in specially designed structures. Buildings shift from their foundations and partly collapse. Underground pipes are broken.
X	Some well-built wooden structures are destroyed. Most masonry structures are destroyed. The ground is badly cracked. Considerable landslides occur on steep slopes.
XI	Few, if any, masonry structures remain standing. Rails are bent. Broad fissures appear in the ground.
XII	Virtually total destruction and general panic. Waves are seen on the ground surface. Objects are thrown in the air.

Seismic Intensity

The oldest method of measuring earthquakes is the **Modified Mercalli Scale of Earthquake Intensity**. This scale provides a crude measurement of observed damage ranging from I, no damage to XII, virtually total destruction. Witnesses to the events produced by the earthquake are often contacted by postcards and requested to mail in to a central location their description of the damage to surrounding structures. A map can be constructed from the returned postcards (Figure 8–9) and the distribution of various levels of damage plotted on a map.

Because the intensity of damage to structures depends on so many variables, including soil type, shaking duration, and details of building construc-

tion, the Mercalli intensity scale is little used in North America at this time. Table 8–2 gives the intensity scale currently in use.

Richter Magnitude

The scale most commonly used to record modern earthquakes is the **Richter magnitude scale.** This scale provides a measurement of the amplitude (magnitude) (See Figure 8–8) of *ground motion*. The Richter magnitude is a number frequently given in newspaper accounts of earthquakes. In practice the reported magnitude ranges from 1 to just over 9.

The number reported is the *logarithm of the observed amplitude of ground shaking*. Since the Richter magnitude is based on a *logarithmic* scale, the change of Richter magnitude by one number, such as a change from 4 to 5, corresponds to a *tenfold* increase (change from 10^4 to 10^5) in amplitude of recorded ground motion. A change of number from 4 to 6 would record a *hundredfold* (10^2 = difference between 10^4 and 10^6) increase in ground motion. The Richter number defines the amplitude or magnitude of the most severe ground motion recorded. The Richter magnitude is a very good measure of the strength of ground shaking.

Sensitive seismographs can record ground motion corresponding to Richter magnitudes of much less than 1; even negative magnitudes have been recorded. Humans rarely feel earthquakes until Richter numbers reach at least 3. When magnitudes are between 3.5 and 5.5, the quake is described as moderate; at values between 5.5 and 7.5 the quake is strong and capable of causing severe damage. Magnitude values over 7.5 describe a great earthquake, one capable of causing catastrophic damage over large areas. The greatest earthquake ever recorded—in Chile in 1960—had a Richter magnitude of 9.5, meaning that its ground motion was more than 1 million times (10^6) greater than one that humans could barely feel.

The Richter magnitude value is a single data point corresponding to the maximum ground motion at the point of earthquake origin. The Mercalli Intensity Scale, by contrast, is commonly used to produce a map showing the extent of damage in the entire area affected by the earthquake.

ENERGY RELEASE

The relation between the Richter magnitude number and the seismic energy released is only an approximate one. Richter magnitude measures ground shaking, which is *not* a direct measure of energy release. Ground motion depends on many variables, only one of which is the amount of energy released.

In a general way, each increase in Richter magnitude by a value of one number corresponds with *an increase in energy release of 32 times*. If we take 30 times as an approximation, an earthquake of Richter magnitude 6 releases 900 times as much energy as a $M_s = 4$ earthquake and 27,000 times as much energy as a $M_s = 3$ quake—one that is barely felt by humans.

The occurrence of an earthquake is a complex physical process. The great majority of the stress stored within the rocks is used to power the growth of earthquake fractures, move the strained blocks of rock (Figure 8–10), and produce both heat and light. People have long described "earthquake lights" when earthquakes occur at night; very large earthquakes may light up the sky in colors of brilliant red and white for several minutes. Of the total energy released by the earthquake, perhaps only 0.1 to less than 0.01 of the available energy is radiated as seismic energy (most of the energy is produced as heat). For those of us who feel the radiation of the energy, we should be grateful that we receive such a tiny fraction of the earth's potential energy. Were all the energy released by a great earthquake expended as seismic energy, we would, no doubt, be catapulted from the earth by the shaking that would be produced.

RECURRENCE INTERVAL

The recurrence interval, averaged over a century or more, is an *inverse* function of the energy released (Figure 8–5). Weak earthquakes occur hundreds of times a day somewhere on earth, while very strong earthquakes having a Richter magnitude greater that 7.0 occur on the average about once a month.

Great earthquakes, those with a Richter magnitude greater than 8.0, occur only once every few years. A few great earthquakes each year release far more stored seismic energy and cause greater offsets of the land surface (Figure 8–11) than all the thousands of weaker earthquakes. One great earthquake may release an amount of energy (approximately 10^{25} ergs, equal to 1340 trillion horsepower) equivalent to the cumulative energy released by *all* other earthquakes in an average year.

FIGURE 8–10
A wooden fence dramatically records shortening of the earth's surface parallel to its length as two geologists continue their investigation of seismic damage. (Photograph courtesy of the U. S. Geological Survey.)

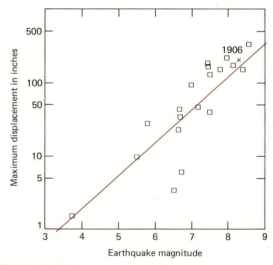

FIGURE 8–11
Relation between the maximum displacement and earthquake magnitude for historic faulting in western North America. The information for the 1906 San Francisco, California, earthquake is shown with a cross and date. Using data from this graph, one can calculate that if 10 inches (25 centimeters) of elastic strain have accumulated, a magnitude 5.5 earthquake would release the accumulated slip. If the average slip rate is 1 inch (2.5 centimeters) per year, a magnitude 5.5 earthquake should have a recurrence interval of about 10 years. (Data from M. G. Bonilla, courtesy of the U. S. Geological Survey.)

The energy released by great earthquakes must approximate the maximum capacity of rocks to store energy. Great earthquakes tell us what the limits of rock strength are in dramatic terms.

Seismic Moment

Several studies have suggested that the energy stored in a unit volume of rock is almost constant, *regardless of the size of the earthquake*. Thus larger and larger earthquakes release the energy accumulated over larger and larger volumes of rock. We can observe that the Richter magnitude is directly related to the length of the fault that ruptures (Figure 8–12).

Because the wavelength, the distance from one

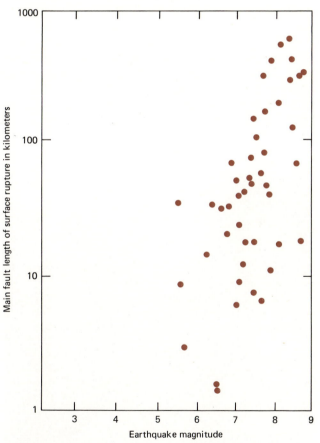

FIGURE 8–12

In this data set accumulated from worldwide and North American earthquakes of all types, there is a correlation between length of rupture produced by surface faulting and earthquake magnitude. If 60 miles (100 kilometers) of fault breaks, the resulting magnitude approximates 8.5. The world's greatest earthquakes may release accumulated slip along distances of several hundred miles or kilometers of an active fault. (Data from H. R. Shaw and A. E. Gartner, 1984, U. S. Geological Survey Open-File Report 84–356.)

wave peak to the next peak, of the ground motion measured to obtain Richter magnitude is rather short (less than 37 miles or 60 kilometers) and because the amplitude of seismic waves represents the energy release from a body of rock whose dimensions are comparable to the wavelength of these waves, the Richter magnitude scale saturates for great earthquakes, which may release the energy stored along faults whose length may exceed 600 miles (1000 kilometers).

For these great earthquakes, very long-period waves (the period is the interval of time between passage of one wave and next), having a period of 200 to 300 seconds, are created. These waves may shake the entire earth for several days. The data measured for Richter magnitude determination does not adequately account for the energy consumed in the generation of these these very long-period waves. Such long-period waves have wavelengths of several hundreds to several thousands miles or kilometers and make a major contribution to the "size" of great earthquakes.

For these reasons, the concept of *seismic moment* (as used here, the word "moment" is a term borrowed from physics) was developed. The **seismic moment** is the product of the surface area of the rupture, the average displacement along the fault plane, and the rigidity of the material along the fault. Seismic moment is one of the most reliable ways to estimate the total "size" of an earthquake, for it is not self-limiting in very large earthquakes and yet parallels Richter Magnitudes determined for smaller earthquakes, which involve motion along smaller fault lengths.

▼

THE EFFECTS OF EARTHQUAKES

▲

Earthquakes create a wide range of hazards to life and property. Among these hazards are surface faulting, ground motion, liquefaction, landslides and rock slides, tsunamis, and seiches. Each of these natural geologic processes pose a risk to those who live in areas where large earthquakes may be expected. Understanding the hazards is the first step in minimizing and understanding the risks of living in earthquake country.

SURFACE FAULTING

Of all the hazards associated with earthquakes, perhaps the most *obvious* is the danger to structures when the ground breaks and moves directly beneath them (Figure 8–13). The breakage and shifting of the earth's surface during an earthquake is termed **surface faulting.**

Surface faulting shifts and breaks the foundations under structures, causing major structural damage and building collapse. The amount of displacement that occurs during an earthquake (Figure 8–11) is related to earthquake magnitude and fault length. Longer faults yield greater magnitudes and greater displacements along the fault. The maximum displacement ever observed in a single earthquake is about 12 yards (11 meters). The horizontal displacement during the 1906 San Francisco earthquake equaled about 6 yards (5.5 meters).

Surface offset may also occur because of seismic creep, the slight chattering movement between blocks separated by a fault. Creep is far less damaging to structures, though walls will slowly be warped and pulled away, windows will pop, and cracks will grow in foundations and walls. The damage produced by creep is cumulative; eventually the structure will be both unpleasant and unsafe.

The history of surface faulting is the single most important tool in evaluating the seismic hazard of an area. In a very general way the risk from earthquakes is directly related to the distance to the nearest active fault.

There is no satisfactory way to engineer any building to withstand several feet (meters) of vertical or horizontal shifting of the ground directly beneath. Displacements in this range characterize even moderately strong earthquakes. *There is no known way that a building built across a potentially active fault can be made safe.* Such buildings are simply disasters ready to happen.

Evaluating the probability of damaging surface faulting is among the geologist's most difficult tasks. Seismic history is full of examples of apparently inconsequential faults that have moved with large displacements after centuries of quiet. A full-scale geologic evaluation would require detailed field investigation, trenching and mapping, examination of boreholes in dangerous areas, and evaluation of the history of seismic events recorded in the rocks. Ideally, instruments for monitoring creep should be installed, along with a host of other instruments, in order to gather data for prediction of potential earthquake activity.

The siting of nuclear power plants has called this issue into sharp focus, for everyone recognizes that surface faulting under a nuclear power plant has the potential for a major catastrophe. Even with the best of information and instrumentation, forecasting which fault zones may move is an extremely difficult task. The lesson is clear: don't build structures near or across a known fault.

GROUND MOTION

In a very general way, the greater the distance of a building from a known, active fault, the safer the building. At some distance from a fault, the major hazard to a building is **ground motion,** described as the sudden acceleration of the earth's surface during an earthquake.

The probable acceleration (Figure 8–14) of a point on the ground surface during an earthquake is a critical factor in correctly designing a building. Along some of the most dangerous areas of southern California, the expected acceleration of the ground surface during an earthquake approaches the value of gravity. That means that the building must be designed to withstand horizontal forces *equal to building the structure standing perpendicular to the vertical face of a cliff so that it hangs out into space supported by nothing.*

Although few buildings must be designed in order to cope with such prodigal forces, many buildings must be designed to withstand horizontal forces equal to half the acceleration of gravity in order to ride out a great earthquake. The costs of such construction are very high, and there is always the temp-

FIGURE 8–13

An example of surface faulting during a moderate 1959 earthquake in western Montana. Try to imagine what would happen to a house built across the area of rupture. (Photograph by I. J. Witkind, courtesy of the U. S. Geological Survey.)

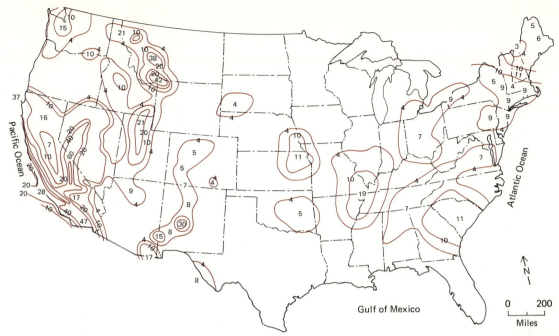

FIGURE 8–14

The map indicates one way of estimating earthquake hazard within the United States. Levels of ground shaking, with 90 percent probability of not being exceeded in 50 years, are shown by *numbered contour lines*. These lines connect all points whose probable acceleration is the numbered percentage of the full force of gravity. As a rule of thumb, if horizontal acceleration is greater than 25 percent of the force of gravity, substantial damage to ordinary buildings results. At levels above 50 percent, damage is essentially total. At levels below 10 percent, damage is slight, but motion is felt by everyone. (Map from S. T. Algermissen and D. M. Perkins, 1976, courtesy of the U. S. Geological Survey Open-File Report 76–416.)

tation to cut corners and hope. At best such deficiencies allow intense and expensive structural damage; at worst the buildings collapse onto those they once sheltered.

The intensity of ground motion depends in part on the type of soil or bedrock beneath the building; the greater the bulk and shear moduli of the soil, the safer the building. The 1906 earthquake in the San Francisco area illustrates the factors affecting ground motion; the study of damage along the more than 250 miles (400 kilometers) of the active zone of the San Andreas Fault, where movement occurred, was extraordinarily thorough. Some of the residents living in homes atop San Francisco's famous hills were not even awakened by the tremor that threw other nearby buildings constructed on soft, wet soils completely off their foundations (Figure 8–15) and caused many buildings to totally collapse. The ground motion in San Francisco lasted one full minute. As one newspaper put it ". . . that was long enough to satisfy everyone." Many buildings finally failed after enduring many tens of seconds of trembling.

FIGURE 8–15

An example of ground failure under a house foundation during the 1906 San Francisco, California earthquake. Such failures will occur again, since the strain along the underlying fault continues to build. (Photograph by G. K. Gilbert, courtesy of the U. S. Geological Survey.)

The lesson from San Francisco was clear. Homes built on solid rock foundations withstood the earthquake well. Homes and structures built on landfills, and reclaimed bay muds and sands, were literally shaken to pieces.

Unsound ground and poorly compacted landfills amplify earthquake waves, much as a bowl of gelatin quivers under the slightest touch. Sixty miles (100 kilometers) away from the epicenter of the 1906 quake, San Jose was totally destroyed by the vibrations, whose intensity probably exceeded those felt immediately adjacent to the fault. San Jose sat on a thick deposit of soft valley-fill material.

LIQUEFACTION AND LANDSLIDES

Another type of ground failure caused by repeated shaking is *liquefaction*. When waterlogged soils are shaken, the shaking may abruptly turn them from solids to liquids; the process of conversion is termed **liquefaction.** Liquefaction occurs because shaking waterlogged soils causes grains to separate from one another and float; the soil loses all cohesion and strength. The process of ground shaking by an earthquake is analogous to stepping into quicksand that looks solid; the more the trapped person struggles and thrashes, the more liquid the quicksand becomes.

The liquefaction of deep waterlogged soils floats the overlying buildings. They may sink into the viscous material that once supported them or heel over into rakish attitudes like sailboats in a stiff breeze (Figure 8–16). Liquefaction can be just as damaging to a building as straddling an active fault. Old riverbeds, deep valley fills, filled delta and marsh areas, filled and diked bay areas, ocean beach and cliff developments, and backfilled estuaries all represent potentially very unstable ground with regard to liquefaction.

Other risks of living near active faults include the possibility that earthquakes may touch off disastrous landslides and rock slides in the hills above you. Building on steep soils at the base of a cliff is triple folly. Not only may the soil fail beneath you, but the cliff magnifies the energy waves and creates still greater shaking. Rocks and soil from above you may also come crashing down as nature smooths and flattens the sloping land.

TSUNAMIS

Tsunamis (pronounced soo-nahm'-ees) or seismic sea waves are high-velocity ocean waves kicked off by earthquakes on the sea floor. Earthquakes abruptly moving the sea floor cause a very large energy release directly into seawater. Since water, like most liquids, is incompressible, the energy is transmitted through the water.

FIGURE 8–16
This aerial photograph taken after the June 16, 1964, earthquake in Niigata, Japan, graphically displays the effects of liquefaction of "solid" ground by prolonged ground shaking. Apartment houses heeled over into a variety of angles. Residents of the central units walked out of the windows and down the side of the outside walls to safety. (Photograph courtesy of NOAA/EDS.)

The wave created in the open sea is of low amplitude, but the pattern moves across the ocean surface at speeds of several hundred miles (kilometers) per hour. As these high-speed waves approach a shallow shoreline, the velocity abruptly slows as the waveform is slowed by friction with the shallower ocean floor. The velocity loss is translated into wave heights of more than 10 yards (9 meters).

The Chilean earthquake of 1960, whose Richter magnitude was 9.5, produced one of the largest tsunamis on record. Chile and Hawaii received most of the damage, but more than eight hours later the same tsunami ravaged the coast of Japan 10,000 miles (16,000 kilometers) away. Thousands of Japanese lost their homes and nearly 200 people died. Property damage alone was about half a billion dollars from this one tsunami.

Low coastal areas may be dangerous places to live. Nearby earthquakes may cause the soils beneath you to shake vigorously and eventually to liquefy. Roaring tsunamis may later clear away anything that is left. One seismic sea wave in Japan 500 years ago destroyed a temple 10 miles (16 kilometers) inland.

SEICHES

A seismic **seiche** (pronounced sáysh) is produced by the sloshing back and forth of water in lakes, reservoirs, rivers, or bays when the rock around them is shaken. Seiches may exceed 10 yards (9 meters) in height, though most commonly they are smaller. Seiches may also be produced by earthquake-generated landslides falling into constricted bodies of water.

In 1958 in Lituya Bay, Alaska, an earthquake produced a rockfall into the bay that caused water to crash against the bay walls more than 600 yards (550 meters) above the former sea level. The combination of water bodies in steep-walled valleys plus earthquake potential can be potentially lethal.

Earthquakes remain for humans an ancient scourge. Within the last few decades, however, the application of modern knowledge has raised a fascinating promise. Might earthquakes be predictable?

THE PREDICTION OF EARTHQUAKES

Among the exciting developments within geology in the 1990s will be the attempt to predict earthquakes based on analysis of precursory phenomena. Precursory phenomena are observable changes in measured earth properties that may dependably occur before an earthquake.

Earthquake prediction, in contrast to the evaluation of seismic risk, attempts to pinpoint the place, time, and magnitude of an impending earthquake. A successful prediction must predict location within a few miles or kilometers, time within a few hours, and magnitude within a Richter magnitude number or two. Only this combination of data could give governmental agencies sufficient time to order the strengthening of buildings and the evacuation of citizens before disaster strikes.

The first successful earthquake prediction in the United States took place in 1973 for a very small earthquake in the Adirondacks. Since that time earthquake prediction efforts have largely shifted to areas where surface faulting is observable and seis-

FIGURE 8–17
The San Andreas Fault (compare with Figure 8–3) is the longest of a network of faults that cuts through the coastal regions of California. The fault is actually a series of interconnected fracture zones that aggregate 600 miles (1000 kilometers) in total length and extend vertically into the earth at least 20 miles (32 kilometers). Blocks on opposite sides of the San Andreas fault move horizontally, the western slice moving northwest with respect to the rest of the United States (note arrows). In some areas, as at Hollister, the fault moves rather steadily by aseismic creep at the rate of 2 inches (5 centimeters) per year. In other areas, such as Fort Tejon and San Francisco, the fault plane is locked; stress has been building up for most of a century.

mic activity is common. Most of the current efforts toward earthquake prediction in the United States are confined to areas adjacent to the San Andreas Fault (Figure 8–17) in California. The San Andreas Fault (compare Figure 7–18) releases much of the seismic energy expended in the United States in an area of very high population density. Nowhere in the United States do more people live in close proximity to a greater source of earthquake risk than in southern California.

Earthquake prediction is a national goal for Japan, Russia, China, and the United States. Every attempt at prediction relies on early signs of changes in physical properties of rocks under stress, including surface tilting and bulging of rocks, changes in water that is circulating underground, changes in velocity of seismic waves from distant earthquakes passing through strained rocks, and changes in electrical and magnetic properties of the strained rocks. Additionally, monitoring of domestic animals has also suggested that several kinds of animals can apparently sense an impending earthquake.

These methods, used separately or collectively to enhance the confidence level of earthquake prediction, are based on recognizing changes in the physical properties of rock that preceded well-documented small to moderate earthquakes. Interestingly, the larger the magnitude of the impending earthquake, the longer the warning interval may be. Such an interval between recognition of something unusual and the occurrence of the earthquake may stretch to several years for great earthquakes.

METHODS OF PREDICTION

A very large number of physical properties of rocks have been studied in an attempt to recognize events that might be dependable predictors of future earthquakes. Among those physical properties are the following.

1. Changes in length of lines surveyed across active faults.
2. Abrupt reversal of changes in elevation of the earth's surface.
3. Tilting of the earth's surface.
4. Changes in the velocity ratio of P and S waves from nearby earthquakes.
5. Abrupt changes in the local magnetic field.
6. Abrupt decrease in microseismic rate, including periods of seismic quiet.
7. Abrupt changes in electrical conductivity of rocks.

8. Abrupt changes in level, color, chemistry, or temperature of underground water.
9. Abrupt increase and sudden decrease in radon-222, a naturally occurring radioactive gas occurring in well water.

When several of these changes in physical properties occur at the same time, the reliability of the earthquake prediction is much improved. We will describe a few of these techniques to give an idea of how the physical changes may be related to precursors to earthquakes.

Local Magnetic Field

Rising pressure within crustal rock appears to raise the intensity of the *local* magnetic field. Pressure-induced change in magnetic properties has been demonstrated experimentally. Where sufficient data on the strength of the local magnetic field have been collected for several years (Figure 8–18), the technique seems to have some real promise.

In some areas a decrease in the strength of the magnetic field seems to signal the rise of magma, which heats the overlying rocks and alters the magnetic field. Continuous monitoring of magnetic field strength over large areas is fairly new. Only time can tell whether this technique will yield dependable results.

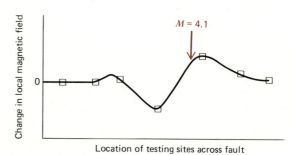

FIGURE 8–18

The data plotted are for simultaneous readings taken at different places along a fault. Each datum point records the *change* in the local magnetic field since some previous reading. This method of earthquake prediction relies on the observation that rocks placed under stress exhibit changes in their local, small magnetic field. The greatest changes through time occur immediately adjacent to the area of slip. A magnitude 4.1 earthquake later occurred just in the area of maximum change shown by the colored arrow. (Data from Malcom Johnston, courtesy of the U. S. Geological Survey.)

Microearthquake Recording and Seismic Gaps

In areas of frequent very, very small earthquakes, called microseismic activity or microseisms, the number of microseisms seems to rise just before a small to moderate earthquake (Figure 8–19). Such increases in microearthquake rate appear to signal an accelerating rate of crack formation that may culminate in fault slip.

In extremely strong rocks, the reverse may be true, with the *lack* of seismic activity indicating that the fault is tightly locked. Stress now builds up without microseismic relief by seismic creep, and a *seismic gap* occurs—an interval of time during which there is little or no small earthquake activity along a particular interval of a fault.

Such periods of quiet are ominous in an area under severe strain, because the quiet forecasts continuing *storage of energy without release* so that the future magnitude of the earthquake may be very large. Such locking and seismic quiet has been typical of both the Fort Tejon and the San Francisco areas of the San Andreas Fault (Figure 8–17) for more than 60 years. Geologists estimate that the stress stored in the San Francisco region is already equal to that released in the 1906 earthquake that devastated a part of central California. If all of this stored energy were released in one great earthquake, that energy release would cause the greatest natural disaster in the history of North America.

Electrical Resistivity

If an electric current is deliberately passed into the ground in one area and the remaining current is measured in rocks on the other side of the fault, data are obtained that measures the *resistivity* of the rock—that is, how much current is lost in transmission from one point to another. The greater the loss, the greater the resistivity of the rock. Dry rock is highly resistive to current flow; most electrical insulating materials are geologic materials like mica and ceramics. On the other hand, water is an excellent conductor of electricity and offers little resistance to current flow.

Resistivity seems to decrease for some rocks (Figure 8–20) before an earthquake, perhaps because naturally circulating underground water moves into dilational cracks as the rock experiences intensive stress. As the rocks become more and more highly saturated with water, their resistance to the passage of electric current falls, as does their resistance to the imposed stress. Rock failure—an earthquake—follows when the saturated shattered rocks can no longer withstand the load placed on them.

Radon-222 Release

If underground water flows through the area undergoing strain, its chemistry may provide another pre-

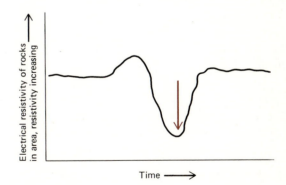

FIGURE 8–20
Substantial changes in the resistivity of rocks to the passage of an electric current suggest that the rocks have gained resistance as cracking occurred. Air is more resistant to the passage of electricity than rock (see Figure 8–4B). As water then flows into the distorted rock (see Figure 8–4C), resistivity falls, because the saturated rock has become a better conductor of electric current. Incoming fluids lubricate the cracks and increase pressure on their tip, leading to amalgamation of cracks and earthquake slip (see Figure 8–4D) along the newly created rupture plane. The colored arrow marks the timing of a predicted event, just as resistivity begins once again to rise.

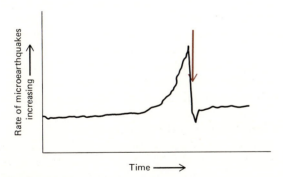

FIGURE 8–19
This chart of a hypothetical area displays the kind of information that has been received prior to several moderate earthquakes. At some time before the main earthquake (shown by the colored arrow) the intensity of microearthquakes in the area increases and then subsides as the strain is relieved by a larger earthquake event. The beginning of subsidence marks the prediction point.

cursory phenomenon. The cracking of rock may release additional amounts of *radon*, a radioactive gas produced by the radioactive decay of trace amounts of uranium in the rock.

Radon is normally present in well water in minute amounts. If the content of radon in the well water is continually monitored, any increase in radon content may signal increased cracking in the rock. The internal cracking caused by dilation that may precede earthquakes gives underground water access to newly shattered rock and its rich harvest of newly released radon. A rise in radon levels (Figure 8–21) may suggest an impending earthquake *if* the well has direct access to a zone of dilation and cracking. If the well water does not have access to highly strained rock, the radon level of the water should remain reasonably constant for many years.

Animal Behavior

The most controversial of all methods of earthquake predictions is the observation of domestic animals. This method, which obviously has nothing to do with the physical properties of rock, has been studied very carefully by Chinese scientists. A wide variety of animals seem to have the uncanny ability to sense when an earthquake is about to occur. Snakes have

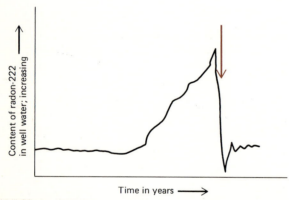

FIGURE 8–21

A chart of the idealized changes in the content of radon-222, a gaseous element formed by the decay of radium (see Appendix B). Radon is commonly present in minute quantities in well water as a result of the radioactive decay of elements in the rocks surrounding the water well. As the rocks near the well undergo severe dilation and cracking, larger amounts of radon are released into the well water. Radon content appears to peak just at or before the earthquake, denoted by the colored arrow. Such changes in radon content have been noted before many earthquakes.

been observed to come out of their holes in wintertime, and ducks, dogs, and cats scurry around in a frenzy. One animal facility observed that chimpanzees took to the ground shortly before a moderate earthquake, and yaks sprawled on the ground. A number of people in San Francisco noted that on the night before the 1906 earthquake occurred horses were *extremely* restless, pawing their stalls and desperately trying to escape.

No one knows what subtle clues animals may pick up before earthquakes. Some scientists doubt whether animals have any premonitory sense at all and chalk up the observed "strange" behavior to selective hindsight after the earthquake. The case for animal observation as one more piece of evidence in earthquake prediction remains unproved, but the circumstantial evidence is often quite convincing. The People's Republic of China has had more experience with this method than any other country and counts it among their more important techniques.

CONCLUSIONS

Earthquake prediction relies on expensive, sophisticated monitoring equipment and a team of individuals to constantly review the incoming data. It also relies on a developing body of data that suggests the prediction of earthquakes will not come easily. The events leading to the final breakage of severely strained rocks must be physically complex, and the trail of data these events will leave behind will not always point clearly to events to come.

There have been a number of successful predictions elsewhere in the world. China has experienced a number of successful predictions since the 1970s, as well as some terrible failures. Probably no country on earth has more to lose than China, with its combination of billions of people and a history of frequent, great earthquakes. Chinese scientists have fully cooperated with western scientists in this important field. The Japanese, too, have had a number of very successful predictions. Although Americans have been a little more cautious, earthquake predictions will likely become an occasional news item in the 1990s. The social implications of a successful earthquake prediction are very large. Perhaps, just perhaps, the ancient affliction will be tamed if sufficient and reliable warnings can be given.

SUMMARY

1. Earthquakes are caused by the locking of fault planes which causes energy to be stored in rocks by elastic distortion. When the fault abruptly

unsticks, the strained rocks undergo elastic rebound. Highly strained rock may also undergo dilation, cracking, fluid diffusion into the cracked rock, and abrupt failure by an aggregation of microcracks.

2. The longer energy is stored, the greater will be the energy release of the resulting earthquake. Great earthquakes recur at longer intervals and release the stored strain from a much larger volume of rock.

3. Energy radiates from the point of first slip on the fault surface and spreads out in waves that travel both along the earth's surface and through the main body of the earth. Energy travels as P, S, and L waves in order of decreasing wave velocity.

4. Decreasing density, increasing incompressibility, and increasing rigidity of rock increase the velocity of earthquake waves as they pass by, elastically distorting the rock.

5. Ground motion is recorded by seismographs as seismograms. The amplitude of ground motion is given by the Richter magnitude number that is an exponent on a logarithmic scale. Mercalli intensity measures the extent of damage, and seismic moment is a number that more adequately represents the energy released by waves whose period ranges from one to several hundred seconds.

6. Earthquake effects include surface faulting, ground acceleration, liquefaction, landslides and rock slides, tsunamis, and seiches.

7. Earthquake prediction depends on long-term measurement of numerous physical properties of the rock in an area where earthquake risk is high. After it has been established what rock properties are normal, lengthy periods of monitoring may turn up anomalies that may precede an earthquake. Techniques of earthquake prediction are in a developmental stage. For an earthquake prediction to be useful, the prediction must specify location to within a few miles or kilometers, magnitude to within a number or two, and time to within a few hours.

KEY WORDS

aseismic creep	bulk modulus
body wave	density
breaking strength	diffusion
dilatancy	seiche
dilatancy-diffusion	seismograph
earthquake	seismogram
elastic	seismologist
elastic rebound	seismic moment
ground motion	shear modulus
liquefaction	S wave
L wave	surface faulting
Modified Mercalli Scale of Earthquake Intensity	surface wave
modulus	tsunami
P wave	wave
Richter magnitude scale	wave velocity

EXERCISES

1. We observe that the relation between earthquake Richter *magnitude* and Mercalli *intensity* is only moderate. Why might this be?

2. How would you find out the seismic risk in the area in which you live?

3. Where is the fault nearest to you that has the potential for moderate to strong earthquakes?

4. If we average the slip observed along portions of the San Andreas Fault, an average rate of motion is about 10 feet (3 meters) per century. Estimated total offset of some geologic units cut and offset by the fault is 350 miles (565 kilometers). How long has the San Andreas Fault been a zone of movement, assuming that the movement rate has been unchanged? Assuming the same rate, how long will it be before San Francisco and Los Angeles—currently about 350 miles (565 kilometers) apart will be a single metropolis?

5. What components are necessary to qualify an earthquake prediction as of value to people in the area in making intelligent choices?

SUGGESTED READING

Bolt, Bruce A., 1978, *Earthquakes—A Primer*, San Francisco, W. H. Freeman, 241 pp.
▶ *Authoritative, modern, inexpensive, this little book is perfect.*

Gere, James M. and Sah, Haresh C., 1984, *Terra*

Non Firma, New York, W. H. FREEMAN, 203 pp.
▶ *Another real winner. Covers it all; well-written.*

RIKITAKE, TSUNEJI, 1982, *Earthquake Forecasting and Warning,* Boston, D. Reidel Publishing Company, 402 pp.
▶ *Updated frequently, this covers the world of prediction.*

WARNER, BRYCE, 1982, *Earthquake,* Alexandria, Virginia, Time-Life Books, 176 pp.
▶ *Lush illustrations, easy reading.*

YANEV, PETER, 1974, *Peace of Mind in Earthquake Country: How to Save Your Home and Life,* San Francisco, Chronicle Books, 304 pp.
▶ *Strongly oriented to western United States, this little book deals with how to evaluate earthquake hazards and make your home more earthquake resistant.*

This locomotive was thrown off the tracks by ground motion near Point Reyes during the 1906 California earthquake. That event marked the beginning of modern attempts to understand the origins of earthquakes. Now the causes are generally understood and the seismic energy unleashed in an earthquake provides a view of the inner earth. Oddly, the two people in the 1906 photograph seem nonchalant—only the dog senses that something is wrong. (Photograph by G.K. Gilbert, U.S. Geological Survey.)

CHAPTER *nine*

The Earth's Framework

What is now proved was once only imagin'd.
                                    WILLIAM BLAKE (1757–1827)

In 1963, a well-known American geologist, M. King Hubbert, suggested that "the acceptance of any proposition by an individual who is not familiar with the observational data on which it is based and the logic by which it is derived is an act of pure faith and a return to authoritarianism." Consistent with that warning, this chapter will not only describe the various component parts, which acting together, form the solid earth, but also outline the evidence and reasoning that has led geologists to this particular view.

This chapter provides an outline of the most important major elements of the earth's outer shell as well as a modern view of the probable characteristics of the earth's interior. It is the capability of inference interacting with observation that allows us to propose reasonable models for the most unknowable part of our planet—the earth within.

THE EARTH'S INTERIOR

The center of our earth is nearly 4000 miles (6378 kilometers) beneath our feet. The deepest well has penetrated 9 miles (15 kilometers) or about one two-thousandth of that distance; the rest can be known only indirectly. Of the various kinds of indirect evidence, laboratory studies of rocks and minerals exposed to extremely high temperatures and pressures have told us something of the physical properties of rocks at depths of many tens of miles or kilometers. Measurements of heat flowing from the earth's interior indicate the energy distribution within the outer earth. Studies of place-to-place variations of the earth's gravitational field allow reconstruction of reasonable models of earth density and constrain the density variations that might be possible within the earth. Measurements of the earth's magnetic field have led to a theory relating its source to properties within one of the earth's innermost shells. Comparisons of the earth's chemistry with that of stars, and especially of meteorites, have provided limitations to what elements and minerals might most reasonably make up the inner earth. Above all else, seismic waves provide "X-rays" of the earth's interior. They outline an earth made of numerous concentric shells, with each shell having physical properties different from the ones above and below.

Earthquakes yield more than violence. The passage of body waves through the earth offers insights into the nature of the inner earth, and the pattern of earthquake locations on the surface of the earth is one link in the evidence for a unified theory of earth dynamics presented in the following chapter.

We have learned that the variation in wave velocity of body waves is a function of the properties of the materials (see Table 8–1) through which they pass. Because hundreds of thousands of earthquakes have occurred and have been studied, the velocity of body waves as a function of distance traveled is quite well known (Figure 9–1). A universal observation is that *earthquake wave velocity increases at varying rates as body waves penetrate more and more deeply within the earth. The more deeply into the earth a body wave has penetrated, the greater its average velocity.* This observation—that velocity changes with distance and depth traveled—forms the basis of Figure 9–1; compare it to Figure 8–1, which describes what happens when velocity is constant.

THE LAYERED EARTH

Since the wave velocity of a seismic wave is increased by increasing incompressibility and rigidity (see Equations 8–1 and 8–2) and decreased by increased density, we may conclude that incompressi-

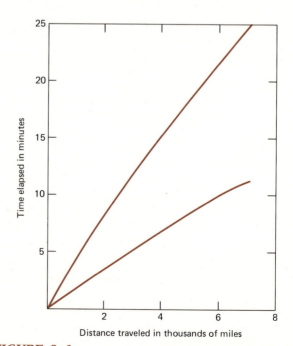

FIGURE 9–1

This is a sample of a time-travel chart for earthquake body waves. Can you figure out which line belongs to the *P* wave and which the *S* wave? Compare to Figure 8–1.

bility and rigidity within the earth rise more rapidly than density. Logic suggests that all three factors should rise under the enormous pressures experienced by rocks many hundreds of miles within the earth, but the slowing effect of increased density is apparently outweighed by the accelerating effect of increased rigidity and incompressibility.

The point on the fault plane where the first slippage occurs is termed the **hypocenter** or **focus** of the earthquake. *Vertically above the focus is the* **epicenter**, *the location on the earth's surface* where the earthquake usually does the most damage (see Figure 8–9). Procedures for determining the exact location of the focus or epicenter are described later in this chapter.

From the focus, energy radiates out in all directions, much as ripples do from a stone dropped into a still pond. *If the velocity of the seismic waves was constant, the ray path taken by the advancing seismic energy would be a straight line*, much as a beam of light passing from headlight to stop sign is a straight line. Since, however, the velocity increases as the waves enter more deeply into the earth, *all of*

the ray paths curve (Figure 9–2) much as a light beam curves when it is slowed by passage through the lens of our eye. *The deflection of the propagating wave is* termed **refraction**, and it is a direct result of the passage of the ray through a material of nonuniform properties. Thus both the acceleration of the time of arrival of the *P* and *S* waves and refraction of the waves tell us that the materials within the earth have different physical properties than those at the surface.

Not only do seismic waves *refract* within the earth, but the arrival of *multiple "copies"* of each *P* and *S* wave on seismograms tells us that many seismic waves *reflect* from surfaces within the earth (Figure 9–12). **Reflection**, or the ricocheting of seismic waves within the earth, can occur only *if the earth is concentrically layered within*, much as an onion is repetitively layered. Seismic waves then bounce off boundaries between shells of inner earth materials

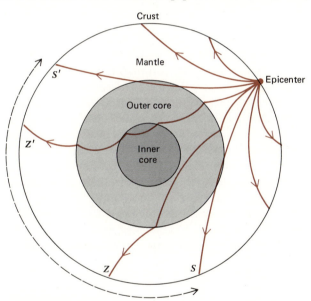

FIGURE 9–2

This sketch depicts the earth sliced through its center. A shallow-focus earthquake yields multiple wave paths. Only a few are shown, and for clarity no reflected waves are displayed. On the other side of the earth from the epicenter, no *P* waves are received over the area Z-Z'; that area is the *P* wave shadow zone caused by intense refraction of the *P* waves by the liquid outer core. Over the area S-S', no *S* waves are received, since the outer core blocks the passage of all *S* waves; this area is the *S*-wave shadow zone. By recognizing and explaining the two shadow zones, and by computing the time-distance curves (see Figure 9–1), a composite picture of the inner earth is created.

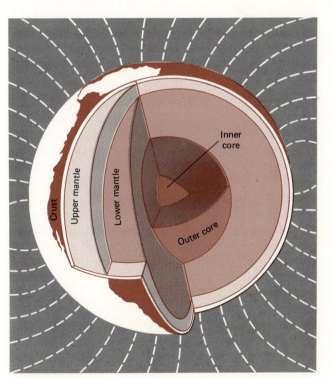

FIGURE 9–3

This three-dimensional rendering shows the major shells within the earth in correct perspective. Radiating out from the earth's surface are the magnetic lines of force produced by the effect of the earth's rotation on the liquid metallic alloys that comprise the outer core. Each of the layers within the earth is recognized by abrupt changes in the velocity of seismic waves as they penetrate the boundaries shown. Compare with Figure 9–10 to see more of the details.

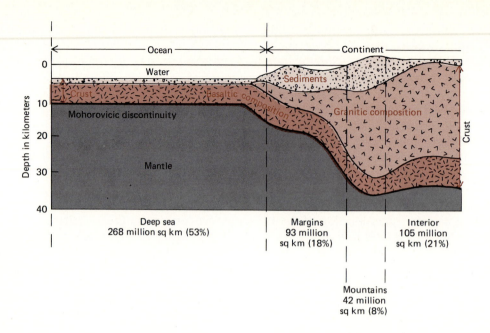

FIGURE 9–4

Continental crust includes margins, mountain (orogenic) belts, and interior or shields. Beneath the continental crust is the Conrad discontinuity, which separates it from the crust of basaltic composition that makes up all oceanic crust. Beneath all crust is the Mohorovičić discontinuity, which defines the base of the crust. (Modified from a sketch by the U. S. Geological Survey.)

having abruptly different physical properties, much as some of the light rays striking the lens of our eye are reflected. Many of us have had the experience of seeing our reflections *on* the lens of a friend's eye that was *at the same time* transmitting *and* refracting our image into their retina. Thus both the reflection and refraction of seismic rays traveling through the earth tell us that areas of gradual change in physical properties are set apart from deeper areas by boundaries displaying *abrupt* changes in physical properties. Our earth is not fully an onion; it is much more like an onion with a generally spherical egg at its center (Figure 9–3).

We live on the outer "skin" of the onion. *The outermost layer of the earth*, comprising only 0.04 percent of the earth's mass, is the earth's **crust.** Easily divisible into a generally granitic continental crust and a wholly basaltic oceanic crust, each of these two subdivisions have three other major distinctive subdivisions, soon to be discussed. Beneath the paper-thin crust is the **mantle**, a region of magnesium- and iron-rich silicates rather similar to the composition of most stony meteorites. The mantle is 1800 miles (2897 kilometers) thick and provides 84 percent of the earth's volume and 67 percent of its mass. Beneath the mantle is the **core,** readily divisible into an

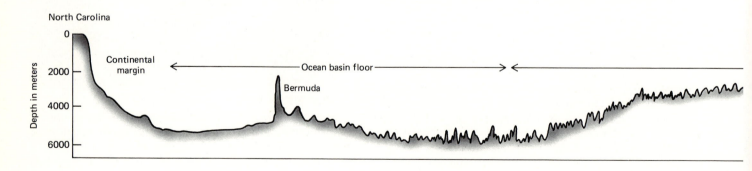

inner and outer core that may be composed of an iron-nickel alloy similar in composition to many metallic meteorites. The core occupies 15 percent of the earth's volume but provides 32 percent of its mass. We will discuss each of these layers in turn.

The Crust

The boundaries defining the principal layers of the earth are recognized by abrupt changes in the velocity of seismic waves. A sudden increase in the velocity of seismic waves takes place when they begin to travel through the mantle. The boundary, defined by an abrupt velocity increase, between the crust and the mantle is the **Mohorovicic discontinuity.**

The boundary is named to honor the Croatian geophysicist Andrija Mohorovičić (pronounced moh-hoh-roh-vee'-cheech) (1857–1936) who first proposed in 1909 that the earth had a layered structure. Studying the seismic waves set up by a Balkan earthquake, he recognized that the waves that penetrated deeper into the earth arrived sooner than those traveling along the surface, even allowing for the difference in total distance traveled. Mohorovičić then proposed that the earth's crust rested on a much more rigid layer within which earthquake waves traveled more quickly. Today we recognize that the abrupt increase in velocity of the P and S waves within the mantle is due not only to the increase in rigidity within the upper mantle but also to the compositional boundary between the silica-rich rocks of the crust and the generally silica-depleted rocks of the mantle (Table 9–1).

The Mohorovičić discontinuity, often abbreviated to the **Moho**, defines the base of the crust. The Moho is generally at a depth of 3 to 5 miles (5 to 7 kilometers) under the oceans but plunges to a depth of 20 to 60 miles (35 to 100 kilometers) beneath the continents. Above the Moho is a thin uniform oceanic crust and a much thicker and more variable continental crust with a generally thin sedimentary veneer (Figure 9–4).

OCEANIC CRUST

The oceanic crust is rather uniform in all its properties. It is created by volcanism at oceanic ridges (see Chapter Ten) where large volumes of a low-potassium basaltic lava are erupted through vents beneath the sea. **Oceanic crust** is the part of the crust that is of basaltic composition and directly overlying the mantle. It may be subdivided into the generally thicker, higher volcanic oceanic ridge (Figure 9–5), yielding high heat flow, and the oceanic floor (Figure 9–5), which is thinner, lower, and yielding diminishing heat flow.

The total heat loss from the Earth is approximately 10^{13} calories/second, an amount sufficient to melt in one year a quarter of an inch of ice covering the entire earth. Of this heat loss, two-thirds is lost through ocean floor volcanism, which is constantly creating new oceanic crust.

The ocean crust consists of a basal layer of gabbro and basalt that averages 3 miles (5 kilometers) in thickness and allows P wave velocities of 6.7 to 6.9 km per sec. Above this generally basaltic oceanic crust is commonly a layer of consolidated sediments and volcanics that averages 1 mile (1.6 kilometer) in thickness with P wave velocities ranging from 4 to 6 km per sec. Above this may be very thin layers of unconsolidated sediment, mostly very fine clays and the shells of tiny marine organisms. The total mass of the ocean crust is estimated at 0.015×10^{24} pounds (7×10^{24} grams). Though the oceanic crust covers 53 percent of the earth's surface, it provides only about 22 percent of the mass of the total crust. Compared to continental crust, oceanic crust is much thinner, much denser, and much more uniform.

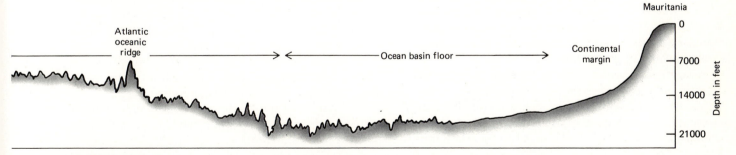

FIGURE 9–5

This is a profile across the Atlantic Ocean from North America east-southeast to Africa. Conspicuous in the center of the oceanic crust is a broad high mountainous area termed the oceanic ridge. Marginal to that ridge and much lower is the deep ocean basin floor, which is marginal to the continents. With the exception of Bermuda, which is of wholly biologic origin, the ocean floor is entirely composed of basalt. (Trans-Atlantic Geotraverse from NOAA.)

TABLE 9–1
Estimates of the Chemical Composition of Continental Crust, Oceanic Crust, and the Upper Mantle

Elements	Continental Crust	Oceanic Crust	Upper Mantle
Silicon	60.2*	48.8	42.9
Aluminum	15.2	16.3	7.0
Total Iron	6.3	8.6	9.3
Magnesium	3.1	7.0	35.0
Calcium	5.5	11.9	4.4
Sodium	3.0	2.7	0.5
Potassium	2.9	0.2	0.003
Titanium	0.7	1.4	0.33

*Elements are given in weight percent of oxides. Data are averaged from several sources.

Oceanic crust is generally thickest, youngest, and hottest nearest the ridges and becomes progressively thinner, older, and cooler with increasing distance from the ridge zones (See Chapter Ten). Rocks on the ocean floor range up to 200 million years in age. The origin of the ocean floor will be much more thoroughly discussed in the next chapter.

CONTINENTAL CRUST

Continental crust is far thicker and more varied than oceanic crust. Because of its much greater thickness, the continental crust provides 78 percent of the crustal mass, even though it covers only 47 percent of the earth's surface.

As noted in Figure 9–4, continental crust has compositional subdivisions, with a basaltic layer at its base, a granitic layer forming its central core, and a thick to thin sedimentary layer covering almost all. The overall composition of continental crust can only be estimated (Table 9–1), since there is no way to sample the lower crust to determine its actual composition. Several models are possible, each of which is consistent with the seismic velocities observed when seismic waves penetrate the base of the continental crust. The model given assumes that the lower crust consists of a layers of granitic and basaltic rocks. Table 9–1 demonstrates the striking compositional differences among continental crust, oceanic crust, and the upper mantle.

A vertical column drilled through continental crust would display striking compositional differences. The *horizontal* subdivision of continents into continental interiors, mountain ranges, and continental margins (Figure 9–4) reveals even more fundamental differences in age, thickness, and origin of the crust.

Continental crust is that part of the crust whose bulk composition is granitic. Large parts (34 million square miles or 93 million square kilometers) of what is geologically a continent are fully or partially

under water, and would therefore be regarded by map makers as ocean. To geologists the *fundamental difference between continent and ocean is defined by differences thickness and composition of the two crusts, not by the temporary location of shoreline.* Continental margins currently occupy 18 percent of the earth's surface and are the part of the granitic continental crust that is currently underwater.

CONTINENTAL MARGINS   The continental margins are areas of sediment buildup (Figure 9–4) marginal to a continent. The continental margin includes the *continental shelf*, an area of gently sloping (1° to 2°) sediment distributed by currents along the edge of the continent. Further out is the *continental slope*, defined by a steeper (3° to 6° slope) that is seaward of the continental shelf.

Still further offshore is the *continental rise*, and even more steeply sloping zone down which soft, water-soaked sediment may cascade during undersea earthquakes. *Turbidity currents* are dense mixtures of water and sediment that may avalanche down continental rise slopes and move hundreds of miles or kilometers out onto the deep ocean floor.

The continental margin described here is typical of those found on the Atlantic Ocean margins and depicted in Figures 9–4 and 9–5. They represent areas of stability, freedom from earthquake activity, and quiet accumulation and dispersal of sediment. These types of continental margins are often termed **passive margins**, or *Atlantic-type* margins.

Along most of the Pacific rim, continental margins are quite different. Here they tend to be extremely narrow, with a shallow sea adjoining the continent with a series of volcanic islands, such as the Japanese islands, further out to sea. Seaward from the volcanic chain is a *deep-sea trench*, a place where the ocean floor abruptly descends to depths of more than 6.9 miles (11 kilometers). These areas are subject to abundant earthquake activity and volcanism and ac-

FIGURE 9–6

A vertical view of the Appalachian Mountains in the Virginia - Tennessee - Kentucky border area taken from Skylab. The long narrow ridge is Pine Mountain; further south is Cumberland Mountain and Cumberland Gap. The Blue Ridge Mountains are at the bottom of the photograph. The Appalachians are a classic orogenic belt, composed of strongly folded and faulted abnormally thick sedimentary rock. The pattern of folds is strikingly displayed by this high-elevation view. (Photograph courtesy of NASA.)

cumulate little sediment. Such continental margins are called **active margins** or *Pacific-type margins*.

OROGENIC BELTS Landward from continental margins may be areas of very strongly deformed, uplifted abnormally thick sedimentary rock formed into a mountain range or an *orogenic belt*. The word **orogeny** comes from the Greek words *oro*, meaning mountain, and *genesis*, meaning formation of, and refers to all processes that make mountains. An **orogenic belt** is a zone that is generally narrow and elongate, such as the Appalachian Mountains (Figure 9–6) of the United States and Canada, and shows evidence of strongly deformed, thick sedimentary rocks, commonly with multiple intrusions of plutonic rocks and associated regional metamorphic rocks.

In the central part of many orogenic belts are huge zones of metamorphic rocks and large batholiths. Examination of the deformed sedimentary rocks as well as the metamorphic and plutonic rocks leaves a distinct impression of extremely strong compressive horizontal forces having acted in the past, turning hard, layered sedimentary rocks into intensely thrust faulted and folded stacks of rock, which grade into even more ductile metamorphic and plutonic rocks. The world's greatest mountain chains seem to have been formed by strong horizontal compressive forces,

which telescoped, folded, faulted, metamorphosed, and melted the rocks caught up in orogenic processes.

The rocks that have been deformed by orogenic forces are those typical of modern *passive margins*. Whatever the source of the forces that make mountains—to be discussed in the next chapter—the rocks involved are the sedimentary accumulations that once formed passive continental margins.

Old orogenic belts are underlain by continental crust with a thickness typical of the continental interiors, but younger belts are underlain by continental crust that is a great deal thicker. The higher the average elevation of the orogenic belt, the thicker the crust beneath the mountains. The top of the world (Figure 9–7), the Himalayas, is underlain by continental crust that is up to 60 miles (100 kilometers) thick.

The relation between average elevation and height is not accidental. It is a relation observed nearly everywhere on earth. To explain this relation, geologists within the last century proposed a concept termed the principle of isostasy. The word **isostasy** comes from the Greek *iso* meaning equal and *stasis*, meaning standing. The **principle of isostasy** holds that the crust floats in the mantle in buoyant equilib-

FIGURE 9–7

This is a Landsat image taken from 568 miles (914 kilometers) above a part of the K'unlun Mountains in northern Tibet. A huge scar running east-west across the center of the snow-capped peaks is the K'unlun Fault, created by the same forces that piled crustal rocks up to total thicknesses of 60 miles (100 kilometers) in this area. (Photograph courtesy of NASA.)

rium, so that the higher the elevation above sea level, the deeper the bottom of the continent. The elevation of any area is proportional to both the mass and the density of the underlying rocks.

Isostasy is a condition of approximate equilibrium. This principle will be discussed in more detail in the next chapter, but it helps to explain why continental crust is thickest precisely where it is tallest.

CONTINENTAL INTERIORS As noted in Figure 9–4, continental interiors have a thin platform of generally horizontal sedimentary rocks resting on much older granitic or metamorphic rocks. The continental interiors are reminiscent of the deep ocean floors, for they are areas of little earthquake activity and great stability covered by a thin veneer of sedimentary rock.

Continental interiors are generally inland from both the orogenic belts and the continental margins. Examination of the underlying ancient metamorphic and igneous rocks shows that they at one time were a part of the central core of many mountain ranges. Often referred to as **basement,** (since these rocks underlie all other continental rocks), these ancient stable rocks form the central foundation for the world's continents. In areas like central Canada, the overlying sedimentary rocks have been largely eroded, and the basement rocks are directly exposed to view. Basement rocks are commonly greatly deformed (Figure 9–8), though their last disturbance may have been billions of years ago.

Also termed **shields** or **cratons,** the basement rocks of the continental interiors occupy 21 percent of the earth's surface area. Where they are exposed by erosion of the overlying sedimentary rock, they tell us that continental basement was formed long ago in vigorous orogenic activity. After prolonged periods of erosion, the shields remain as roots of former mountain chains. After very lengthy periods of erosion, the shields or cratons are covered by very thin layers of much younger sedimentary rocks (Figure 9–9). In this condition they remain as stable continental platforms, generally devoid of earthquakes and volcanism, covered by extensive areas of gently tilted or flat-lying sedimentary rocks exposed to modern erosion.

The central part of North America is cratonic, including all of central Canada and everything in between the Rockies and the Appalachians. Cratonic rocks are exposed to view in areas as the Hudson Bay region; northern Michigan, Minnesota, and Wisconsin; the Black Hills; the Adirondacks; the Llano Uplift; the Wichita Mountains; and in the bottom of the Grand Canyon. Elsewhere we can only sample them when deep wells penetrate these rocks and bring up specimens for us to see.

Most of the basement appears to have formed by successive additions of ancient orogenic belts welded and deformed together. The igneous and metamorphic rock within the cratons average 20 to 24 miles (35 to 40 kilometers) thick and allow *P* wave velocities in the 6.0 to 6.9 kilometers per second range. When covered by sedimentary rocks, they may be a mile (0.6 kilometer) thick and allow *P* wave velocities in the 2 to 4 kilometers per second range.

Heat flow from continental crust varies as a func-

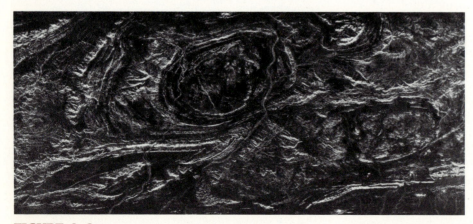

FIGURE 9–8

This image is formed by radar waves bouncing off a basement surface in the Hammersley Range in far western Australia. The image was acquired by the SIR-A experiment aboard the Space Shuttle Columbia. The circular pattern is of eroded folds surrounding a prominent granite dome. The rocks are approximately 1.5 billion years old and are part of the craton of Australia. The image covers an area of 30 by 60 miles (50 by 100 kilometers). The Hardey River runs down the center of the image. (SIR - A image courtesy of NASA.)

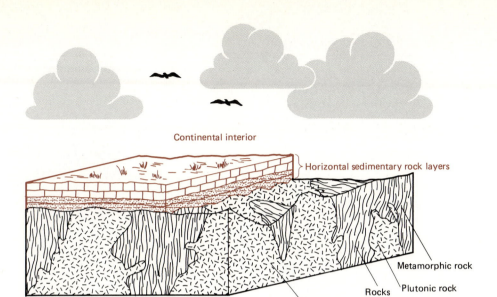

Continental interior

Horizontal sedimentary rock layers

Metamorphic rock

Plutonic rock

Rocks

Basement

FIGURE 9–9

The typical continental interior consists of a series of thin sedimentary rocks that are generally horizontal or gently dipping and lie on an unevenly eroded surface of ancient igneous and metamorphic rocks. If the sedimentary rocks are eroded away, as in the right part of this sketch, the craton is exposed (see Figure 9–8) as a group of strongly metamorphosed and plutonized rocks. If the sedimentary cover remains, the earth's surface looks like the American midwest.

tion of age of the rocks, since the younger the rock's recrystallization, the greater the amount of radioactivity displayed by radioactive minerals in the rock. Within the ancient rocks of the shield, heat flow is less than half of that measured in young orogenic belts. Newly deformed, highly compressed young orogenic belts are areas of relatively higher heat flow, as are areas of continental extension, such as the Basin and Range area of the western United States. The Basin and Range area has a rather thin crust, and a variety of structures suggesting that continental crust in this region is being extensively stretched and thinned. Here the anomalously higher heat flow seems to emanate from the mantle beneath the stretched and thinned crust.

The Mantle

Since the mantle lies beneath the zone of direct observation, we must infer its probable composition and structure from a variety of data. The profile of density within the mantle is constrained by what we know of the pattern of slow deceleration of the earth's rotational speed and the wobble of its rotational axis in space. The spinning earth is like a gyroscope, with its stability and diminishing spin rate direct functions of the pattern of density distribution within.

Abrupt variations in seismic velocities within the mantle suggest that the mantle itself is strongly layered in shells of different physical properties. In particular, no earthquake *foci* have been recognized below a depth of 400 miles (670 kilometers). This depth divides the mantle into an *upper mantle* and a *lower mantle*.

Laboratory studies have suggested what the probable mineral content of the mantle is; they have also defined very sharp zones where minerals should adopt new, higher-density forms under the crushing load of the rocks above. The zones where minerals change their crystal structure under pressure are termed *phase-change zones*. Numerous phase-change zones have been predicted on the basis of experimental laboratory work, and most of them have been identified. As very accurate measurements of seismic wave velocities allow recognition of the predicted zones, they are marked as areas of abrupt velocity increase.

Some violent volcanic eruptions have brought materials to the surface that appear to have come from the mantle; the best known of these are *kimberlite pipes*, areas composed of a vertical cylinders filled with a type of brecciated dark volcanic rock in which the high-pressure mineral diamond may be found. More ordinary mineral pairs, including very high-pressure forms of olivines and pyroxenes, have also been recognized in some igneous inclusions in other rocks. Thus certain violent events have brought us inadvertent mantle samples, which have largely served to confirm our theoretical predictions.

THE UPPER MANTLE

The upper mantle begins at the Mohorovičić discontinuity at the base of the crust, and continues down to 400 miles (670 kilometers). Studies of the inclusions carried up into the crust as well as studies of basaltic magma generation and theoretical studies of phase transitions all suggest that the upper mantle is composed largely of magnesium-rich olivines and pyroxenes with small amounts of high-pressure garnets and spinels.

The uppermost 45 miles (70 kilometers) of the upper mantle is a strong, rigid zone, allowing P wave velocities averaging 8 kilometers per second and S wave velocities of around 4.6 kilometers per second. The upper mantle and the entire crust together form a rigid, strong unit that ranges from 60 to 100 miles (100 to 160 kilometers) in thickness; this composite layer is the **lithosphere.** As we will discover in the next chapter, the lithosphere is the fundamental *structural* unit of the earth's surface. It floats in buoyant isostatic equilibrium in the layer just beneath it.

THE LITHOSPHERE AND ASTHENOSPHERE

The layer beneath the lithosphere, called the **asthenosphere,** is a zone of abnormally low seismic velocities, muffled S waves, and other characteristics that indicate that it *is partly liquid.* The long-term fluidity of the asthenosphere not only allows the lithosphere above it to be in *buoyant equilibrium,* but also permits the lithosphere to be in *both horizontal and vertical motion.*

The asthenosphere reaches down to depths of 200 miles (350 kilometers), making it about 100 miles (150 kilometers) in thickness. Within the asthenosphere P wave velocity drops from the 8.0 kilometers per second typical of the lower lithosphere to 7.6 kilometers per second, whereas S wave velocities drop from 4.6 to 4.2 kilometers per second. The S waves also suffer a great deal of muffling. These properties suggest that the asthenosphere may contain up to 10 percent fluid. It should probably best be thought of as a crystalline mush, having some of the physical properties of an asphalt road on a blistering hot day.

From the base of the asthenosphere (Figure 9–10), seismic wave velocities remain at almost the same level down to a depth of nearly 250 miles (400 kilometers), at which point the phase transition of olivine to spinel causes an abrupt increase in rigidity. At this boundary, P wave velocities increase from 7.6 kilometers per second to 8.5 to 9.6 kilometers per second, and S wave velocities move from 4.5 kilometers per second to 5.5 kilometers per second.

Beneath the 250-mile phase transition zone, rigidity must slowly increase as seismic velocities slowly increase down to the base of the upper mantle,

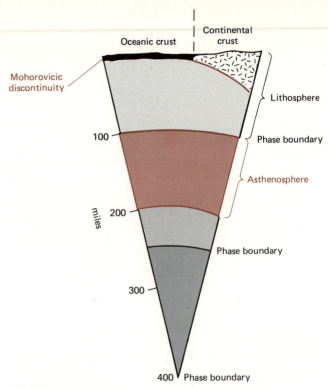

FIGURE 9–10

The upper mantle and crust include the uppermost 400 miles (670 kilometers) of the earth. The upper 60 to 100 miles are termed the *lithosphere,* since this block of combined uppermost mantle and crust acts as a strong, rigid unit. Beneath the lithosphere is the asthenosphere, which is partially molten and allows the lithosphere to be in isostatic equilibrium. Beneath the asthenosphere is the rest of the upper mantle, whose base is defined by the lowest depths at which earthquake foci have been recorded..

which is at 400 miles (670 kilometers), the maximum depth at which an earthquake has occurred. This boundary is marked by another phase change, perhaps coincident with the repacking of the mineral spinel's components into the still denser mineral *perovskite.* This change is recognized by still another abrupt increase in seismic wave velocity.

The abrupt stoppage of all earthquake generation suggests that downward movement of discrete units of rocks has become impossible at this depth, since they meet resistance so strong they cannot move deeper. This boundary also suggests that the upper mantle is isolated from the lower mantle, with little transfer of material across this boundary; not all geologists would agree with this interpretation. *By whatever description, the 400-mile discontinuity marks a fundamental change in the earth; it is the boundary between upper and lower mantle.*

THE LOWER MANTLE

The mineralogy of the lower mantle is poorly known. Theoretical considerations suggest that the silicate phases of the upper mantle should transform into extremely dense phases, but there are too few constraints to allow direct prediction of what the lower mantle is like.

Many geochemists have argued that the earth's crust was formed through the fractionation (see Chapter Three) of the upper mantle into two phases: a lower-density phase that rose to the surface and continues to float there as continental and oceanic crust and a higher-density phase that remained below the lower-density crust. According to geochemical arguments, the entire crust could have formed from fractionation of only the upper mantle, leaving the lower mantle as primitive material little changed since the days of earth origin, billions of years ago.

Given that we have little knowledge of the temperature within the lower mantle, the amounts of certain elements that are present, and phase transitions at extraordinary pressures, we must restrict our models to the already limited data. We may argue that the lower mantle becomes increasingly rich in metals and depleted in silica, but probably retains its character as a magnesium- and iron-rich silicate crystalline material similar in composition to many stony meteorites.

Within the lower mantle *P* wave velocities rise to their maximum velocity, reaching 8.5 mi/sec (13.7 kilometers per second) at the **Gutenberg discontinuity,** the boundary that marks the base of the mantle at 1800 miles (2920 kilometers). The Gutenberg discontinuity is named after Beno Gutenberg (1889–1960), a German-American geophysicist who recognized in 1913 the existence of the earth's core, and with it the core–mantle boundary, which was named after him (Figure 9–11).

The Core

The Gutenberg discontinuity is the second vertical break in the pattern of a constant increase in velocity as seismic waves penetrate deeper into the earth. At this discontinuity, the density nearly doubles, from 5.7 gm/cc to 10 gm/cc, which is over one million times greater that atmospheric pressure. The core is 32 percent of the mass of the earth and can be readily divided into an inner semi-solid core and an outer liquid core.

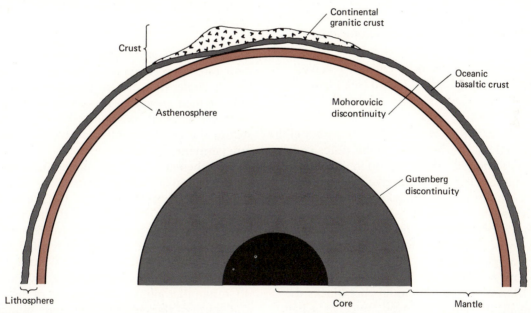

Continental granitic crust

Crust

Asthenosphere

Oceanic basaltic crust

Mohorovicic discontinuity

Gutenberg discontinuity

Lithosphere   Core   Mantle

FIGURE 9–11

This cutaway diagram reveals the relation between the crust and the rest of the earth. The diagram is not to scale; if it were the entire crust would be a single thin line. In terms of *chemistry*, the earth is separated into a granitic continent, underlain by basaltic crust, underlain by a magnesium-rich silicate mantle, underlain by a nickel-iron-sulfur alloy core. In terms of *seismic boundaries*, the granitic crust is underlain by the Conrad discontinuity, the basaltic crust is underlain by the Mohorovičić discontinuity between crust and mantle, the mantle is underlain by the Gutenberg discontinuity between mantle and core, and the outer core is underlain by the Lehmann discontinuity between inner and outer core. Each of these discontinuities is a sharp zone of abrupt change in seismic velocities. In terms of the large scale *structures of the earth*, the earth is composed of a rigid lithosphere resting on the fluid asthenosphere. The asthenosphere is both a magma source to the crust and a zone that allows isostatic compensation.

The temperature within the core may range between 6900° to 8100 °F (3800° to 4500°C). The velocity of seismic waves, the earth's rotational characteristics, the earth's density distribution, and the cosmic abundance of elements are all consistent with an overall core composition of a iron–nickel alloy, with small amounts of sulfur mixed in. This composition is also consistent with the composition of many metallic meteorites.

The overall density of the core is approximately 11.5, a density similar to that of lead. P wave velocities in the core average 8 to 10 kilometers per second, with an abrupt temporary decrease at the boundary between the inner and outer core.

THE OUTER CORE

As seismic waves strike the Gutenberg discontinuity, four things happen simultaneously:

1. S waves are stopped as they enter the core, suggesting that the outer core is liquid (see Figure 9–2).

2. Many P waves bounce off the core - mantle interface, so that less compressional energy enters the core (Figure 9–12).

3. P wave velocity precipitously falls from 13.7 kilometers per second to 8 to 10 kilometers per second.

4. Because of the abrupt velocity change, P waves are strongly refracted into the core (see Figure 9–2).

THE SHADOW ZONE Two directly observable effects are created by the liquidity of the outer core. One effect has to do with the distribution of areas on the earth's surface that fail to receive transmitted P and S waves from distant earthquakes. Locations that fail to receive seismic waves are said to be in one of two **shadow zones,** the areas on the earth's surface (Figure 9–2) where the core blocks the receipt of *direct* seismic energy.

One shadow zone (S–S' on Figure 9–2) is termed the S wave shadow, where the liquid core blocks the transmission of *any* direct S waves, much as a ball held in front of a flashlight blocks out part of the light. The other shadow zone (Z–Z' on Figure 9–2) is much less extensive. Termed the P wave shadow, it is the zone where no direct P waves are received because of the intense refraction of the P waves created by the abrupt decrease in velocity across the Gutenberg discontinuity (Figure 9–2).

Shadow zones outline the outer core as surely as an X-ray outlines bone against muscle. They are one

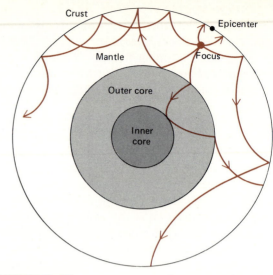

FIGURE 9–12

This sketch shows a few of the *reflected* ray paths as seismic energy bounces off both the discontinuities at the surface of the inner and outer core. Other waves bounce repetitively off the crust, shown here in true scale as the outer line of the largest circle.

of two kinds of evidence that the earth's outer core is liquid. The other is related to the origin of the earth's magnetic field.

THE EARTH'S MAGNETIC FIELD Lines of magnetic force (see Figure 9–3) extend outward from the earth's center in a pattern reminiscent of a bar magnet oriented nearly north–south subparallel to the earth's rotational axis. But the high temperatures within the earth would rule out any solid retaining any permanent magnetism.

That the earth has a permanent magnetic field has been known for four centuries. William Gilbert (1544–1603), court physician to Queen Elizabeth I, showed in 1600 that the earth itself was a great spherical magnet, to whose magnetic poles our compass needles point. He also showed that a magnetized needle points downward toward the earth at an angle of inclination that varies with latitude, the lines on a map that allow measurement of distances north and south of the equator. The angle of inclination of a numerous magnetized needles placed on the surface of the earth outline the lines of magnetic force that come from the earth (Figure 9–3).

It remained for Walter Elsasser (1904–), a German–American physicist, to predict in 1939 that the earth's rotation would set up eddy currents in the liquid nickel–iron outer core of the earth. The eddying flow of a magnetic metal creates small electric currents; these in turn form a magnetic field.

This model has since been reasonably confirmed, for it precisely explains slight variations in the magnetic field, as well as the tendency for the orientation of the magnetic north and south poles to wander slightly in an orbit centered about the earth's rotational pole. *Thus the earth is an electromagnet, its magnetic field produced by the feeble electric currents caused by rotation of the outer core.*

THE INNER CORE

The inner core, defined by the Lehmann discontinuity at 3200 miles (5200 kilometers), transmits seismic waves as if it is partially molten. At the Lehmann discontinuity, *P* wave velocities abruptly drop back to 5 mi/sec (8 kilometers per second) and then rise rapidly, increasing to nearly 6.9 mi/sec (11 kilometers per second).

The inner core is slightly less than 5 percent of the total volume of the entire core. It is a tiny kernel, whose statistics stagger our imagination. Pressure in this region is estimated at nearly four million times atmospheric pressure, with temperature estimated at 4500 °C, a temperature similar to that of the surface of the sun. The center of the earth is 3963.5 miles (6378.4 kilometers) from its equatorial surface. Having arrived at that locale, we terminate our review of the layered earth (Figure 9–13).

EPICENTRAL PATTERNS ON THE EARTH'S SURFACE

There is order to the concentric shells within the inner earth. Each layer is increasingly denser than that above it. The mantle is more metal-rich than the crust, and the cores are more metallic than the mantle. An earth layered in concentric shells of increasing density suggests that at some early time the earth was at least partially molten, allowing denser materials to sink in the fluid and lighter materials to rise. This is an idea to which we will return as we look at models of earth formation in Chapter Eighteen.

If we return to the earth's surface and look at the patterns of earthquake epicenters and foci, we once again find order, and in this order we find clues to a revolutionary view of the forces that form the largest features of the earth's surface. An outline of these ideas forms the next chapter.

Our first task is to see how to locate an earthquake epicenter. Since the earthquake epicenter is vertically above the earthquake source region (the focus), we might expect that the epicenter would be easily located as the area sustaining the maximum damage. Because of the effect of types of soils and foundations, the epicenter does not, unfortunately, always coincide with the area of greatest building damage.

EPICENTER DETERMINATION

Epicentral location is obtained by direct cooperation among at least three seismographic recording stations. For maximum accuracy, the stations should be at widely varying distances and directions from the epicenter (Figure 9–14).

The principle is simple. If two racing cars cross the finish line exactly a minute apart, with car A averaging 5 miles (8 kilometers) per minute and car B averaging 4 miles (7 kilometers) per minute, how far away was their starting point? A bit of arithmetic will show that their starting point had to be 20 miles (32 kilometers) distant, *for this distance is the only one*

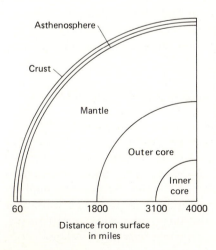

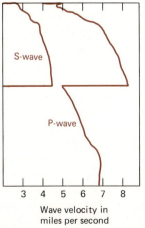

FIGURE 9–13

This diagram relates body wave velocity to the concentric shells within the earth. Both *S* wave and *P* wave velocity increase abruptly at the crust - mantle boundary (the Mohorovičić discontinuity or the "Moho") and then increase steadily through the mantle. At the core - mantle boundary (the Gutenberg discontinuity), the *S* wave is absorbed and the *P* wave velocity greatly slowed. On through the core, *P* wave velocity gradually increases into the center of the earth.

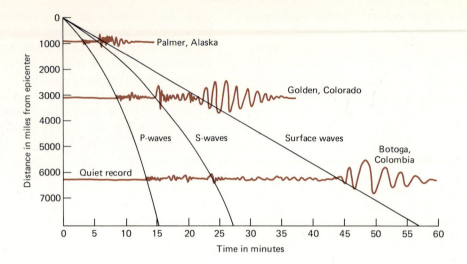

FIGURE 9–14

The colored lines in this graph are seismograms of the same earthquake recorded by three progressively distant seismographs. The sweeping diagonal lines connect *all* points on the seismograms at which *P* waves, *S* waves, and surface waves *first* arrive at the station. Notice that the *difference between the arrival times at any one station is a* *direct function of the distance to the epicenter*. If several seismographic observatories plot the distance to epicenter as a radius, the intersecting arcs define the epicenter (see Figure 9–15) (Diagram courtesy of the U. S. Geological Survey.)

that could produce the observed difference in their arrival times.

Similarly, when a seismograph records the arrival of the faster *P* wave, and then later that of the slower *S* wave, the time difference (Figure 9–14) between the arrivals—since the velocity of each wave is well known—is a direct function of the distance to their starting point. Each seismographic station is then able to directly determine the distance, but not the direction, to the epicenter.

If three or more seismographic observatories co-operatively exchange the information each has gathered about distance to epicenter, they may plot all three epicentral distances as radii of circles (Figure 9–15), *and the intersection of these radii is the only unique point at which the epicenter could be.* In practice the calculations are more difficult than here described because of a variety of corrections, but the technique is as described. With the advent of high speed computers and data communications, approximate epicenter determinations may be available within a few minutes.

The location of the focus or hypocenter requires much more detailed calculations, but its determination allows an understanding of the vertical pattern of earthquakes over a period of time (see Figure 9–19). This vertical pattern, in turn, may allow much greater understanding of the origin of the stresses that are relieved by the earthquake. Another important piece of information obtained from an earthquake is the direction of first slip along the fault; this too requires more detailed calculations but allows a much better understanding of the strain patterns that relieve stresses.

In an average year, perhaps 1 million earthquakes are recorded by sensitive seismographs around the world—an average of approximately two earthquakes per minute. Before the advent of high-speed computers, the calculation and plotting of epicentral and focal data was a time-consuming, laborious task. Since the 1960s, the advent of computers has revolutionized **seismology**, the branch of geology that deals specifically with earthquake activity. High-quality seismographic stations, linked worldwide by telecommunications networks, compute and plot epicentral location data and focal positions for thousands of earthquakes each year. The ability to plot huge amounts of computer-derived data on a worldwide basis, combined with other kinds of information, has revolutionized our view of our not-so-solid earth.

AN UNEXPECTED ORDER

A cumulative plot of many years of epicentral and hypocentral data reveals striking three-dimensional patterns of energy release. In Chapter One we defined science as the search for repetitive patterns, which must then be explained. Earthquake data on a worldwide basis certainly provide an intriguing puzzle; whatever the forces that cause earthquakes, *those forces are not randomly distributed because earth-*

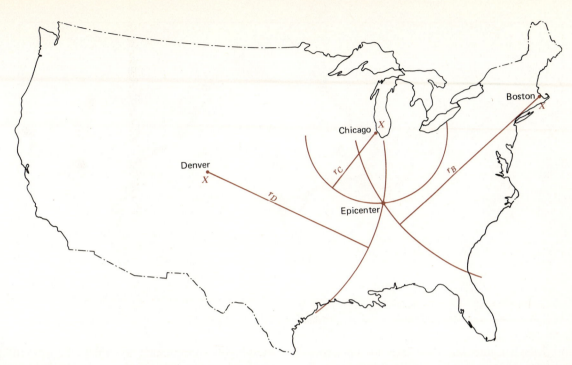

FIGURE 9–15

The simplest method of locating the epicenter is to find the time interval between the P wave and S wave arrivals at several seismographic stations. The distance from each station is then plotted using time-travel curves (Figure 9–1) and plotted as a radius from each station ($r[npD]$, $r[C]$, r●). The arcs intersect at a common point, which is the

epicenter. In this example, seismographic observatories in Boston, Chicago, and Denver have located an earthquake epicenter in southeasternmost Missouri. In 1811 - 12 a prolonged series of earthquakes in this same region constituted probably the most powerful earthquakes experienced in the United States in historical time.

quakes themselves are not randomly distributed on any scale.

Three narrow zones, strikingly displayed on a world map (Figure 9–16) *release more than 95 percent of the world's seismic energy.* Having searched, we find order.

The *earth's lithosphere is divided by epicentral patterns into irregularly shaped elements called plates, whose margins are defined by narrow zones of abundant earthquake activity.* In the central regions of plates, earthquakes are relatively uncommon; at the margins of plates, earthquakes are, unfortunately, extremely common. Earthquakes not only originate most often at plate boundaries, but they also form three-dimensional patterns of earthquake zones that in cross section slant steeply into the outer mantle (Figure 9–19). In some areas, earthquake foci may reach to 400 miles (670 kilometers) depth—defining the deepest boundary of a plate.

A **plate** is an irregularly shaped block of lithosphere bounded by zones of abundant earthquakes. Pacific plate margins may be defined by planes that contain foci dip or slant downward; other types of margins are common. Since all plates are composed

of lithosphere, *they float on the asthenosphere, though the earthquake zones that define them may penetrate the entire upper mantle (Figure 9–10).*

Among the largest plates is the one encompassing most of the Pacific Ocean (Figure 9–16). Defined by swarms of epicenters along the western coasts of the Americas, along the Aleutians and Japanese islands, in a double zone along the western Pacific, stretching southeast through Indonesia and New Zealand and then skirting the Antarctic continent, the margins of this plate *release 75 percent of the world's seismic energy.*

TYPES OF SEISMIC BELTS

There are three types of seismic belts, linear zones of intense earthquake activity. The longest single belt coincides with the ridges of the oceanic crust. The second belt stretches in a series of curves eastward across the Mediterranean and Caspian Seas, across the Himalayas, and then south into the Bay of Bengal. The third belt is the seismic zone that encircles

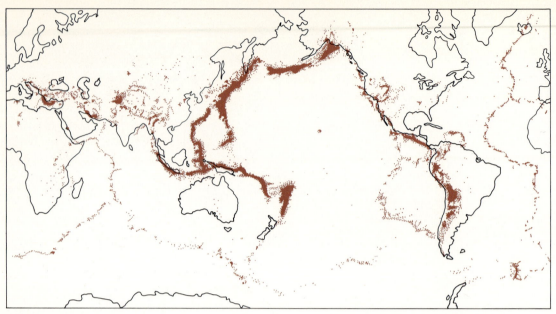

FIGURE 9–16

Plot of worldwide epicenter locations; each dot represents one epicenter for the years 1961 - 1969. One sparse band of dots defines the oceanic ridges. Another band of more concentrated dots leads from the Mediterranean Sea east across central Europe to the Himalayas; this is the Mediterranean - Himalayan seismic zone. The band of most highly concentrated dots circles the western, northern, and eastern margins of the Pacific Ocean; this zone of extremely intense earthquake activity defines the Pacific Rim seismic zone. (NOAA/EDS, courtesy of the U. S. Department of Commerce.)

the Pacific Ocean. These three belts release most of the total seismic energy released by the earth in any one year. The annual release is estimated at 9,000,000,000,000,000,000,000,000 ergs; this enormous number is slightly greater than 1 million trillion horsepower.

OCEANIC RIDGE BELT

The world's longest single seismic zone coincides with the oceanic ridges, previously mentioned in this chapter. The **oceanic ridges** are plate boundaries marked by

1. Abundant basaltic volcanism from very shallow magma chambers along the ridge axis; this volcanism creates the ridge.
2. Frequent low- to moderate-magnitude shallow-focus earthquakes along the ridge axis, which create
3. Grabens or rifts (Chapter Seven) along the crest of the ridge.
4. Offset of the ridge axis by faults trending perpendicular to the ridges.
5. Release of 5 percent of the world's seismic energy.

As we have already noted it is volcanism from the oceanic ridge system that has created the basaltic oceanic crust. The oceanic ridge system stretches nearly 50,000 miles (80,000 kilometers) around the world. In the Atlantic Ocean the ridge runs exactly down the center of the ocean basin and surfaces to form Iceland; in the Indian Ocean the ridge is about halfway between Africa and India and extends up the Red Sea; in the Pacific Ocean the ridge runs along the southern margin of the ocean basin to form a complex pattern off South America that runs northerly into the Gulf of California (Figure 9–16).

Studies of first motion or slip directions along this seismic belt reveal a movement pattern consistent only with *tension*. The oceanic ridge zone forms within the ocean basins because *forces tend to pull the oceanic lithosphere apart*, much as you would pull on a string to break it. Pulling in opposite directions on the lithosphere causes the center of the ridge to constantly break and collapse, creating a central *rift* or *graben*. A **rift** or **graben** is (Figure 9–17) a down-dropped area where crustal material under tension has broken and slipped downward along two normal faults dipping toward each other when the neighboring higher blocks are moved away from one another.

The worldwide rift zone in the center of the oceanic ridges suggests that forces within the earth

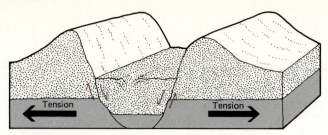

FIGURE 9–17
Tensional forces (the black arrows) may pull the lithosphere apart, creating a downdropped (the colored arrows) central depression, which is known as a *graben* or a *rift*. Along oceanic ridge systems, a central rift zone is common, so that the oceanic ridge is actually a system of parallel ridges with a central rift zone. Other grabenlike areas include the Dead Sea of Jordan, the East African Rift zone, the Purcell Trench in British Columbia, and Death Valley in California, U.S.A.

act to pull the lithosphere apart along a tear nearly 50,000 miles (80,000 kilometers) long, which causes shallow earthquakes and eruption of lava along the resulting rift zones. These rift zones are not exclusively oceanic. The Gulf of California, the Red Sea, and the East African Rift Zone are all in continental areas that are being pulled apart by the same mysterious forces.

Whatever the origin of the forces (to be discussed in the next chapter), logic suggests a puzzle. If the lithosphere is being pulled apart along a 50,000 mile long tear, it must be converging somewhere else, *or*, the lithosphere is growing in size. The resolution of this puzzle leads to a whole new understanding of how the earth works. That understanding forms the subject for the next chapter.

THE MEDITERRANEAN - HIMALAYAN BELT

Stretching east–west along the Mediterranean and Caspian Seas to the south margin of the Himalayas, the Mediterranean–Himalayan seismic belt has

1. Shallow- and intermediate-focus earthquakes, predominantly across strike-slip and reverse faults.
2. Earthquake foci extending well into the lithosphere.
3. Compression in a general north–south direction, which creates
4. Intensely deformed sedimentary rocks in young orogenic belts trending roughly east–west.

5. Release of 15 percent of the world's seismic energy

Studies of the slip direction associated with the faults on this easterly trending zone suggest strong compression from the convergence of Eurasia and Africa, the Arabian peninsula, and India. Numerous high, young mountain chains including the Alps and the Himalayas document what happens when continent slowly rams into continent. Both intermediate- and shallow-focus earthquakes are produced by the forces moving these continental landmasses toward one another.

THE PACIFIC RIM BELT

An extremely distinctive—and deadly—seismic zone, the Pacific Rim has

1. Epicentral patterns on the earth's surface that form great looping arcuate shapes (Figure 9–18).
2. Shallow-, intermediate-, and deep-focus earthquakes largely in seismic zones that slant down under the continents.
3. Deep-focus earthquakes that stretch down to the bottom of the upper mantle.
4. Seismically active, narrow continental margins, with abundant volcanism.
5. Strong horizontally-directed compression perpendicular to the length of the seismic belt.
6. Release of 75 percent of the earth's seismic energy; the Pacific Rim is home to more than 20 great earthquakes (see Chapter Eight) in an average year.

The Pacific Rim seismic belt, with a rare exception, is the only place on earth where *deep-focus* earthquakes occur. These earthquakes may occur at depths down to 400 miles (670 kilometers), the deepest earthquakes ever recorded. This depth defines the base of the upper mantle. Deep-focus events are rare, comprising about 3 percent of all earthquakes, and probably do not originate by the same elastic or dilational (see Chapter Eight) processes that seem to trigger shallow-focus earthquakes.

Both deep- and intermediate-focus earthquakes must involve processes other than simple shearing, which should be impossible at the high pressures that are typical of depths beyond the asthenosphere. Laboratory experiments as well as the waveforms produced by intermediate- and deep-focus earthquakes suggest that they may originate because of

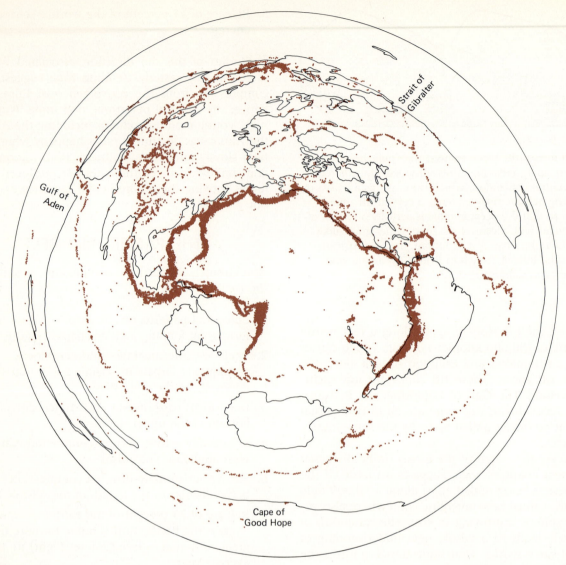

FIGURE 9–18

This map projection is centered on the Pacific Ocean, with each dot representing the epicenter of a single earthquake during the period 1963 - 1973. This plot clearly reveals the enormous concentration of earthquake foci around the rim of the Pacific. (Plot by NOAA/EDS, courtesy of the U. S. Department of Commerce.)

sudden volume loss as minerals such as olivine shift its atomic packing to a denser form (perovskite).

However, when viewed in three dimensions, deep earthquake hypocenters are typically at the base of a sloping seismic zone that dips under the continents (Figure 9–19). Such a pattern suggests that the compression between oceanic and continental margins may be accommodated by movement on great faults that dip under the continents, as though the Pacific Ocean Basin were being stuffed *under* the granitic and andesitic rocks that rim it. In such circumstances, earthquakes may be caused by grinding and shearing as continental crust slides over oceanic crust.

In still other areas, shallow-focus earthquakes are produced by horizontal slip along faults whose fault planes are near vertical. With motion along these seismic belts nearly horizontal, lithospheric plates act like cars that sideswipe one another as they head in opposite directions. One such seismic belt is the San Andreas Fault of southwestern California (see Figure 8–17). Along this terribly dangerous fault system, energy built up over a century or more along locked sections of the fault may find its release in a few devastating minutes.

Volcanism is not prominent along the San Andreas Fault. It appears that the fault plane does not extend deeply enough to open any easily available

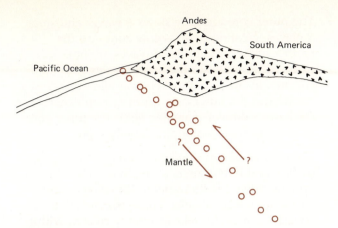

FIGURE 9–19

In this highly schematic cross section west to east from the Pacific Ocean basin to the western coast of South America, earthquake hypocenters or foci form a downward-sloping seismic zone that may extend to depths of 400 miles (670 kilometers). Such downward-sloping seismic zones are typical of most of the Pacific Ocean margin, and suggest that the continents are overriding the Pacific Ocean floor. The sketch is not to scale; the Andes reach not quite 5 miles (8 kilometers) above sea level.

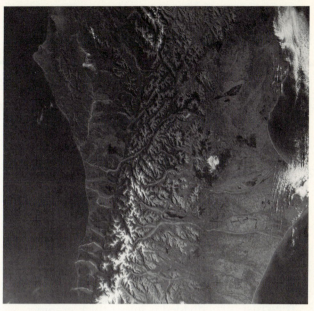

FIGURE 9–20

This Skylab photograph of South Island, New Zealand, has clearly captured the effect of a large strike-slip fault (the Alpine Fault) on topography, as the fault forms a straight line running north-northeast across the west (left) quarter of the island. The western part of the island is moving northeasterly along the fault; notice the offset stream courses and the widening of the stream channels. The city of Christchurch is just under the clouds on the center right margin. The distance from the top to bottom of the photograph is about 175 miles (290 kilometers). (Photograph courtesy of NASA.)

magma source. In other areas of the earth, where magma may lie at shallower depths, strike-slip faulting and volcanism may be closely related (Figure 9–20).

The association of volcanism with the compressive margins of the Pacific Basin, where enormous seismic zones dip under the continental crust, has earned the area the informal name as *The Pacific Ring of Fire* (see Chapter Three). The volcanoes occur in great arcuate patterns across the western and northern Pacific. Associated with them are deep oceanic trenches that parallel the pattern of volcanoes on the surface.

These oceanic trenches, also termed **foredeeps,** appear to be linear downfolds of the ocean floor that cause the ocean floor to flex to depths in excess of 6 miles (10 kilometers). The foredeeps are themselves the site of shallow-focus earthquakes, with deeper-focus earthquakes progressively landward from them.

Along these systems of foredeeps and arcuate islands, we see yet another expression of the earth's dynamism. It appears that oceanic crust is being stuffed under continental crust along the foredeep seismic belts. The volcanic islands mark *the edge of the granitic continent*, that then extends under the marginal sea between it and the mainland.

The Pacific Rim is the most complex of the three major seismic zones. It combines seismic zones due to intense horizontal compression with sliding strike-slip fault zones like the San Andreas system. The Pacific Rim exhibits shallow-, intermediate-, and deep-focus earthquakes, as well as great arcs of foredeeps and volcanic islands that mark the continental edge. The axes of many of these arcuate systems split, breaking the Pacific plate into smaller micro-plates, each of which has its own history.

In spite of the complexities, the patterns along the Pacific Rim and other seismic belts allow us a view of an earth whose lithosphere appears to be in constant motion. The edges of the plates jostle one another in head-on collisions, grinding sideswipes, and 50,000-mile-long splits. Such a revolutionary view of the earth comes from the combination of sensitive, quivering seismographs, worldwide computer and telecommunication links, and patient human beings asking questions of the earth. The answers to some of these questions fill the next chapter, a review of the earth's dynamic processes on their grandest scale.

SUMMARY

1. The center of the earth is approximately 4000 miles (6378 kilometers) beneath us. Models of our earth's interior are constrained by available data on earth's average density (5.5), heat loss, gravitational and magnetic fields, the chemistry of stars and meteorites, rotational characteristics, and the way the earth's interior affects earthquake body waves.

2. The velocity of earthquake body waves increases unevenly with their depth of passage through the earth. Using data from seismic wave velocities, geophysicists divide the earth into a two-part crust, a two-part mantle, and a two-part core. Each of the three major subdivisions is separated by major changes in probable composition and related changes in seismic wave velocity.

3. The crust is subdivided is into a thick granitic continental crust and a thin basaltic oceanic crust. The continental crust can be further subdivided into active and passive continental margins, orogenic belts, and continental interior, cratons, or shields. The oceanic crust is split by long, linear oceanic ridge systems with a central rift. The ridge system is flanked by the deep ocean floors.

4. The mantle can be subdivided into an upper mantle, defined as the deepest zone within which earthquake foci or hypocenters occur, and a lower mantle, which is seismically quiet. Within the upper mantle, which is 400 miles thick, is the asthenosphere, a partially liquid layer that may be 100 miles thick and whose upper boundary lies from 60 to 100 miles beneath the crust.

5. The uppermost mantle, above the asthenosphere, plus all of the crust forms a rigid unit that is from 60 to 100 miles thick and is termed the lithosphere. The lithosphere appears to float in buoyant equilibrium in the asthenosphere, so that thick, low density granitic crust stands high to make continents and thin, high density basaltic crust stands low to make ocean basins. The principle governing this equilibrium is termed isostasy.

6. The lithosphere is subdivided by narrow seismic belts into a dozen or more plates. The high seismicity associated with the plate margins indicates that the plates are in continuous horizontal motion, colliding, stretching and tearing as well as grinding past one another along their margins.

7. The outer core abruptly slows P waves and stops S waves, creating two shadow zones on the earth's surface where these waves are not received. Rotation of the earth creates eddy currents in the liquid nickel–iron alloy that constitutes the outer core, causing it to generate the earth's dipolar magnetic field. The inner core, of the same composition, is largely solid.

8. The pattern of earthquake epicenters on the surface and hypocenters at depth outlines the oceanic ridge, Mediterranean–Himalayan, and Pacific Rim seismic belts. These narrow belts release most of the seismic energy created within the earth.

KEY WORDS

active margin	mantle
asthenosphere	Moho
basement	Mohorovičić discontinuity
continental crust	
core	oceanic crust
craton	oceanic ridge
crust	orogenic belt
epicenter	orogeny
focus	passive margin
foredeep	plate
graben	reflection
	refraction
Gutenberg discontinuity	rift
hypocenter	seismology
isostasy, principle of	shadow zone
lithosphere	shield

EXERCISES

1. If the earth's core slowly became solid, what effects would that have on the earth?

2. If the continental crust (density = 2.6 gm/cc) and oceanic crust (density = 3.3 gm/cc) were the same thickness, would their upper surface have a common elevation? Why or why not?

3. Since the ratio of densities of continental crust to oceanic crust is approximately 4:5, what would happen to the elevation of a point on the

continent from which 1000 feet of rock was removed? What would happen to the elevation of the ocean floor if the 1000 feet of rock removed from the continent were placed on it?

4. Since oceanic rift zones tell us the lithosphere is being stretched and torn, what evidence would we look for to see whether the lithosphere elsewhere is being shortened by an equal amount? What evidence would we see if the earth's surface were gaining in surface area?

5. Scientists think that meteorites are fragments of an ancient planet that exploded in collision with other objects. Most meteorites have compositions similar to that proposed for our mantle and the remainder have compositions similar to that proposed for our core. What does this evidence from space tell us about the earth?

6. In terms of plates, continents are indifferent passengers in a moving plate. What determines whether a continental margin is active or passive?

7. The fact that all rotating astronomic bodies larger than a particular size are spherical tells us what about their strength?

8. The heat flow or heat loss through the surface of the earth tells us something of what's going on in the crust. Why should older and older orogenic belts release progressively less and less heat? Would you expect continental interiors to be areas of high or low heat flow? Why?

9. If the earth's overall vertical density stratification was reversed, that is, if the densest rocks were at the surface, how would that effect the current rate of rotation of the earth? (Hint: think about what happens to spinning ice skaters who pull their arms in toward their body.)

10. One model suggests that the earth's lower mantle is the source of lava for continental volcanics, whereas the upper mantle is the source for oceanic ridge eruptions. How would one test this hypothesis?

11. High-quality seismic reflection profiling by COCORP has suggested that the Moho is moved to high crustal levels by crustal extension in areas like the Basin and Range; how could such a flat, young Moho be formed?

12. A minority of geophysicists currently argue that slabs of oceanic crust penetrate the lower mantle to depths exceeding 750 miles (1200 kilometers). This speculative idea would require the lower and upper mantle to intermix by convection, making heat transfer to the crust more efficient. How could one test such a proposal?

13. If the convective flow of heat in the mantle is driven by heat emanating from the core, what can we make of recent studies that suggest the core–mantle boundary surface is irregular on the scale of 6 to 10 miles (10 to 16 kilometers)?

SUGGESTED READINGS

BOLT, B. A., 1973, The fine structure of the earth's interior, *Scientific American*, March 1973.

O'NION, R. K., et al., 1980, The chemical evolution of the earth's mantle, *Scientific American*, May 1980.

PARKER, E. N., 1983, Magnetic fields in the cosmos, *Scientific American*, August 1983.
► *Three examples of why Scientific American is unexcelled at explaining science—in this case the earth's interior.*

BRUSH, STEPHEN G., 1984, Inside the Earth, *Natural History*, February 1973.
► *Delightful research into the history of ideas. Wonderful reading.*

WESSEL, GREGORY R., 1986, *The geology of plate margins*, Geological Society of America Map and Chart Series, Map MC-59.
► *Well-illustrated introduction into the complexities of margin types.*

WYLLIE, P. J., 1976, *The Way the Earth Works*. New York: John Wiley & Sons, Inc., 288 p.
► *Delightful and authoritative. Science at its best.*

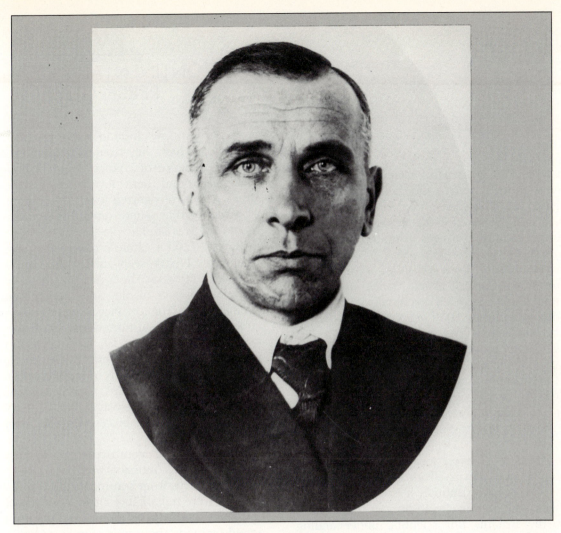

This photo of Alfred Wegener, taken sometime in the 1920s, captures the spirit of a scientist who looked for the forces that move mountains and continents. His ideas regarding continental displacement were the fertile ground from which the theories of plate tectonics emerged. (Photograph courtesy of United Press International.)

CHAPTER *ten*

Plate Tectonics

To see things, it is necessary to believe them possible.

M. BERTRAND (1891)

All great truths begin as blasphemies.

GEORGE BERNARD SHAW (1856–1950)

New ideas are the irritants of science. Because science is very much a human endeavor, new concepts challenge comfortable ideas. Progress, however, can come only from challenging the prevailing ideas. Scientific understanding proceeds in uneven, often bitterly contested steps through time.

Understandings of the ways things are may last a long time. When Copernicus and Galileo proposed a sun-centered universe, they eventually displaced a model that was nearly 2000 years old, and thereby upset the intellectual and religious worlds. When Hutton and Lyell proposed that the world was cyclic, uniform, and ancient beyond the chronologies of Moses, the catastrophists and neptunists fought a losing battle for nearly two centuries. Some of that sense of loss is still reflected in modern day creationism, an idea that lost its empirical foundation centuries ago.

There is a time for every idea. Even a correct but revolutionary view that is advanced before the prevailing world view can support it is doomed to scathing attack and attempts by authorities to ignore or suppress the irritant. But scientific understanding grows only as old ideas are displaced by more fruitful ones. Each new model may explain a series of known facts more acceptably than the one preceding it and forecast phenomena yet to be discovered.

This chapter is concerned with two models about the possibility of horizontal and vertical motion of continents and ocean floor. The earliest model is termed *continental drift*. It and its ancestors can be traced back to Sir Francis Bacon (1561–1626). Bitterly resisted or ignored for three centuries, the idea was replaced in this century with the concept of *plate tectonics*. At first plate tectonics was regarded as even more a heresy than continental drift, but within ten years of its proposal it became the orthodox view, and remains so.

Although this chapter may, at first, seem to deal with plate tectonics, the central explanatory theory of geology, it is also a study of scientific ideas. It looks at how ideas gain acceptance, where they come from, and the intensely human story of some of the scientists who spawned the ideas and others who spurned them.

CONTINENTAL DRIFT—A PERSPECTIVE

From earliest times human beings regarded the continents on which they lived as permanent and fixed, yet the evidence of change could be found in the rocks all around them. Five centuries before Christ, Greek writers recognized fossils in sedimentary rocks as evidence that where once there was marine life there now was land.

In A.D. 1200 a Chinese observer wrote: "In high mountains I have seen shells. They are sometimes embedded in rocks. The rocks must have been earthy materials in days of old, and the shells must have lived in water. The low places are now elevated high, and the soft material turned into hard stone." Leonardo da Vinci (1452–1519) speculated on the antiquity of the earth two centuries before Hutton and correctly interpreted fossils in sedimentary rocks now lifted into high mountains. Clearly land and sea did not remain fixed. Some way had to be found for land and sea to exchange places vertically from time to time.

By the late nineteenth century still another puzzling aspect of continents needed to be explained. Fossil life forms from continents far distant from one another were remarkably alike. Why was there such constancy among fossil animal and plant species separated by many thousands of miles of ocean?

THE CONTRACTION MODEL

The predecessor to the idea of continental drift was termed the *contraction hypothesis*. This model was the orthodox model for geology in the late nineteenth and early twentieth centuries. It attempted to explain the rise of seafloor to become continent by proposing that *the early earth had been molten and was cooling and contracting, losing volume.* Because of the contraction, the "skin" of the earth wrinkled upward to form mountains, much as wrinkles appear on the skin of an old dried-up apple.

Critical to understanding this model was the assumption that the outer layer of the earth was everywhere identical. There was no difference between seafloor and continent. The ocean basins were simply parts of the earth's surface that had sunk low along gigantic fault zones while the continents rose high along the same faults. If the ocean basins sank still more deeply, water would drain off the continents, and former seafloor would now be exposed as *land*

FIGURE 10–1
Born in Belfast, William Thomson (Lord Kelvin) was educated at the University of Glasgow, where he later received a faculty appointment and became Chancellor. Lord Kelvin's work on the earth's mantle proved how fruitful an error can be in establishing truth. (Painting by Frank McKelvey, courtesy of the Ulster Museum, Belfast, Ireland, used with permission.)

bridges. Along these bridges animals would wander back and forth from continent to continent at will. In this way similar life forms would be maintained on all the continents. As sediment filled the ocean basin, the water level would rise again and conveniently cover the land bridges.

The contracting earth model was successful *in the light of the knowledge of the time*. The ocean basins were essentially unexplored, and the earth was obviously losing internal heat in the form of volcanism. The earth was apparently cooling, and therefore the earth's skin was shrinking.

The assumption that the earth had originally been molten was an old one; calculations by Irish mathematician and physicist William Thomson, better known as Lord Kelvin (1824–1907) (Figure 10–1), attempted to assess *how long* the earth had been cooling as a means of establishing the age of the earth.

Kelvin had *assumed* a temperature for the "original" earth and then *assumed* a probable cooling rate; his calculations proposed a date for the formation of the originally molten earth that was so young (no more than 100 million years) that it outraged uniformitarian geologists. Kelvin, one of the most prestigious scientists of his day, replied "If you cannot measure, your knowledge is meagre and unsatisfactory!" His reputation and great strings of numbers supported the contraction hypothesis, for a time.

It remained for the English physicist Lord Rayleigh (1842–1919), the co-discoverer of argon, to recognize that uranium, present in small amounts in many rocks within the earth, provided a *continuing* source of heat. The earth, with its own internal heat source, was not simply cooling from a fiery beginning to become a cold clinker. The discovery of radioactive heat production alone seemed to knock Kelvin's strings of assumptions and formulae into the wastebasket of discarded ideas. However, Kelvin's model of a juvenile, steadily cooling earth was not only flawed in ignoring radioactivity, but it also contained an additional erroneous premise.

We now know that the earth (see Chapter Nine and discussion of convection later in this chapter) is dominantly a *convecting* body, more fluid and cooling much faster than Kelvin could imagine. Instead, he had assumed the earth was solid and lost its internal heat entirely by *conduction*, a far less efficient process of heat transfer within rock. The discovery of radioactivity did not solve the problem of the heat flow-age paradox, especially in the oceans, where only a small fraction of the heat production can be explained as heat produced by radioactivity. A convecting Earth is *required* to satisfy the observed relation between heat flow and age of the earth. The age proposed by Kelvin was the thermal time constant of the *lithosphere*, not the whole Earth. With the destruction of Kelvin's model of a solid, juvenile, conductive earth, the contraction model collapsed, though for decades there was little to replace it.

EARLY SPECULATION ABOUT CONTINENTAL DRIFT

The contraction model was slowly replaced by the far more controversial theory of continental drift. Briefly, the theory of **continental drift** suggested that all modern continents came from the breakup, millions of years before, of a gigantic supercontinent, whose pieces drifted through the ocean basins to their current locations like ships drifting through warm tar. The continental shapes in map view reminded a few insightful observers of pieces of some ancient jigsaw puzzle, separated, and then strewn across the surface of the earth by some unknown force.

The first person to note the mirror-image shapes of the shorelines of all continents bordering the Atlantic was Sir Francis Bacon. In 1620 Bacon noted the congruent Atlantic shorelines on one of the first reasonably accurate maps of the world. Bacon did not, however, speculate on how these shapes might have originated.

The first person to have suggested a breakup of the Americas from Europe and Africa was Antonio Snider-Pelligrini in 1858. He developed his concept just at the time when catastrophism was under severe attack, and the breakup and dispersal of a supercontinent, as he proposed it, was certainly a catastrophic idea. His contribution was quickly ignored.

In 1893 Eduard Suess (1831–1914), an Austrian geologist with a keen interest in the origin of mountains, suggested that mountains were formed from the crumpling of the earth's surface as crustal material moved away from poles that were flattening. Suess had thereby provided a plausible mechanical force to drive continents horizontally. In 1907 Bailey Willis, a distinguished British geologist, proposed that the mountains of Asia, including the Himalayas, were produced when the Pacific and Indian ocean basins underwent "northward spreading or underthrusting of the deep-seated mass beneath Asia."

In 1910 Frank B. Taylor, an American geologist,

published a lengthy paper in the *Bulletin of the Geological Society of America* on "The Origin of the Earth's Plan". In it he laid out the whole idea of continental drift, describing Greenland as recently rifted from Canada, and the Mid-Atlantic Ridge as "a line of parting or rifting—the earth crust having moved away from it on both sides. . . the two continents on opposite sides of it have crept away in nearly parallel and opposite directions."

Taylor's "crustal creep" was continental drift. His 1910 paper laid out the mechanisms, most of the evidence, and a general statement of how his theory explained the opening and closing of ocean basins and the collision of continents. Taylor's paper was the first clear statement of continental drift, and it fell into the hands of continental drift's greatest exponent, Alfred Wegener.

WEGENER AND CONTINENTAL DRIFT

Alfred Lothar Wegener (pronounced vay'-guh-ner) (1880–1930) was a German meteorologist (see chapter frontispiece) who spent the last 20 years of his short life marshaling the evidence for and advocating the idea of continental drift. He may have read the

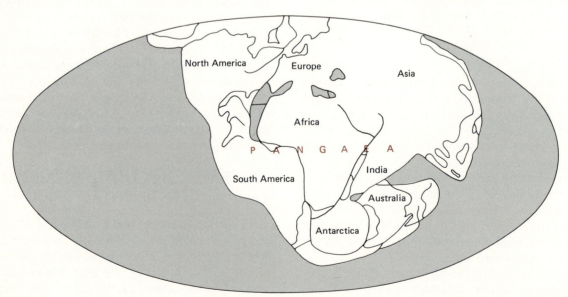

FIGURE 10–2

The ancient supercontinent of Pangea incorporated all the world's landmass, including the continental shelves. Wegener's reconstruction, slightly modified here from that

shown in the 1929 edition of his book has been challenged by the intervening 60 years of research, but it is still generally correct. (Courtesy of Dover Publications, Inc.)

earlier work by Suess, and most certainly he read Frank Taylor's 1910 article, having referenced it in his earliest brief statement on continental drift. More than anyone else, Wegener's name is associated with this once extremely controversial idea.

Though others preceded him in stating the idea, Wegener's fame comes from his dogged determination to advance the concept and to supply increasingly sophisticated supporting evidence. Starting in 1912 Wegener published his ideas in journals and four successive editions of his famous book, *The Origin of Continents and Ocean Basins*. The last edition was published in 1929, the year before his untimely death in an accident while doing research in Greenland. More than any other person, Wegener made the revolutionary idea of continental drift the common property of educated people.

Wegener proposed that all the world's landmasses must once have been grouped together as a single supercontinent to which he gave the name **Pangea**, (also spelled Pangaea) from two Greek words *Pan* meaning ''all'' and *gaea* meaning ''land.'' In 1929 Wegener estimated that Pangea had broken apart 30 million years ago (we now know that that estimate is off by 200 million years), and proposed that the continents have drifted to their current locations in the intervening time. Figure 10–2 is a slightly modified version of a sketch taken from the last edition of Wegener's book showing the continental reconstruction into Pangea (''all land'') that he proposed.

Wegener pointed to five main lines of evidence supporting continental drift:

1. Detection of *modern* continental motion by repeated surveys.
2. Congruence of the Atlantic shorelines and correspondence in geologic history of predrift rocks in the continents now separated by the Atlantic Ocean.
3. Correspondence of fossils in predrift rocks in now separated continents.
4. Similar distribution of ancient climates as revealed by predrift rocks.
5. Isostatic equilibrium of continents and ocean basins, leading to a new view of the differences between continent and ocean floor.

We now know that Wegener's *mechanisms* for continental motion were totally wrong, but continents *do* move, just as Wegener said they must. We now examine each of these five points in some detail. They provide a splendid example of how a controversial idea is proposed, defended, and modified.

MODERN CONTINENTAL SEPARATION

If repeated surveying were to show that the latitude and longitude of the same spot were changing systematically through time, the horizontal motion of a continent would be proved. By inference, past horizontal motion of the continent would have to be accepted.

Wegener presented the results of repeated surveys showing that Greenland was moving in a predictable direction through time. His critics attacked the accuracy of the surveys, and in hindsight, they were correct. Long-distance surveying was still a primitive art at the turn of the century, and Wegener's data have been shown to reflect the systematic errors made in these early surveys.

Very accurate data on the latitude and longitude of a point through a long period of time *would* prove or disprove the idea that continents are in horizontal motion. Such data are now being gathered by orbiting satellites, and in their preliminary form seem to indicate clearly horizontal continental motion. Laser ranging (the transmission and reflection of laser light by a specially designed satellite back to fixed points on earth) to and from *Lageos,* the Laser Geodynamics Satellite, has shown that six of seven measured points are moving in the direction currently predicted by plate tectonics models. Had the marvels of our technology been available to Wegener, he would have had one piece of data very difficult to argue against. Instead, the technology of his time did not allow the precision required.

CONGRUENT ATLANTIC SHORELINES AND GEOLOGIC HISTORY

Wegener pointed to the jigsaw-puzzle match of the Atlantic coastlines, suggesting that they reminded him ''of the pieces of a cracked ice floe in water. The edges of these. . . blocks are even today strikingly congruent.''

Critics suggested that the mirror-image shoreline ''fit'' was accidental and dismissed his evidence. Today, computer estimates of the probability that such congruent shapes originated by accident run one billion to one. Thus the shoreline fit is a strong link in the overwhelming evidence that Europe, Africa, and the Americas separated ''like pieces of a cracked ice floe in water.''

Geologic mapping of western Europe, western Africa, and the eastern coastlines of North and South America had revealed to Wegener startling similarities in predrift histories. Restored to predrift positions, our Appalachian Mountains align perfectly with the Caledonian ranges of western Europe and Greenland and record a similar predrift history in detail. As another example of similarity, Alexander du Toit, a South African geologist who was a contemporary of Wegener, once remarked that while standing in eastern Brazil he had "great difficulty in realizing that this was another continent, and not some portion of one of the southern districts in the Cape of South Africa."

Documented cases of correspondence only met with more criticism. The similarities were not as great as alleged; the patterns were far more complex than recognized, and even if they were similar, that did not imply that the continents were once joined—and on and on. Orthodoxy dies slowly, and Wegener was challenging fundamental beliefs in the permanence of continents.

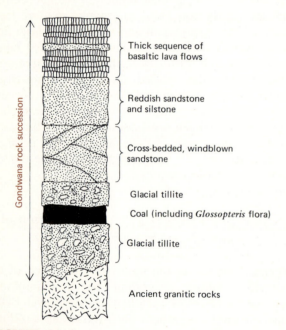

Thick sequence of basaltic lava flows

Reddish sandstone and silstone

Cross-bedded, windblown sandstone

Glacial tillite

Coal (including *Glossopteris* flora)

Glacial tillite

Ancient granitic rocks

Gondwana rock succession

FIGURE 10–3

A highly schematic composite columnar section of Gondwana rocks. This distinctive vertical sequence of rocks—with glacial tillites on the bottom resting on much older rocks of many kinds, topped with sequences of unusual basaltic rocks—is found on the widely scattered continents of Antarctica, South America, southern Africa, India, and Australia. That such a nearly identical sequence of rocks should form during the same interval— 100 to 300 million years ago—on five widely separated continents is improbable; a more likely explanation is that the Gondwana strata formed on what was at the time a single continent, which later fragmented. The coals between the tillites indicate interglacial periods.

DISTRIBUTION OF FOSSILS

Wegener noted that the fauna and flora of separated continents were remarkably alike before the breakup of Pangea, but had become increasingly unalike since that time—just what one should expect if the breakup had really occurred. As noted, *paleontologists*, geologists who specialize in the study of fossil life, explained these same facts by the disappearance of alleged *land bridges*, which had connected continents in the past and had now conveniently foundered out of sight.

Wegener, arguing on isostatic principles, showed that land bridges could not be expected to sink into the seafloor, since the seafloor was denser than the land bridge! He took paleontologists and others to task for not keeping up with progress in other sciences—an even greater problem today. Wegener, whose formal training was in astronomy, whose profession was meteorology, and whose lasting contributions have been largely in geology was in the perfect position to ask scientists to listen to one another.

The paleontologists did not listen, however. As Wegener quoted one paleontologist: "It is not my job to worry about geophysical processes." Paleontology did not mix well with geophysics and isostasy, and land bridges remained as solutions for paleontologists long after Wegener had shown it was physically impossible to make them sink and disappear. Wegener's faunal evidence was shrugged off by paleontologists as being inconclusive.

THE ROCK RECORD OF ANCIENT CLIMATES

Among the unchanging facts about our earth is the general latitudinal distribution of sunshine and overall climate. High latitudes always mean colder climates, lengthy periods of total darkness lasting for months, with animals and plants adapted to this colder climate. Moderate latitudes always mean moderate climates, except for the abundant mid-latitude deserts; four definite seasons; unequal length of day and night, changing with the seasons, but with some sunlight year around; and plant and animal life adapted to these more moderate climates. The lowest latitudes always mean tropical climates, equal length of day and night, and a plant and animal community adapted to tropical climates.

Wegener recognized that the climate at the time the rocks were forming is faithfully recorded in the rocks and the fossil life they contain. IF a continent had been moved north or south for a great distance, its older rocks would have recorded a climate largely unlike that it currently enjoyed. This older rock record would prove that continents had moved in the past.

The Gondwana Strata

Working primarily with rocks from the southern continents and from rock *that predated Pangean breakup*, Wegener was able to identify a group called the **Gondwana strata** (Figure 10–3). These rocks, now scattered across India, Australia, Antarctica, southern Africa, the Arabian Peninsula, and parts of South America *record a similar geologic history for a period of about 200 million years.*

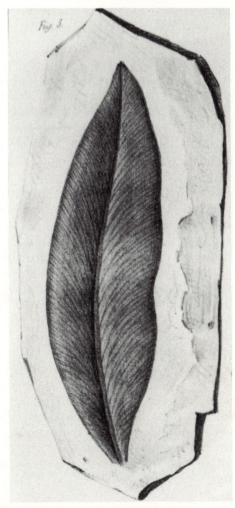

FIGURE 10–4

This leaf fossil from the seed fern *Glossopteris* is one bit of evidence for the former existence of Gondwanaland, on whose land mass it was a common plant. (From Alexandre Brogniart, 1857, courtesy of the History of Science Collections, University of Oklahoma Libraries.)

The Gondwana strata contain a record of intense glaciation. These glacial deposits, a highly unusual group of sedimentary rocks known as *tillites*, indicate a period of major glaciation about 300 million years ago. But what are glacial deposits doing in continents that are now close to the equator? To Wegener, the answer was simple. At the time of glaciation, 300 million years ago, these southern continents *clustered around the south pole.*

Interspersed in the middle of the glacial deposits is a distinctive series of coals that seemed to have formed in interglacial periods, much like the period of time in which we live today. Within these coals are abundant plant fossils—termed the *Glossopteris flora.* The seeds of *Glossopteris,* (Figure 10–4) a seed fern, are found as fossils in the six continents and subcontinents that are now separated by many thousands of miles of ocean. How could seeds be spread across the distances between each of the continents? Such distances are so great that wind dispersal seems out of the question. Wegener postulated, instead, that at the time of coal formation all six of the southern continents and subcontinents were joined.

The *Glossopteris* flora present another challenge. How could great huge trees and seed ferns have survived *if* the southern continental grouping called **Gondwanaland** had been at the south pole? How could coal, that most tropical of all sedimentary rocks, have formed near the south pole? Clearly the formation of coal swamps and tropical seed ferns are incompatible with six months of polar darkness and bitter cold.

The *Glossopteris* flora appear to record a brief movement of the southern continents north to a more temperate climate. During this interval of time, coals formed, along with fossils of seed ferns and cold-blooded reptiles. Indeed fossils of seed ferns and cold-blooded reptiles were found in the 1960's within a few hundred miles of the south pole. Wegener reasoned that clearly Antarctica has not always been buried under ice; millions of years ago its location and its resultant climate were subtropical.

Other Environmental Reconstructions

In much the same way, Wegener was able to locate abundant evidence of past deserts forming during Gondwana time in continental areas that currently have cool temperate climates. He was even able to summarize the information and draw in the location of the ancient equator by reconstructing past climates in a normal order by latitude (Figure 10–5). Wegener felt that *surely* no one could argue against ancient glacial deposits in continents now adjacent to the

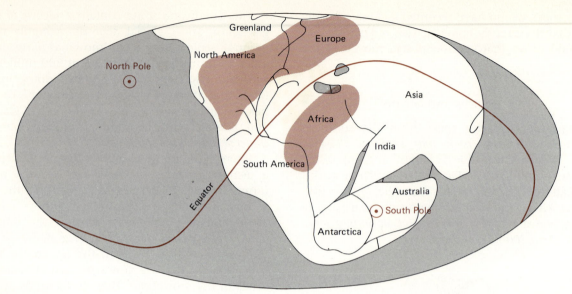

FIGURE 10–5

In the colored areas, the rock types suggest that the climate at the time of their deposition was a desert climate. Deserts on the modern earth are generally at about 25° north and south latitude, a situation that depends only on the position of the earth's rotational axis with respect to the sun. Along the equator, whose position was inferred by Wegener, coals were common, as are swamps in tropical regions today. Near the only pole that intercepted landmasses in the southern continental group, evidence of glaciation in the southern continental group is widespread. (From *The Origin of Continents and Oceans, 1929, courtesy of Dover Publications, Inc.*)

equator or ancient coals, seed ferns, and cold-blooded reptile fossils found in the bitter cold of the Antarctic.

The critics, as usual, were less than kind. The matching of glacial features after more than 200 million years of erosion was ridiculed. The environmental reconstructions were ignored. Professor R. T. Chamberlain (1843–1927), well-known American geologist and glaciologist, remarked that "Wegener's hypothesis in general is of the footloose type. . . and is less bound by restrictions or tied down by awkward, ugly facts than most of its rival theories."

ISOSTATIC EQUILIBRIUM

In the face of his many critics, Wegener proposed still another test of his theory, and to Wegener the test was decisive. Wegener noted that when a random sampling of the elevations of points on the globe were plotted by frequency of occurrence (Figure 10–6A, B) *two maxima* were evident.

The most abundant higher elevation parallels the average elevation of the surface of the continental crust at 300 feet (92 meters) above sea level. The most abundant low elevation parallels the average elevation of oceanic floor at 15,000 feet (4600 meters) below sea level. Other elevations on the earth are *far* less common. As Wegener phrased it "For

uplift and subsidence . . . we know only one law: the greater, the rarer."

As Wegener pointed out, the assumption held by most geologists at the time—that the earth's crust was a *uniform* layer moved *randomly* up or down to make continents or ocean basins—was totally *inconsistent* with the well-known data on distribution of elevations on earth. If a common elevation were to be moved randomly up or down, the distribution pattern would be the *colored* line in Figure 10–6A. This distribution is termed a *Gaussian* or *statistically normal* distribution, one that would result from a truly random subsidence and elevation.

To Wegener, the actual distribution of elevations on earth meant that *continents were fundamentally different than oceans.* Continents must be made of less dense stuff and stand higher, whereas ocean basins must be made of higher density stuff, and therefore sink lower. In Wegener's own words, ". . . the two layers behave like open water [analogous to the ocean floor] and large ice floes [analogous to the continents]."

Now Wegener presents the final argument. Here it is in Wegener's own words:

". . . the whole isostasy theory depends on the idea that the crustal underlayer has a certain degree of fluidity . . . if this is so and the continental blocks really do float on a fluid, even though a very viscous

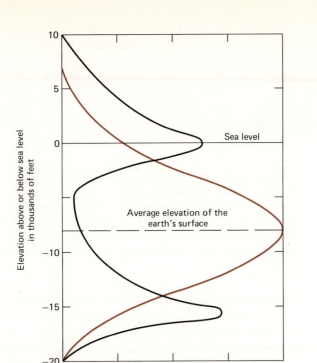

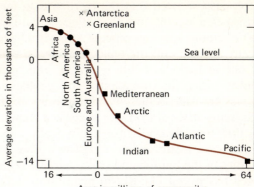

(B) *Here is another way of looking at Wegener's old information. This chart compares average elevation of each continent with its surface area, which increases in both directions from zero on the horizontal axis. The greater the area of continent, the greater its average elevation. The larger the ocean basin, the lower its elevation. Only Greenland and Antarctica are unusual. Their abnormal elevations are due to the enormous mass of glacial ice on them. If their ice were removed, their elevations would be appropriate for their area. (Adapted with permission from The Science Teacher, September 1963, with the courtesy of Dr. Ned Ostenso.)*

FIGURE 10–6

(A) If the topographical features of the earth's surface were formed by uplift and submersion of *identical* continental and oceanic crust, a random distribution of elevations centered about an average—*the pattern shown by the colored line—would be expected, and continents would be much lower and smaller than they actually are. In fact, two levels predominate: the continental platforms just above sea level and the ocean floor, 15,000 feet (4600 meters) below sea level. (Adapted from Alfred Wegener, The Origin of Continents and Oceans, 1929, courtesy of Dover Publications, Inc.)*

one, there is clearly no reason why their movement should only occur vertically and not also horizontally, provided only that there are forces in existence which tend to displace continents. . . . That such forces do actually exist is proved by the orogenic compressions. . . . The forces which displace continents are the same as those which produce great fold-mountain ranges. . . . Continental drift, faults and compressions, earthquakes, volcanicity . . . are undoubtedly connected causally on a grand scale.''

Wegener had built his argument step by step. The distortion of the spherical earth, with its polar flattening and equatorial bulge was *exactly* what would be expected from a very viscous stiff fluid rotating at the earth's known speed. The distribution of elevations on earth showed that the continents and oceans float in a viscous fluid beneath, with each block reaching the overall elevation appropriate to its

density and size (Figure 10–6B). This indicated that the continents could move up and down when they were loaded or unloaded with rock, supported by the viscous fluid beneath them, much like a boat.

Since the continents could move up and down within a fluid, driven by vertical forces, they could clearly move sideways, plowing through that same fluid—*if* it could be demonstrated that there are natural forces in a horizontal direction. The existence of a force to make them move is obvious, for major mountain chains reflect the application of large amounts of horizontal force to the crust.

THE COUNTERATTACK

Sir Harold Jeffreys (1891–1984), an English astronomer and geophysicist, attacked Wegener precisely where his ideas were weakest. Wegener had been unable to provide a satisfactory mechanism to move the continents *across the earth's surface through the mantle.* Jeffreys showed with the mathematical certainties of physics that the mechanisms tentatively proposed by Wegener to induce continental motion were too weak to overcome the rigidity and strength of the mantle. Though little was actually known about the mantle, it was *assumed* in Jeffrey's time to be as strong and rigid as high-grade steel.

Jeffrey also showed that the earth was in far less perfect isostatic equilibrium than Wegener would admit. The whole driving process was dismissed as physically absurd, and the biologic evidence was rejected out of hand as being imprecise and open to numerous other interpretations. The geologic evidence was set aside as biased and imprecise. In short, though many lines of evidence suggested that continental drift was plausible, it was physically impossible. Given the assumptions of Jeffrey's time about the rigidity of the mantle and crust, Jeffrey's criticism was well based, but like Kelvin before him, Jeffrey's *assumptions* about the earth would turn out to be wrong.

So great was Jeffreys' prestige that few asked the next obvious question. Many phenomena, including gravity, electricity, and magnetism, were accepted and used long before the mechanisms behind them were thoroughly understood. As we noted in Chapter One, data provide certainty without explanation, and theory provides explanation without certainty. Even if the mechanisms in the theory tentatively proposed by Wegener were not quite right, how *could* the data reported by Wegener be explained? For many decades after Wegener's death, there was no answer.

▼
THE SEARCH FOR A MECHANISM
▲

Before his death, Wegener saw a few responsible geologists around the world begin to accept his idea of continental drift, because it pulled together many otherwise poorly explained pieces of information. Continental drift explained many puzzling aspects of geology much more satisfactorily than the cumbersome collapsing land bridges, disappearing continents, and rhythmically rising and falling ocean floors that had been proposed. As more research on the ocean floors began, Wegener's conjectures on the distinctive nature of oceanic crust were strongly validated. However, a fundamental concern still remained. What *mechanism* could drive the continents through the mantle?

Among the people to come to an early acceptance of Wegener's ideas was Arthur Holmes (1890–1965), an English geologist whose specialty was the application of the newly recognized phenomena of radioactivity to geology. Holmes was deeply interested in whatever role the heat produced within the earth by radioactive decay might play in driving earth processes.

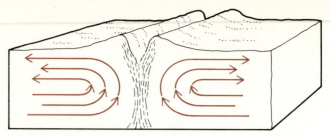

FIGURE 10–7
Colored lines represent the flow of heat as it rises from the hotter mantle beneath an oceanic ridge - rift system. As hot material rises and flows away, the neat flow from the inner earth to the surface is increased, and the surface is placed under tension. The patterns of heat flow by convection in a pan of water being heated on a stove are similar.

It had long been known that hot, low-density fluids transfer their heat to colder, high-density fluids by a process known as *convection* (Figure 10–7). **Convection** is the transfer of heat by the upwelling of hotter, lower-density fluid replacing colder, denser fluid, which sinks. Physicists had worked out the physics of heat transfer by convection long before. Holmes's proposal was simple. The mantle was being heated by the radioactive decay of some of its elemental constituents (primarily potassium, uranium, and thorium), causing the heated rock to expand, and thereby lowering its density. The result was a *convection cell,* in which heat was spread by rising and falling currents of fluid (Figure 10–7).

The convective circulation, once begun, would cause the rise and spread of hot mantle material. As this material struck the thin ocean floors, it might crack and break them, causing mantle material to erupt through the surface (Figure 10–7). Moving underneath the continents, convection currents might be just the force needed to push them sideways and initiate their drift.

Significantly, Holmes had to await far more information on the nature of the world's ocean floor before his ideas could be accepted or rejected. It remained for the oceanographic research of the 1950s and 1960s to verify that Holmes's speculations on *convection currents as the driving force for horizontal continental motion were one reasonable explanation.*

Slowly researchers began to recognize that long ridges across the ocean floor were sites of abnormally high heat loss, volcanism, and shallow-focus earthquakes. The ocean floors turned out to be far more complex than previously imagined, with long linear fault systems cutting across the ridges. Data from the ocean floor revealed extraordinary changes in the earth's past magnetic field; similar data from continental rocks firmly established past horizontal continental motion as fact. Research directed toward a better understanding of the earth's magnetic field led in the 1960s to a revolution in understanding the earth's dynamism.

▼

THE EARTH'S MAGNETIC FIELD

▲

Characteristics of the earth's magnetic field suggest that at least 95 percent of it is created in some manner within the earth. That discovery was made in 1835 by J. K. F. (Karl) Gauss (1777–1855), an extraordinarily gifted German mathematician who set up the world's first observatory designed to study the earth's magnetic field. He was also the first to calculate mathematically the location of the magnetic poles and to create units for the measurement of magnetic phenomena. The units of magnetic intensity he created were eventually named the *gauss;* the earth's magnetic field is about 0.3 gauss unit.

Rotation of the entire earth plus core convection currents apparently combine to cause sluggish flow within the earth's outer core. The flow of metallic materials sets up a weak electrical current in the conductive liquid metal. This flow of conductive metal, much like electrical current flowing through a wire, in turn creates a magnetic field. Most of this field is trapped within the core, but a very small amount of it exits the crust and can be measured hundreds or thousands of miles into space. The core of the earth, about a third of the earth's mass, acts as an extremely large electrical and magnetic field generator or dynamo. *The magnetic field produced, however, is two hundred times weaker than that of a child's toy magnet.* Once created, such a magnetic field is self-sustaining, as long as the slow motion within the outer core continues.

The rotating earth is its own dynamo, creating an electrical current that generates its own magnetic field. The field is *dipolar,* that is, it is formed in a line between the north and south poles, and it has a definite polarity aligned along an axis approximately parallel to the earth's rotational axis (Figure 10–8A). The "plus" or "positive" end of the field is nearly coincident with the earth's rotational pole near the north pole, and the "negative" end is near the earth's south pole. We observe this polarity when we use a magnetic compass, for one end of the needle always points northerly, i.e., toward the "plus" pole of the field, which is nearly coincident with the north pole.

Because the outer core is an extremely stiff fluid, the rotational motion of the outer core does not keep pace with the rotation of the solid material above it. The result of this lag in motion is that the axis of the field as it exits the earth's crust drifts westward across the earth's surface at a rate of about 1° longitude every five years. The axis of the field wanders over

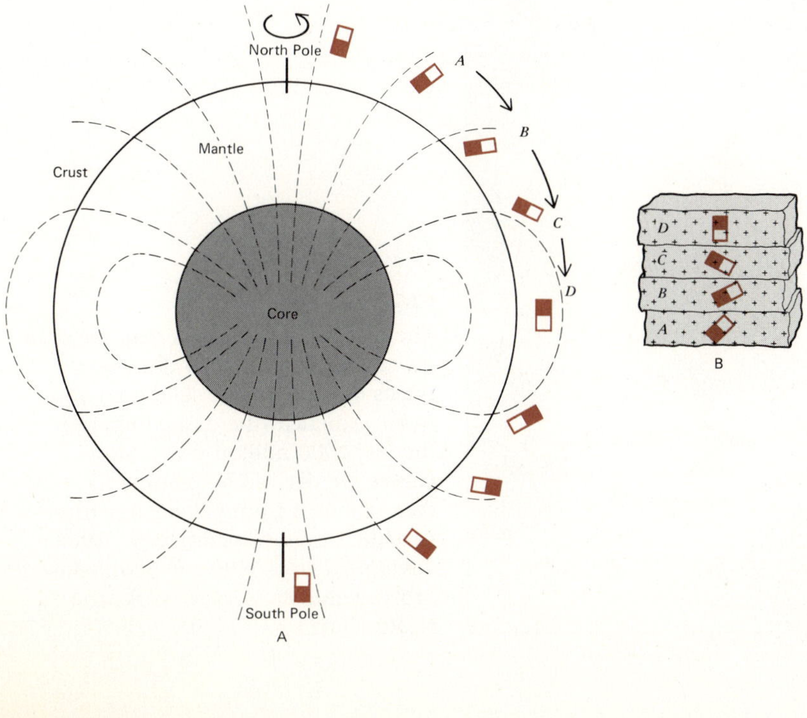

FIGURE 10–8

(A) A cross section of the earth showing the magnetic lines of force coming out of the core. Over any length of time the magnetic pole is coincident with the rotational pole, as is indicated. Shown arrayed around the earth are blocky crystals of magnetite, with their "north" or "plus" end colored. As each mineral orients itself with the magnetic field, its *orientation is a direct function of its* latitude. These crystals in southern latitudes point *upward,* whereas those in the north point *downward.* Those at the equator have no dip. (B) Imagine an area of the earth subject to occasional lava flows that *moves through time from points A to D.* In the stack of lavas shown in sketch B, the magnetite grains in each layer faithfully record the latitude they were in when they formed. The record of diminishing inclination downward shown by the vertical sequence of lava flows is a record of the *change in position* of the area through time from points A to D.

time in other directions as well, but it always maintains an *approximate* parallelism with the earth's axis of rotation. Thus the *orientation of the earth's magnetic field with respect to lines of latitude is essentially unchanging.*

PALEOMAGNETISM

Notice that in Figure 10–8A, grains of magnetic minerals orient themselves in cooling lava or a watery sediment *in a direction that is coincident with latitude north or south.* If we later recover the solid rock containing these magnetic minerals and carefully measure their orientation, *we will measure the ancient latitude* of the place where the rock formed.

If the rocks were never moved from their original location, such an exercise would be futile. We would simply recover the *original* latitude of the rock's original location and discover that it is the same as the rock's *current* location. *But, if the rock has been moved from the location where it formed* (Figure 10–8B), then the latitude recorded in the orientation of its magnetic minerals will be *different* than the latitude where the rock currently resides.

Paleomagnetism is the study of the orientation of the ancient, faint magnetic fields preserved in the oriented magnetic minerals in rocks. If rock samples only a few million years old are tested, their magnetic minerals point to the *modern* position of the

magnetic and rotational poles. This is additional evidence that magnetic and rotational poles have changed their orientation little over the last several millions of years.

Older rock samples tell a different story. The older a rock is, the greater the difference between its ancient latitude of formation and the latitude of its current location. If we then plot the *apparent* position of the north pole as constructed from the latitudes obtained from a series of rocks of varying age, we create an *apparent polar wandering curve* (Figure 10–9). Such a curve, a continuous plotting of the *apparent* position of the north pole through time, at first suggests that the position of the magnetic north pole has progressively changed.

Yet we know that the magnetic north pole has not experienced a *large* departure from its current position *near* the earth's rotational north pole for two reasons.

1. As just mentioned, we have positive evidence that the orientation of the axis of the earth's magnetic field has nearly paralleled the rotational axis of the earth for the last several million years.

2. Since the earth is an extremely large, rapidly rotating body, very much like an enormous top, its orientation in space cannot be changed. Therefore the orientation of the axis of both the magnetic field and rotation must have remained

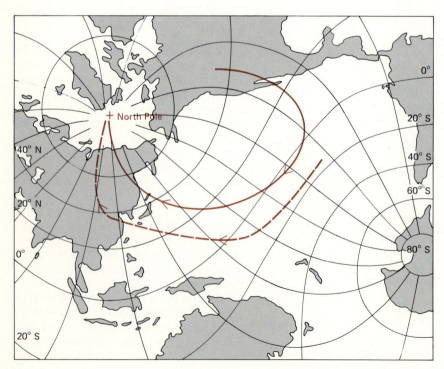

FIGURE 10–9

The colored lines on this map are *polar wandering curves,* which describe the tracks of Europe (the solid line) and North America (the dashed line) for the last 2000 million years. Such curves are derived by plotting location data obtained from paleomagnetism. (Adapted from Runcorn, S.K., 1962, *Continental Drift,* with the permission of Academic Press, New York, and S. K. Runcorn.)

parallel for all the earth's history of billions of years.

Yet we have data plots that *appear to show* that the earth's magnetic pole has wandered all over the earth's surface. Once again, data and explanation seem to conflict. How can the apparent polar wandering curves be explained?

Another way to view these curves is to say that the *continents have moved through time,* so that the points on the earth's surface where we took the rock samples have moved from the latitude where the rocks originally formed. If movement of the continents can be allowed, *the apparent polar wandering curve we see does not trace the movement of the poles in relation to fixed continents, but instead traces the movement of the continents in relation to a fixed pole location.*

Apparent polar wandering curves plot the movement of continents over the earth's surface, tracing the changing latitudinal information frozen into the orientation of magnetic minerals in the moving rocks (Figure 10–9). The sampling of the orientation of ancient magnetic fields in rocks of differing age is like playing back a movie of the history of continental motion through time. The results of thousands of paleomagnetic determinations are among the most convincing lines of evidence that continents have in the past moved horizontally for long distances.

FIELD REVERSAL

Anyone who has attempted to use an old map and a magnetic compass for navigational purposes knows how variable in orientation the earth's magnetic field is. Not only is the field orientation irregular from place to place at any one time, but the orientation of the magnetic north-south axis also drifts westward, making old maps increasingly inaccurate. In addition, the location of the magnetic pole also moves, looping around the position of the rotational pole in a period of perhaps 10,000 years. All the variation *in field orientation* can be accounted for by assuming that rotation of the molten fluid in the outer core is lagging behind the rate of rotation of the earth's surface by about 50 yards (meters) per day.

The *strength* of the field itself can also vary by half over a period of several thousand years. What could be causing a decline in the *strength* of the field? Nothing in the knowledge of geophysicists in the 1950s could account for the observation that the field strength is currently in a period of long decline, decreasing at the rate of less than 1 percent per decade. If the current rate of decline were to continue—and we do not know whether it will—the field strength would reach zero in 1200 years. Then what?

According to current theory, at that point the field would have about 1 chance in 20 of rebuilding its strength with a *polarity opposite that observed now, so that the ''north end'' of compass needles would point south!* Such a change in the polarity of the earth's magnetic field is termed a **field reversal**, or polarity reversal.

Field reversals naturally recorded in rock were first recognized by French physicist Bernard Brunhes, who in 1906 was working with volcanic rock from the Massif Central of France. His study showed that the rocks of various ages showed two basic polarities, each 180° from the other (Figure 10–10). Motonori Matuyama at Kyoto University found the same results in 1929 and concluded that the magnetic field of the earth reversed itself about a million years ago, a time coincident with the onset of major continental glaciation.

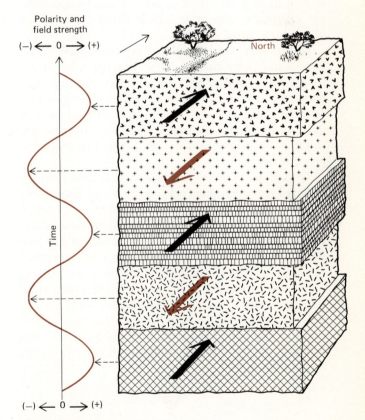

Polarity and
field strength

(−) ← 0 → (+)

North

Time

(−) ← 0 → (+)

FIGURE 10–10

If we carefully examine a vertical pile of lava flows, all deposited and cooled within the last few million years, we will find *field reversal* recorded in the magnetic minerals contained in the rocks. Each individual layer records the field polarity *at the time it cooled*. Such magnetic reversals are fairly common, though they occur *far* more randomly in time than shown.

As recorded in rocks, the earth's magnetic field appears to have reversed itself from time to time in the remote past. Such field reversals have occurred at intervals ranging from a few thousand to several million years, with the time between reversals averaging very roughly half a million years. A detailed record of reversals shows that they are completely random events.

Periods when the magnetic field is north-seeking, as it is today, are termed *normal epochs*. Periods of time when the magnetic field is south-seeking are termed *reversed epochs*. Within each of these larger epochs are still smaller periods of reversal (Figure 10–11), termed *events*.

By noting the patterns of events and epochs recorded in a series of rocks of unknown age and noting the apparent pole location produced by the inclined magnetic minerals within the rock, we can use the pattern of reversals and events to assign rock age. This technique is similar to using the unique pattern of tree rings from a tree of known age to figure the age of another tree whose rings demonstrate a part of the same unique pattern.

In this way, field reversals have been used to create a "magnetic clock." Now that the relation between patterns of reversals and known rock age have been correlated for millions of years, the data on field reversals have become an effective tool to provide dates of formation on rock sequences otherwise unsuitable for more conventional age determination processes.

One of the obvious places to look at field reversals in detail is the basaltic rocks of the ocean floor. Since basalt commonly contains *magnetite*, a mineral strongly oriented by the earth's magnetic field as lava cools, it is among the best rocks to study for this purpose. Using both airborne and seaborne *magnetometers*, devices which measure the intensity and polarity of the local magnetic field, various governmental research agencies began in the 1950s to study the details of the magnetic field recorded in the basaltic rocks of the ocean floor. The results were stunning and led to a wholly new understanding of the origin of the seafloor.

SEAFLOOR SPREADING

We know from observation that molten basalt is continuously pouring out of the ocean floor along the rifts in oceanic ridges. The volume is huge; *from two-thirds to three-quarters of all magma that escapes through*

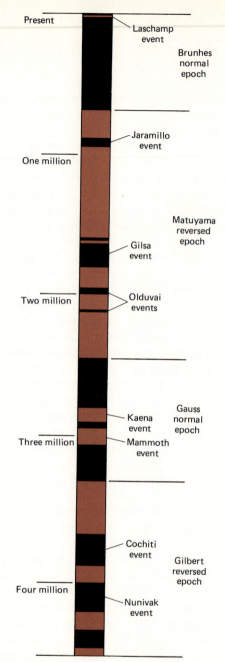

FIGURE 10–11

This chart records the known paleomagnetic events within the last 4 million years. The black stands for periods of normal (north-seeking) polarity, and the color for periods of reversed polarity (south-seeking). The timing of events is random, but field reversals averaged over a long period of time take place every 500,000 to 600,000 years. (Modified from National Science Foundation.)

the earth's surface each year comes from spreading centers. When that lava pours out on the seafloor it quickly cools, and magnetic domains within minerals in the lava align themselves with whatever the po-

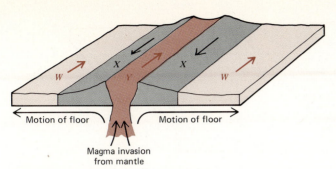

FIGURE 10–12

In this three-dimensional sketch of a part of an ocean ridge, modern extrusion of lava (*Y*) splits *X*, a slightly older flow, and shoves it aside. *X*, in turn, had once split the still older *W* flow which had occupied the rift zone. Layers *W* and *Y* record epochs when the polarity of the earth's field is normal, while *X* was intruded while the field was reversed. Note that each strip of rock successively farther from the ridge is *older, colder, and thinner* because of the volume loss on cooling. (Compare with Figure 10–22.)

larity of the magnetic field is at the time. When at some time the polarity of the field reverses, the newly formed lavas record the reversed field, and a boundary between rocks recording normal and reversed polarity is created.

When magnetometers were towed across the oceanic ridges perpendicular to their long axis, an incredibly symmetrical pattern (Figure 10–12) resulted. When the first pattern (the ELTANIN-19 profile) was displayed at a scientific meeting in 1966, it created an immediate sensation. Those involved in research in paleomagnetism realized that the perfectly symmetrical pattern of boundaries between normal and reversed basalt flows paralleled the ridge axis for thousands of miles, *and the linear bands of magnetized rock on one side of the ridge were the mirror image of those on the other side* (Figure 10–12). The variation in width of bands was equal on both sides of the ridge.

It was quickly recognized that these strange, perfectly symmetrical patterns were the product of a newly recognized process termed *seafloor spreading.* **Seafloor spreading** is a process that is now believed to create the entire ocean floor by *magmatic intrusion at rift zones in the ocean floor repetitively splitting and filling the rift.* It had been known for several decades that the ocean basins contained large ridges; the recognition that those ridges were commonly bisected by rifts, along which volcanism and dike injection were intermittent, came as a result of oceanic research in the 1950s and 1960s.

By intermittent intrusion of basaltic lava filling the rifts, the ocean floor grows wider by the welding on of one strip after another to the older rocks of the ocean floor. *The basaltic seafloor is believed to have been formed by repetitive magma intrusions that split and spread the older seafloor, causing the old seafloor to be moved away from the ridge in a nearly continuous horizontal spreading motion.* The oceanic ridges and their central rifts have been more recently termed **spreading centers,** because from these centers the ocean floor is poured out and split into a series of parallel stripes of rock of the same age and polarity. Each stripe is moved or spread horizontally from the central rift zone. Equal amounts of each new pulse of lava are added to both moving edges of the spreading rift.

As an example, imagine that layer *W* in Figure 10–12 was formed by intrusion of basalt along the central rift of an oceanic ridge. As it cooled, its magnetic minerals recorded the normal north-seeking polarity of the time. Layer *W* was intruded in time *W*, split and shoved or pulled aside in both directions perpendicular to the ridge. Then lavas and dikes erupted in time *X*, a period of south-seeking polarity. They filled the rift and recorded reversed (south-seeking) polarity. Still later, in time *Y*, new intrusions split *X* in half, shouldered or pulled it aside, and filled the rift with lavas and dikes which, recorded normal (north-seeking) polarity. As the process repeats itself time after time, the essentially continuous eruption of lava forms "stripes" of basaltic and gabbroic rocks parallel to the ridge, with each stripe recording the alternation of polarity through time. *The ocean floor lava eruptions turn out to be a gigantic tape recorder, recording the changes in polarity for the last 200 million years.*

TRANSFORMING THE OCEAN FLOOR

Further research on the events surrounding oceanic ridges established that the spreading centers are not continuous. Instead, the ridge and central rift system is cut by *numerous* faults whose trend is perpendicular to the long axis of the ridge. *These faults connect the axes of offset spreading centers and thus transform the spreading motion between the two ridge segments into a sliding motion along the fault.*

J. Tuzo Wilson, a well-known Canadian geologist, suggested that such faults formed a new class of fractures unique to spreading centers. He termed these **transform faults,** since they *transform* the continuous spreading of lava from eruptive centers into sliding motion perpendicular to the ridge axis. Transform faults are unique in that earthquakes occur *only* along the parts of the fault (area *T* in Figure 10–13) where sea floors are spreading in opposite directions. In areas *A* and *B* of Figure 10–13, the sea floors are

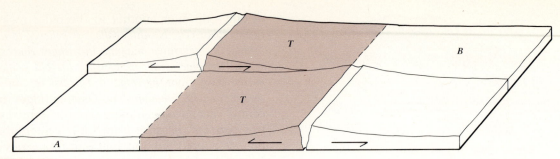

FIGURE 10–13

This sketch shows a portion of an oceanic ridge, offset and connected by a transform fault. As the seafloor is spread away from the two central rift zones, *only* in areas *T*—shaded in color—*are the two areas of ocean floor sliding in opposite directions.* Along such areas, the grinding of rocks sliding in opposite directions creates abundant shallow-focus earthquakes. In the *unshaded* areas *A* and *B*

of the diagram, *sea floor on both sides of the fault are sliding in the same direction at the same speed.* In areas *A* and *B* the fault is passive, and there are no earthquakes. Prove this to yourself by adding arrows to points *A* and *B*; compare the movement directions you draw to those of the adjacent seafloor across the fault.

moving in the *same* directions at similar speeds, so that relative motion along these *fracture zones* does not occur.

The existence of transform faults had been predicted by the Swiss mathematician Leonhard Euler (1707–1783), who had shown two centuries earlier that if the surface of a sphere were set into sliding motion, transform faults would have to arise to conserve surface area and separate zones that spread at different velocities. Transform faults serve this exact function, *since surface area is neither created nor destroyed along a transform fault.*

The driving forces for seafloor spreading remain somewhat speculative. Since it is increasingly clear that *all dynamic earth processes are driven by internal differences in heat energy and density,* it is reasonable to assume that seafloor spreading is the result of convection currents in the mantle (see Figure 10–7). A search for a more detailed understanding continues,

but it is clear that the forces driving the ocean floors away from the ridge systems that created them must, in some way, be the surface expression of the earth's vast reservoir of heat, both conducted and convected from deeper sources (Figure 10–14).

MODERN EXAMPLES

Better understandings of seafloor spreading may come from examining three areas on earth where it is currently in operation and easily visible. The requirement for visibility limits our search to areas where spreading centers interfere with continents. Three areas that meet these criteria are.

1. the East African Rift Zone (Figure 10–15),
2. the Red Sea between northeastern Africa and the Arabian Peninsula,

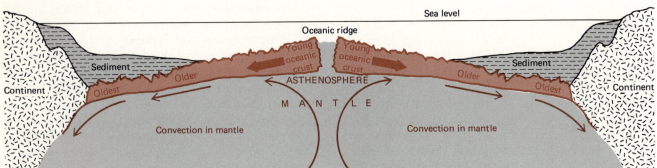

FIGURE 10–14

Seafloor spreading creates basaltic ocean floor from erupting mantle-derived material from the oceanic ridge, which is then conveyed sideways by convection currents in the mantle to become basaltic ocean floor, a process

known as seafloor spreading. The lithosphere is thinnest over the oceanic ridge and thickest under the continent. As seafloor cools it grows progressively older, denser, and thinner, and is moved farther from the ridge.

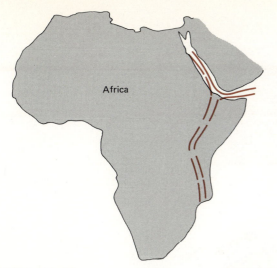

FIGURE 10–15

The East African Rift Zone is one arm of a three-legged rift or *aulacogen* between eastern Africa, the Indian Ocean, and the Arabian peninsula. Owing to crustal tension, the area has abundant earthquakes, volcanism, hot springs, and high heat flow over stretched and thinning crust.

3. the San Andreas Fault zone of southwestern California.

The East African Rift Zone

A continent rifts when a spreading center moves under it and begins to pull it apart. The signs of incipient spreading are much elevated heat flow accompanied by stretching and thinning of the crust (Figure 10–16A). Volcanism begins to occur and earthquakes become locally very abundant.

The East African Rift Zone (Figure 10–15) appears to be such a place. East Africa is breaking up along a line marked by long, linear lakes such as Lakes Nyasa, Tanganyika, and Victoria; boiling hot springs; towering volcanoes, such as Mount Kilimanjaro and Mount Kenya; and frequent earthquakes. The seismic belt follows the rift zone all the way north to where it joins the Red Sea.

Cross sections across the African rift zone and the Mid-Atlantic oceanic ridge (Figure 10–17) reveal how alike these two spreading centers are. The Atlantic spreading center is at least 200 million years old and has created the entire Atlantic Ocean basin in that time. The spreading center under Lakes Nyasa and Tanganyika has been operating for perhaps one-tenth of that time and is still in the process of fragmenting the continent above it.

As noted in Figure 10–15, the East African Rift Zone is actually one "arm" of a "three-legged rift zone" termed an *aulacogen* that is affecting all of eastern Africa. The East African Rift Zone is the youngest "arm" in a triangular system that first interfered with the African continent and began spreading it apart about 20 million years ago. It separated the Arabian Peninsula from Africa and created the Red Sea and the Gulf of Aden. If current spreading rates and orientations continue, East Africa will split from Africa and become an island, much as Madagascar has already done.

The Red Sea

The Red Sea appears to represent the second stage in continental separation, in which the spreading center breaks the thinned crust. A wedge-shaped block drops downward as the crust on either side of it moves outward. Volcanism occurs along the fault zones pouring lava into the collapsed graben (see Chapter Seven). The downdropped area is filled with water and sediment, and a small, narrow, linear sea has been created (Figure 10–16B). Were it to continue to grow by spreading over hundreds of millions of years, a new major ocean basin like the Atlantic would form.

The Red Sea (Figure 10–18) at its northern end appears to be somewhat younger than at its southern end, so that we can easily imagine that in a few million years there will be a *continuous* body of water separating all of the Arabian Peninsula from all of Africa. The spreading center in the Red Sea floor is a zone of shallow-focus earthquakes as well as a zone where abnormally hot, metal-rich saline water erupts. Water temperature is one-half of its boiling point, with a measured salinity among the highest ever recorded in natural waters. The water has an abnormally high content of heavy metals; iron, zinc, copper, silver, and gold are deposited in the fractures in seafloor rocks and in the overlying sediment. Similar zones of hot, briny vents are associated with oceanic rifts around the world. We may well be seeing a modern example of one of the major processes that formed ore deposits in ancient rocks; the same processes continue to enrich younger rocks on the seafloor.

Beneath the young floor of the Red Sea are 6000 yards (5500 meters) of salt and other evaporites. Clearly the ancestral Red Sea was once a closed basin produced by advancing rifting. This basin must have experienced periods of prolonged evaporation, a history shared with the Mediterranean Sea to the north.

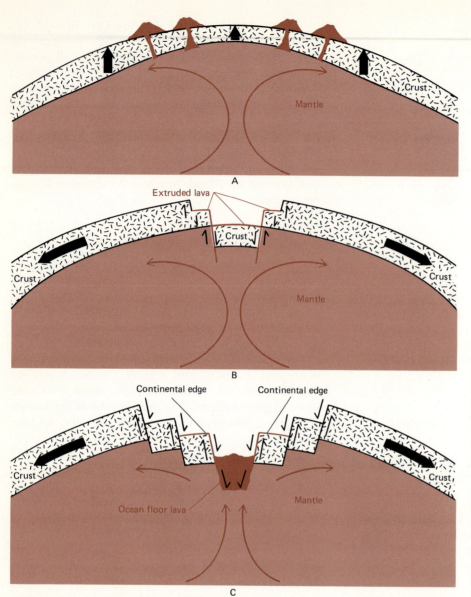

A

Extruded lava

Crust

Crust   Crust

Mantle

B

Continental edge   Continental edge

Crust   Crust

Ocean floor lava   Mantle

C

FIGURE 10–16

(A) The continental crust is first stretched, thinned, and intruded owing to high heat flow from a spreading center beneath it. (B) The crust breaks, and the central keystone block has collapses to form a graben whose surface and margins may be covered with new lava flows. (C) The collapse is complete, and the central rift is being filled with basaltic ocean floor. The continental margins will, by now, be covered with layers of sedimentary rock which will obscure the process; for the sake of clarity this diagram omits these materials.

The San Andreas Fault

A somewhat more controversial example of geology affected by spreading centers is the San Andreas Fault of western California, first discussed in Chapter Seven. Horizontal movement along a part of this fault in 1906 moved parts of the western block seven yards (64 centimeters) northwestward relative to the eastern block and produced one of the most devastating earthquakes in North American history.

Many geologists believe that the San Andreas Fault is a transform fault that connects the Juan de Fuca and East Pacific spreading centers (Figure 10–19). Perhaps the only thing unique about this particular transform fault is that it happens to include a sliver of the American continent that has overrun an oceanic spreading center called the East Pacific Rise. With continuing spreading southwestern California will be detached and become an island, which will head toward a collision with the Aleutian Islands far to the north; that collision will take place in about 30 million years.

Not all geologists are convinced that the San Andreas Fault is a transform fault, but it does satisfy most of the geometric requirements, connecting as it does the displaced ends of the East Pacific spreading center with other spreading centers to the north. For California, the spreading that has produced the Gulf of California has had devastating consequences.

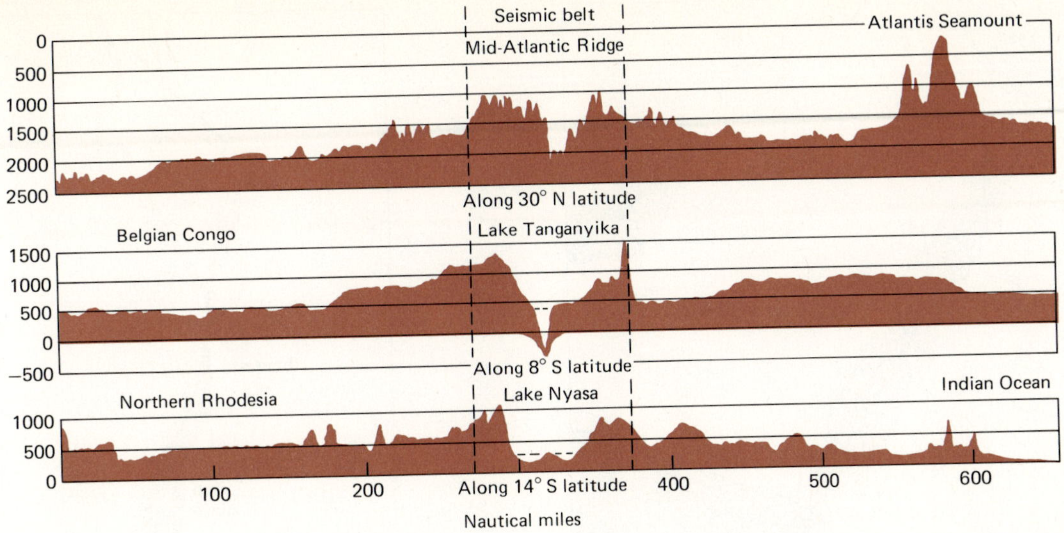

FIGURE 10–17

Two profiles of the African rift valleys are here compared with a profile at the same scale across the Mid-Atlantic Ridge. Notice the great similarity in form of the three rift valleys. Similarity in form implies, but does not prove, similarity in origin. (From Runcorn, S. K., 1962, *Continental Drift*, with permission of Academic Press, New York, and S. K. Runcorn.)

The explanatory power of the idea of seafloor spreading is well established by the diversity of structures it can explain. At first glance, no one would have thought that the African rift zone, with its majestic volcanoes and linear lakes; the Red Sea with its hot, briny mineral wealth and extremely young ocean floor; and the San Andreas Fault, bounding a sliver of California, share a common explanation founded in a common origin.

AGE RELATIONS ON THE SEAFLOOR

Among the most powerful tests of the theory of seafloor spreading would be the establishment of the age of the seafloor. As noted in Figure 10–14, if the seafloor is formed and spread outward from the oceanic ridges, then the age of the seafloor should *increase directly with distance from the ridge.*

There are three ways to obtain the age of the rocks on the ocean floor.

1. Chemical analysis of seafloor basalts using radiometric dating techniques, to be discussed in Chapter Seventeen.

2. Measuring the age of sediment overlying the basalt by analyzing tiny marine fossils trapped in the sediment.

3. Using paleomagnetic techniques to determine the pattern of polarity reversals in seafloor basalts.

All these methods have been employed, each serving as a check on the others. Most of the samples have been obtained by an internationally supported deep-sea drilling program carried on from two specially modified ships. This process is ongoing, with increasingly sophisticated technology allowing deeper and deeper probes into the crust.

The results of analyzing these data are given in Figure 10–20. It is easily seen that the direct relation between rock age and distance from ridge totally supports the theory of seafloor spreading. That 70 percent of the earth's surface could be explained by geometries as simple as layers spreading laterally from a single continuous ridge is one of the astounding results of the study of thousands of cores of recovered oceanic rock by scientists around the world.

Recognition that sediment is nearly absent from the ridges (Figure 10–14), and that it becomes increasingly thicker and older with greater distance from the ridge, was a second result of this same study. The record in the sediments of the ocean floor confirmed the record in the basaltic rocks beneath it.

A brilliant confirmation of the theory of magnetic polarity reversals was a third result of studying cores from the basaltic ocean floor. Basaltic ocean rocks of the same age from all around the world yielded similar patterns of polarity reversal, stretching back more than 100 million years.

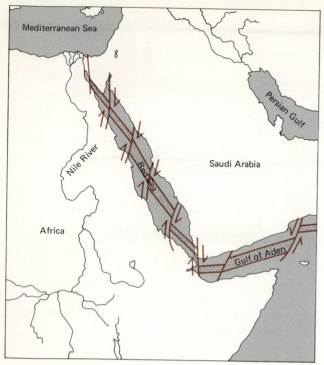

FIGURE 10–18

In this map of the Red Sea area (compare with Figure 10–27), double lines represent spreading centers of the Red Sea rift zone and the Indian Ocean Ridge which trends through the Gulf of Aden. The single lines are diagrammatic transform faults that offset the spreading centers. The Red Sea, like the Gulf of California, is underlain by quite new oceanic crust; both of these seas are ocean basins caught in the process of formation.

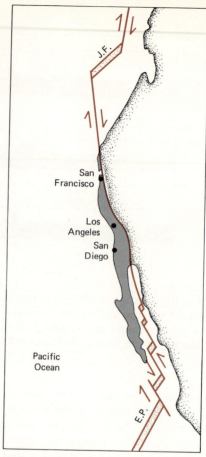

FIGURE 10–19

Double lines are spreading centers, including the East Pacific (E.P.) spreading center and the Juan de Fuca (J.F.) Ridge. Single lines are transform faults connecting the two spreading centers. Part of the San Francisco Bay region and all of Los Angeles and San Diego are part of the Pacific plate. The rest of the United States is part of the American plate.

Isochrons

Lines that connect points of equal age on maps of the sea-floor (Figure 10–21) are termed **isochrons.** The isochrons on these maps stretch parallel to the ridge system of the world's oceans like grooves left by a garden rake. Though offset by numerous transform faults, they tell us that rocks of equal age are in strips parallel to the ridge that formed them.

Parallel to the Mid-Atlantic Ridge, the isochrons are closely spaced. Those parallel to the Indian Ocean Ridge are somewhat broader. The isochrons that parallel the East Pacific Ridge are much more widely spaced.

The width of the ocean basin between isochrons is a function of the *rate of spreading*. The more rapid the spreading rate, the greater the width of ocean basin created in the same unit of time.

Comparison of the width of ocean floor between equal isochrons has revealed a substantial difference in spreading rates of the world's ocean basins. The average spreading rate in the Atlantic is lowest, averaging about 1 inch (2.5 centimeters) per year, whereas that in the Pacific is highest, averaging about 5 inches (13 centimeters) per year along the East Pacific Ridge (Figure 10–21).

SIMPLICITY AND PARADOX

The regularity of the magnetic patterns and associated isochrons suggests that the ocean floors move away from the ridge systems as remarkably rigid plates whose areas include many thousands of square miles or kilometers, offset only by the long

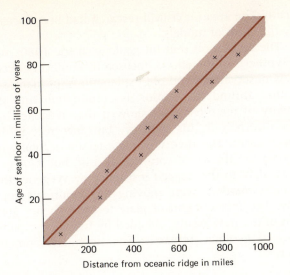

FIGURE 10–20

In this generalized chart the data points describe a simple linear relation between distance from ridge center and seafloor age. What is the spreading rate in this area? Was this graph prepared using Atlantic or Pacific data?

transform faults that connect short segments of the ridge systems. The whole ridge pattern (Figure 10–21) reminded one geologist of "the stitching on a baseball."

The apparent simplicity of the ocean basins, born from their own ridge and rift zone, complicated only by transform faults perpendicular to the ridges, and carried away at spreading rates of a few inches per year, is astounding. But the very simplicity of this highly successful model created *two* paradoxes. Like all paradoxes, they consist of two statements that are believed to be true but that appear to be mutually contradictory. Resolution of these paradoxes was the final key to creating a model of the world that both unified and revolutionized the geologic view of our earth and other planets.

The first paradox can be stated as follows. Rocks of the ocean floor tell us that the earth has had ocean basins for only the last 5 percent of all its history, yet continental sedimentary rocks tell us that the earth has had oceans for all its history. Each set of data pointed to opposite conclusions. Either the earth always has had ocean basins, or it has not.

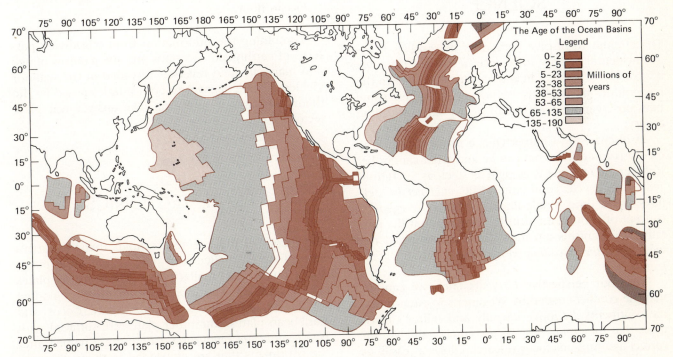

FIGURE 10–21

Age patterns on the ocean floor. The youngest portion of the floor coincides with oceanic spreading centers. Notice that stripes of rock of equal age tend to be wider in the Pacific, suggesting that the Pacific has the highest rate of seafloor formation. Notice that in the Pacific the Americas have consumed large parts of the ocean floor. With permission from Geological Society of America, Inc. Copyright © 1974, and Walter C. Pitman III, from 1974, *Age of the Ocean Basins*.

The second paradox has been briefly mentioned before. Since new area and materials are being constantly *added to the seafloor*, either

1. The earth is increasing its total size, its skin growing in area like the surface of a balloon being blown up. Yet we have no evidence that the earth has gained or lost volume since it formed, or,

2. The earth is maintaining its volume, and the crust being created at the ocean ridges is being destroyed in equal amounts somewhere else. Yet, the earth yielded no evidence in the early 1960s of any processes that destroyed the seafloor.

What was going on? Either the earth was expanding, or it was consuming its own surface; yet *neither* explanation made sense in terms of what was then known about the earth. The stage was set for a revolution.

THE ROLE OF PARADOX IN SCIENCE

The attempt to resolve paradoxes or conflicts is *not an incidental activity in science; it occupies the center of scientific effort.* Most new discoveries begin with the recognition of conflicts, which occur when some new observation does not fit neatly into prevailing theory.

As discussed in Chapter One, a major role for any theory is to decide what can be and what cannot be. *If* the earth's mantle is as rigid as steel, as Sir Harold Jeffreys believed, continental drift *could not be.* Jeffreys, for a time, single-handedly buried Wegener's concept of continental drift by showing that the tentative mechanisms advanced by Wegener could not work *given the prevailing theory about the rigidity of the mantle.*

From our perspective 60 years later, we can see that the conflict created by Wegener's insistence that the continents *did* move laterally and Jeffrey's insistence that they *could not* was extremely valuable. Incorrect assumptions focus research efforts as easily as correct ones. As geologic evidence mounted that horizontal continental displacement was probable, the prevailing assumption that the mantle was as rigid as steel forced a detailed reexamination of the evidence for mantle rigidity and recognition of the mantle's semifluid *asthenosphere* within a decade of Wegener's

death. The area for critical research had been focused by the clash of models.

The revolution that hit geology in the 1960s was a synthesis of the efforts, focused by conflict between new observations and old explanations, of many scientists around the world. Its culmination into the theory of plate tectonics showed that Wegener had been correct in his geology, but incorrect in his mechanism. The theory of plate tectonics totally replaced the theory of continental drift, but it owes a great debt to the dogged insistence by Wegener that *some mechanisms* were moving the continents laterally. The introduction of plate tectonic theory grew out of the new results obtained by two new capabilities in geologic research. The new developments were

1. The capacity, made possible by the advent of high-speed computers, to plot thousands of earthquake epicenters and foci onto a world map.

2. The capacity, made possible by modern oceanic research ships capable of drilling deeply into the ocean floor, to obtain abundant data on patterns of magnetic field reversals and the age of the ocean floor.

Combining the information from worldwide patterns of earthquake locations and the patterns of age and magnetic field reversals recorded in the seafloor yielded a single answer to both puzzles posed by the theory of seafloor spreading. The answer was plate tectonics.

PLATE TECTONICS

The theory of **plate tectonics** holds that the earth's lithosphere is divided into nine or ten large plates and a number of smaller plates. There are six plates whose surface is *largely oceanic crust;* they are named the Pacific, Philippine Sea, Scotia, Nazca, Cocos, and Caribbean plates. The other plates are surfaced with *both oceanic and continental crust* and are named after the continents that are embedded in them; they are the Antarctic, African, Arabian, Eurasian, Indian-Australian, North American (including Greenland), and South American plates (Figure 10–22).

The plates of lithosphere are approximately 5 to 60 miles (8 to 100 kilometers) thick. Their margins are defined by narrow zones of high seismicity; the

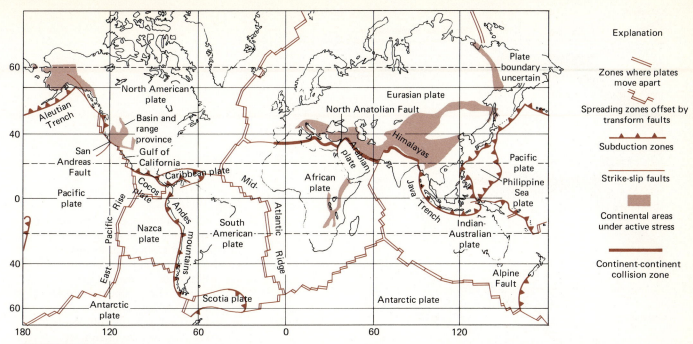

Explanation

Zones where plates move apart

Spreading zones offset by transform faults

Subduction zones

Strike-slip faults

Continental areas under active stress

Continent-continent collision zone

FIGURE 10–22

Plates and plate boundaries of the earth's lithosphere. Data for this map were compiled and adapted from numerous sources and much simplified in complex areas

(Modified from Warren Hamilton, U.S. Geological Survey.)

world's seismic belts define plate margins. Plates are rigid and extremely strong. The plates float on the semifluid *asthenosphere* within the upper mantle. The boundary between plate and asthenosphere is not a compositional one but rather a mechanical one where rigid lithosphere encounters semifluid asthenosphere. The plates are in constant relative motion, moved by poorly understood forces at rates of less than an inch up to 5 to 8 inches (1 to 20 centimeters) a year. The average rate is about the same rate as the growth of your fingernails.

The term tectonics comes from the Greek word *tekton,* meaning "builder." Plate tectonics (the term was coined in 1968) is the process that builds the overall architecture of the earth, creating ocean basins and enlarging and modifying continents.

The shape of the continents bears little relation to the shape of plate margins; only along the continents fronting the Atlantic Ocean do the shapes of continental and plate margins coincide. This, of course, is what we should have expected, since the continents fronting the Atlantic were split by the Atlantic spreading center hundreds of millions of years ago, just as the Indian Ocean spreading center is currently tearing East Africa and the Arabian Peninsula apart.

The continents ride on the plates like passive logs stuck in ice. Unlike the ocean floors that are constantly being spread from oceanic ridges, *the conti-*

nents are permanent features of the earth's crust. Because they are less dense but thicker than oceanic crust, they are buoyant and cannot be destroyed.

Continents can, however, be split into fragments, as in East Africa. The separated continental fragments can be sent on their own individual path, to eventually crash into and become a part of another continent somewhere else on earth. *Continental crust cannot be consumed, but it can be fragmented and welded to other continental crust.*

When plates containing continents interact, there are only four basic possibilities.

1. When plates containing two continents collide, the two colliding continental margins may be crushed into narrow orogenic belts like the Appalachians.

2. When plates containing a continent collide with a plate margin composed of oceanic crust, the buoyant continent overrides and consumes the oceanic crust, creating mountains like the Andes.

3. If a plate containing a continent overruns an oceanic spreading center, the buoyant continental crust is thinned, torn apart by continuing spreading beneath it, and split into smaller fragments like the Arabian Peninsula.

4. When plates containing continents slide past one

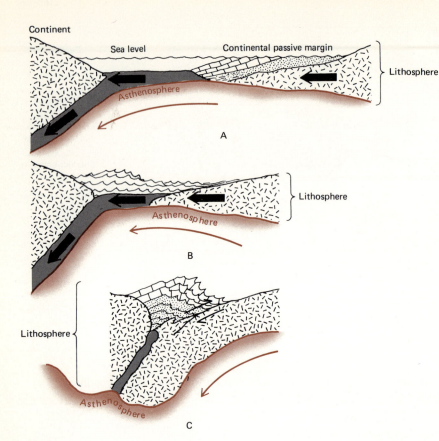

FIGURE 10–23

(A) One continent approaches another, being passively borne along on the plate in which it is embedded. Its passive continental margin is out in front. (B) The collision of the passive margin occurs, and these sedimentary rocks begin to be strongly deformed and arched upward. (C) The collision is complete. The two continents are sutured together with an intervening piece of oceanic crust, and huge thrust faults define the passive margin, which is now high in the sky above an abnormally thickened lithosphere.

another, pieces of continent may be sliced off, as along the San Andreas Fault.

The movement of plates can rearrange continental fragments, but it cannot destroy them. Examples of the four types of interaction occur along plate margins.

PLATE MARGINS

If we imagine fragments of a giant ice sheet floating in water, we will have the correct mental image of the behavior of plates. *As each individual fragment moves, its various margins may simultaneously be involved in (1) continent-continent collision; (2) continent-ocean collision; (3) rifting due to tension; and, (4) grinding and sliding past one another.* These four types of interchange exhaust the potential interactions along plate margins.

Continent–Continent Collision Margins

Collisional margins may also be termed **convergent margins**, which may include both continent-continent and continent–ocean types. Modern examples of continent–continent collision include the Alps,

formed when Italy rammed north into Europe a few tens of millions of years ago, and the lofty Himalayas, produced when the continental fragment we now call India, traveling northeast across the Indian Ocean for 35 million years, slammed into the passive continental margin on Asia's southern margin (Figure 10–23).

The result of this collision of continent with continent was the trapping of the wedge of passive margin sediments between the two advancing continents, followed by intense thrusting as the margins of the two continents were welded together. As the continental crust grows thicker and thicker owing to both folding and thrusting, it rises higher and higher and its growing "root" moves deeper and deeper into the mantle, an example of *isostasy* (see Chapter Nine).

Ultimately the deeply buried crust becomes both metamorphosed and melted, leading to both volcanic and plutonic activity. The two masses of continental crust weld into a single mass, plate motion stops, and global plate motions become reorganized. The total volume of the continent has been increased by the addition of continental fragments, volcanic rocks, and the metamorphosed passive margin sedimentary rocks. After hundreds of millions of years of erosion,

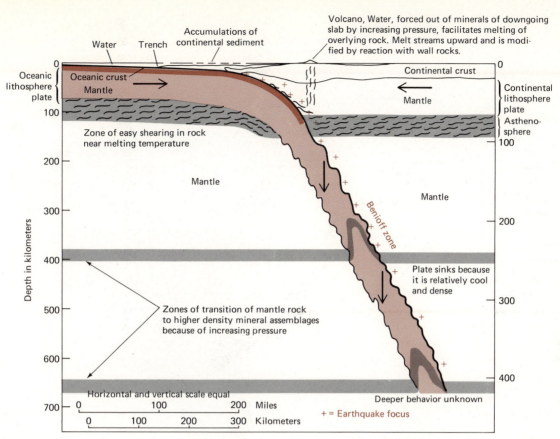

FIGURE 10–24

Cross-section through a subductional margin, showing a Benioff zone. Oceanic lithosphere is forced beneath the continental plate. Vertical and horizontal scales are equal so there is no vertical exaggeration. Details and dimensions are those for western Java and the Java Trench system, but other continental systems are similar. (Modified from Warren Hamilton, U.S. Geological Survey.)

the intensely deformed metamorphic and igneous rocks at the base of the former mountain range may be exposed to view; *at that time we would call these rocks a part of the continental interior, craton, or shield,* as defined in Chapter Nine.

The ultimate result of continental collision is the welding on of a new orogenic belt to the margin of the continent, increasing total continental volume as well as size. *Thus continents grow in size through time by the welding on of orogenic belts to their active margin.*

Continent–Ocean Collision Margins

When thin, dense oceanic plate margins run into a thick, low-density continental margin, the result is as predictable as the collision of a subcompact with an "18-wheeler." The thin oceanic crust is flexed downward and forced under the continental crust where it may be wholly or partially melted (Figure 10–24).

Collisional margins of this type encircle most of the Pacific plate (Figure 10–22). The ocean floor is buckled downward, forming deep oceanic trenches or *foredeeps,* as described in the previous chapter. In cross-section, foredeeps are deep infolds of the ocean floor caused by overriding continental crust.

Bands of earthquake foci (see Figure 9–19) dipping under the continent define a downward-sloping seismic zone which may reach depths of up to 400 miles (670 kilometers). These downward-sloping zones of *inclined earthquake foci are unique to areas of ocean - continent collision* and are termed **Benioff zones** after Hugo Benioff, the geologist who first adequately described their geometry. *Earthquake foci at depths of 400 miles (670 kilometers) are confined to Benioff zones.*

When the slab of oceanic lithosphere is carried deeply into the mantle, it encounters temperatures that will cause it to melt completely or partially. The fact that there are no earthquakes at depths greater than 400 miles (670 kilometers) is taken as evidence

Pacific plate

Cocos plate

East Pacific Rise

Pacific Ocean

Nazca plate

Antarctic plate

Peru-Chile Trench

Andes mountains

South America

FIGURE 10–25

The Andes are the result of the consumption and partial melting of the Nazca plate as South America moves west over it. The Peru - Chile Trench and the Andes mountains are mirror images of each other.

that all slabs of oceanic lithosphere are melted or softened by the time they reach this depth, though very recent research is challenging this view. As the material melts, it may become part of the mantle again, though partial melting will often send its less dense components up to the overlying continent as lavas and magmas of intermediate to silicic composition. Along Benioff zones, the descending slab of oceanic lithosphere is consumed by complete or partial melting.

The products of melting rise to form a continental margin magmatic belt termed a **magmatic arc.** The Andes on the west coast of South America are a good example of a magmatic arc composed of intermediate, andesitic volcanoes, underlain by intrusions

of granitic magma and metamorphosed rocks formed in a high-temperature, low-pressure environment. Such a magmatic arc is typically 180 to 240 miles (300 to 400 kilometers) inland from the foredeep.

Seaward from the magmatic arc is the **mélange**, a complex mixture of deep-water sediments, sedimentary material formed from oceanic landslides from the continental rise, and ocean floor basalts scraped off the advancing margin of the subducted plate. All these types of rock are metamorphosed in a high-pressure, low-temperature environment and are intricately folded and faulted. It is in mélange zones that *blueschists* (see Chapter Six) may form as well as metamorphosed basaltic ocean floor plus its sedimentary cover; this complex of blueschists, sedi-

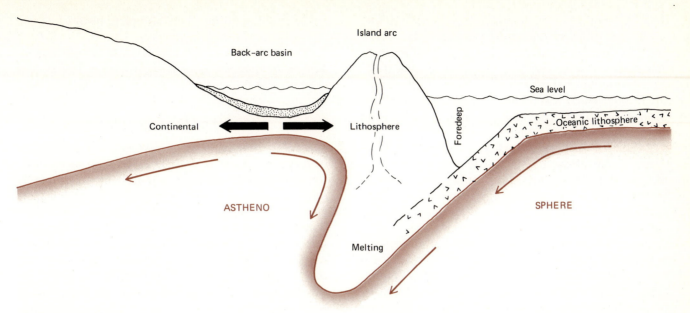

FIGURE 10–26

Stretching behind the volcanic arc extends the crust and forms a back-arc basin, which fills with sedimentary rocks from the continent. Oceanic lithosphere plus sediments accumulating in the foredeep are melted and puncture the

mentary rocks, and metabasalts form distinctive, complexly sliced suites of rocks termed **ophiolites**.

The combination of paired belts of mélange zones and magmatic arcs are the signature of continental–ocean plate collision. An example in western North America is the blueschist mélange of northern California which is paralleled by the Sierra Nevada magmatic arc. This paired zone marks the ancient (approximately 110 million years ago) collision of the North American and Pacific plates.

SUBDUCTION OF OCEANIC LITHOSPHERE

Geologists refer to zones of oceanic plate consumption as **subduction zones.** The verb *subduct* comes from the Latin verb *subducere,* which means *to lead away'' or ''draw under.''* Geologists have chosen the unfamiliar word *subduction* to convey a precise image. Many geologists believe that the newly formed oceanic crust is *pushed* away from the ridges while it is hot and *pulled* down into the mantle by its increased density when it is cold.

Volcanism is a common companion of subduction, for seafloor is partly melted and returned to the earth's surface as volcanoes in many settings. Though subduction zone volcanoes cover less than 1 percent of the earth's surface, they are by far the most visible. *They account for 84 percent of the volcanic eruptions known since recordkeeping first began.*

A modern example of subduction of oceanic lithosphere directly beneath a continent is furnished by the collision of the Nazca and South American plates

lithosphere forming islands whose arcuate map pattern parallels the offshore foredeeps. The line of island arcs defines the edge of the geologic continent.

(Figure 10–25). The Andes Mountains form the western mountain spine of all of South America and are paralleled, offshore, by the Peru–Chile Trench—a typical oceanic trench or foredeep. Nine miles (15 kilometers) of elevation separate the volcanic rocks at the top of the Andes from those at the bottom of the adjacent trench.

The ocean floor basalts of the Nazca plate are being forced under the western margin of South America at a rate of 2 inches (5 centimeters) per year in a process that may have begun nearly 200 million years ago. As the ocean floor that was stuffed beneath South America reached the melting point of its lowest-melting components, these lower-density siliceous materials melted and rose toward the surface. Erupting as *andesite,* a volcanic rock named for its abundance in the Andes, the andesitic lavas have continued to erupt and thicken the west coast of South America over many millions of years. The thickened low density continental crust began to *rise.*

Areas like the East African Rift Zone and the Red Sea (Figure 10–27) show us what a spreading center beneath a continent can do. Hidden beneath shallow ocean water or stretched crust, spreading centers can tear continents in two. Perhaps somewhat less than 150 million years ago, part of what is now South America was pulled apart from Africa as what we now call the Mid-Atlantic Ridge began to thin and split the overlying continent.

The line of ancient parting parallels both the Mid-Atlantic Ridge and the west coast of Africa. Heading west, the motion of the fragment that would become South America accelerated the rate of subduction of Pacific Ocean crust over which it rode. The great thicknesses of sediment lying along the ancient western continental margin were also formed into a mélange zone which was folded, faulted, and partly metamorphosed; this zone too, became a part of the increasingly thickened Andes.

As the subduction of the Pacific Ocean floor continues today along the Peru–Chile Trench (Figure 10–25), the South American continent continues to grow at the expense of the Pacific Ocean. Once again, *plate tectonic processes cause the growth of both continental volume and area through time.*

In a different example, the young, hot lithosphere bordering the *island arcs* of the Pacific is forced down under arcuate chains of andesitic volcanoes. Behind them (Figure 10–26) a small sea may be created by what is termed *back-arc spreading*. The *andesitic island arcs mark the edge of the geologic continent.* It is no accident that in the western Pacific bands of island arcs are paralleled by great foredeeps, both of which form looping arcuate chains on a map of the Pacific.

THE RESOLUTION OF PUZZLES

Earlier in this chapter, we pointed out that the phenomena associated with seafloor spreading led to two puzzles.

1. Why do the ocean floors record only the last 5 percent of geologic history, whereas the continents record essentially all of it?

2. What happens to the enormous volume of material added to the seafloor from spreading center volcanism?

Subduction zones provide the answer to both puzzles. The seafloor records only the last 5 percent of the earth's history because sea floors are destroyed by subduction, melting, and addition to the continents within no more than 200 million years of their formation at an oceanic ridge. Being of high density, seafloor is doomed to destruction by being forced under a continent and remelted. The sea floors do, however, provide a splendid record of the last 200 million years of earth history.

Since continents are of low density, they can never be subducted. Since they cannot be consumed, they record almost all the earth's total history. Since they can, however, be eroded, the record *at any one location* is fragmentary, so that the scraps of history from many areas must be pieced together in order to arrive at a composite, more complete history.

The second puzzle is as easily solved. *The rates of seafloor destruction by subduction equal the rate of seafloor production.* Sea floor is destroyed as fast as it is created, and the earth's total area remains the same.

Pull-Apart or Divergent Margins

Spreading centers, also termed pull-apart zones or **divergent plate margins,** have high heat flow, shallow, abundant seismic activity, and an extremely thin stretched and rifted lithosphere. Divergent margins display abundant basaltic volcanism of a low-potassium, olivine-deficient type of lava. Approximately two-thirds of the annual lava eruption on the earth's surface is along the oceanic ridges. The process has been responsible for creating 70 percent of the earth's surface area within the last 5 percent of the earth's history.

Transform faults allow the spreading motion between offset ridge axes to be transformed into a sliding motion perpendicular to the ridge axis. In this way transform faults permit conservation of surface area. Since transform faults define the direction of sliding or spreading, any relative motion of the plates that is not parallel with the transform fault will inevitably cause either divergence (separation) or convergence (collision) of plates.

Rates of spreading may vary from essentially zero to rates (measured in the Pacific) of nearly 8 inches (20 centimeters) per year. If this Pacific rate were to be long continued, it would carry a plate all the way around the earth's surface within a period of time equal to the last 5 percent of earth history.

On the average, oceanic ridges stand about 2 miles (3 kilometers) above the general level of the ocean floor, since the newly erupted rock is much hotter and less dense than its older counterparts. As seafloor ages it *slowly cools, becomes denser, contracts and subsides as it is moved away from the ridge.* The end result of this simple relation is that the seafloor *grows both thinner and deeper as a direct function of increasing age* (Figure 10–13).

One of the few places on earth where an oceanic spreading center can be examined up close is Iceland, which is directly astride the Mid-Atlantic Ridge (Figure 10–22). It is formed from tremendous effusions of lava along the Atlantic spreading center.

Volcanism and earthquakes in Iceland are extremely common; *all* of Iceland having been formed within the last 1 percent of earth history. Bubbling hot springs are everywhere, and their heat is tapped to heat homes and greenhouses. The energy that feeds the enlarging Atlantic Ocean is also used to

FIGURE 10–27
A photograph taken from Gemini XI shows the Gulf of Aden and the Red Sea, two arms on a spreading center between the Arabian Peninsula to the right and Africa to the left. (Photograph courtesy of NASA.)

FIGURE 10–28
Old Faithful is the best known example of a geyser in North America. Recent earthquakes in the region have somewhat disturbed its usual clocklike eruptive cycle.

grow orchids on the Arctic Circle. One bubbling, erupting spring, named *the Great Geysir*, has given its name to worldwide springs of this sort (Figure 10–28).

The eastern part of Iceland moves eastward, while at the same time the western fragment moves westward. The rift between them is a series of grabens that have been filled by basaltic fissure eruptions over the last 16 million years. One of those grabens provided the valley ampitheatre in which the world's first parliament, the *Althing*, met beginning in A.D. 930. Iceland continues to grow in size, an emergent product of the processes that make seafloor.

Sliding Boundaries

Plate margins marked by sliding are also termed **conservational boundaries,** since sliding boundaries neither *create* surface area as do divergent boundaries, nor *destroy* surface area as do convergent boundaries.

Examples of this type of plate margin include the Alpine Fault of New Zealand (Figure 9–20), the North Anatolian Fault of Turkey (Figure 10–22), and the San Andreas Fault of California (Figure 10–18). Boundaries of this type commonly exhibit no volcanoes. Instead, continuing large displacements in hor-

izontal directions over millions of years are accompanied by shallow focus earthquakes that are occasionally of high magnitude.

The North Anatolian Fault has been the site of very large earthquakes for the past 400 years. These earthquakes have brought misery and death to the residents of northern Turkey. Horizontal displacements up to 13 feet (4 meters) have accompanied some of the quakes. This fault is approximately 800 miles (1300 kilometers) long and forms a small part of the southern boundary of the Eurasian plate (Figure 10–22). The northerly segment of this fault has been moving easterly at the rate of 1 inch (2.5 centimeters) a year for at least the last 15 million years. The total displacement along the plate boundary is several tens of miles or kilometers.

A small part of the boundary between the Pacific Plate and the North American Plate causes much concern to Californians, since the San Andreas Fault marks the sliding boundary between these two plates (Figure 10–22). The San Andreas Fault may have originated 100 million years ago, although most geologists suggest its age is somewhat less than that.

Like the North Anatolian Fault, if you were to stand on the ground on one side of the San Andreas Fault and look across the fault plane, the ground on the other side of the fault will always move to your right (Figure 10–29). The cumulative displacement over the many tens of millions of years of its existence may be as much as 300 miles (480 kilometers). Geologists are able to match up older geologic features now offset by the fault, but it requires 300 miles of backward motion along the fault to bring these offset features together again (see Exercise One). Current rates of relative movement along the fault average 1 to 3 inches (2.5 to 7.5 centimeters) a year. Extended over a possible 100 million year life-span, displacements of 300 miles (480 kilometers) are not unreasonable.

If the motion along the San Andreas Fault should continue in the same direction at the same rate, Los Angeles will be a western suburb of San Francisco in about 20 million years. Separation from the mainland will occur not long after that, and a sliver of North America will begin an existence as a long island moving northwest across the Pacific Ocean floor. Tens of millions of years later, what once was southern California will collide with the Aleutian Islands as a small part of the continuing "demolition derby" we call plate tectonics.

SUMMARY

1. Dissatisfaction in the early part of this century with the prevailing theory of a contracting earth led to a reexamination of an old idea—continental drift.

2. The evidence for continental drift consisted of the congruent Atlantic shorelines, similar fossils and histories in rocks now far separated, ancient climates recorded in rocks that were now tens of degrees of latitude from where the climate could have occurred, direct measurements of horizontal continental motion, and direct plotting of apparent polar wandering paths. Jeffreys challenged and nearly destroyed Wegener's concept of continental drift by insisting that, given the knowledge of his time, the mantle was far too rigid to allow continents to plow through them.

3. In the 1960s evidence from several decades of oceanographic research led Harry Hess and others to propose seafloor spreading from oceanic ridge–rift systems. The evidence for seafloor spreading included the patterns of magnetic reversal recorded on the basaltic rocks of the ocean floor, accompanied by linear age–

FIGURE 10–29

In this aerial photo along the San Andreas Fault, a stream flowing toward the camera is offset to the left along the fault line. Such offset of modern stream courses is direct evidence of continuing horizontal motion along the fault. (Photograph by R. E. Wallace, U.S. Geological Survey.)

distance to ridge relation, and a mechanical understanding of the role of transform faults.

4. With the acceptance of seafloor spreading, it was recognized that the seafloor records only the last 5 percent of the history of an earth that has had ocean basins for all of its recordable geologic history. Plate tectonics consumes the marine record of its past. The theory of plate tectonics, developed in the 1960s, proposed that seafloor is steadily created at spreading centers, destroyed in subduction zones, and conserved along transform boundaries.

5. Plate tectonics proposes that the earth's lithosphere is broken into plates whose margins are marked by epicentral and focal patterns creating narrow seismic belts. Convergent or subductional boundaries between ocean and continent are marked by downward-sloping Benioff zones, abundant intermediate volcanism and magmatism, mélange zones, abundant high-magnitude earthquakes, orogenic activity accompanied by foreland folding or back-arc spreading accompanied by strong ocean floor metamorphism. Continent-continent collision boundaries are accompanied by thickening and welding of the continental crust, extensive thrust faulting and intense folding accompanied by compression of the former continental margins, deep-seated metamorphism and plutonism, little volcanism, and high-magnitude, shallow- to intermediate-focus earthquakes. Convergent boundaries cause continents to grow in size through time.

6. Sliding boundaries yield high-magnitude earthquakes, little volcanism, strike-slip faulting, and no limits to the amount of movement that can be accomplished over millions of years. Sliding boundaries are conservational boundaries.

7. Divergent boundaries display high heat flow, thinned, rifted lithosphere, abundant shallow low- to intermediate-magnitude earthquakes, prolific eruption of oceanic crust, and spreading. Spreading centers beneath oceanic crust cause the crust to arch, thin, and collapse along a series of grabens, accompanied by eruption of oceanic basalts, hot springs, and abundant low-magnitude, shallow earthquakes. Fracturing and spreading of a continent may take millions of years to complete.

8. Plate tectonics is currently the best theory explaining the earth's past and its future. Widespread acceptance of plate tectonics has placed geology in its most fruitful period of continuing research into the mysteries of our dynamic earth.

9. Plate tectonics is the mechanism that cools the earth, which currently loses heat at a global average heat flow of about 80 milliwatts per square yard or meter of the earth's surface. That is enough heat to melt a layer of ice 0.25 inch (0.6 centimeter) thick in a year.

KEY WORDS

Benioff zone	mélange
conservational margin	ophiolite
continental drift	paleomagnetism
convection	Pangea
convergent margin	plate tectonics
divergent margin	polar wandering curve
field reversal	seafloor spreading
Gondwana strata	spreading center
Gondwanaland	subduction zone
isochron	transform faults
magmatic arc	

EXERCISES

1. The following table contains some information of rocks that are now separated by the San Andreas Fault, with part of the rock on each side of the fault. The data are real.

Rock Formation	Age in Million Years	Amount of Offset in Miles (Kilometers)
A	25	200 (320)
B	20	175 (280)
C	17	130 (210)
D	13	78 (126)
E	3	19 (30)

(a) Plot a graph of relation between offset and rock age.

(b) What is the *average* rate of movement along the San Andreas Fault?

(c) Is the rate of fault motion constant?

(d) What factors might control the rate of motion?

2. If plates are pulled down into Benioff zones by their own density or driven sideways from spreading centers, is lava filling the oceanic rifts the *cause* or the *effect* of spreading? Or is it both?

3. The majority of transform faults are subparallel the equator while the majority of the length of spreading centers runs north-south. Would you expect this relation? Why?

4. It has been observed that there is a systematic *decrease* in the amount of heat emitted through the ocean floor as the floor grows older. How do you explain that relation?

5. If continental crust continues to thicken with time, will that have any effect on the rates of plate motion? Is there any limit to how thick continents may become?

6. Might there be plate tectonics on other planets? What would you look for to recognize plate tectonics, say on Venus?

SUGGESTED READINGS

Hsu, K. J. (ed.), 1984, *Mountain Building Processes,* New York, Academic Press, 263 pp.
► *Broad technical review of the topic.*

Lewis, T. A., (ed.), 1983, *Continents in Collision,* Alexandria, VA, Time-Life Books, 176 pp.
► *Beautifully written and illustrated; nice introduction.*

Understanding Landscape

On many other planetary bodies, the dominant agent forming landscapes is meteorite impacts. In this view of the lunar farside surface, the entire landscape is a regolith of material churned by the impact of countless meteorites. Above the lunar surface hangs the crescent shape of earth at earthrise. (Apollo 17 photography, courtesy of NASA.)

CHAPTER *eleven*

Introduction to Landscapes

*Standing here in the deep brooding silence
all the wilderness seems motionless, as if the
work of creation were done.*
   *But in the midst of this outer steadfastness,
we know there is incessant motion and change.*

JOHN MUIR

Landscapes form a stage for human beings and seem
almost changeless. A few rocks may fall, floods may
churn, stinging winds may shift the sands, and waves
may attack the shore, but the land seems to endure,
little changed. Yet each scene offers its past for examination. The record of the distant past is frozen
into the rocks themselves, but the landscapes carved
into these rocks record nearly modern events. Landscapes reflect perhaps the last one-twentieth of the
earth's history.

Each landscape results from a delicate interplay
among the life forms, rocks, and soil that compose it
and the unending forces that change it. The diversity
of landscapes is great, but understanding them starts
with a study of only two very basic processes:

1. **Erosion,** a collective term for all processes that
   remove earth materials *from* the earth's surface,
2. **Aggradation,** a collective term for all processes
   that deposit materials *on* the earth's surface.

THE ORIGINS OF LANDSCAPES

A geologist once suggested that all that was needed
to understand the earth was "one disgracefully simple idea . . . first it goes up, and then it comes
down!" That bit of wry humor is not only correct
but also the key to understanding the landscapes
around us.

Landscapes are the surface features of the earth
seen at all scales from mountain range to smallest
gully. The interaction of erosional processes, which
lower the earth's surface, with any combination of
aggradational, orogenic, and isostatic processes,
which collectively raise the earth's surface, creates
landscapes. This chapter and the three that follow it
outline the multitude of more specific processes that
create the rich variety of landscapes on planets, especially on Earth.

Two centuries ago James Hutton (see Chapter
One) recognized that the land surface was always in
a state of reconstruction. As he phrased it "This

earth, like the body of an animal, is wasted at the
same time that it is repaired." The wasting away of
the earth is the most easily *visible* half of the history
of the earth's surface. We are conscious of the bits of
loose soil that streams carry away as well as the sand
whipped into the air during sandstorms. Erosional
forces are easy to see as they do their work.

EROSIONAL PROCESSES

All processes that cut and carry away earth materials
are erosional. The verb *erode* comes from the Latin
erodere, which means to "gnaw away." Erosional
processes gnaw away at both the solid and loose materials of the earth and carry the eroded materials
generally downward away from their source. Earth's
erosional processes reflect the presence of a thick atmosphere and hydrosphere (Figure 11–1). The atmosphere, unevenly heated by solar radiation, is in
constant motion, bringing water from the hydrosphere onto the land; that water running back to the
sea is a powerful erosional agent.

The effect of erosional processes is to modify the **relief,**
*the difference in elevation, on the earth's surface. Erosion
also redistributes and spreads earth materials to lower elevations.*

There are four major agents of erosion. The dominant one in any area depends *largely on climate,*
though one form of erosion is at work in every climate at every instant. That omnipresent agent is
gravity. The role gravity plays in erosion is reviewed
in this chapter; the remaining erosional agents are
discussed in subsequent chapters. The four forms of
erosion are

1. **Mass wasting**—erosion caused solely by the
   effects of gravity overcoming the cohesion within
   loose materials.
2. **Stream erosion**—erosion caused by the
   dissolving, cutting, and transporting power of a
   stream pulled down a slope by gravity.
3. **Wind erosion**—erosion caused by the abrading
   and transporting power of wind driven by
   differences in atmospheric pressure.
4. **Glacial erosion**—erosion caused by the
   abrading and transporting power of glaciers,
   including their meltwater streams.

One other erosional agent should be mentioned.
On many other planetary bodies, the *predominant*
means of erosion is the *impact of meteorites* (see Chapter Frontispiece). The landscape on these planets is
dominated by circular impact structures; on Earth

FIGURE 11-1

The earth's oceans or hydrosphere and its atmosphere with swirling cloud patterns are obvious in this photo of earth from space. Even at this distance, the energy within the earth is also evident; the Gulf of California, created by seafloor spreading, is in the center of the photograph, and the southern Rocky Mountains and the San Joaquin valley of California can be seen. The interaction of the earth's external and internal energy creates landscapes. (Apollo 10 photography, courtesy of NASA.)

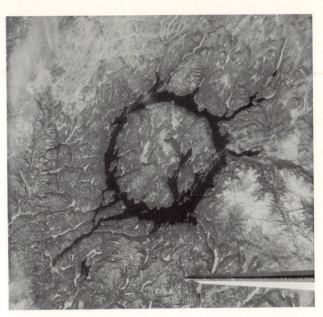

FIGURE 11-2

Because earth is protected by a dense atmosphere, the impact of meteorites is a rare event, and the dynamism of our planet means that the evidence of the impact will be eroded in time. The Manicouagan crater, in northern Quebec, Canada, is an example of a impact scar that has yet to be erased; it is also proof that Earth can be a target. (Space Shuttle photography, courtesy of NASA.)

(Figure 11-1) the main relief forms are linear—mountain chains, and oceanic ridges and trenches.

Because the earth is protected by a dense atmosphere, bombardment of the earth's surface by large meteorites is a rare event, separated by thousands or millions of years of quiet. Much smaller meteoritic remnants pass through the earth's atmosphere and strike the earth in a continual rain of fragments and dust.

Earth's topography is generally young, since the earth's surface is constantly being resurfaced by plate tectonics and active erosional processes. New crust is continually being formed and reabsorbed while erosional processes denude the planetary surface at an extremely active rate. *The average erosion rate for earth has been estimated at 0.001 inch (0.003 cm) per year,* with estimated rates for other planetary bodies averaging one-one hundred thousandth to one-millionth that of Earth.

Consequently, impact craters are rarely preserved on Earth (Figure 11-2). On the surfaces of other planets that are not protected by an atmosphere and are not resurfaced by either plate tectonics or active erosion, the record of billions of years of bombard-

ment still remains. The surface of the earth is different from that of any other planet; only Venus, a planet whose size, mass, density, and internal energetics are similar to earth, may share with Earth some similarities in landscape and landscape processes.

AGGRADATIONAL AND OROGENIC PROCESSES

A very large number of processes build up the elevation of the earth's surface. Sediments are deposited by *mass wasting*, rivers, glaciers, and wind, and mountains are formed by volcanic processes. Volcanism is an important aggradational process, building up large mountainous areas on the surface of the earth from reservoirs deep within or beneath the crust. It is the <u>only</u> aggradational process that involves one-way transfer of materials, from the mantle to the crust (Figure 11-3).

The effect of aggradation is to increase the elevation of the surface on which the material is deposited. In this way aggradational processes have an effect opposite that of erosional processes, sometimes called *degradational* processes. Although aggradation and degradation are natural antagonists, it is com-

FIGURE 11–3
The black volcanoes in the center of the photograph have formed within the Afar Triangle in southeastern Ethiopia, where the rift systems forming the Red Sea, Gulf of Aden, and East African Rift Zone join. Basaltic lavas forming these volcanoes come from beneath the crust and are deposited on it as the crust breaks up. (Shuttle imagery, courtesy of NASA.)

mon for erosion at high elevations to send the eroded materials downward so that they become aggradational deposits at lower elevations. Thus, except for volcanism, most aggradation is due to degradation at higher elevations. The *construction* of volcanic landforms on the earth's surface is commonly one result of deformation.

Compared to many other planetary bodies, the rates of volcanism and deformation on earth are extremely high. The areas of greatest relief are along plate margins. The highest elevations are along convergent boundaries where passive margins have been compressed and elevated between colliding plates. The lowest elevations are within subduction zones, where oceanic crust has been infolded into foredeeps. The total relief on earth is about 12.5 miles (20 kilometers) or 0.003 of its radius; the earth's surface is smoother, for its size, than a billiard ball (Figure 11–1).

Orogenic processes form linear landscapes concentrated along plate margins. The major orogenic belts of the modern earth are around the Pacific Rim—the Andes, Rockies and Alaskan ranges as well as the Aleutian, Japanese, Indonesian, Philippine, and New Zealand areas—and in the Mediterranean–Himalayan belt, which extends from the Alps, to Turkey, to Iran, to the Himalayas, and to Indonesia.

Though the driving force for the formation of orogenic belts is poorly understood, clearly in many of them the dominant force is a strong, long-continued horizontal compression caused by plate collision (Figure 11–4). The landscapes formed by orogenic forces are rugged, narrow linear belts of high elevation.

The highest elevations, because they are the most rapidly eroded, are the rarest. Why should the earth's surface be so smooth? Why should the earth be able to sustain only as its tallest point an area projecting up one-thousandth of one percent of its radius?

ISOSTATIC AND EPEIROGENIC PROCESSES

Epeirogenic movements, unlike the orogenic events confined to narrow belts, are a gentle, broad upwarping of large crustal areas. The word comes from the Greek *epeiros*, meaning "mainland" or "continent," and may refer to the slow uplift on an entire continent. More commonly, epeirogenic movements are of large areas of continental interior.

Among the most common causes of epeirogenic movement is crustal uplift or sinking in order to maintain isostatic compensation in the area. As we have learned, the concept of **isostasy** proposes that crust is buoyed up in direct proportion to its thickness and in inverse proportion to its average density. This theory suggests that continents and ocean basins alike are like ships in water. When loads are added to them, they sink a little bit, and when loads are removed, they rise a little bit.

Since continental and oceanic crust are buoyed by floating on the denser mantle beneath them, their surface elevation depends both on their density and on their thickness. The denser and thinner they are, the lower their surface elevation. The thicker and lower their density, the higher their surface elevation.

The theory of isostasy proposes that mountains and continents are supported by low-density crustal "roots," and that oceanic crust is much thinner and denser. This concept requires that the elevation of the surface of the crust rise or fall when its floating equilibrium is disturbed. Because the mantle is so thick, it may take thousands of years for equilibrium

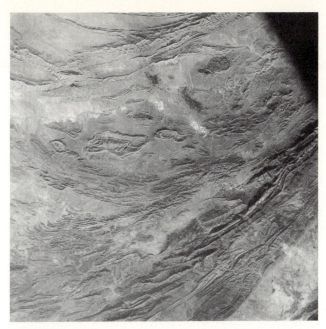

FIGURE 11–4

The continuing collision of India with Asia forms these twisted swirling patterns on the surface of the earth. The Sulaiman Range in west Pakistan, photographed from 108 nautical miles by the Apollo 7 crew, was created by strong orogenic forces. (Image courtesy of NASA.)

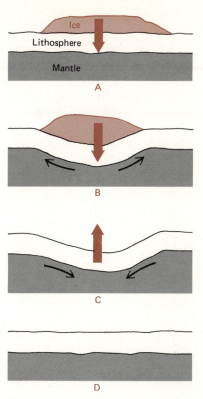

FIGURE 11–5

(A) The advance of glacial ice imposes a load on the asthenospheric rocks sluggishly move away. (C) The ice has melted back, and isostatic rebound begins; mantle material slowly moves into the depressed area as it elevates. (D) Isostatic equilibrium has been regained.

or compensation to take place after an older equilibrium has been disturbed.

Orogenic processes commonly interfere with isostatic equilibrium. Continent–continent collision may cause large increases in crustal thickness and force masses of low-density continental crust deep into the mantle. Continent–ocean collision with subduction may force thin layers of oceanic crust *far* into the mantle, creating substantial disequilibrium. Once orogenic stresses stop, both oceanic and continental crust forced into the mantle will slowly rise, until their elevation is once again appropriate to their thickness and density.

As mentioned in Chapter Nine, the addition of heavy loads of glacial ice to a continent may also force it below the level of isostatic equilibrium. Even the weight of water in a lake may cause the crust to depress temporarily. About 16,000 years ago, during a time of high rainfall coincident with continental glaciation, a large area of central Utah was covered by a lake, Lake Bonneville. As the climate has warmed and dried, the lake has grown far smaller, leaving the Great Salt Lake as its remnant. In 16,000 years the Lake Bonneville region of Utah has risen 210 feet (64 meters), for an average uplift rate of 0.01 foot (4 millimeters) per year.

The observation of what is termed **isostatic rebound,** the increase in elevation caused by unloading from an area (Figure 11–5), was no surprise. Postglacial rebound was first recognized by the Swedish bishop of Abo in 1616. A careful study of the coast of Maine by Maine surveyor N. S. Shaler in 1874 revealed the same pattern. Shaler, who recognized marine sea shells attached to rocks a hundred or more feet above sea level, pointed out that the *increase in elevation was greatest precisely where the glacial ice had probably once been thickest.* The addition of a mile or two of ice had caused the asthenosphere beneath to flow like an extremely sluggish fluid away from the source of extra mass; once the glacier had melted, the asthenosphere sluggishly flowed back under the area and it rose once again.

The laws of an isostatic earth are simple. *Mass excess above sea level must be balanced by mass deficiency beneath it; mass deficiency above sea level must be balanced by mass excess beneath it* (Figure 11–6).

Areas of high elevation on earth are largely compensated for or supported by low density rocks beneath them. Low elevation areas are largely compensated for or supported by dense materials beneath. Most of the compensation occurs within the upper lithosphere. In comparison to other planets, Earth is generally in isostatic equilibrium.

When material is removed from a high area to a low one, the eroded areas will rise while the aggraded areas will sink (Figure 11–6). The loss of mass from the high area will be compensated for by mantle material slowly moving in to replace the mountain's diminishing "root." The mantle material will be displaced from the area of deposition or aggradation and transferred to the area of erosion.

With this simple model we see that eroding the earth *causes uplift,* and uplift *accelerates erosion.* The model we are using is still another version of Archimedes' principle, first encountered in Chapter Two. *The wasting away of the earth's surface is balanced by uplift.* The earth's continents are rising at about the same rate as their elevation is being lowered by erosion. If this were not true, we would all be fish, for the continents would have worn down to the level of the sea many billions of years ago. If the buoyant forces were to cease today, the continents would be at sea level in only a few tens of millions of years.

MASS WASTING

The incessant pull of gravity draws everything on earth downward. Fluids possess no strength, so a cup of soup and the Pacific Ocean are alike; their surfaces are leveled by the same force that pulls rocks down from the mountains. Rocks, unlike fluids, possess some strength and for a time resist being pulled down toward the earth's center. But rocks resist only for a time, because the destiny of any material on earth is to be moved downward by gravity. In time, rocks tumble, too, brought to their lowest level, just like sea water. Everything is brought down with time. As someone has said, "The oceans are the graveyards of the land."

Mass wasting is the term for all erosional processes in which gravity is the active agent. Because of the pull of gravity, any elevated object has the potential to move downward and do work. The measure of the work it could do in moving from a high to a lower place is its **potential energy.** The amount of potential energy depends on only two factors.

1. The mass (experienced on earth as weight) of the object.
2. The elevation difference (relief) between its current location and its final location.

A very heavy rock perched high above a valley floor has a large amount of potential energy, *the energy of position.* If it begins to move, its potential energy is instantly converted to **kinetic energy,** the energy of motion. As the rock rumbles down a slope, it does *work,* the expenditure of energy (Figure 11-7). When the rock comes to rest, its kinetic energy drops to zero, and its potential energy has been lowered. Only when a rock reaches the lowest point on the surface of the earth can its potential energy reach zero; place it anywhere else and it has the capacity to do work as it falls farther down.

Slopes are nature's highways. They provide a surface that is steep enough to allow the products of weathering to be carried down it, whether driven by gravity alone, or by its allies—running water or glacial ice. Since the eventual resting place for everything on the continents is the sea, most slopes are inclined toward valleys, which in turn are sloped toward the sea. Slopes and canyons are funnels leading seaward.

Decrease the slope, and the weathered and eroded material will aggrade or deposit, *until the slope angle becomes steep enough to start downslope movement again.* What factors control slopes, and what factors control slope failures? These are more than academic questions, since the failure of slopes accounts for property damage and loss of life every year.

FORCES AND SLOPES

Any sloping surface is being acted on by two forces. The pull of gravity tends to flatten the slope, and cohesive forces tend to hold the material in place. **Cohesion** is the tendency of loose particles to stick together and resist shearing.

Gravity tugs at all particles equally, but the cohesive forces of resistance arise for two different reasons. The cohesive forces in clays and fine silt come from *small electrical forces among the grains* which cause them to be attracted to one another. The charges, similar to static electricity, are small, but they may collectively make a large clay mass quite cohesive. The cohesion among coarser grains, such as coarse silt, sand, and gravel, comes from *friction among the angular grains.*

The **angle of repose** is the *steepest* stable slope for a given material. For most natural materials the an-

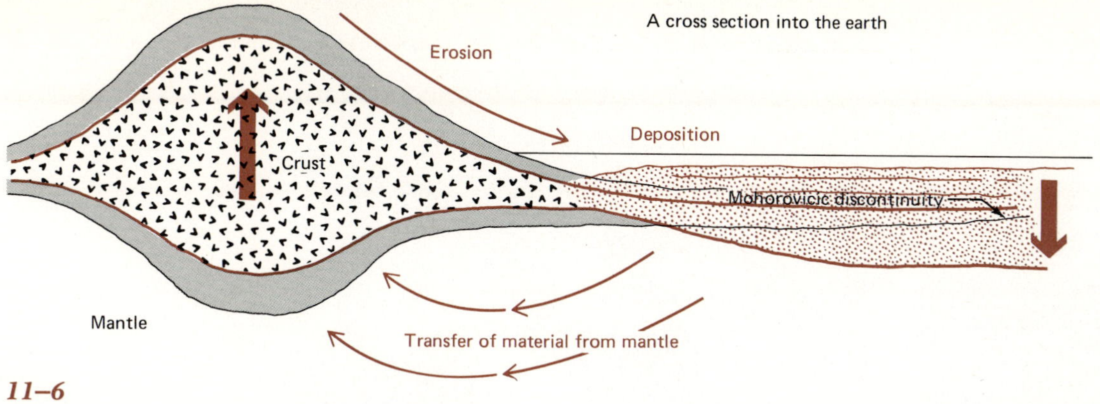

A cross section into the earth

Erosion

Deposition

Crust

Mohorovicic discontinuity

Mantle

Transfer of material from mantle

FIGURE 11–6

The scene prior to erosion is shown in black, that after erosion in color. Erosion of material from the mountain top is partly compensated for by the rise of the crust as weight is removed. As the crust rises, denser materials from the mantle move in to take the place of less dense crust, and the depth to the Mohorovicic discontinuity diminishes. Because of the difference in density, 4000 feet (1220 meters) of mantle will "replace" 5000 feet (1500 meters) of eroded continental crust, so the actual elevation loss is only a 1000 feet (305 meters). The opposite would be true of adding lower density sediment to a higher density oceanic crust, which would cause the M-discontinuity under the ocean to sink as that under the continent rose.

gle of repose ranges from 20° to 40° from horizontal. Dry sand blown into a sand dune will form a stable slope near 30° from horizontal; cinders in cinder cones may approach 50°. The most spectacular example of cohesion is, however, furnished by **loess,** a wind-deposited silt of glacial origin, whose angle of repose is often *vertical* (90°) (Figure 11–8). With loess, the electrical attraction among its fine silt grains and the extreme angularity of the silt grains combine to form an unconsolidated material of high cohesion.

Moist soils may have temporary strong cohesive forces. Such cohesion is due to the effect of *surface tension*, at the boundary between grains, air, and wa-

ter. The phenomenon of surface tension is familiar to anyone who has noticed that the contact between a drinking glass and the water it contains is slightly curved, owing to the attraction between the dipolar water molecule and the glass pulling the water up the side. The same force causes the sand grains in a moist sand castle to cling tightly together.

The cohesive forces caused by surface tension are temporary. As the water dries out in the pores between the sand grains, the sand castle will collapse. Conversely, if too much water is added, air is driven out of the pores, the grains become buoyantly supported, and the sand–water mixture will liquefy and flow, the water serving now as a lubricant. Sand that

M

h

Still

FIGURE 11–7

The scene before the boulder moves is depicted in black; the aftermath, in color. The potential energy of the boulder is the product of its mass (*M*) and its height (*h*). When the boulder starts rolling downhill, its potential energy is converted to kinetic energy which flattens the still. *Moral:* Potential energy is still-dangerous.

FIGURE 11–8
This photograph taken in 1928 is
of a road cut in loess near
Jerseyville, Illinois. Notice that the
road cut remains vertical.
(Photograph courtesy of the U.S.
Geological Survey.)

has water filling *all* its pore space is called **quicksand;** its cohesion is zero.

Solid rock is as strong as its weakest natural plane. If rocks on an incline have natural planes, such as bedding, foliation, flow banding, or fractures and joints, *that are inclined down the slope,* the potential for the rocks to separate and slip along these planes may be very high. The addition of excess water acts as a lubricant, and a small shock may be all that is needed to set the whole slope in motion.

SLOPE FAILURE

In order to make a slope fail (Figure 11–9), we must

1. Steepen the slope angle, increasing the *effect* of gravity, or,
2. Increase the height of the slope, increasing the *effect* of gravity, or,
3. Increase the load on the slope, increasing the *effect* of gravity, or,
4. Saturate the slope with water, diminishing the material's cohesion as excess water buoys grains, or,
5. Accelerate failure along natural fractures slanting downslope.

An Example

Let us look at one example of a recent slope failure. Notice how many factors often combine to lessen the cohesion of both rock and soil. The failure of slopes, unfortunately for us, is absolutely natural.

The Vaiont River is a tributary to the Piave River in the highest part of the Italian Alps 80 miles (130 kilometers) north of Venice. The Vaiont River valley is a narrow deep canyon (2000 to 3000 feet of relief), its walls cut into clay-rich limestones whose bedding planes on one side of the valley slant strongly downward toward the valley floor. Alpine glaciation a few thousands of years ago severely eroded the valley, and isostatic rebound after the glacial ice melted away left many additional zones of natural fractures parallel to the valley walls. Water draining into the river has enlarged these fractures, some of which were the gliding plane for one prehistoric landslide. One could hardly imagine a more dangerous combination of natural elements than those already present in this steep, narrow valley.

Added thoughtlessly to this area in 1959 was the world's highest thin-arch dam, which backed the Vaiont River water upstream for 6 miles (10 kilometers) to depths of 2,200 feet (680 meters) (Figure 11–10). A year after dam construction, a large mass of soil and rock slid into one side of the reservoir as it was filling; it was an omen of what was to come. As the reservoir continued to fill, lake water under pressure of the many tons of overlying water was forced into the clay-rich heavily fractured rock in the lowermost valley walls. Four years after dam construction in September 1963, slope movement greater than 16 inches (40 centimeters) a day was recorded. A month later in October 1963, nearly 300 million cubic yards (230 million cubic meters) of rock and soil roared down the canyon wall into the reservoir.

Displaced water stormed over the dam in waves more than a hundred yards (90 meters) above dam level, accompanied by blasts of explosively compressed air that flattened every living thing. As the

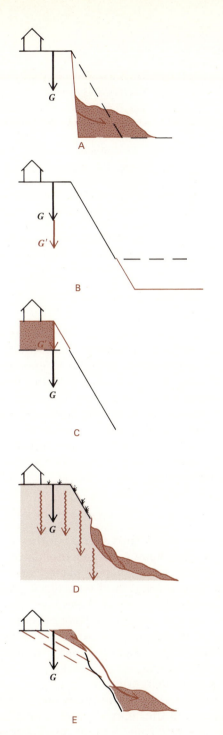

FIGURE 11–9

The slope before failure is shown in black, after failure in color. (A) Oversteepening the slope causes it to fail by debris flow. (B) Increasing the height of the slope weakens it, making it unstable. (C) Adding a load on top of slope also leads to an unstable slope. (D) Overwatering permeates the soil, causing a slump and debris flow. (E) Natural planes, weakened perhaps by water and vibration, fail by landsliding. The house, as in all five of these sketches, is now in imminent danger.

water surged upvalley and roared downstream, several thousand people died, and property damage was in the many millions of dollars. The dam, incredibly, survived the onslaught—an engineering triumph which was a monument to geologic ignorance. The dam is still in use today, but with water confined to a much lower, less efficient, but safer level.

Everything had worked together to produce disaster. The loose rock was clay-rich and heavily fractured in zones inclined down valley. Water forced into the fractures by both the lake and rain further weakened the rock and destroyed the forces that held it together. Gravity did the rest.

TYPES OF MASS WASTING

Although slope failures of the Vaiont type are spectacular, mass wasting causes the detachment and downslope transport of soil and rock materials in a variety of ways. All that is needed for mass movement is deformable loose material with potential energy. Mass movement occurs wherever these conditions obtain; landslides underneath the Atlantic Ocean, on the moon, and on Mars have been documented. One Martian landslide displaced 3.5 million cubic feet (100 cubic kilometers) of martian soil.

Figure 11–11 provides an outline of one of the many possible classifications of mass wasting. Material may move down a slope in one of five modes. It may

1. *Fall* vertically,
2. *Slide* along multiple planes,
3. *Slump* and rotate along a curved surface,
4. *Flow* with the assistance of air, water, or ice,
5. *Creep* slowly down a gentle slope.

Falls

In **falls** rock or debris drop down cliff faces or steep slopes to bounce and shatter on contact with the bottom of the slope. **Rockfalls** are the free fall of loose rock down vertical to steep slopes, whereas **debris falls** are the direct free fall of *regolith,* as loose surface material is called (see Chapter Four).

Rockfalls and debris falls more commonly occur in arid to semiarid climates, where the dry environment allows steep slopes and cliffs to form. In a more humid climate weathering may be so intense that slopes are eroded to form more gentle profiles. The fall of rocks is a simple process, often kicked off by minor earthquake tremors, vibrations, vehicular traffic, blasting, thunder, and the like.

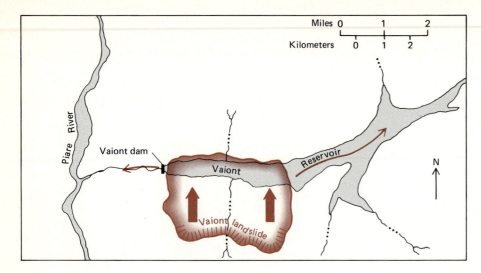

FIGURE 11-10
The scene before the landslide is shown in black; the aftermath in color. As the Vaoint landslide occurred along a water-weakened slope, 300 million cubic yards (230 million cubic meters) of rock and soil roared down the steep canyon walls into the reservoir. Water was driven high over the top of the dam. Huge waves were driven in all directions; several thousand people were drowned.

If the fallen material accumulates, it collects at an angle of repose typical of the angularity of its fragments (Figure 11–12) to form a **talus** (pronounced "tay'-luhs") slope. Talus slopes caused by rockfalls are commonly steep, since the broken rock fragments are often extremely angular and therefore have high intergrain friction.

Rockfalls constitute a major hazard to structures built near the base of a cliff or at the bottom of a narrow canyon. Free-falling boulders are capable of obliterating anything in their way.

Slides

Slides are rapid downslope movements of detached rocks or soils along a steeply inclined plane. **Rock slides** are masses of rock that have broken loose along natural surfaces; in **landslides** solid rock and soil are mixed. **Debris slides** consist of wholly unconsolidated soil (Figure 11–13). In practice, these three types of slides may become difficult to separate, since the sliding mass may begin to break up almost as quickly as it begins to move.

Slides are commonly initiated along natural planar surfaces that dip toward the valley floor, as in the Vaiont River dam disaster. Loading the fractures with water, undercutting or oversteepening the slope, vibrating the area, and the presence of slippery soils all contribute to the likelihood of sliding.

Once the suspended slide breaks along the failure planes beneath it, its downslope motion commonly becomes turbulent and chaotic. Air may be trapped beneath the moving mass, causing the mass to move on an almost frictionless carpet of violently compressed air. Downslope speeds in excess of 60 miles (96 kilometers) per hour have been estimated for

some landslides.

The largest landslide on earth within the last million years may have occurred about 400,000 years ago on the north coast of Molokai, one of the Hawaiian islands. As much as *half* of the island may have disappeared into the Pacific, creating the steep cliff on northeast Molokai that runs 3300 feet (1 kilometer) down to the sea. As much as 17 million cubic feet (500 cubic kilometers) of volcano may have been sheared away, perhaps because of movement on the seafloor.

Slumps

Slumps grade into debris slides, but they are distinguished by the slipping of unconsolidated material along multiple spoon-shaped planes, accompanied by the backward rotation of each block (Figure 11–11). Slumps occur wherever slopes have been undercut, as by stream valleys, or have become unstable for a myriad of reasons. Failure occurs along a curving, parallel scarp (Figure 11–14) in the natural slope as the material rotates backward and downward as a semicoherent unit.

Near the base of a slump moisture may begin to accumulate, and the slumped material may begin to move as an *earthflow*. Slumps produced when a highway is cut into a slope, removing its support, are particularly dangerous. The slump that may result will quarry away not only huge volumes of material but also the highway itself.

Slumps are a common hazard associated with earthquakes. Prolonged shaking may loosen unstable materials and send them on their natural path. The Good Friday earthquake near Anchorage, Alaska, in 1964 is one sad example.

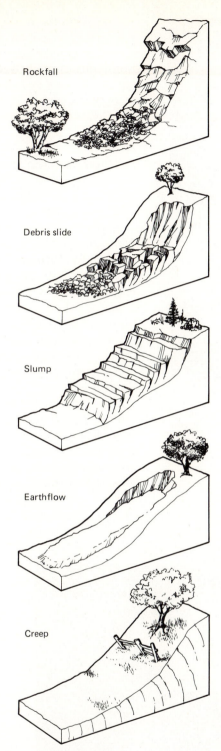

Rockfall

Loosened rock and debris fall from a cliff through air, bounce down a slope, and come to rest as a talus pile.

Debris slide

Loosened rock and debris move downslope by sliding on a surface that underlies the deposit. Instability is increased by fluid pressure along planes of potential failure and by vibration.

Slump

Coherent masses of soil or poorly consolidated rock move downslope by rotational slip on surfaces that both underlie and penetrate the soil.

Earthflow

Soil and poorly consolidated rock move downslope in a manner similar to a viscous liquid, forming lobate ridges on the valley floor.

Creep

Soil and poorly consolidated rock move extremely slowly downslope, causing flexing of vertical layers and bending of fences, tree trunks, etc.

FIGURE 11–11
Five kinds of mass movement. (Sketch adapted from Tor Nilsen, U.S. Geological Survey.)

The Turnagain Heights area of Anchorage was an area of upscale homes built on a promontory overlooking the Bering Strait. The views were lovely, but the homes rested on a series of recently uplifted soft ocean floor sediments, tipped strongly toward the sea. Within the sediment were numerous layers of soft, slippery clays, sometimes called "quick clays," which are notoriously slippery and unstable. Geologists had warned 20 years earlier that the combination of steeply dipping, young, soft, poorly consolidated clays in an area with high earthquake potential was a disaster just waiting to happen.

FIGURE 11–12
Glacial scenery high in the Uinta Mountains of Utah is augmented by massive talus piles at the base of the cliff. Such talus piles are formed by repeated rockfalls. (Photograph courtesy of the U.S. Geological Survey.)

The warning was ignored, however, and homes were constructed on Turnagain Heights. When the great earthquake occurred, the area collapsed along numerous curving fractures downward toward the sea. Millions of dollars of homes and lifelong dreams were destroyed.

Flow

In **flows** unconsolidated earth material moves as a viscous liquid, with the assistance of water or ice. Flow types include *earthflows, mudflows, solifluction,* and *rock glaciers.*

Many flowage phenomena occur in conjunction with slumps, since the two phenomena are quite similar. **Earthflows** and **debris flows** are a slow movement flow of earthy materials lubricated with water (Figure 11–15); earthflows can be studied over a period of time. Such studies suggest their flow mechanism is similar to that of a glacier. **Mudflows** are an extremely rapid movement of earthy material mixed with large amounts of water. The necessary ingredients are loose soil material, water, and a slope.

Mudflows may start as small soil slips on slopes saturated with rain. The mixture of soil and water may be stable until it is vibrated by a passing truck, a sonic boom, trees swaying in the wind, or a very low-magnitude earthquake. Such soils are said to exhibit **thixotropy,** the phenomenon in which semi-

FIGURE 11–13
A recent debris slide in the San Juan Mountains of Colorado has left a large area unforested. The toe of the slide must have partially dammed the stream at one time; the stream has now gone around it and cut through it. (Photograph courtesy of the U.S. Geological Survey.)

FIGURE 11–14
The upper part of a slump is defined by a series of curving small cliffs, each marking the separation and rotation of a slump block. (Photograph courtesy of the U.S. Geological Survey.)

solids turn to fluid when shaken. Ordinary house paint is deliberately thixotropic, being sufficiently thick to *not* flow when placed on a vertical surface, but capable of instantly becoming fluid when a paint brush or roller disturbs its equilibrium. Quicksand is another thixotropic substance, as are most clay-rich soils when wet. Anyone who has repeatedly tapped saturated sand at a beach or shore to watch it turn liquid has caused thixotropism.

Mudflows may range from the consistency of stiff mortar to that of a very muddy stream. As they move downslope and pick up more and more material,

FIGURE 11–15
A typical small debris flow. The upper surface is defined by a cliff along which separation first occurred. The mass moved downslope as a viscous fluid, leaving a toe area of bumpy topography. (Photograph courtesy of the U.S. Geological Survey.)

FIGURE 11–16
High in the Rocky Mountains, a rock glacier moves very slowly down a former stream valley. A mixture of ice and rock, the surface of a rock glacier is often furrowed and broken by fractures; a rock glacier is an extremely viscous liquid. (Photograph courtesy of the U.S. Geological Survey.)

their density and speed commonly increase. Ultimately mudflows may pick up and carry house-sized boulders, homes, railroad engines—anything at all in their path. The classic mudflow is created by heavy rains on poorly vegetated soil, loading the soil with water more rapidly than the water can leach downward into the subsoils. The mass moves at first as a slip along a discrete surface, but almost immediately undergoes a thixotropic transformation into a viscous fluid.

One specialized type of mudflow is the **lahar,** a flow of volcanic ash and water created by heavy rains on an erupting volcano. This phenomenon, described in Chapter Three, is a self-accelerating process, since the eruption of huge volumes of ash into the air creates the necessary nucleii to cause rain, which then falls on the fresh slopes covered with loose ash. Such lahars on the slopes of Vesuvius buried Herculaneum 19 centuries ago, were common on the slopes of Mount St. Helens in the 1980 eruption, and continue to be one of the major geologic hazards affecting cities such as Tacoma, Washington.

Another specialized form of mudflow is **solifluction,** a type of mudflow unique to subarctic regions where the soil is permanently frozen at some depth year around. Solifluction mudflows occur when spring thaws unfreeze the upper part of the soil, while the deeper soil, termed *permafrost,* remains frozen. The upper soil becomes waterlogged, because its water content cannot trickle downward into the permafrost beneath, and it becomes saturated, thix-

otropic, and poised for flow. Solifluction is a major construction hazard in large parts of the subarctic world. In the Arctic, where soil remains frozen even at the surface, solifluction does nor occur. Any structure in subarctic areas must have its foundation well below the zone of thawing; otherwise foundation and structure alike may flow downslope during a thaw. Some structures have been literally built on "stilts," allowing solifluction to occur and harmlessly flow *under* the structure.

Still another landscape produced by flow is the **rock glacier,** (Figure 11–16) a mixture of ice and rock found in the highest part of mid-latitude mountain ranges. A rock glacier may range from being mostly ice with some rock material (often talus) to a talus pile having a small amount of ice between the rock fragments. The whole mass moves ponderously down steep slopes, very much like a true glacier. Its movement is that of an extremely viscous fluid. Because of its high viscosity, its surface is often deformed into ridges by uneven flow patterns within the mass.

Creep

Creep is the slow, even imperceptible downslope movement of earth materials under the direct influence of gravity. Soil creep creates few distinctive landscapes of its own, but it modifies essentially all landscapes developed on slopes.

Creep is largely caused by temperature differences

FIGURE 11–17
The vertical layering in these rocks once extended all the way up to the land surface. The movement of material along the slope has caused the layering to be flexed downslope along a railroad cut in Schuykill Valley, Pennsylvania. (Photograph courtesy of the U.S. Geological Survey.)

between night and day. As water freezes at night, frost may nudge rock and mineral grains up, *perpendicular to the hillslope*. When the slope warms up the next day, the ice melts, and the fragment sinks *vertically downward*. In this way a gentle seesaw motion occurs for each grain, and the net result of years of this motion is downhill creep.

Even solid rock is bent by the incessant pull of gravity (Figure 11–17). Rock fragments may slowly move downslope by *thermal creep*, the daily process by which rocks warm and expand, then cool and contract. The drag of snow creeping downhill, animal traffic which dislodges soil from slopes, frost action, alternate wetting and drying of expansive clays, even the filling of burrows by loose material from upslope are agents that contribute to rock and soil creep.

The results of creep can be recognized in hilltops with thin soils, as well as leaning fences, trees, telephone poles, and gravestones. When careful measurements have been made year after year, the rate of creep is highest at the soil surface, decreasing steadily downward for several feet or meters.

SUMMARY

1. Landscapes are the expression of a series of variables and usually reflect the last one-twentieth of earth history. The dominant variable is *climate;* no other variable has so much effect. Other variables include rock resistance, orogenic forces, and former landscapes.

2. Three fundamental forces interact to form landscapes: erosion, aggradation, and isostasy. Slopes are the fundamental units of a landscape and may be created and modified by both erosion and deposition. Isostatic uplift increases the potential energy available to erosional agents.

3. Erosional agents include mass wasting, glaciers, wind, and running water.

4. Mass wasting is a term for all erosional processes in which gravity is the sole active agent. Gravity on a slope is opposed by the cohesive forces among regolith and the strength of solid rock. Any processes that lower the strength of solid rock or the cohesive forces among particles will increase the probability that the slope will flatten at rates ranging from imperceptible to disastrous. Mass-wasting processes include falls, slides, slumps, flows, and creep.

5. Falls are defined as free fall of material down vertical or steep slopes. Slides move at high speeds along preexisting planar surfaces. In slumps unconsolidated material slips and rotates backward along a spoon-shaped curving slope. In all flows water or ice is added, which turns regolith into a viscous fluid; flow speeds range from glacial slowness to high, downslope velocities. Creep is imperceptible downslope movement of regolith.

6. Mass movement is generally associated with moist soils and slopes of intermediate inclination. Steep slopes are rapidly stripped by water erosion, but on gentle slopes gravitational forces are ineffective. Mass wasting is most common in areas of high relief, broken or weakly layered rocks, and seismic activity.

KEY WORDS

aggradation	debris flow
angle of repose	debris slide
cohesion	earthflow
creep	epeirogenic
debris falls	erode

erosion	potential energy
falls	quicksand
flows	relief
glacial erosion	rock fall
isostasy	rock glacier
isostatic rebound	rock slide
kinetic energy	slides
lahar	slumps
landscape	solifluction
landslide	stream erosion
loess	talus
mass wasting	thixotropy
mudflow	wind erosion

EXERCISES

1. The landscapes of earth reflect perhaps the last one-twentieth of its history, whereas those of many other planetary bodies reflect most of their history. What is it that makes earth different from, say, the moon?

2. If erosion removed 1000 feet (305 meters) of material having a density of 2.8 from the top of a continental mountain, and mantle having a density of 3.5 flowed underneath to replace the lost mass, how much elevation decrease would actually be observed, given that the area was isostatically balanced?

3. We observe that mass wasting is much less common in areas of low relief. Why?

4. Recent studies in semiarid regions show that mass wasting is much more common on north-facing slopes than on south-facing slopes. Can you suggest a reason why?

5. If the density of continental rock was increased by 50 percent, how would this affect the elevation of continents?

6. Why is it important for a retaining wall to have drain holes in it?

7. If you live on a hillside and the neighbor above you maintains a lush yard by overwatering it, what problems might you expect?

8. You are considering buying a home on a hillside lot that has trees and retaining walls both tilted about 10° to 20° from vertical. Hairline cracks finger out at 45° from the corners of windows and doors, and some of the windows and doors are very difficult to close. Outside, the concrete stairs have pulled away from the walls. What processes would make your dream home a nightmare? How might the current owner limit the problem?

SUGGESTED READINGS

CHORLEY, R. J., et al., 1984, *Geomorphology*, New York, Methuen, 606 pp.
► *See Chapter 10; excellent overall reference for landforms*

DENNEN, W. H., AND MOORE, B.R., 1986, *Geology and Engineering*, Dubuque, Iowa, W. C. Brown Publishers, 378 pp. and appendices
► *See Chapters 11 and 13 for treatment of slope failures.*

HAYS, W. W., (ed.), 1981, *Facing Geologic and Hydrologic Hazards*, Washington, D. C., U. S. Government Printing Office, U. S. Geological Survey Professional Paper 1240-B, 110 pp.
► *See Chapter 4. Extremely well-written and illustrated.*

ZARUBA, QUIDO, AND VOJTECH, MENCL, 1969, *Landslides and their Control*, Amsterdam, Elsevier, 341 pp.
► *Comprehensive discussion of mass movement.*

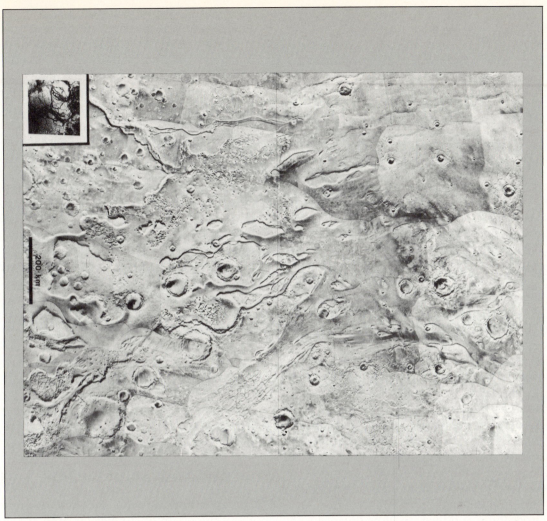

The outflow channels on the surface of Mars suggest that among the last recorded events was catastrophic flooding. The sinuous shapes of the landforms and the stream patterns are similar to the forms produced by running water on planet Earth. (Viking mosaic, image courtesy of NASA.)

CHAPTER twelve

Stream-Dominated Landscapes

*Following the path of least resistance is what
makes men and rivers crooked.*

ANONYMOUS

It is stupid to sleep in the floodplain.

DON BARNETT, MAYOR OF RAPID CITY,
AFTER THE 1975 FLOOD.

▼

THE HYDROLOGIC CYCLE

▲

A little over 200 years ago, Benjamin Franklin (1706–1790) made what could be described as an incisive geologic statement: "Rivers are ungovernable things, especially in Hilly Countries." Centuries later we continue to learn that Franklin was correct, though rivers in many countries are now much more "governed" by dams strewn along their length.

Rivers exist to drain water from the land into the ocean. This drainage is a part of a large pattern of water movement termed the **hydrologic cycle** (Figure 12–1). The cycle is powered by solar energy, which evaporates water from the world's oceans, as well as lakes, rivers, ponds, and plants, and fills the atmosphere with water vapor. The uneven heating of the atmosphere drives storm systems that drop an annual average of 3400 million cubic feet (96,000 cubic kilometers) of precipitation on continental surfaces. Of this amount, two-thirds is evaporated back into the atmosphere or directly used by organisms. A little less than a third runs off the land surface in streams which slope toward the sea. The remaining

small amount soaks deeply into the earth and moves very, very slowly toward lower elevations within the outer mile or two of the earth's surface as *groundwater,* the topic of Chapter Fifteen.

If all precipitation onto the land's surface were to stop, the amount of water in streams (estimated at 300 cubic miles or 1200 cubic kilometers) would keep flowing for about two weeks; this water is about 0.65 percent of the earth's total supply. Additional freshwater storage on land is provided by glaciers, which store 2.15 percent of the total amount of water on earth. The remainder—97.2 percent—of our earth's water is stored in its ocean basins, which hold 317 billion cubic miles (1.32 billion cubic kilometers) of water.

The hydrologic cycle depicts the continuous movement of water from the earth's ocean to the atmosphere to the land and back to the sea. The cycle was an early part of folk wisdom, beautifully described in Ecclesiastes 1:7 "All the rivers flow into the sea, yet the sea is not full. To the place where the rivers flow, there they flow again". Though it was observed that rivers ran to the sea and did not cause them to fill up, the connection between rain and snow and the formation of streams, and stream valleys was not understood until the end of the eighteenth century!

STREAMS AND STREAM VALLEYS

To sixteenth-century people, there was little question about the origin of streams, or of their valleys. Since the time of Aristotle, everyone "knew" that there was too little rain to serve as the source for springs

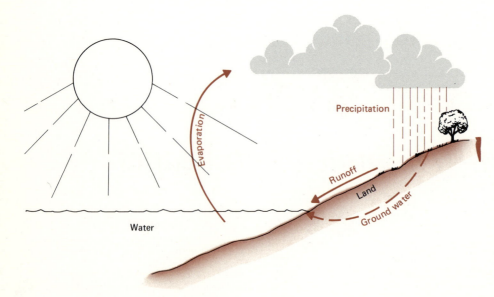

FIGURE 12–1
The hydrologic cycle is powered by the sun. Water runs off the land, creating valleys, and seeps seaward underneath the land. The higher the land elevation onto which precipitation falls, the more work water will do on its way to the world's ocean.

and rivers. Rain could only augment the flow into streams from the oceans of water *within* the earth (Figure 12–2)

Rather, streams and springs were supplied with water from gigantic caverns within the earth, and they flowed down valleys created by catastrophic cracking of the earth's crust by great earthquakes. Four centuries later, these ideas seem ludicrous, but we have depended on rivers much longer than we have understood them.

In 1580 Bernard Palissy first recognized that water circulating underground was the source of springs, and that this underground water came only from rain seeping into the earth. The skeptics remained unconvinced. Almost a century later, Pierre Perrault carefully measured the amount of rain that *fell* in the area surrounding the Seine River in France and also

FIGURE 12–2

This seventeenth-century engraving depicts the understandings of the time. Water was thought to be sucked from the ocean (creating whirlpools) into underground caverns. Caves, in turn, siphoned the water into streams, which emerged from underground. From there, the water flowed back to the sea. (From A. Kircherii, *Mundus subterraneus*, 1665, courtesy of the History of Science Collections, University of Oklahoma Libraries.)

recorded the amount of water *leaving* the area in the Seine River. He found that *six times* as much rain fell as left in the river. There was actually far *more* rain than needed. An idea that dated back more than 1900 years finally died, the victim of patient measurement and human curiosity.

It remained for eighteenth-century scientists to establish that streams carved their own valleys. Resistance to this idea had been based on the assumption that the earth was only a few thousand years old, and anyone could figure out that rivers would not have been able to cut their deep canyons in so short a time. In 1746 French geologist Jean Etienne Guettard (geh-tahrd') (1715–1786) described in great detail the erosive work of running water. Fifty years later John Playfair, Hutton's friend and fellow naturalist, recorded a number of observed facts about stream valleys that were irreconcilable with the theory that streams occupied preexisting, earthquake-formed chasms. These observations included the following.

1. Tributary streams join main streams at a *common* elevation, which would be unlikely if main streams flowed in canyons created by catastrophe.

2. Stream valleys are rarely straight in map view for any great distance, again an unlikely happenstance if streams followed ready-made fractures.

3. Valley width was generally proportional to the amount of water flowing in the stream. Again, this relation would be unlikely if the canyon was there first.

4. Valleys become narrower as their elevation increases, never the other way around.

Playfair cast the first of his observations into the **principle of accordant stream junctions,** which states that tributary streams always join main streams at a common elevation. If streams had simply followed preexisting fractures caused by earthquakes, stream junctions would commonly be waterfalls; instead waterfalls are the scenic exception.

Valley Widening

Today we say that a stream *cuts* its own valley, but this statement is only partly correct. Consider (Figure 12–3) what a valley would be like in cross section if streams were the only force forming and modifying their valley. If streams only cut their canyons, all stream valleys would be narrow "slot" canyons, no wider than the stream itself, and stream valleys would look like saw cuts into a board. Instead, as shown in Figure 12–3, the stream *cuts straight down,* and the unsupported stream walls collapse into the stream *by various mass wasting processes.*

Seen in this way, streams not only drain the land but *function as conveyor belts* at the bottom of the valley, carrying away *the material fed to them by gravity.* Streams are superb transportation agents, but they are generally poor cutting tools. Downcutting in solid rock takes an immense length of time—time in which the valley walls slowly collapse into the stream to give the stream valley its characteristic notched or ∨-shaped cross section.

Streams are in many climatic zones the dominant agent of erosion. Only in the most intense of deserts or within huge areas of continental glaciation is evidence of stream erosion largely absent. Elsewhere, the evidence of stream erosion and deposition is commonly the dominant aspect of the landscape. What is the source of all of this erosive power?

STREM ENERGY

Water or snow falling at high elevations acquires potential energy. As water moves down a slope, its potential energy is converted into kinetic energy, the energy of motion. Water always seeks its lowest level and continues to travel downslope until it reaches

the sea. Sea level is termed **ultimate base level,** the elevation below which streams can no longer erode. At this level their potential and kinetic energies have dropped to nearly zero.

Stream energy is a direct function (Figure 12–4) of only two quantities,

1. The mass or *volume* of water flowing per unit of time, and
2. The slope or *elevation difference per unit of horizontal length* down which this mass moves.

The elevation difference is an expression of the potential energy of each drop of water, and the mass is the total weight of all drops. Computing the energy of a stream is analogous to finding the product of an elevated rock mass and the elevation difference down which it moves, in order to compute the total *potential energy* of the boulder.

The energy expended by water running off the land is used *to carve* and *to carry* rock and rock fragments to the sea. The results are impressive. The Mississippi River alone carries 230 million tons (207,000 kilograms) of sand, silt, and mud past New Orleans each year (Figure 12–5). The estimated total loss of materials from the continents to the sea is 2 billion tons (18 trillion kilograms) every year.

STREAM DISCHARGE

The *volume* of water moving past a point per unit of time is the stream's **discharge,** which is usually defined in terms of cubic feet or cubic meters per second. The discharge of a small stream may be only a few hundred cubic feet per second (*cfs*), the discharge

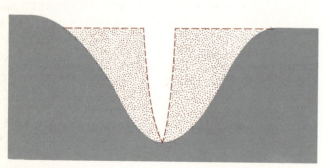

FIGURE 12–3

If canyons were formed *only* by cutting, they would have the cross-sectional profile shown in color. Instead, once a stream has notched the earth's surface, the unsupported walls slowly collapse into the stream valley, and the valley becomes wider. All the material brought onto the valley floor by mass wasting is removed gradually by stream erosion. The result of extensive mass wasting is a wide gv-shaped canyon, as shown in gray.

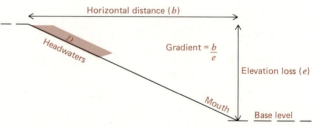

FIGURE 12–4

A stream's energy is the product of its total gradient from headwaters to mouth and the total water mass or volume. The greater the volume of water and the greater the elevation loss, the greater the energy expended.

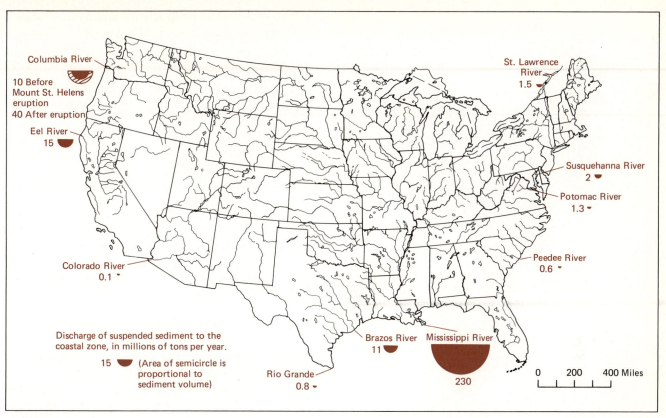

Columbia River
10 Before
Mount St. Helens
eruption
40 After eruption

Eel River
15

Colorado River
0.1

Discharge of suspended sediment to the
coastal zone, in millions of tons per year.

15   (Area of semicircle is
        proportional to
        sediment volume)

Rio Grande
0.8

Brazos River
11

Mississippi River
230

St. Lawrence
River
1.5

Susquehanna River
2

Potomac River
1.3

Peedee River
0.6

0   200   400 Miles

FIGURE 12–5
This compilation of recent stream data by the U.S.
Geological Survey indicates that the Mississippi River
carries three-quarters of the total sediment load delivered
to the continental shelves by all streams in the United
States. The load of the Mississippi has been cut in half in
the last decade or two by the installation of more dams
upstream. These dams catch sediment that would
otherwise have made it to the Gulf of Mexico. (Map
courtesy of the U.S. Geological Survey.)

of major rivers in flood stage several million. A record of stream discharge is maintained at about 7000
stream gauging stations in the United States. At these
stations data on discharge, load, and velocity are au-
tomatically recorded.

The discharge of a stream varies continuously. The
base flow of a stream, its average lowest volume of a
stream in a year, is commonly furnished by the seepage of *ground water*. Additional water in the stream
channel is furnished by runoff from the land's surface. Variations in the amount of surface inflow into
the stream causes the discharge to vary from time to
time. If the stream runs year around, it is said to be
a **perennial stream.** If the stream dries up during
part of the year, it is an **intermittent stream.**

Since most streams are fed by tributary streams,
the discharge increases downstream, as more and
more tributaries feed water into the stream. Since
discharge ≅ width × depth × velocity, both the

width and depth of a stream increase downstream in
order to handle the increasing volume of water.

Drainage Basins and Stream Ordering

Any stream is an open physical system. Rock material and energy both enter and leave at many points
within the system. The upper part, or **headwaters,** of
any stream collects water, the middle part transports
water and sediment, and the lower part, or **mouth,**
of a stream disperses the sediment and water. The
physical limits of the area drained by a stream are
the **drainage divides,** the areas of highest elevation
that separate one drainage basin from another. A
drainage basin is an area, bounded by drainage divides, whose runoff water and regolith are collected
by a stream and its tributaries (Figure 12–6).

The average discharge of a stream increases in direct proportion to the area of the drainage basin. Dis-

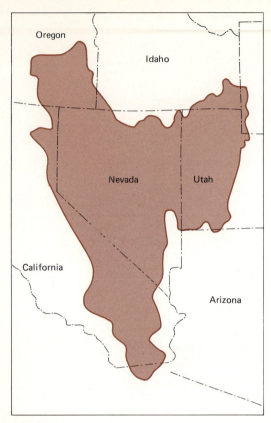

FIGURE 12–6

The Great Basin, shown here in color, is an extremely large drainage basin. About 210,000 square miles (544,000 square kilometers), it is a region of interior drainage. The colored line bounding this area defines the drainage divide—the ridge of land that separates water flowing in one direction from water flowing in the opposite direction. (Modified from a map by G. K. Gilbert, 1890, U.S. Geological Survey Monograph I, Pl. II.)

charge is directly proportional to the total available source area; this means that the larger the overall drainage basin, the longer the overall stream and the greater its discharge. The relation between the discharge of each individual tributary, its stream length, and the area of its individual drainage basin is also relatively constant throughout the whole stream system.

Other simple relations have also been observed for the stream system itself. Some of those relations are

1. The size of the stream valley increases downstream in proportion to discharge.
2. The average length of tributaries becomes geometrically greater downstream.

3. The slope of the main stream and its tributaries exponentially decreases downstream.
4. The number of tributaries decreases downstream in a constant ratio.
5. The area drained by each individual stream segment increases downstream according to a constant area ratio.

All these direct mathematical changes in streams imply that streams are highly organized *systems* of energy distribution. Rivers are systems that both input and output water and sediment; the slope and size of a stream valley are adjusted so that input matches output along any stretch of the stream. The stream system is at equilibrium. The physical systems we call streams obey one fundamental law: *changes in the stream valley that tend toward equilibrium persist; other changes are transient.*

STREAM GRADIENT

As defined in Figure 12–4, a stream's **gradient** is the *slope* of the stream valley, or the elevation loss per unit of horizontal distance. Stream gradients are usually expressed in terms of feet per mile or meters per kilometer. Stream gradients vary from less than a foot per mile to nearly a thousand feet per mile in some mountainous valleys.

If we plot many points representing the the elevation of a stream from its headwaters to its mouth, we create a smooth, concave-upward curve known as a **longitudinal profile** (Figure 12–7). When longitudinal profiles from streams of different size and climatic zones are compared, they exhibit very strong similarities. The overall shape of a stream's longitudinal profile is essentially the same all over the earth.

A feature so general must express relationships among the basic dynamics common to all rivers. The headwaters of a stream, point *A* in Figure 12–7 is an area of high gradient but low discharge. The midstream of a stream (point *B*) is an area of moderate gradient and moderate discharge. The mouth of a stream (point *C*) is an area of low gradient but high discharge. In each of these three different areas of a stream, the *product of discharge and gradient—stream energy—remains relatively constant.* The stream develops its concave upward longitudinal profile because that profile *yields constant energy expenditure from headwaters to mouth.* Nature is conservative with energy; slopes steeper than necessary are soon flattened by erosion, while gentler slopes are steepened by deposition.

The slope along any stretch of the stream valley is

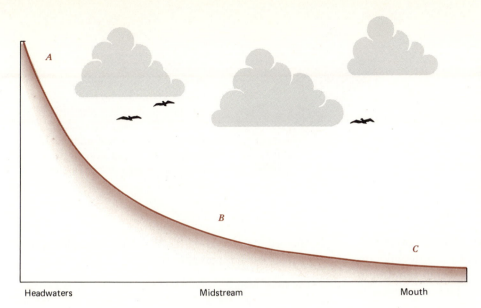

FIGURE 12–7
A longitudinal profile connects all points along the gradient of a stream valley from head to mouth. If we could look from the side at the entire notch cut by the stream, this is the profile we would see. Sometimes entire notch cut by the stream, this is the profile we would see. Sometimes termed a *profile of equilibrium,* the longitudinal profile reflects even expenditure of energy over the whole length of the stream.

itself the product of conflicting processes. Stream *erosion*, which cuts downward, increases slope, and adds sediment to the water, is offset against stream *deposition* which builds up the streambed by depositing these sediments and thus decreasing the slope. The interaction of these two forces develops a profile of equilibrium—the longitudinal profile.

The longitudinal profile allows an output of water and sediment at the river mouth equal to the input of water and sediment over the rest of the river. The gradient developed at any one point is *just sufficient to carry the load supplied to it by the drainage basin.* On many levels, we note that a stream is a system at equilibrium.

STREAM VELOCITY

If isostatic or orogenic uplift increases the gradient of a stream, the **velocity,** the stream's rate of flow, will increase. The velocity of a stream is commonly expressed in miles per hour or meters per second. Since the gradient of a stream diminishes from headwaters to mouth, we might reasonably expect that the velocity of a stream would also decrease in the same way. We would be wrong.

Careful measurements in many streams have shown that the velocity in most streams remains relatively constant from headwaters to mouth. The decreasing slope tends to slow the current, but increasing downstream discharge pushes the water along, offsetting the effect of diminishing gradient. Decreasing friction downstream—because less of the water in a deeper, wider stream is in contact with the val-

ley walls—also offsets the effect of diminishing gradient. The net effect is to keep stream velocity little changed throughout the stream's length. *Downstream, rivers carry larger quantities of finer- and finer-grained material at a higher discharge rate over a lower and lower slope.*

Stream Competency

The *largest* particle a stream can move is termed the stream's **competency,** measured in inches or centimeters. The stream's competency is related to stream velocity in the following way. If the stream velocity doubles, twice the discharge hits each particle with twice the force. We observe that *competency is roughly proportional to the square of velocity.* Tripling the stream's velocity increases its competency by roughly ninefold.

Not surprisingly, swiftly moving streams in flood stage are capable of lifting and moving ridges, houses, and other objects in their path. As a result, streams transport almost all their annual load during brief intervals of flooding.

Stream Loads

The total *weight* of material, expressed in tons or metric tons, carried by a stream is its **load** or its **capacity.** As noted before, the load carried by the Mississippi River system averages 230 million tons/year. *Load is proportional to the square of discharge.* Increase the discharge by five times, and the load that *can* be carried will increase by twenty-five times *if* the sediment is available.

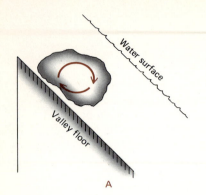

A

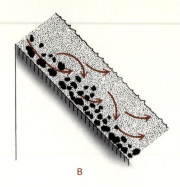

B

FIGURE 12–8
Stream A rolls a single large boulder; though highly competent, it is carrying little of its capacity. Stream B moves a dense swarm of fine-grained material in suspension; it has low competence but it carries a heavy load.

The capacity of a stream is only loosely correlated with competency, even though both are dependent variables of velocity and discharge. A high mountain stream of steep gradient may have high competence but little capacity (Figure 12–8), if the stream is surrounded by bare rock walls and little sediment is supplied. Table 12–1 reviews the relation between the dynamic factors of a stream along the longitudinal profile.

Streams carry their total load using a combination of three methods.

1. **Rolling** coarser fragments along the channel bottom. This load is termed the stream's **bed load.**

2. **Floating** finer fragments within the turbulence of the current. This load is the stream's **suspended load.** Once the current stops, the suspended load will fall to the bottom and the water will be clear.

3. In **solution** as chemicals dissolved in the water. This load is termed the stream's **dissolved load.**

In order to recover the dissolved load, the water must be totally evaporated.

The total load or capacity is the bed load plus the suspended load plus the dissolved load. Most of the load of a stream in a humid region is dissolved load; streams in semiarid regions carry most of their load in suspension.

STREAM-DOMINATED LANDSCAPES

Streams form patterns on the land that are well delineated from a low-altitude orbit of the earth (Figure 12–9). The most common pattern for streams flowing on homogeneous material is the **dendritic pattern**, a continuous branching of each of the tributaries into smaller and smaller tributaries (Figure 12–9) much like the branching pattern of a tree. As we

Downstream Changes in Stream Characteristics

Characteristic	Observed Change	Effect on Stream
Gradient	Decreases	Lowers velocity
Channel width and depth	Increases	Increases discharge and decreases friction
Discharge	Increases	Increases velocity
Competence	Decreases	Deposition of coarser load
Total load	Increases	Increases friction
Channel roughness	Decreases	Decreases friction
Velocity	Variable	Remains relatively constant
Distance to stream divide	Increases	Less mass wasting into main stream valley
Tributary length	Increases	Increased discharge
Number of tributaries	Decreases	Decreases size of drainage basin

FIGURE 12–9

A dry wash, known in Arabic as a *wadi*, is well developed on the Hadhramaut Plateau in Yemen on the southern edge of the Arabian Peninsula. A dry wash is a stream valley along which water rarely flows. Small rills join to form a *dendritic* pattern cut into the plateau rocks. The intensity of drainage lines in this dry area is very fine, typical of an arid climate. The number of tributaries increases upstream in a fixed geometric ratio. (Image courtesy of NASA.)

have mentioned, such branching follows rigid geometric rules—rules that also apply to root systems, blood vessels, the bronchial tubes, and all other natural systems of fluid distribution. Mathematically the patterns are classic fractals.

The Effect of Slopes

Dendritic drainage patterns are common because they represent the most effective way to collect water with the least expenditure of energy. The branching pattern of blood vessels and capillaries in animals is analogous, representing the pattern that most efficiently expends the energy of the pumping heart. Dendritic drainage patterns are common even in areas of steep slope (Figure 12–9).

Regions of steep slope, like that shown in Figure 12–10, suggest an arid to semiarid climate, resistant rock, and a landscape dominated by mass movement. Steep slopes are also common in high mountainous areas, where glaciation and stream erosion, resistant rock, and mass wasting all combine to produce rugged topography. Gentler slopes suggest softer rock, a more humid climate, and erosional processes dominated by stream erosion.

A century ago, three empirical laws of stream erosion were described by American geologist Grove Karl Gilbert. Each of these laws relates to the effect of slope on stream flow. These principles are

1. Steepness of slope is inversely related to the quantity of water.
2. Steepness of slope is directly related to rock resistance.
3. Steepness of slope is directly related to distance to the stream divide.

FIGURE 12–10

In this engraving of a butte within the Grand Canyon of Arizona, notice that the drainage pattern on the steep slopes is *dendritic*. This pattern is almost universal if the rock is relatively homogeneous and not disturbed by structural events. (Engraving courtesy of the U.S. Geological Survey.)

Erosion is most rapid where slopes are steepest, and deposition is most probable where slopes are gentlest. In this way, each stream generally apportions water collection and both headward and vertical erosion to the headwaters, stream migration and both transportation and deposition of load to the midstream, and both water dispersion and load deposition to the mouth. Let us look at the landforms and processes typical of each of these areas along the course of the stream.

HEADWATERS AREA

The headwaters area is the part of the stream closest to the drainage divide. The dominant activity within this area is water collection, downcutting, and headwater erosion.

Runoff Collection

Where streams are just starting, gradients are steep, discharge is low, and the full longitudinal profile has yet to form. Stream courses are marked by rapids, waterfalls, and a highly irregular valley floor. Tributaries are short, abundant, and closely spaced, pro-

FIGURE 12–11
Here is a typical ∨-shaped stream cross section. Notice the coarse boulders on the valley floor. They have been brought down the steep canyon walls by mass wasting and have been rounded by stream erosion. (Photograph courtesy of the U.S. Geological Survey.)

viding a dense network of small channels which collect water from the upper drainage basin.

The formation of small channels, termed *rills*, concentrates the erosional energy of running water, changing the overall flow pattern from uniform sheetlike flow across the land surface into flow confined to a narrow area. Stream valleys *concentrate* the potential energy formerly dispersed in sheet flow into channel flow; *it is the concentration of energy dispersed over a wide area into volumes of water moving through a narrow channel that gives streams their erosive power.*

Downcutting

A cross section of the stream reveals a ∨ shape (Figure 12–11), typical of the headwaters part of any stream. This profile results from downcutting—an important process for any stream near its source. As the stream cuts directly down, the unstable valley walls break away and fall onto the narrow canyon floor; mass wasting becomes a dominant process in widening the canyon.

In areas undergoing rapid uplift, the stream, energized with increasing potential energy, may downcut so rapidly that extremely narrow "slot" canyons are formed. The Royal Gorge of Colorado is the best known example of this phenomenon, but narrow canyons are the rule in areas that underwent rapid land uplift while they were being formed.

Steep gradients, including waterfalls and rapids, mean higher stream velocity and increased stream erosion. In time the concentrated effects of stream erosion selectively remove rapids and waterfalls and move the whole stream toward a profile of equilibrium, with energy evenly distributed along the full reach of the stream. In this way, the hydraulic forces in a river tend to eliminate concentrations of erosional power.

Headward Erosion

Waterfalls and rapids are not the only sites of extreme gradients and rapid erosion. The uppermost edge of the stream valley is an area where a channel terminates in a sloping land surface drained by *sheet flooding*, the transport of water in smooth sheets across a planar land surface. At the point where the channel or gully wall contacts the gentle slope above it, water entering the channel experiences a vertical "gradient," so that erosion along the edges of the channel is extremely rapid. This type of erosion (Figure 12–12) is termed *headward erosion*, for it is the process by which the stream *eats its way upslope, extending the overall length of the stream.* Over a very long period of time, the gradient at a fixed point on a stream steadily diminishes as the stream grows in length.

Anyone involved in the control of erosion knows that *spreading* water uniformly over the land surface limits erosion; in contrast channelizing water accelerates erosion. The point at which accelerated erosion first occurs is the lip of the gully or channel; the extent of erosive power in this small areas is striking (Figure 12–12).

Erosional Tools

Streams are generally weak cutting tools, effective at quarrying impediments in the stream course only over very long periods of time. The major erosional processes by which streams actively deepen and lengthen their channels are

1. **Solution**. The only common rock which slowly dissolves as streams run across it is limestone; in areas where the bedrock is limestone, solution is an important erosional process.
2. **Abrasion**. The sharp-edged rock and mineral particles carried by a stream will slowly abrade

FIGURE 12–12
A gully marks the change from water moving down slopes as a continuous sheet to its concentration into a channel. This is the earliest stage of stream valley formation and marks the extreme upper end of a stream's headwaters area. (Photograph courtesy of the U.S. Geological Survey.)

the rock over which the stream flows. Abrasion during floods may be a significant process, when large angular fragments are bounced along the channel floor.
3. **Cavitation**. The trapping of air bubbles in natural cracks in rocks tends to set up an air-hammering effect, splitting and dissolving the rocks.

THE MIDSTREAM AREA

Characteristics of the middle part of the channel slowly replace those of the headwaters area as the stream extends its length by headward erosion. What once was the headwaters now forms the midstream.

The trunk or main stream is well defined and is more distant from drainage divides. Gradients and discharge are both moderate, and the main stream has fewer, longer, well-integrated tributaries. The stream is dedicated to the processes of valley filling and widening, stream migration, and carrying the load.

Valley Filling and Widening

Swollen by the discharge of many tributaries, the main stream receives a load of sediment too coarse and too abundant to be carried all at once. Most of the load is temporarily deposited as **alluvium,** the stratified sediment deposited by running water.

Alluvium, deposited on the widening channel floor, becomes the surface over which the stream flows. Unlike the headwaters area, in midstream the water rarely flows across bedrock. Instead, a stream moves across material it once carried, and will carry again during storms. Alluvial material is always, on the scale of centuries, a temporary deposit (Figure 12-13) saved for a rainy day.

Valley widening continues; once canyon wall is exposed to open air, mass wasting never stops. Undercut by the river as it strikes the canyon wall, the valley sides are inherently unstable. They slowly pitch into the stream, their remnants swept downstream in flood.

The valley floor, built of alluvium, is commonly flat, having been formed by sudden deposition of load as the flooding stream spilled out of its channel. As the stream spills out of its confined channel, its width dramatically increases. In order for the stream discharge to remain constant, the velocity must quickly slow as width increases. As velocity slows, the stream's competency, which is proportional to the square of velocity, rapidly falls, and deposition occurs.

FIGURE 12–13

In this scene from the Canyon de Chelly area of Arizona, resistant sandstones form vertical cliffs in a semiarid climate. At the base of those cliffs, talus piles grade outward into an alluvial fill of a widening valley. An intermittent stream flows across its own floodplain. During spring and late summer floods, the floodplain is churned by flood waters, and a small fraction of the total valley fill is moved downstream a few miles. (Photograph courtesy of the U.S. Geological Survey.)

An alluvial plain, built by flooding, is called the **floodplain.** *The development of a floodplain signals the transition from headwaters to midstream* along the downstream longitudinal profile.

Stream Migration

Another distinctive pattern that separates the headwaters area from the midstream region is the *beginning of lateral stream migration.* Imprisoned between bare rock walls, the headwaters area stream follows a generally erratic course along joints or zones of pulverized rock along faults, but has numerous straight reaches. The generally straight stream course may be split by tributaries, but the stream course never wanders across the landscape.

As a stream enters its midstream stage, it characteristically begins to wander back and forth across its floodplain. The river ceases to run straight downslope, but rather develops a looping pattern (Figure 12–14). A stream loop is called a **meander,** after the ancient Maiandros River in Turkey; a stream with an overall looping pattern is called a *meandering stream.*

Meandering is an extremely general pattern, seen in atmospheric turbulence, ocean currents, streams of glacial meltwater, and some lava flows. The meander pattern is a wave form, analogous to the wave forms produced by seismic energy, and like any wave form, meanders are stable forms which travel. Mean-

der patterns slowly travel downstream as well as shifting from side to side across the floodplain.

Just as the longitudinal profile allows a uniform expenditure of energy over the length of a stream, *the meandering pattern is formed because it, too, achieves uniform potential energy loss for each unit of stream length.* Water piles up in the channel as it is slowed by passage around the meander curve.

The meander bend coincides with the deepest part of the river, where the downchannel slope of the water surface would normally be gentlest. But because water piles up, thereby steepening the downchannel slope of the stream's surface just where it would ordinarily have been gentlest, the meander pattern *maintains a constant slope for the water surface,* and thus achieves a uniform expenditure of energy

Because rivers meander in their midstream and downstream sections, *over a period of time, the river's channel will occupy every possible location between the confining valley walls.* The floodplain has often been called a *meander belt* (Figure 12–15) since the floodplain is formed by deposition in the waning stages of a flood and is modified by erosion as the stream migrates across it.

A meandering pattern lengthens the total distance water must travel from head to mouth, effectively lowering the gradient within the middle and lower reaches of the stream. By lowering the gradient, meanders cause deposition within the channel, which intensifies meandering.

FIGURE 12–14
The Milk River in northernmost Montana displays a typical meandering pattern across its floodplain. The outer part of each meander bend is an area of erosion, and the inner part is an area of deposition. Trees have taken advantage of the favorable environment furnished by deposition. (Photograph courtesy of the U.S. Geological Survey.)

Capacity

Because the tributaries furnish more and more water to the midstream portion, the total load or capacity of a stream is very much increased in its midsection. The widening of the channel associated with in-creased discharge means that less and less of the water is in contact with the valley floor, so that friction is lessened, and velocity is slightly increased.

This increase in velocity is countered by the increase in load, which increases friction and turbu-

FIGURE 12–15
The circular scars across the floodplain of the Laramie River in Wyoming mark the site of former meander loops. At some time the stream has occupied essentially every position on the floodplain. (Photograph courtesy of the U.S. Geological Survey.)

lence. As noted, the net effect of the many factors affecting a stream is to cause velocity to stay nearly constant throughout the entire length of the stream.

The capacity of many streams remains relatively constant, but it rises in periods of high discharge. It has been calculated that most streams carry 95 percent of their load during 5 percent of the year. Anyone who has seen a small stream in full flood cannot help but be impressed with its capacity to scour its channel severely and to move alluvial materials that may have lain dormant for decades.

The total time an individual particle resides within a stream valley has rarely been measured. The few measurements available and some reasonable estimates suggest that, for example, for a sand grain to travel from the headwaters of the Mississippi in Minnesota to its mouth south of New Orleans may require 2 million years. Obviously most particles spend very little of their time in active transport.

THE MOUTH AREA

As a stream nears base level, the gradient continues to decline. Two major activities remain: the deposition and dispersal of the total load and the disposal of the water collected within the drainage basin. The fate of both the load and the water depend on what kind of base level is encountered by the stream.

Which Base Level?

The majority of streams are tributaries to still larger streams. For each tributary stream, *the elevation of its junction* with the main stream is its **temporary base level.** Such a base level is temporary because as the main stream erodes its channel, *the elevation of the junction must also decline.*

Streams may also enter lakes; for these streams the lake level is a temporary base level, which will rise as the lake fills with sediment. The floor of a basin may be the temporary base level for a high-gradient stream exiting a mountain range (Figure 12–16). As the gradient abruptly disappears, the streams drop their sediment load as **alluvial fans,** deposits of stratified sediment which form an apron at the base of many mountain ranges.

The temporary base level of a tributary stream is always slowly dropping, so that the *rate of downcutting by any stream is controlled by the rate of downcutting of the stream into which it flows.* Each stream controls the erosional rate of the stream above it and is controlled by the erosional rate of the stream beneath it. Each stream is part of a larger interlocked system

whose whole job is to drain the land surface and remove its weathered products. The rate at which this work can be done is set by a whole series of delicately balanced stream systems, each interacting with the other.

The system's rate of operation can be slowed or accelerated by natural events. If sea level, the ultimate base level, rises, all the world's streams lose potential energy, and the rate of downcutting declines. If sea level falls, or the land is uplifted, or both, streams gain energy and begin to downcut vigorously. A sudden increase in the erosional power of a stream is termed **rejuvenation.** When an area is uplifted, the meandering patterns typical of lower gradient streams may be quickly carved into the rising land surface as $\vee$-shaped canyons (Figure 12–17). The incising of stream patterns typical of low gradients is classic evidence of recent land uplift and stream rejuvenation.

Sediment Dispersal

As a stream nears its mouth, the floodplain widens and deepens. The width of the meander belt may approach tens to hundreds of miles or kilometers. The floodplain of the Mississippi River (Figure 12–18) in the United States includes large parts of several southern states. The total area of the floodplain of the Mississippi River encompasses about 36,000 square miles (90,000 square kilometers).

Lateral migration of the stream across the floodplain spreads and sorts the deposits by size. Coarser material is deposited as sand or gravel bars within the channel, while finer material fills in the slack water areas between the bars. As a river rises in flood stage, it leaves its channel and spills out onto the wide, flat floodplain, suddenly and dramatically increasing its width. Since discharge $\cong$ width $\times$ depth $\times$ velocity, the sudden increase in width for a constant discharge just as suddenly decreases stream velocity. Competency drops with velocity, so that deposition rapidly occurs.

The coarsest materials are dropped at the margins of the natural channel, where the velocity drop is most rapid. This material often builds up **natural levees,** low mounds of material along the main channel margins that serve as dams to help confine the river within its channels during flood stages.

The stream that sorts and deposits alluvium in one area is eating it away from another. For areas where erosion is currently active, the process may be a disaster. Perhaps nowhere do the shifting whims of the Mississippi cause more trouble than at Natchez, Mis-

FIGURE 12–16

This view looks across the salt pans in front of the Panamint Range in California to the extensive alluvial fans at the base of the range. The distributary channels evident on the surface of the fans have a dendritic pattern. The fans have coalesced into a broad apron of gravel and sand. (Photograph courtesy of the U.S. Geological Survey.)

FIGURE 12–17

The Colorado River within Canyonlands National Park has tightly compressed meander patterns cut into well-jointed limestones and sandstones. Such meandering patterns do not generally form in the same environment that causes streams to downcut rapidly; the combination of looping meanders and deep \/-shaped channels suggests rapid rejuvenation of a meandering stream. (Photograph courtesy of the U.S. Geological Survey.)

sissippi. There river erosion and landsliding are eating away at the highest bluff overlooking the river. Composed of loess, the bluff supports pre-Civil War mansions and the most historic part of Natchez, property valued at millions of dollars. The normal processes of channel widening by mass wasting here interfere with human treasures. The bluff margins have retreated as much as 90 feet (27 meters) in the last century, and many of the finest mansions are literally on the brink.

Sediment dispersal is not confined to floodplains. The most distinctive landform built at a river's mouth is the **delta,** a stratified deposit of silt, sand, and mud. In map view, deltas are approximately equilateral triangles (Figure 12–19); their name comes from their resemblance to the Greek letter *delta* Δ.

Deltas are the lake or oceanic equivalents of *alluvial fans* (Figure 12–15), deposits of similar overall shape fronting many mountain ranges. Deltas and alluvial fans form for the same reason—the streams traveling through mountains or over floodplains lose the power to transport their loads. The finest materials are deposited farthest from the river mouth, whereas the coarsest materials fall nearest the mountain or shoreline, their deposition creating the last of the stream's slope or gradient.

FIGURE 12–18

This vertical view is of a part of southeastern Louisiana occupying an area of more than 500 square miles (1300 square kilometers). The Mississippi River meanders southward from Baton Rouge, at the top of the image, to New Orleans at the bottom right. The light gray areas adjacent to the meanders are long-lot farm tracts established by the early French settlers. Such tracts also outline an abandoned channel on the left. Everything in the photo is part of the floodplain of the Mississippi. (Image courtesy of NASA.)

FIGURE 12–19

Egypt's Nile delta forms the entire triangular dark area in this view from the Space Shuttle. The Nile River enters at the bottom of the image and fans out to form the delta. About 45 million people, almost all of Egypt's population, live within the 7500–square-mile (20,000 square kilometer) area of the delta.

The distribution and dispersal of sediments across a delta's surface are accomplished by sluggish, meandering streams, aptly called **distributaries**; they migrate across the sediment–water interface at the top of a delta (Figure 12–20). The sediment being carried into the sea (Figure 12–20) forms bars within the channel mouth, diverting the stream flow into ever more convoluted patterns.

The sediment once deposited on a delta has still not finished its journey. Vulnerable to attack by ocean waves and tidal currents, the soft sediments of a delta are commonly moved seaward by oceanic processes. The building of a delta is a constant fight between the rate of sediment deposition from the river and the rate of sediment erosion by the ocean. Consequently a delta's size is determined by a dynamic equilibrium between wave erosion and stream deposition. There is another limitation to delta size.

As deltas grow out to sea, their distributary gradient becomes too low for water to flow, and the flow shifts laterally to another distributary channel of higher gradient.

No matter what its size, the delta is the end of the road for a stream. Having drained the land and carried its load to the sea, the stream simply disappears. All that is left is a delta, a reminder that "the oceans are the graveyards of the land."

RIVER FLOODING

Floodplains are the part of a river valley floor that is subject to sporadic flooding and that is built up by that flooding. Approximately 16 percent of the nation's urban areas are already located on floodplains. They are attractive places to live. Transport of goods along adjacent navigable streams is cheap, the soil is often rich and easily used for both construction and agricultural purposes, and water sports can be readily enjoyed.

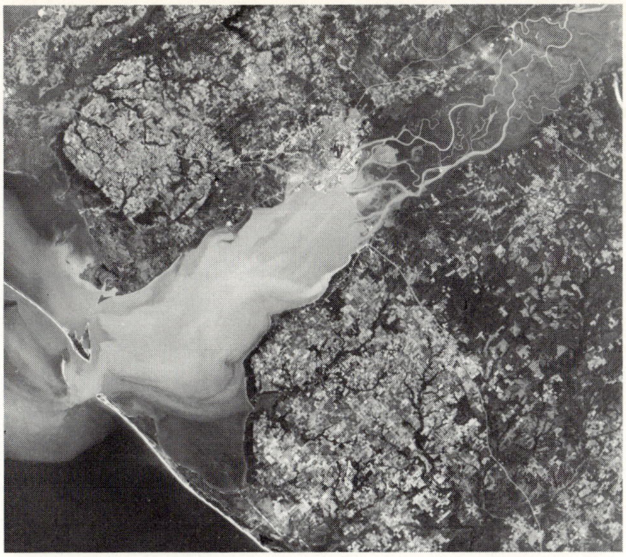

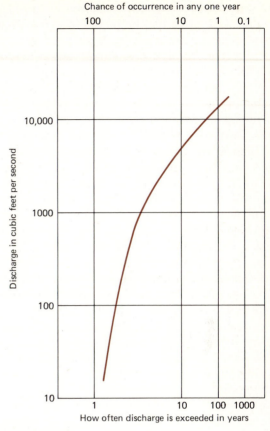

FIGURE 12–20

The Mobile Bay area of Alabama is in the center of this photograph. North (up the long axis of the bay) of Mobile Bay, the Tensaw and Mobile rivers form distributaries out onto the advancing delta. The flow of sediment is well marked by plumes of off-white turbulent currents, which fill all of Mobile Bay and spill out into the Gulf of Mexico. (Skylab EREP photography, courtesy of NASA.)

FIGURE 12–21

Graph showing the discharge versus recurrence interval for Orestimba Creek, near Newman, California. According to 42 years of recorded data, the 100–year flood should be about 11,000 cubic feet per second (310 cubic meters per second). Such a flood has a 1 percent chance of occurring in any year. (Graph modified from *Guidelines for Determining Flood Flow Frequency*, U.S. Water Resources Council, 1976.)

The rate of urban expansion onto floodplains averages 2 percent per year. At this growth rate nearly one-third of our urban areas will be on floodplains within the next 40 years. Increased development onto floodplains has both benefits and risks. The benefits have been listed; how can we analyze the risks? We already know that urbanization increases risks, as concrete and asphalt divert water into runoff that would have been partially absorbed on open ground. But can the risks be quantified?

The discharge of over 7000 of America's streams has been recorded for many years. Each level of discharge can be plotted against its *historical recurrence interval*—how often that discharge has occurred since records have been kept. Such a record (Figure 12–21) gives us the history of discharge along any stream.

Given such historical information for a stream, the *probability* of flooding in any single year can be easily computed. For the example given, a discharge of 100 cubic feet/second (nearly 3 cubic meters/second) occurs *on the average* every two years. This means the probability that this discharge will occur in any one year is 1 in 2, or 50 percent. A discharge of 7000 cfs (200 cubic meters/second) will occur, *on the average* every 10 years; the probability that that discharge

will occur in any one year is 1 in 10 or 10 percent. A discharge of 11,000 cfs (312 cubic meters/second) occurs *on the average* once a century; the chance of this discharge occurring in any one year are 1 in 100 or 1 percent. It is not possible to say when this discharge will recur. It is only possible to say that the odds are 1 percent that it will happen next year.

Such a discharge, equaled or exceeded on the average only once a century is termed the **hundred-year flood.** For most engineering purposes, the standard flood that structures should be designed to withstand is the hundred-year flood. If a structure can be protected against an event with a 1 percent chance of occurring, it has *reasonable* protection against known risk.

There are still greater floods. Both 500-year and

1000-year floods have been witnessed within historical time, and there will be others to come. Protection of areas against events with a risk of 0.002 and 0.001 percent, respectively, would simply cost too much for the benefits expected. Life consists of limiting risks, not insuring oneself against every conceivable event.

If only the probabilities of recurrence can be obtained, how can we protect our lives and property from floods? The construction of dams along streams allows nearly empty reservoirs—in the ideal situation—to absorb much of the extreme runoff from major storms and to level out the discharge of a river system over a longer period of time. Many people living below flood-control dams have learned that guessing the vagaries of weather is indeed a hazardous occupation, for heavy storms all too often find reservoirs already brimming with water. Dams impose a hazard of their own, as dam failure imposes catastrophic risks to those who live below them.

Upstream from the dam, the lake created is rapidly filled in with sediment that would have once gone on its way downstream. Such filling limits the lifetime of the dam and causes substantial expense if dredging and pumping have to be employed to remove sediment. Downstream from a dam, the sediment-free water now runs at much higher velocities, accelerating the rate of erosion. Evaporation losses from the lake mean that total discharge is much diminished, so that tributaries now furnish the main stream with coarse sediment it can no longer move. This sediment armors the streambed and causes navigational problems.

Building levees along a river bank may limit flooding, but it also artificially increases stream velocity and erosive power causing a greater discharge to remain within a channel rather than spilling over channel walls onto the floodplain. Channelizing streams and straightening their course also brings much increased velocity and erosive power.

There are few solutions to flooding that do not bring their own set of risks, hazards, and costs. Perhaps the best solution would be to use floodplains only for purposes compatible with the peril they pose. Floodplains provide superb agricultural areas, make excellent parks and playgrounds, and supply logical sources for sand and gravel quarries—in short, they are compatible with any use that is nonintensive and that does not require large, expensive permanent structures. Every city of any size has a map in its planning department showing floodplain limits for 50-year and 100-year floods; consulting this map before you buy or build could save your property or even your life.

SUMMARY

1. All streams slope toward the sea, their ultimate base level. Drainage basins define the area drained by an integrated drainage system, which consists of a headwaters collecting area, a midstream transportation area, and a downstream dispersal area.

2. The longitudinal profile of a stream develops to bring the expenditure of energy along the stream into equilibrium with the work that needs to be done at any point along the stream. The meander performs the same function on a lesser scale by smoothing the slope of the water surface.

3. The headwaters area of a stream extends itself by headward erosion and deepens itself by abrasion, cavitation, and solution. The cross-channel profile is $\vee$-shaped because mass wasting occurs along the unsupported valley walls.

4. The stream in midsection widens its valley by undercutting and by continued mass wasting. Meander belts form as the stream migrates across floodplains composed of alluvium.

5. Near its mouth, the stream pattern is composed of wide, looping meander bends over a floodplain whose width may be one hundred times that of the stream flowing over it. Sediments are distributed and deposited during lateral migration of the stream; the job is completed by the formation of distributaries and a delta.

6. The potential energy of a stream is the product of its elevation and mass or volume. This potential energy is converted to kinetic energy along the stream's gradient. A stream system is a series of interconnected streams, each controlled by the one beneath it and controlling the one above it. Increased discharge, lowered sea level, uplifted land, or a wetter climate lead to rejuvenation of the stream; rejuvenation is recognized by the incision of downstream patterns into resistant rocks.

7. The probabilities of flooding can be estimated, depending on the adequacy of the historical record. Control of flooding is usually attempted with networks of dams which collectively try to even out discharge over a year or more.

KEY WORDS

accordant stream junctions, principle of

alluvial fan

alluvium

base level

bed load

capacity

competency

delta

dendritic pattern

dissolved load

discharge

distributary

drainage basin

drainage divide

floodplain

gradient

headwaters

hundred-year flood

hydrologic cycle

intermittent stream

load

longitudinal profile

meander

mouth

natural levees

perennial stream

rejuvenation

suspended load

temporary base level

ultimate base level

velocity

EXERCISES

1. What features would you look for to separate an alluvial fan from a talus pile? From a debris flow? From a delta?

2. According to a tongue-in-cheek description by Mark Twain, if we straightened out the Mississippi River, its mouth would be over 1000 miles or 1600 kilometers out into the Gulf of Mexico. What was he talking about?

3. Meander patterns have been observed etched into the surface of Mars, which currently has no evidence of liquid water on its surface. How would you explain this?

4. What limits how much loose sediment, all of the same size, a river can pick up at any one time?

5. Over one millennium (1000 years) what will happen to an unmaintained dam? Give its total history in outline form.

6. Short of moving the river, what could be done to render the bluffs of loess in Natchez, Mississippi less susceptible to landsliding?

SUGGESTED READINGS

CHORLEY, R. J., 1985. *Geomorphology*, New York, Methuen and Co., 607 pp.
▶ *See Chapters 12 to 14; excellent bibliography.*

DINGMAN, S. L., 1984, *Fluvial Hydrology*, New York, W. H. Freeman and Co., 403 pp.
▶ *Technical but comprehensible.*

"Floodplain Delineation Map" (or similar title) in your hometown.
▶ *Vital. Every town has one.*

LEOPOLD, L. B., AND LANGBEIN, H., 1966, River Meanders, *Scientific American*, (June 1966), pp. 60–70.
▶ *Classic article, superbly written and illustrated.*

SHAW, H. R., AND GARTNER, A. E., 1984, *Empirical laws of order among rivers, faults, and earthquakes*, U. S. Geological Survey Open-File Report 84–356, 51 pp.
▶ *A superb synthesis on stream energy. Thoughtful.*

SULLIVAN, W., 1984, *Landprints*, New York, New York Times Book Co., Inc., 384 p.
▶ *Engaging writing about landscapes; easy reading.*

The 330 yard (300 meter) high crescent-shaped sand dunes of the Namib Desert in Namibia, southwest Africa, are interrupted by a dry streambed that occasionally carries flash flood waters from the adjacent highlands into the desert. Away from the dry wash, the crescent-shaped dunes give way to longitudinal dunes whose axes are parallel to the Namibian coast. (Image courtesy of NASA.)

CHAPTER *thirteen*

Arid Landscapes

*I love all waste and solitary places
where we taste the pleasure of believing
what we see is boundless. . .*

PERCY BYSSHE SHELLEY (1792–1822)

Deserts offer a uniquely fascinating scene, one made familiar by generations of Westerns filmed among the beauty of southwestern landscapes. Stark scenery, bare rock, rippling sand dunes, and alkali lakes—the words create visions of desolation and a hostile land. Yet the recent population shift in the United States toward the southwestern "Sun Belt" suggests many find the warmth, sunny living, and splendid blue skies a welcome relief from the gray, snowy winters elsewhere.

Deserts are not easy hosts. Nowhere on earth have deserts supported large permanent populations for centuries. Still, one out of eight people today lives in an arid or semiarid climate—defined as an area receiving less than 16 inches (40 centimeters) of annual precipitation. In most of the world, life in the desert is a perpetual struggle to find enough water. The desert dwellers in the American Southwest, supplied with swimming pools and abundant fruits and vegetables from nearby irrigated fields, are simply living on borrowed water. The reserves stored underground for millennia are currently being depleted after less than a century's use. The essence of deserts is that they are where water is not.

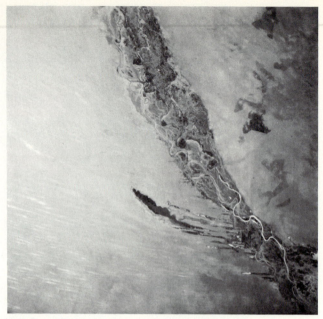

FIGURE 13–1

This image is of an area approximately equal to the state of Maryland and spans the boundary of Senegal and Mauritania. The image was taken from an altitude of 125 nautical miles during the Apollo 6 mission. The Senegal River bisects the scene. A typical desert river, it is intermittent and has a highly braided stream pattern with multiple channels. (Image courtesy of NASA.)

ARID LANDS

Geographers most commonly define a desert as an area that receives less than 10 inches (25 centimeters) of rain in an average year. *Defined in this way, deserts occupy 30 percent of the continental land surface.* Still other climatological definitions have classified deserts as areas where annual evaporation exceeds annual precipitation.

WHAT IS A DESERT?

Geologists would prefer still a different definition— one which depends on observable geologic processes. Geologists define a **desert** as an area *within which all drainage is internal.* So little rain falls on a desert that runoff does not reach the sea. Stream courses are short, stream flow is intermittent (Figure 13–1), and drainage is always into local basins. Deserts are areas lacking external drainage.

We cannot, however, escape nature's complexity by making definitions. Like most definitions, ours has exceptions. Several deserts are bisected by large streams which manage to make it across the desert to the sea. These streams typically originate in adjacent highlands. Examples include the Colorado River system, a group of about 50 streams that drain one-twelfth of the United States and a small area of Mexico. Formed primarily from melted snow that once fell among the highest peaks in western Colorado, nearly 2000 miles (3200 km) of permanent stream slashes through the heart of the southwestern deserts of the United States. The Nile (see Figure 12–19) exits the highlands of Ethiopia and makes it across the Nubian and Arabian deserts on its rush to the Mediterranean.

With these exceptions, we can alter our definition slightly. Of the rain that falls *within* a desert, none of it leaves. Unlike more humid areas, in deserts there are *no* integrated drainage systems that feed runoff into streams which in turn funnel it to the sea. Almost all the meager precipitation is evaporated; the little that remains does not flow very far.

FIGURE 13–2

The pattern of light-colored streamlines from each of the smaller impact craters suggests the prevailing direction of atmospheric flow in this cratered part of Mars. The wind direction on this planetary surface must be remarkably constant. The material being redistributed by the wind is fragmental debris from the impact event. Wind transportation is clearly effective on this surface, but the sharpness of the impact crater margins suggests that wind abrasion is an insignificant factor. (Viking Orbiter image, courtesy of Jody Swann, U.S. Geological Survey.)

We have learned only recently that humankind may enlarge existing deserts as a direct result of overpopulation. Within the last 50 years more than 400,000 square miles (1 million square kilometers) of formerly tillable land have been lost to the desert. Overcultivation of land, overgrazing by livestock, and wholesale collection of firewood remove vegetative cover, so that the soil is left to crumble and dry, exposed to intense wind erosion. Such practices also reduce the soil's water-holding capacity, remove dark-colored vegetation, thereby increasing the land surface's reflectance, and remove the atmospheric water nuclei whose presence may precipitate rain. Denuding the ground appears to accelerate these processes collectively termed **desertification**. Desertification converts arable land into intense desert.

WHAT CAUSES DESERTS?

Having defined deserts as lands of little rain and no outflow tells us nothing about *why* they are *where* they are. There are four types of natural desert, each with a slightly different origin. Those desert types are

1. Polar deserts,
2. Mid-latitude hot deserts,
3. Continental deserts,
4. Rain-shadow deserts.

Polar Deserts

Polar deserts encompass areas near the north and south pole in which *liquid* water is largely absent. Areas where the climate keeps water year around in its *solid* form are as dry and barren as the hot deserts of the mid-latitudes. The earth's poles are among the driest spots on earth, analogous to the deserts typical of many other planetary bodies (Figure 13–2) where wind is the only geologic agent modifying the landscape dominated by meteoritic impact.

Another process is also at work. The poles are regions of stable, high atmospheric pressure, where bitterly cold, dry air descends from the atmosphere (Figure 13–3). Air that is extremely cold can hold very, very little moisture, so little moisture ever reaches the polar regions. In fact, these regions receive less precipitation than any other area on earth. Though covered by snow, the addition of *new* snow is a rare event in these dry, cold regions.

Mid-latitude Deserts

The hot deserts of the world are a *mid-latitude phenomenon*. The location of these deserts coincides with worldwide atmospheric circulation patterns dominated by high pressure (Figure 13–3), and descending hot, dry air. As air descends, it is compressed, which causes it to grow warmer. As air is warmed, its capacity to hold moisture is *increased*, and so these deserts are areas of hot, dry, evaporating air currents.

Evaporation greatly exceeds precipitation. In some of these deserts, fewer than 2 inches (5 centimeters) of rain fall in an average year, yet the intense heat would allow evaporation of ten times this amount of moisture.

These mid-latitude deserts are shown in a stipple pattern on Figure 13–4. They include the Tanami, Great Victoria, and Great Sandy deserts of western Australia, the Kalahari and Sahara deserts of Africa, the Sonoran desert of northern Mexico and southwestern United States, the deserts of the Arabian Peninsula, the Great Indian desert of northwest India, and the Atacama desert of western Chile.

Continental Deserts

The deserts of central Asia owe their existence partly to their great distance from the ocean, a source of

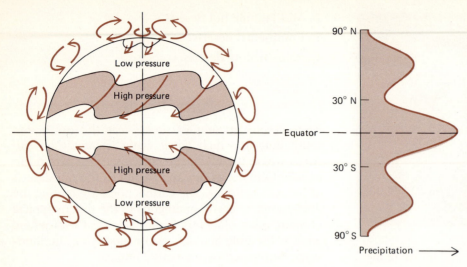

90° N
30° N
Equator
30° S
90° S
Precipitation ⟶

FIGURE 13–3

The stable high-pressure zones of the poles and the mid-latitudes both coincide with regions of low precipitation, and therefore with regions in which deserts are likely. The equator is a trough of continuous low pressure within which winds are often calm and the atmosphere extremely humid. The mid-latitudes at 25708 to 35708 N and S are narrow high-pressure belts sometimes termed the *horse latitudes*.

moisture. These *continental deserts* are denied access to water because almost all the atmospheric moisture has been removed before storm fronts penetrate that far into the large continental land mass. The Gobi desert of Mongolia is the best example of a continental desert.

Rain-Shadow Deserts

Rain-shadow deserts form on the leeward side of mountains. As moisture-laden air rises to get over the mountain range it expands and cools. As it cools, it is able to hold *less* moisture, a phenomenon familiar to anyone who has breathed on a cold mirror. The cooler air drops its rain or snow on the windward slope of the mountain.

The air that tops the mountain peaks has been dried and cooled. As it descends to the leeward side (Figure 13–5), it is compressed, and as it is compressed, it is heated. Hot air is able to hold more moisture than cold, and so the descending air is a hot, dry, evaporating air mass. The hot air pulls moisture from the ground, creating an arid environment in the leeward region downwind of mountain masses.

The climatic effects of migrating air masses are seen on many scales. A comparison of natural vegetation around any hilly area will reveal a "dry side" on even the local scale. On a worldwide scale (Figure 13–4), Death Valley and the Mojave desert of California, the Great Basin deserts of the American West, and the Patagonian desert of Argentina are splendid examples of rain-shadow deserts. The deserts of central Asia, including the Gobi, are intensified by the rain-shadow effect of the Himalayas and associated ranges.

In summary, polar deserts form because bitterly

cold air can hold almost no moisture, and what little moisture is present is in solid form. The world's hot deserts form in regions of descending, hot, dry air currents, which may be intensified by remoteness from a source of moisture and/or rain shadows in the lee of mountain masses.

Just as hot deserts are a mid-latitude phenomenon today, they have presumably always been so. The distribution of worldwide atmospheric circulation (Figure 13–3) is little affected by the presence or absence of continental land masses. Areas at the poles, and at latitudes of 25° N and S have probably always been regions of extensive deserts. We have encountered the *principle of uniformity* again; understanding the present is a key to understanding the past.

If the characteristics of ancient rocks indicated that they had been formed in deserts, we would have a good clue to their location on the earth's surface at the time they were forming. As we have seen, such thinking led Alfred Wegener (Chapter Ten) to a bold reconstruction of past continental positions and the theory of continental drift, forerunner to modern plate tectonic views. If we are to understand the rocks forming in modern deserts, we must ask some familiar questions. What are the products of desert processes? What processes are unique to deserts?

DESERT PROCESSES AND LANDFORMS

The desert is a place of fascinating deception. Because the desert world appears to be geologically quiet, space and time seem endless, as though

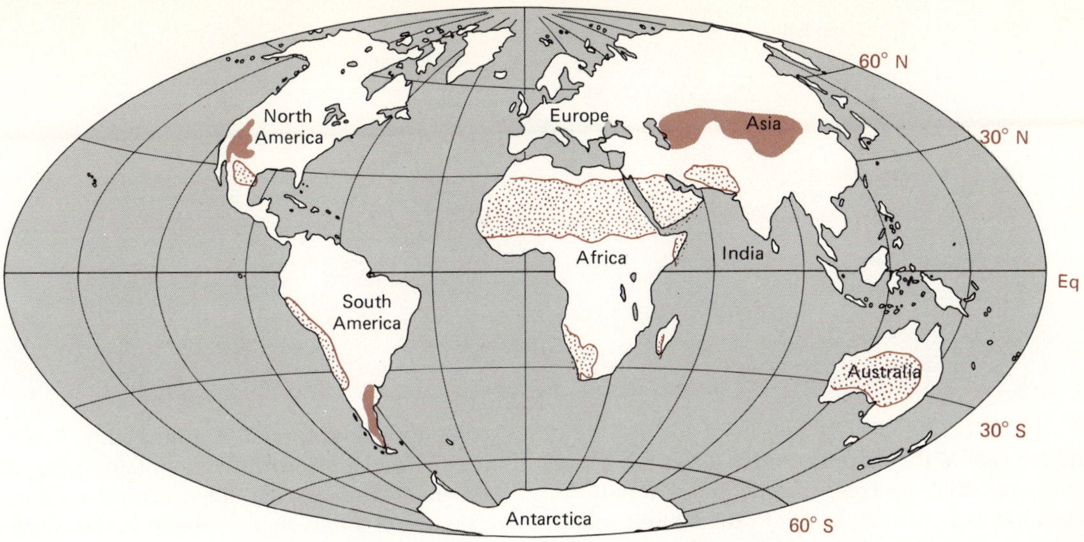

FIGURE 13–4

Mid-latitude deserts are shown with a stippled pattern in color; rain-shadow and continental deserts are in solid color. Polar deserts occupy the regions north and south of the Arctic and Antarctic circles. The deserts of the world amount to one-third of the world's land area.

change were frozen. The quotation that opened this chapter is a common view of those who do not understand deserts. Deserts are not endless wastelands. Rather they are areas where the familiar geologic processes — which elsewhere produce soft, rolling landscapes covered with green — now operate within very different time scales.

The scenery is bizarre for those, like Shelley, who knew only rolling hills of green. Craggy, splintered mountains split the crystal-clear sky only to end abruptly at the margins of huge, gently sloping desert basins (Figure 13–6). Few plants hide the geologic scene; slopes are steep, stark, and rugged. Sunlight is intense. Soils are poorly formed, thin, and gravelly, enriched in the soluble alkalies that would be washed away in a more humid land.

WEATHERING PROCESSES

As discussed in Chapter Four, the central agent in almost all weathering processes is water. Deserts, lacking water, are little weathered. Compared to humid areas, the weathering processes are painfully slow. In the absence of roots or ice, mechanical weathering depends only on alternate heating and cooling of rocks, which may have little effect. Salt weathering may be locally quite important. The occasional root can still split rocks, and in high-elevation deserts frost does the same. In sum, the processes of mechanical weathering operate, but at an imperceptibly slow pace.

In the absence of water, chemical weathering is also unhurried. Rock surfaces are often colonized by *lichens,* a "plant" that is really an intergrowth of fungi and algae. Looking more like dry off-color scales on a rock surface, the lichen grows at a snail's pace, but the life activities of this organism provide natural acids which slowly leach the rock of nutrients, an example of microorganic chemical weathering.

The most distinctive result of arid chemical weathering is **desert varnish,** a glossy brownish-black, paper-thin surface finish that develops on many rocks in desert environments. Analysis of desert varnish shows that it is a mixture largely of clays with iron and manganese oxides. One proposal suggests that dew formed on the rock surface may slowly leach the components from the outer centimeter or two of rock and bring the insoluble clays and oxides back to the surface, where they dry as a thin film. Desert varnish is found worldwide and is extremely common. Once on the surface it is chemically stable, some varnish is likely to be more than a 100,000 years old.

MASS WASTING PROCESSES

Slope steepness is inversely related to annual precipitation, so that desert slopes are often quite steep. Because of this, *rockfalls* are a dominant mass wasting process in the desert, producing *talus slopes* that maintain their position at the base of a slope for thousands of years.

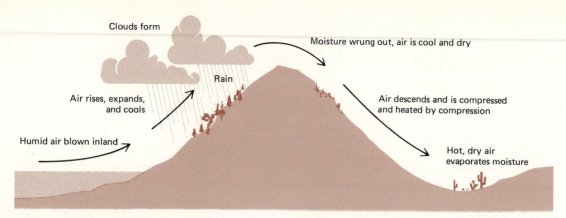

Clouds form

Moisture wrung out, air is cool and dry

Rain

Air rises, expands, and cools

Air descends and is compressed and heated by compression

Humid air blown inland

Hot, dry air evaporates moisture

FIGURE 13–5

The windward side of the mountain mass forces the humid air up, causing it to cool, expand, and dump its water as precipitation. On the leeward (downwind) side of the mountain, the dried air descends, and is compressed and heated. A rain-shadow desert is created on the leeward side of the mountain; if the intervening mountain were moved, the desert would disappear. An example of a rain-shadow desert in the western United States is the Death Valley region of eastern California.

Mudflows are another important mass wasting process. Two distinguishing characteristics of desert precipitation are that it is *sporadic* and it is *intense* when it does fall. The loose, sparsely vegetated desert soils are easy prey for the occasional sporadic thunderstorm. Runoff is rapid. Valleys that may have been dry for decades may experience flash flooding, and sheet flow across desert slopes may soon produce mudflows.

In summary mass wasting operates at moderate intensity within a desert, but it is infrequent. A rock may fall now and then, and a sporadic cloudburst will produce an occasional mudflow, but most of the time little is happening.

STREAM PROCESSES

Stream processes include sheetwash, gullying, and the development of integrated but intermittent drainages. Deserts very commonly contain evidence that stream activity was more vigorous in the past and that modern stream processes are episodic and often catastrophic. Desert scenery is formed by brief periods of catastrophic restyling that punctuate long periods of little change.

Alluvial Fans

As a desert stream spills out of a mountain range onto an adjacent basin—its temporary base level—it often creates an **alluvial fan.** Alluvial fans are a low cone of stratified boulders, gravel, sand, silt, and mud radiating from the mouth of a stream valley onto the floor of a desert basin (Figure 13–7). Many years of study have shown that the area of an alluvial fan is directly related to the size of the drainage basin that feeds it and inversely related to the overall fan slope. The more resistant the rock material is to weathering, the smaller and steeper is the average fan.

The surface of the fan is crisscrossed by dry stream channels, remnants of mudflows, and occasional plant debris. The fan slopes more steeply near the mountain and more gently near the basin. Each area of slope is neatly adjusted to the approximate size of the particles composing it. Steep slopes form in order

FIGURE 13–6

This photograph taken from the top of the House Range, Utah, is typical of desert scenery in the western United States. The contact between the wide, flat basin and the mountain range is a long, gentle slope. (Photograph courtesy of the U.S. Geological Survey.)

FIGURE 13–7
An alluvial fan near Salt Lake City, Utah. The original engraving was drawn by W. H. Holmes. The human figures at the base of the dry fan suggest scale. (From G. K. Gilbert, 1890, U.S. Geological Survey Monograph 1, pl. XLIV.)

to provide enough incline to carry the coarser bed load deposited near the mountain front as finer debris is swept progressively along gentler slopes near the basin floor.

Finally, as temporary base level is reached, the gradients become too slight to carry anything, and even the finest-grained material in suspension is deposited. The fan has progressively sorted the stream load by size.

Playas

Farther out on the basin floor, the stream exiting the alluvial fan deposits its solution load by evaporation. Most of the year, the area of deposition is a dry lake bed, termed a **playa.** Most commonly the lake is bone dry, with only glistening white material evaporated from the stream to mark the location. Mud cracks, sometimes several feet or meters deep, cover the lake floor, indicating the extent of contraction as the muddy, alkali-rich sediments dried and evaporated (Figure 13–8).

Examining the chemistry of the lake bed materials is like averaging the chemistry of all the mountainous region surrounding the lake. The chemistry of playa sediments is a *direct* reflection of the kinds of rock above the lake, as well as of the weathering processes that released soluble materials into the streams that drained into the lake.

The total profile from the top of the alluvial fan to the shimmering dry lake is a **profile of equilibrium,** analogous to the longitudinal profile of a stream in a more humid area. Nature is again conservative with energy; slopes steeper than necessary

FIGURE 13–8
A playa in Death Valley, California. The peaks in the distance wring the moisture out of the atmosphere before it moves over Death Valley. The polygonal lines in the dry lake bed are large mud cracks, formed when the lake bed dried out a few days after an infrequent rain. (Photograph courtesy of the U.S. Geological Survey.)

FIGURE 13–9

A pediment surface in front of the Growler Mountains in southeastern Arizona. The pediment is covered with a few feet (up to a meter) of loose rock and soil; desert scrub vegetation is on top of it all. The contact between the mountain front and the pediment is abrupt. (Photograph by James Gilluly, U.S. Geological Survey.)

are soon flattened by erosion, and gentler slopes are steepened by deposition. Each slope is just sufficient to do the work imposed on it by the load reaching that slope.

Playa deposits may have economic value. Not only may they furnish sparkling mineral specimens of rare, soluble minerals to collectors, but they may also contain natural deposits of other minerals useful in the chemical industry including compounds of potassium and other alkali metals. Borax is perhaps the best known of the materials quarried from dry lake beds; the borax deposits of dry lakes near Death Valley, California are the only known economic source of this valuable material, which is used in glass, drugs, welding compounds, and porcelain enamels. It is probably most easily recognized when it forms a component of hand soaps and laundry powders.

Pediments

Pediments are bedrock surfaces of very low slope at the foot of mountains. They commonly surround a mature mountain range on all sides. Often covered with a very thin veneer of gravel (Figure 13–9), the pediment surface is extensive and forms an *abrupt* contact with the adjacent mountain front. Moving toward the adjacent basin, the pediment gradually grades into an alluvial fan (Figure 13–9).

The pediment, first described by the American ge-

ologist Grove Karl Gilbert in 1877 in the Henry Mountains of Utah, is a nearly universal landform in arid and semiarid regions. They arise in environments where both soft and hard rocks are the norm, and they occur in a variety of sizes.

The origins of pediments are controversial. Overall they seem to originate from sheet flooding across bare rock surfaces during sporadic torrential rainstorms. Both their position and their overall character suggest that the *pediment is a surface of transportation,* across which large volumes of sediment are moved toward the basin beyond the mountain. The sharp angle at which pediment surfaces meet the mountain front indicates that, unlike streams in humid areas, which eat their way *down* into bare rock, streams flowing across pediment surfaces eat their way *back* into the adjacent mountain.

As desert slopes retreat *back,* gentler slopes replace steeper ones at the mountain front (Figure 13–10). Because the slopes become progressively gentler, *the land surface gradually becomes buried under its own debris.*

The history of desert scenery is that of mountainous areas being carved back into long, sloping pediment surfaces. In time, the remnants of the mountain mass form isolated knobs, centered within radiating sheets of debris that rest on pediment surfaces.

Such a landscape is stable and little changing, as if frozen in time. Slopes have become too low to allow movement of materials across them. The rate of change becomes almost nil. The landscape is at equilibrium, barring isostatic or orogenic uplift or changes in climate.

Stream Valleys

Streams in deserts are intermittent and overloaded with bed load. Their pattern is typically braided (Figure 13–1), with load in excess of the ability of the stream to transport it. The stream deposits are channel sequences of aggrading sand bars and laterally shifting meander deposits (Chapter Twelve).

With little vegetation to hold loose materials in place, stream valleys become major sites of aggradation (Figure 13-11), which may then be heavily scoured during intermittent periods of intense flooding. The dry streambeds in desert areas of the American Southwest are known as *arroyos* and as *wadis* (Figure 12–9) in the Arabic world.

Major stream valleys in deserts are widely spaced since rainfall is so rare. Streams rapidly lose most of their flow by evaporation and by soaking into the porous, gravelly materials that make up their valley

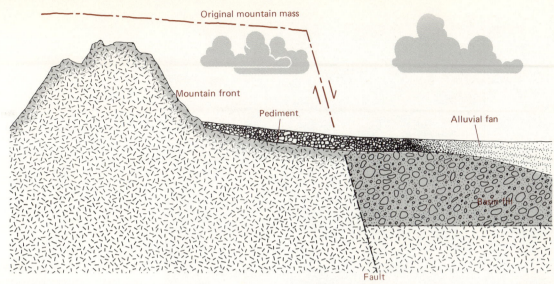

Original mountain mass

Mountain front

Pediment

Alluvial fan

Basin fill

Fault

FIGURE 13–10

This cross section is typical of much of the desert of the American West. The mountain mass was originally (dashed line) twice the size of the remnant now left at the surface. The edge of the uplifted mountain is back of but parallel to a buried fault (its downward continuation is the dashed line) that separates the mountain range from the adjacent basin. The pediment, which meets the mountain front at a sharp angle, serves as a surface of transportation for the material removed from the mountain front. The pediment grades basinward into an alluvial fan.

floor. Rarely do streams flow more than a few miles or kilometers from where they originate. The streams empty into desert basins, blocked by mountains, mudslides, lava flows, and alluvial fans. The lakes are short-lived and evaporate to form dry lake beds within a few days or weeks.

Mesa and Butte Topography

Among the distinctive landforms of arid and semiarid regions are the **butte,** a small flat-topped hill, and the **mesa,** a larger butte. These distinctive landforms are created by stream erosion which cuts into layered sedimentary rocks of uneven resistance.

FIGURE 13–11

A dry wash meanders across a semiarid area of brush in northwestern New Mexico. The arroyo is dry, which is its typical condition most of the year. The channel is poorly defined, heavily braided, and waiting for a flash flood to do a day's work.

FIGURE 13–12
The effect of a resistant horizontal rock layer is easily seen. The master stream has cut through the resistant cap, and isolated remnants stand as buttes and mesas, skirted by aprons of less resistant soft rock.

sedimentary rocks of uneven resistance.

As soon as the stream has breached a resistant rock layer, it encounters soft, easily erodible rock beneath and quickly cuts a deep channel. Since stream erosion is now concentrated into channels, the remainder of the resistant rock layer is "saved" from erosion and is exposed largely to sheetwash, an erosional process less effective against resistant rock layers. As streams cut their way back into resistant areas, blocks of resistant rock become isolated by the erosion into free-standing buttes and mesas (Figure 13–12).

Once a mesa has been created, continuing erosion of the softer material beneath the resistant caprock causes undermining of the caprock, which soon fails by rockfall. In this way, both rockfall and stream erosion reduce the size of a mesa until at some stage it begins to be referred to as a butte. As it grows still smaller, it may be called a *spire, and finally it, too, disappears, a victim of the law that steep slopes concentrate stream erosion.*

EOLIAN EROSION

For wind to be an active erosional agent, there must be an atmosphere capable of being disturbed and surficial materials that are loose and available for transport. On Earth, Venus, and Mars (Figure 13–2) these conditions apply in some areas.

Twenty percent of the earth's surface is arid and largely devoid of vegetation; about one-quarter of this arid land surface is covered with mobile sand. This sand has come from intermittent stream channels, mudflows, older dunes, coastal sands, and

FIGURE 13–13
In the northeast end of the Saudi Arabian desert called the Rub' Al Khali (Empty Quarter) is the great sand dune field known as the Ash Shaiba. The dunes rise to great heights, with an extremely steep *slip face.* Any disturbance of the base can bury a person under a cascade of sand. The dunes are called *barchan* dunes.

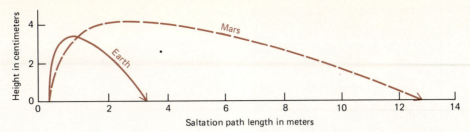

FIGURE 13–14

This graph compares saltation on Earth with that on Mars, with its low atmospheric density. Notice that a particle on earth may bounce to a height of 3.6 centimeters (1.6 inches) and be carried a horizontal distance of 3.5 meters (11.5 feet) at the wind speed of the experiment; the same particle will travel four times as far on Mars. (Data from NASA Contractor Report 3788.)

weathering of sandstones and is deposited as **ergs** or *sand seas*. These huge expanses of shifting sand are shaped into a variety of landforms by the wind. The largest erg on earth is the "Empty Quarter" of Saudi Arabia where the Rub´al Khali (Figure 13–13) occupies nearly a quarter of a million square miles (575,000 square kilometers) of the southern Arabian Peninsula.

In areas of loose sand and sparse vegetation, **eolian** or *wind-driven* processes are dominant. In these areas, wind is capable of forming distinctive depositional and erosional landforms by both vigorous erosion and rapid transportation.

Deflation

Deflation is a term used to describe the *removal* and transportation of material by wind. As measured in repeated studies, the majority of material deflated during windstorms is fine- to medium-grained sand which is lifted and carried within a 3 to 4 feet (1 meter) zone directly above the ground surface. The sand is moved in two related processes.

1. **Saltation.** A process by which sand grains are both lifted and bounced into the layer of air moving above the desert surface. The grains move by repetitive bouncing (Figure 13–14), much as does bed load along a channel floor. The word saltation comes from the Latin *saltare,* which means "to leap"; saltation puts about four-fifths of the total load into temporary suspension above the land surface.
2. **Surface Creep.** A process that moves grains too large to saltate by rolling them directly along the land surface. The energy driving this motion is provided by the shearing force of wind across the surface, aided by the impact of saltating grains. The force from the wind shear drops to zero at a height above the land surface equal to one-

fortieth of the mean grain diameter. Thus a rough surface including large gravel, cobbles, or boulders will protect smaller grains from being transported by limiting surface creep.

If the desert surface is composed of material of widely varying size, the finer material may be deflated in time, leaving behind a surface covered by a concentrated mix of coarse material. Because a surface of coarse interlocked grains limits wind velocity at ground level, such a surface is immune to further deflation.

Many desert areas expose such a surface, variously termed **desert pavement, deflation armor, or lag gravels** (Figure 13–15). Once protected by this mosaic of coarse materials, only flash flooding can tear through the armored surface and expose finer-grained materials to the deflating effect of wind again.

Deflation of loose material is an important process on other planets having a minimal atmosphere. In the absence of liquid water, the surfaces of several of the rocky planets could reasonably be described as deserts. Many images of other planets taken from exploratory spacecraft have depicted scenes indicative of strong deflation. The first Mariner mission to Mars arrived just as that planet was shrouded in reddish-orange dust stirred by a planetwide dust storm. The storm continued for a month or more, a spectacular confirmation of the importance of deflation as a geologic agent on Mars.

Mars furnishes a particularly interesting comparative example, since the atmospheric pressure on Mars is only about 2 percent that of Earth. Even though the atmosphere is sparse, it is largely composed of carbon dioxide, a gas much denser than our own atmosphere, so that uneven heating and cooling of the planet can still drive large windstorms. Deflationary surfaces (Figure 13–16) were recorded at high resolution by Viking landers, which returned

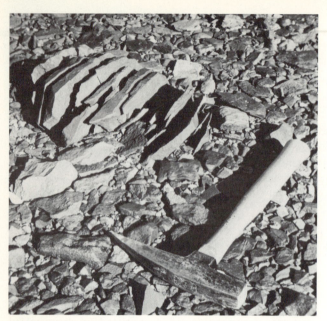

FIGURE 13–15

Lag gravel on the surface of Death Valley. Notice how the rocks fit together almost like a mosaic to protect the surface with a complete deflation armor. For scale, the hammer is about 1 foot (30 centimeters) long. (Photograph courtesy of the U.S. Geological Survey.)

FIGURE 13–16

This was the first color picture ever taken on the surface of Mars. The image, from the Viking 1 Lander, shows an area of volcanic rock littered about on a strongly deflated surface. The horizon is about 1.8 miles (3 kilometers) from the camera. Other features located on the Martian surface include sand seas, dune fields, *yardangs*, deflation basins, and crater streams (see Figure 13–2). (Image courtesy of NASA.)

scenes reminiscent of many desert areas on Earth.

Similar data from Venus also reveal a surface stripped of its fine materials by turbulence in the dense atmosphere of the planet. Since Venus is essentially the same size and density as Earth, comparison of geologic processes on the two planets has been an important part of international research efforts. Figure 13–17 displays some panoramas of Mars, taken by the American Viking landers, and of Venus, taken by the Russian Venera landers. The uppermost image will show clearly defined dune forms on the surface of Mars. The lower Venera images are severely distorted, but still reveal a significant amount of information about the surface of a planet perpetually shrouded from view.

Either of the two sets of images could have been taken in one of Earth's deserts. Once again, the principle of uniformity connects product to process, even though the processes affected planets millions of miles or kilometers distant from Earth.

Wind Abrasion

Wind, armed with sharp, angular particles of sand, would seem to be a formidable agent of abrasion. Anyone who has ever had a windshield frosted in a single violent windstorm knows firsthand how effective wind abrasion can be.

Studies of sandstorms show that wind velocities rise steadily as the air moves above the retarding effects of friction with the land surface. At the same time, the *amount* of sand in a given volume of air drops rapidly the farther above the desert surface it is. The result of these two factors interacting is that most wind abrasion occurs within a foot (30 centimeters) of the land surface. For instance, armoring the lowest fraction of a wooden telephone pole in an erg makes it a reasonably permanent structure; if the pole is not armored, it will be sandblasted in two in a year or less.

The scouring effect of sand-laden wind is most effective when it is applied on a fairly soft surface, where wearing away of material is easiest. Patterns of wind scour exist at all scales, from grooves a few inches long cut into hard rocks, to surfaces combining both deflation and abrasion (Figure 13–18), to a variety of polished and fluted hard-rock landforms.

Among the more interesting examples of wind-scoured landforms are hills streamlined by wind erosion into highly symmetrical oval shapes whose long axes parallel the prevailing wind direction. Such hills, called *yardangs*, have formed in many desert areas on Earth and have also been observed on Mars.

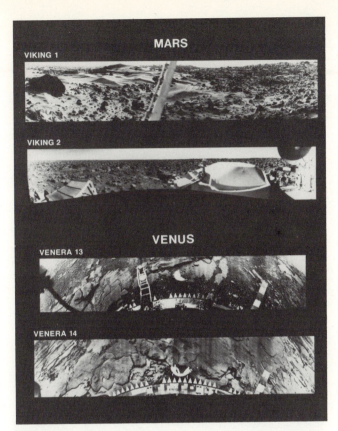

FIGURE 13–18

The evidence of wind abrasion and deflation is obvious in this photograph of a scoured surface in the American Southwest. The surface is weakly consolidated sand, a material easily scoured and deflated by windstorms. (Photograph courtesy of the U.S. Geological Survey.)

FIGURE 13–17

These panoramas of the Martian landscape by the Viking orbiters reveal a rock surface typical of the desert surface many places on Earth. Because of the imaging technique used by the Russian Venera landers, the Venusian horizon is severely distorted, but the available image suggests an extremely arid landscape. (Image courtesy of NASA.)

EOLIAN DEPOSITION

The ergs are aptly called "sand seas," for their appearance from space may be as monotonous as a view of the oceans. Stretching for miles or kilometers in every direction is seemingly endless sand. Such sheets of sand are themselves a distinctive landform, best seen from the reaches of space.

Within many sand seas the most distinctive landform is the **dune,** an accumulation of sand that forms originally in the lee of a rock, a fence, some vegetation, or any other obstacle that slows the flow of wind near the earth's surface, causing the suspended and saltating sand to deposit. Once a dune has begun to form, it traps saltating grains and so begins to grow.

Most dunes are hills of loose sand, with a gentle slope on the windward side, the side facing the wind, and a steeper slope on the leeward side, the side fac-

ing *away from* the wind (Figure 13–19). They form by wind shear around an obstacle and grow by accumulating saltating sand. Dunes migrate generally downwind by the movement of saltating sand grains up the windward slope and then collapse down the leeward slope. Usually the leeward side maintains a slope near the 30° to 35° angle of repose (see Chapter Eleven) for sand. Because sand slips, small sand avalanches, are the common mode of mass wasting on the leeward face this side of the dune is often called the **slip face.**

Dunes may form anywhere abundant particles and wind interact, so they are not restricted to deserts. Indeed, snowdrifts are a form of dunes, as are the dunes formed in volcanic ash. Coastal dune fields, composed of beach sand and produced by strong onshore winds, are common. Dune forms have been recognized on other planets as well (Figure 13–17).

The smallest "dune" is an individual *ripple* on the surface of loose sand. Though not termed dunes, ripples obey the same mechanical principles as their much larger analogs. They migrate downwind and behave in every way like the wave form that they and dunes truly are.

Seen from space (Figure 13–20) dune fields look like nothing more than giant ripples with wavelengths of several hundred feet or tens of meters. They are among the most repetitive of all landforms.

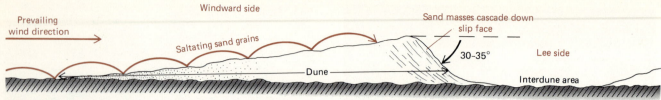

FIGURE 13–19

This cross section of a dune indicates the dominant activities of an unstabilized dune. The windward side is a surface of transportation by both saltation and surface creep; it is commonly intensively rippled. The slip face is an active zone of sand cascades which collapse as soon as the slope is oversteepened beyond the stable angle of repose. In interdune areas deflation continues until bedrock or desert pavement is reached.

There are numerous classifications of dunes, based primarily on overall shape. The variety in shapes of dunes is very, very great, and classification schemes can become quite numerous and tedious. We adopt for our purposes the following classification.

1. **Longitudinal** or **Seif** Dunes. Ridges formed parallel to prevailing wind.
2. **Transverse Dunes.** Ridges formed at right angles to prevailing wind.
3. **Barchanoid Ridge Dunes.** Transverse dunes with wavy ridge crests.
4. **Barchan Dunes.** Crescent-shaped dunes whose points face downwind.
5. **Star Dunes.** Dunes of complex shape owing to shifting wind directions.

Longitudinal Dunes

Longitudinal dunes are sometimes termed *seif* dunes from the Arabic word for "sword." The swordlike shape in map view is formed by long streamers of sand subparallel to the prevailing wind direction (Figure 13–21A).

Found in areas of abundant sand supply, longitudinal dunes appear to be formed by spiraling wind currents whose central axis is parallel to the ground and the length of the dunes. The wind currents form a vortex (whirlwind) that sends sand grains spiraling up the double slip faces characterizing this dune style. Longitudinal dunes may extend for distances of tens of miles or kilometers; they make up about a third of the world's dunes. The area between the dunes (Figure 13–22) is deflated down to a bare rock surface. In crosssection these dunes are an inverted V (∧), whose crests may be straight or slightly sinuous.

Transverse Dunes

Transverse dunes are defined by long ridges of sand *perpendicular* to the prevailing wind direction (Figure 13–21B). The sand is usually somewhat coarse and is moved only by strong, prevailing winds. These dunes appear to arise when disturbances to the laminar ground flow of air produce turbulent eddies that

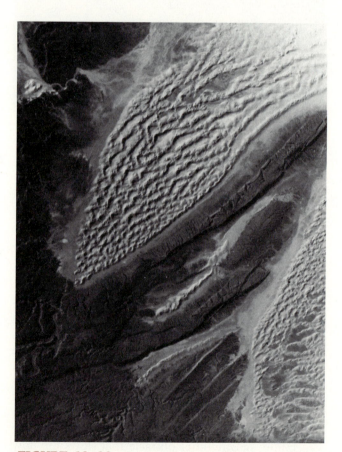

FIGURE 13–20

This photo of Algeria's Tifernine dunes, trapped within a structure in the Tassili n'Ajer Mountains in southeasternmost Algeria, was taken from the Space Shuttle Columbia. The dunes are a *barchanoid* ridge type, and they form a part of the Issaouane Erg. The dunes are in excess of 1000 feet (305 meters) high. This view is from an orbital altitude of 160 nautical miles (296 kilometers) above the land surface. (Shuttle photography courtesy of NASA.)

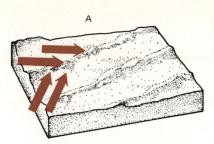

Longitudinal or seif dunes. Arrows show
probable dominant winds.

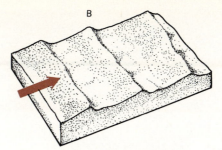

Transverse dune. Arrow shows
prevailing wind direction.

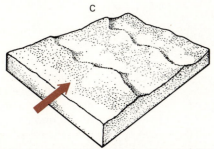

Barchanoid ridge. Arrow shows prevailing wind direction.

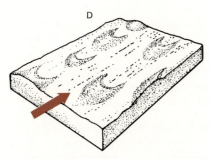

Barchan dunes. Arrow shows prevailing
wind direction.

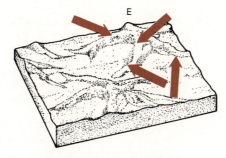

Star dunes. Arrows show effective
wind directions.

FIGURE 13–21

These sketches portray five of the
dominant dune types. (Modified
from E. D. McKee, U.S.
Geological Survey Professional
Paper 1052.)

cause deflation in what will become the interdune
area; deposition occurs just downwind to form the
initial ridge. Once the pattern of dune - interdune is
set up, it is self-sustaining.

Transverse dunes migrate by saltation of sand
grains up the windward side and sand slips down the
leeward slip face. Movement of the entire dune form
downwind at rates of several tens of yards or meters
a year have been recorded. If the sand supply becomes somewhat limited, or if wind directions become somewhat more complex, transverse dunes
pass progressively into *barchanoid ridges* (Figures 13–
20, 13–21C, and 13–23) and then into *barchan dunes*
(Figures 13–13, 13–21D, and 13–24).

Barchanoid Ridge Dunes

Increasing the sinuosity of the ridge crests gives a
transverse dune the shape typical of a *barchanoid ridge
dune* (Figure 13-23). Horizontal eddies in the wind

currents, alternately driving parts of the ridge crest
downwind at higher rates than the rest, increase the
concavity of the ridge crest downwind.

The ridge crest may also become convex downwind for the same reason, until the ridge crest becomes extraordinarily complex in map view. Some of
the patterns resemble fish scales and similar herring-
bone patterns. Barchanoid ridges usually suggest that
wind patterns have become more complex or that
their deflationary capacity exceeds available sand
supply.

Barchan Dunes

The term barchan comes from the Turkish word *bar-
kan* meaning "sand hill." Barchan dunes display a
crescent-shaped slip face oriented downwind (Figure
13–24) and terminating in two distinct points. The
gentle slope of the windward side is convex into the
wind, while the lee side is steep or abrupt.

FIGURE 13–22
This photo shows wind ripples on the slip face of a small seif dune. The interdune area in the foreground has been deflated to yield materials for the dune. (Photo by G. K. Gilbert, U.S. Geological Survey.)

Transverse dunes, barchanoid ridges, and barchans are transitional stages in dune formation caused by diminishing sand supply, so that the wind has excess deflationary capacity. If the capacity of the wind increases, or sand supply diminishes further, the floor of the desert will be swept clean and dunes will not form.

Star Dunes

Dunes that have irregular shapes and multiple slip faces are collectively called star dunes (Figure 13–21E). They commonly have a high central peak, from which three or four "arms" radiate outward, giving the appearance in map view of a poorly drawn star.

Star dunes are little more than defined piles of sand. Their internal structures are nearly chaotic, composed of lenses of sand piled on top of and cross-cutting one another. Star dunes tend to grow vertically rather than migrating horizontally. They are a common dune form, making up about one-fifth of all dune forms on earth. They originate in areas with highly variable wind directions, with each shift in wind direction creating a new "arm" on the star from the destruction or modification of the other arms.

SUMMARY

1. About a third of continental land masses receive so little liquid water that drainage from the high elevations does not reach the sea. Deserts form because of the bitter cold at the poles, the stable zones of descending hot air in mid-latitudes, the remoteness of continental interiors from seawater, or the rain-shadows in the lee of mountainous areas.

2. Geologic processes in deserts tend to be both intermittent and intense. In the near absence of water, very steep slopes are the norm, and weathering is unhurried.

3. Mechanical weathering processes are dominated by salt weathering and thermal expansion. Chemical weathering is limited to acid attack by lichens and the production of desert varnish by dew. Mass wasting is limited to rockfalls and mudflows.

4. Erosional processes are dominated by stream erosion, which tends to be extremely sporadic and vigorous. Gullying and sheetwash are both common, with intense scouring and very short-range transport being the norm. Desert washes tend to be areas of braided streams with intermittent flow regimes. Taken together, landforms from the heads of alluvial fans at range margins to playas in the center of the basin form an overall profile of equilibrium.

5. Deflation becomes a dominant process where abundant, loose fine- to medium-grained sand is present. Wind abrasion is limited to within a foot or so (about 30 centimeters) of the land surface.

FIGURE 13–23
Barchanoid ridges in the San Luis Valley of Colorado. The steep slip face is in shadow, and the gentler windward slope dips to the right. The prevailing wind in this area must blow from right to left; the dune forms themselves migrate in the same direction. (Photograph courtesy of the U.S. Geological Survey.)

Continuing deflation may create a lag gravel which serves as a deflation armor.

6. Distinctive arid zone landforms include pediments meeting cliffs at abrupt angles covered by angular talus slopes; mountain remnants surrounded by enormous pediment sheets and alluvial fans; poorly integrated intermittent stream valleys or arroyos; playas, deflation basins, mesa and butte topography; and a wide variety of dunes within sand seas.

KEY WORDS

alluvial fan	barchanoid ridge dune	deflation	mesa
barchan dune	butte	deflation armor	pediment
		desert	playa
		desertification	profile of equilibrium
		desert pavement	saltation
		desert varnish	seif dune
		dune	slip face
		eolian	star dune
		erg	surface creep
		lag gravel	transverse dunes
		longitudinal dune	

FIGURE 13–24
Barchan dunes near Biggs, Oregon. The scale is given by the telephone poles just visible over the tops of the dunes. The slip faces are in shadow; the windward profile slopes gently to the left. (Photograph courtesy of the U.S. Geological Survey.)

EXERCISES

1. Since sand dunes move downwind, if you were foolish enough to buy a house just downwind of a large, moving dune, what would you do to attempt to stabilize the dune?

2. Extremely well-formed, integrated $\vee$-shaped stream valleys have been found beneath the sands of the Sahara. These landforms suggest that the Sahara desert in the past was a region of flowing streams and abundant rainfall. What factors can you suggest that might account for such a dramatic change in climate?

3. It has been said that too many human beings makes too much desert. Given your knowledge of *desertification*, what does this remark mean?

4. The dust storms of the mid-1930s in the United States left 150,000 square miles (390,000 square kilometers) of land devoid of topsoil. Individual dust clouds stretched for 1,350,000 square miles (3,496,230 square kilometers) and were up to 3 miles (5 kilometers) thick. The estimated total loss was 4.2 *billion tons* (3.8 billion metric tons) of soil. A combination of prolonged drought and loss of soil moisture by repeated plowing set the stage for dust storms. What steps would you have taken as a government to try to reclaim this disaster area and return the Great Plains to the breadbasket they are today?

5. It has been estimated that wind-blown dust deflated from deserts amounts to 500,000,000 tons (450,000,000 metric tons) annually. How does this compare to the total sediment removed from just the North American continent by streams (see Chapter Twelve)?

SUGGESTED READINGS

BAGNOLD, R. A., 1941, *The Physics of Blown Sand and Desert Dunes*, London, Methuen, 524 pp.
► *The classic study, by a British officer who used Model T Fords and rolls of chicken wire as aids to desert research in the 1930s.*

MCKEE, E. D. (ed.), 1979, *Global Sand Seas*, U.S. Geological Survey Professional Paper 1052, Washington, D.C., U.S. Government Printing Office, 284 pp.
► *Beautifully illustrated introduction to eolian deposition.*

PÉWE, T. L., 1981, *Desert Dust: Origin, Characteristics, and Effect on Man*, Geological Society of America Special Paper 186, 124 pp.
► *The definitive study of deflation around the world . . . and Mars.*

UNITED NATIONS, 1977, *World Map of Desertification*, Scale 1:25,000,000, New York.
► *The global outlook for desert intensification; available from FAO and UNESCO.*

Patterns of crevasses in the Agassiz glacier, Alaska dwarf the climber (note the figure in the lowest part of photograph). Glacial ice, whose beauty is well displayed here, is among the most vigorous agents of erosion on earth. (Photograph courtesy of the U.S. Geological Survey.)

CHAPTER *fourteen*

Glacial Landscapes

Ice is the silent language of the peak . . .
CONRAD AIKIN (1889 - 1973)

For the moving of large masses of rock, the most powerful engines . . . which nature employs
are the glaciers, those lakes or rivers of ice.
JOHN PLAYFAIR (1802)

MODERN GLACIATION

In the same way that about 30 percent of the earth's land area is too arid for external drainage to the sea, *approximately 10 percent of the land is too cold for water to run freely;* water moves instead as **glaciers,** *accumulations of snow and layered ice which migrate under the direct influence of gravity.* Essentially all the area covered by glaciers today is accounted for by the glaciers of Greenland and the Antarctic. Within the last few million years, however, nearly one-third of modern continental surfaces were buried under glacial ice.

Ice is far more viscous than water so that

1. The flowage of glaciers is ponderous and slow.
2. Ice accumulates to thicknesses of up to nearly 3 miles (4.8 kilometers) on continents. Ice on continental margins floats out onto the ocean to form ice packs.
3. Glaciers store three-quarters of the world's fresh water.
4. The upper part of a glacier is composed of layered deposits of snow, analogous to a sedimentary rock, and the lower part is composed of foliated ice, analogous to a metamorphic rock (Figure 14–1).
5. Glaciers have extreme competence and extreme capacity; they carry and deposit huge loads whose size range is commonly very, very large.
6. The landforms produced by sluggishly flowing ice are unlike landforms produced by running water or by wind.

Physical properties of ice also have direct effects on glaciation. Since ice is the only common solid that is less dense than its liquid form, it floats on the oceans, making icebergs up to 12,000 square miles (31,000 square kilometers) in area, bigger than Belgium. Because 90 percent of floating ice is below its

waterline, the bottoms of coastal glaciers are able to quarry coastal stream channels well below their base level.

Melting ice requires a very large amount of heat. The same amount of heat required to simply melt a block of ice would raise the temperature of the same amount of water by 144° Fahrenheit (62° Celsius). The heat energy required to melt the north polar ice pack, which covers 5.5 million square miles (14.2 million square kilometers) and is larger than the continental United States, exceeds the available heat energy of all fuel sources available to humankind.

SNOW LINE

A glacier can be considered as a system that *accumulates, transports, erodes, deposits, and loses mass.* A glacier *accumulates* snow and *recrystallizes* it under pres-

FIGURE 14–1
Snow under the pressure of the mass of snow above it converts to granular recrystallized ice and becomes *foliated* like a metamorphic rock. Both the subhorizontal foliation and an intense subvertical cleavage can be seen in this photograph. The human figure at lower left gives the scale. (Photograph courtesy of the U.S. Geological Survey.)

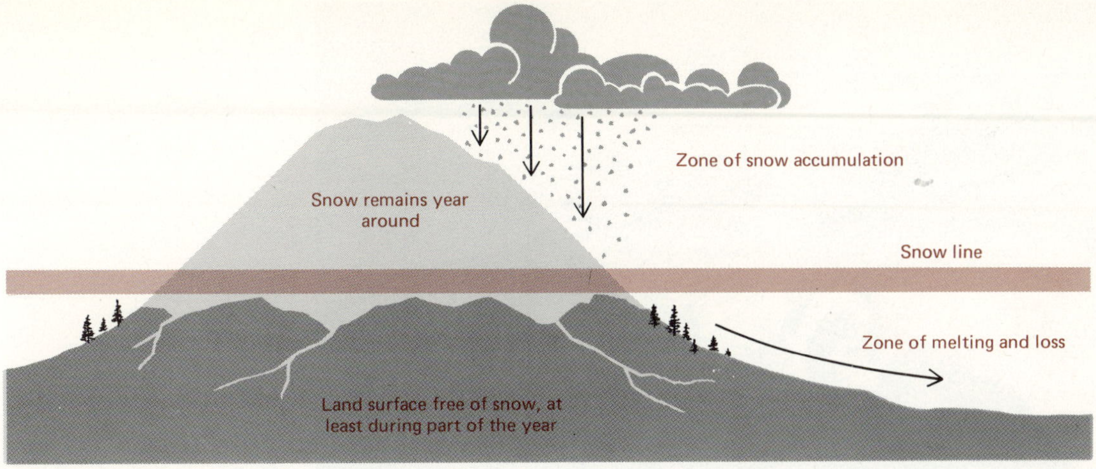

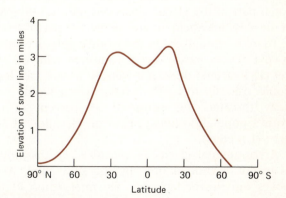

Snow remains year around

Zone of snow accumulation

Snow line

Zone of melting and loss

Land surface free of snow, at least during part of the year

FIGURE 14–2

The snow line, shown in color in this sketch, is the *equilibrium plane*. Above it, snow accumulates; below it, snow melts. As climate warms, snow line elevation rises; as climate cools, snow line elevation falls.

sure (Chapter Six) into ice. It then transports both ice and the broken rock within it in response to gravity. The glacier vigorously *erodes all surfaces it contacts by abrasion, ice wedging, and freezing. It deposits ice fragments, rock fragments, and meltwater. Finally, it loses mass by melting and depositing rock material caught within it* by means of meltwater streams and by mass wasting from melting ice margins.

Snow line is the average elevation *above which* snow accumulates and *below which* glaciers melt (Figure 14–2) over several centuries. Snowline separates the zone of accumulation above it from the zone of wastage below it.

The elevation of snow line varies, depending on climate, latitude, and the moisture supply. The chief determinant is latitude (Figure 14–3). The snow line reaches nearly 20,000 feet (6000 meters) near the equator and drops to sea level at both the Arctic and Antarctic circles. In midlatitudes the snow line averages 10,000 feet (3000 meters), dropping on the average of 3000 feet (920 meters) for every 15° of increasing latitude.

The effect of moisture is more subtle. Snow line rises with increasing distance downwind from the ocean where the air is drier; it falls in regions of high humidity, such as near the equator. Areas of generally dry atmosphere have anomalously high snow line elevations. The snow line is a *direct measure of climate*. Snow line rises with increasing temperature and falls with declining warmth.

Nearly everyone is aware that the earth has recently experienced intense glaciation, but few realize how widespread this glaciation has been within the last few tens of thousands of years. Twenty-four thousand years ago, small glaciers occurred as far south as Mexico City. The worldwide snow line dropped 3000 feet (920 meters), equivalent to decreasing the latitude of an area by 15°. Glaciers moved toward the equator 15° to 20° of latitude, while worldwide average temperature declined by 5° C. Almost one-third of the earth's land area was under glacial ice. Glaciers were of two types, *continental* and *valley*.

CONTINENTAL GLACIERS

Continental glaciers are the giants of the glacial world. A **continental glacier,** sometimes termed an

FIGURE 14–3

The relation between average snow line elevation and elevation and latitude on earth. Can you suggest why the snow line is highest at 20 degrees north and south of the equator? *Hint:* Compare with Figure 13-3.

FIGURE 14–4

An aerial view of northeastern Victoria Land, Antarctica. Floating sea ice is to the right. Volcanoes within the Hallett volcanic province make up the high ground within a small part of the largest continental glacier on earth. (Photograph by U.S. Navy.)

ice sheet, is a broad dome of ice (Figure 14–4) which totally buries preexisting topography and flows out from its central area as a sheet. It flows under its own weight and occupies much of a continental area.

The Antarctic ice sheet is an example of a continental glacier. It covers more than 90 percent of the Antarctic continent, includes nearly 5 million square miles (13 million square kilometers), and may reach 13,000 feet (4000 meters) in total thickness. The continent has subsided under the enormous weight of the overlying ice, so that parts of the land surface are 8200 feet (2500 meters) *below sea level.* An additional part of the glacier floats out into the Ross Sea as the Ross Ice Shelf, an area about the size of Texas.

There is enough ice in the world's polar ice sheets to cover every part of every continent with ice 600 feet (183 meters) deep. If it were all to melt, sea level would rise 300 feet (91 meters), which would seriously threaten the estimated 38 percent of the world's population living below an elevation of 300 feet (91 meters).

A continental glacier is *unconfined by topography.* Its extent is limited only by the snow line. As it moves into lower latitudes, the intersection of the rising snow line with the land surface marks the maximum extent of glaciers. Based on our understanding of the last few tens of thousands of years, glaciers cover a third of the earth's land surface when long-term climate turns colder by 9° F (5° C).

The effect of a continental glacier on a preexisting landscape is *to smooth the surface, lowering relief by filling in the valleys and planing off the uplands.* Left behind after the glacier melts back is a land surface typical of northwestern Europe, central Canada, and the American Midwest with bare rock surfaces, abundant lakes, gravelly, bouldery soils, and gently rolling topography.

VALLEY GLACIERS

Valley glaciers, sometimes termed *alpine glaciers,* are smaller glaciers confined between valley walls produced by stream erosion. They form above snow line in high mountains and flow *downslope* (Figure 14–5) like an extremely stiff and sluggish stream, which melts as it moves below snow line. ''Smaller'' is a relative term; Alaska's Malaspina glacier, flowing from Mount St. Elias, is larger than Rhode Island.

A valley glacier, *confined* by valley walls, exerts its erosional power by roughening the preexisting topography, actively cutting the stream valley into a new cross-profile and deepening the valley. Much of the erosional power of a glacier is exerted through the large basal area of ice-rock contact, termed the *sole,* which is full of jagged rocks. As these rocks are forced against the solid rock surface beneath, the moving glacier cuts and carves its valley like a giant rasp.

After the passage of a valley glacier, drainage di-

FIGURE 14–5
Terminus or snout of Mer de Glace glacier, Chamonix Valley, Switzerland. Terminal and lateral moraines are both prominent as a part of this valley glacier. (Photograph courtesy of the U.S. Geological Survey.)

vides are sharpened, and the whole mountain mass acquires a much more rugged appearance. The new scenery is typified by the Alps, the Coast Ranges of British Columbia, and the highest parts of the Rocky and Sierra mountain ranges of the United States.

GLACIAL FLOW

Glaciers flow because of their weight, which creates shear stresses that cause ice particles to slip past one another. This slippage internal to the glacier is often augmented by slippage between the glacier and the surfaces beneath it, which are often flooded with meltwater.

Continental glaciers flow because their great thickness and weight cause crystalline slip within the glacier as well as slippage along the glacier's water-lubricated base and sides. Valley glaciers flow because the gradient or slope of the stream valley down which they move sets up stresses causing both internal crystalline slip and slip along meltwater-lubricated surfaces (Figure 14–6).

The upper surface of all glaciers is under less pressure, and may deform like a brittle solid and crack, creating **crevasses** (Figure 14–7). These open fractures form when the brittle upper surface is forced to extend, as when the glacier rides over an obstruction or turns a corner. Under the increasing weight of overlying ice, the glacial mass slowly changes downward from a brittle solid to a plastic, then to a vis-

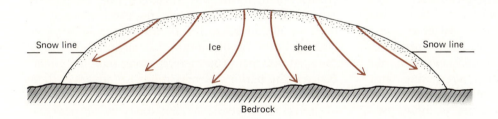

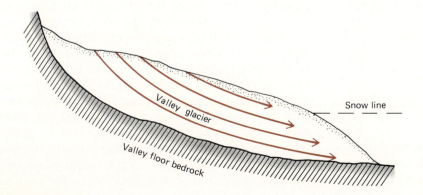

FIGURE 14–6
The colored arrows indicate the flow lines of particles within the glacier. Glaciers are like streams in that their greatest velocity is in their center, but movement is retarded in the area where they contact the valley wall. Both the ice sheet and the valley glacier are sliding over their bedrock base. The two sketches are not to the same scale; continental glaciers are thousands of times larger than most valley glaciers.

FIGURE 14–7
Crevasses are prominent in the surface of the Skolai glacier, Alaska. They reflect the stretching that occurs in the upper brittle part of a glacier as the material beneath it flows. Note the quantity of dark rock material on the surface of the glacier. As such material falls into a crevasse, it may eventually work its way down to the base of the glacier, following one of the paths outlined in color in Figure 14-6.

cous solid. Crevasses then close downward into zones of solid flow (Figure 14–1), marked by strong foliation in the ice and conversion to textures typical of a foliated metamorphic rock (Chapter Six).

Mechanisms of glacial flow are affected by ice temperature and are extremely sensitive to the level of shear stress. Very high levels of shear stress, as for example at the base of a glacier, render the ice extremely viscous and promote rapid flow. The abundance of meltwater at the base of a glacier also enhances flow, as does melting of ice under pressure. Thus a glacier advances by a combination of internal crystalline creep, folding, faulting, crevassing, and both internal flowage and slippage at its base.

Not surprisingly, because there are so many processes contributing to glacial movement, the rates of glacial flow vary widely. Glaciers sometimes have a *surging* flow, with velocities briefly up to 100 times "normal." The range of normal velocities is approximately 10 to 100 feet (30 to 300 meters) *per year,* with a few glaciers moving faster than that.

▼

CAUSES OF GLACIATION

▲

There have been four *major* periods of massive glacial advances in the geologic past. The earliest was about 600 million years ago, the next about 450 million

years ago, the next from 300 to 250 million years ago. The latest one, still in progress, began about 15 million years ago in Antarctica. During the majority of its history, the earth has been largely free of ice over continental land masses.

In addition to these periods of large-scale ice advance, there have been many other periods during which smaller-scale ice advances can be proved. There have been, for example, no fewer than ten periods of ice advance within the last million years. During times of maximum ice advance, sea level dropped 300 feet (100 meters) lower than it is today. *Between* the periods of maximum ice advance, there have also been numerous warmer times lasting tens of thousands of years. We live today in an **interglacial epoch,** a brief period of relative warmth within an otherwise glacial climate.

The contributors to climate change are numerous. Volcanic eruptions that throw huge volumes of ash and acids into the atmosphere may limit the amount of sunlight reaching the earth, creating a colder climate for a very few years. Careful study of the sun reveals that its heat and light output are not absolutely constant, as was previously thought. Variations in solar output correspond reasonably well with the known 11-year sunspot cycle, but there may be longer variations as well. A decrease of less than 2 percent in solar output over several centuries would turn the world from its current state into one in which arctic cultures would stretch as far south as the Mediterranean Sea and Kansas. Still other proposals for the causes of widespread glaciation have been put forward; these include changes in atmospheric chemistry, opening and closing of oceans, and changes in oceanic currents.

Until recently, no one could determine whether any of these variables could bring on *prolonged* periods of *major* glaciation. Though the evidence recorded in the rocks for past glaciations was overwhelming, no one could propose a *cause*. The theory of glaciation was stuck where continental drift (Chapter Nine) was in the 1930s; the evidence for it was overwhelming, but it was not accepted because a plausible *mechanism* for bringing on glacial ages was unknown.

EFFECTS OF ORBITAL AND TILT CHANGES

While climatologists struggled with the complexities of the earth's climate in efforts to find a simple causal theory, geologists struggled with the complexities of the sedimentary record of past climates. Astronomers also labored over with the intricacies of orbital motions to assess the role of the earth-sun interaction in causing ice ages. Many scientists of different persuasions reasoned that ice ages were brought on by changes in solar radiation reaching the earth—but what caused the changes?

In the 1920s the Czechoslovakian astronomer Milutin Milankovitch proposed that orbital variations in the earth's attitude and path around the sun led to changes in incoming solar energy that were cyclic. These changes allegedly caused glacial ages to develop during the periods of minimum receipt of solar energy. Milankovitch felt that three overlapping alterations in earth's orientation and orbital location, with corresponding variations in the amount of sunlight received, combined to force changes in the climate. Those changes are

1. In a cycle lasting approximately 26,000 years, the earth advances in its elliptical orbit so that its closest approach to the sun occurs at different times of the year. Currently, the earth is closest to the sun in January and farthest away in July, which makes winters warmer and summers cooler than they will be about 12,000 years from now, when the earth will be closest to the sun in July. During that distant period winters will be colder and summers warmer than now.

2. In a cycle lasting approximately 41,000 years, the earth's spin axis wobbles so that the earth's rotational axis becomes more nearly perpendicular to the sun than it is currently. If the axis were to become precisely perpendicular to the sun, there would be no seasons, and high-latitude ice would never melt, which could help bring on an ice advance. Currently the tilt is diminishing, which should lead to less change from season to season.

3. In a cycle lasting approximately 93,000 years, the earth's orbit about the sun changes from being perfectly circular to more notably elliptical. When the orbit is nearly circular, it diminishes the effect of the 26,000-year cycle. As the orbit becomes more eccentric, it enhances the effects of both the 26,000 and the 41,000 year cycles.

These three orbital changes cause the amount of sunlight received by the total earth to vary only slightly. But, *the amount of sunlight received at high northern latitudes in the summer* may vary by as much as 20 percent when all three orbits coincide. If less sunlight is received at high latitudes in the summer, snow will tend *not to melt* and will, over a long period of time, accumulate. The accumulation of snow because it does not melt in summer causes continental ice sheets to begin to form, which accentuates the coolness of the area. Earth has entered an ice age.

TESTING THE THEORY

For 50 years after its proposal, the Milankovitch theory was considered untestable. Geologists could not provide a really precise record of temperatures over the last half million to million years against which to test Milankovitch's mathematical model. More recently geologists studying lengthy, undisturbed deep-sea sediment records were able to provide a precise record of climate for the last half-million years. When this record was matched with the appropriate predictions of Milankovitch, it was a good fit. The Milankovitch theory became the accepted explanation within a short time.

The earth experiences glacial ages as a function of changes in orbital and axial tilt. These changes are further modified by the effect of *isostasy*. As snow continues to accumulate, the surface of the snow rises for a time, attracting still more snow. When finally the land sinks under the weight of the ice, the elevation of the snow surface declines, which favors melting. Glaciologists have noted that glacial periods seem to start with a long period of snow accumulation, followed by a much briefer interval of meltback. Perhaps the slow movements of the mantle affect the timing of glacial ages.

Investigations have shown that shifts to a colder

FIGURE 14–8

Louis Agassiz is remembered not only for his contributions as a zoologist but also as the scientist who firmly established proof of recent major glacial advances in mid-latitude areas. (Courtesy of the History of Science Collections, University of Oklahoma Libraries.)

climate have been remarkably regular, with larger glaciations at intervals of approximately 100,000 years and lesser glacial advances at periods of 19,000, 23,000, and 42,000 years—corresponding to the times predicted by the Milankovitch theory. We also recognize that plate motions may move continental plates into or out of higher latitudes, effectively "turning on" or "turning off" the potential for glaciation during a Milankovitch climatic cycle.

As still more precise and detailed records of world climate are attained, we will no doubt find that other variables are involved in glaciation. Quite recent research has suggested that intervals of glacial advance correspond with periods of low carbon dioxide content of the atmosphere. Our burning of enormous quantities of hydrocarbon fuels may be inadvertently warming our climate. Research on the multiple cause of glaciation continues.

ANCIENT GLACIATIONS

Jean Louis Rodolphe Agassiz (pronounced agg'-uh-see) (1807–1873), a Swiss-American naturalist (Figure 14–8), was born in a time of scientific, social, and political revolution. In the early part of the eighteenth century, as we have seen, the earth was thought to be, at most, a few thousand years old and little changed since the flood recorded in Genesis. Then, in the late 1700s, James Hutton had maintained that the earth was incredibly old, yet constantly changing. He had seen ". . . no vestige of a beginning—no prospect of an end." Shortly after that Baron Georges Cuvier (1769–1832), perhaps one of the greatest naturalists of all time, found fossil evidence that giant elephants and other "monsters" had roamed the area around Paris thousands of years ago. In 1837, after studying glaciation in the Swiss Alps with friends, Louis Agassiz, whose specialty had been the study of fossil fish with Cuvier, startled Europeans by announcing that much of Europe had only recently been submerged under vast sheets of glacial ice. It was all too much to accept!

AGASSIZ AND THE ICE AGE

Others at the beginning of the nineteenth century gad begun to discern that glaciers had the power to quarry and move rocks over limited distances. It was John Playfair, Hutton's friend, who in 1802 vividly described the erosive power of glaciers (see quotation as chapter opener). Agassiz was not even the first to propose that glaciers had covered large parts of Europe in the past. Herrn A. Bernhardi suggested widespread European glaciation in 1832, and two friends of Agassiz, Ignace Venetz and Jean de Charpentier, presented a scientific paper in 1834 suggesting that the Alps had once been covered by much more extensive glaciers.

Agassiz spent a summer with his friends in the Alps, intending to dissuade them of their ideas. Instead, it was Agassiz who was persuaded to accept widespread glaciation. Evidence of past glaciation was everywhere. Swiss mountaineers had long recognized that

1. Valley glaciers *moved* downvalley at very slow rates.

2. The moving glaciers gouged and scratched the rocky valley floors over which they flowed.

3. The moving glaciers carried quantities of boulders with them, leaving piles of boulders behind as they melted back. These piles of boulders, locally termed *erratics,* were often many miles or kilometers from their place of origin.

Agassiz was to Venetz and de Charpentier what Playfair was to Hutton, and what Wegener was to Taylor (Chapter Ten). He took their ideas, clarified them, expanded them, and devoted several decades of his life to the forceful exposition of an *extension* of their theory. The concept he championed was so enormous in scope that Agassiz is remembered 150 years later not only as a distinguished zoologist, but also as the man responsible for the worldwide acceptance of widespread continental and valley glaciation.

From his initially cautious acceptance of limited extensions of glaciers in the Alps, Agassiz later recognized the evidence of glaciation over large parts of Europe. When he came to North America in 1846, he found the same kinds of evidence widespread and proposed still another extension. Here is own description of the essence of his theory.

In my opinion, the only way to account for all these facts and relate them to known geological phenomena is to assume that at the end of the [recent] geologic epoch . . . the earth was covered by a huge ice sheet . . . .This ice sheet filled all the irregularities of the surface of Europe . . . the Baltic Sea, all the lakes of Northern Germany and Switzerland. It . . . even covered completely North America and Asiatic Russia.

Agassiz presented his ideas more fully in 1840 in a book entitled *Studies on Glaciers.* The first chapter was a very fair review of the work of others who had preceded him, including de Charpentier, who had taught him the core of glacial geology. The middle chapters described the glaciers of the Alps and Jura mountains and included many ideas on the biologic control of glaciation and the physics of glacial movement, now known to be erroneous. In the final chapter Agassiz extended his theory of glaciation to propose a recent **Ice Age,** a period of time that we now know encompasses the last 1.6 million years during which glaciers covered much of the northern world.

Adam Sedgwick (1785–1873), a well-known English geologist, spoke for many scientists of the time when he provided the following succinct reaction to Agassiz's ideas, especially the proposal in Agassiz's last chapter about a worldwide Ice Age. ''I have read his Ice-book. It is excellent, but in the last chapter he

FIGURE 14–9

This caricature of Professor Buckland by one of his students proves that university humor was still alive and well a century ago. Buckland, attired for the study of modern glaciers, did present a ponderous image. The cartoon none too gently suggests that his was an ''IMMENSE BODY'' not unlike a glacier, and perhaps capable of scratching grooves in rocks. (Courtesy of the History of Science Collections, University of Oklahoma Libraries.)

loses his balance, and runs away with the bit in his mouth.''

The battle was on, and it raged for several decades. To those comfortable with an earth only a few thousands of years old, recently deluged by the flood of Noah, Agassiz's ideas were a frontal assault. The boulders strangely out of place, the grooved and polished rocks, the huge strips of gravel that Agassiz pointed to as evidence for glaciation had always been taken as *signs of the biblical flood,* not some sheet of ice that had covered most of Europe! To smug citizens, staggered by Hutton's insistence on an earth with no beginning or end and Cuvier's Parisian monsters, Agassiz's glaciers was yet another intrusion into a comfortable, static world.

Dr. William Buckland, English geologist, clergyman, and confirmed diluvialist, decided in the 1840s to go to Switzerland and attack Agassiz's ideas at their source. Instead, he came away—as Agassiz had—a confirmed believer in a recent period of extensive glaciation. He spent his remaining years as a vigorous proponent of glacial theory and accompanied Agassiz on several expeditions in England and Scotland, where glacial features abound. Figure 14–9 is a caricature of Buckland, dressed for field studies, drawn by one of his students. College humor was just as lively then as now.

In time, Agassiz's meticulous accumulation of evidence for worldwide glaciation won the acceptance of the world's scientific community; he is remembered as the father of the theory of worldwide glacial ages. His concepts were accepted slowly, and after much debate, but time was on his side. He had "seen what others had seen, but dared to ask what no one else had ever asked."

His later years were devoted to the study of zoology at Harvard, where the Museum of Comparative Zoology still bears his name. His grave, at the Mount Auburn cemetery in Cambridge, Massachusetts, is marked by a boulder from the glacier of Aar in Switzerland, where it all started.

NORTH AMERICA AND THE ICE AGE

The last 1.6 million years of history have been marked by intermittent episodes of glacial advance from the polar ice sheets far into central North America (Figure 14–10). This period of time is informally termed the *Ice Age* and more formally termed the **Quaternary Period.**

The Quaternary Period is subdivided into the **Holocene Epoch,** which includes only the last 10,000 years, and the **Pleistocene Epoch,** which includes all 1.6 million years of Quaternary time except the last 10,000 years. This text uses the terms Ice Age and Quaternary interchangeably to mean the last 1.6 million years of earth history, in which continental and valley glaciation have been intermittent but dominant events in North America, Europe, the United Kingdom, northern Eurasia, Greenland, and the Antarctic.

The general term for all glacially deposited sediment is **drift,** a term dating from the time when such deposits were assumed to have <u>drifted</u> into

place during the flood of Noah. As geologists have mapped drift and other glacial deposits throughout North America, it has become apparent that there are *several different layers of drift*. The oldest layers, on the bottom, have well-developed soils containing fragments of plants that can grow only in warm climates. Above these soils are still more layers of drift, with soils at *their* top containing fossils of warm-climate plants. Such evidence, pieced together from all across North America, affirms that between each period of glacial advance there were *long periods of warmth*.

By carefully checking a wide variety of glacial features, geologists were able to recognize *four* periods of major glacial advance within the *Pleistocene* deposits of North America. These periods of glacial advance were named *Nebraskan, Kansan, Illinoian, and Wisconsin* in order of *decreasing* age. The Quaternary Period can then be subdivided in the following way.

Quaternary Period

▶ Holocene Epoch—last 10,000 years.
Coincides with current *interglacial time*.

▶ Pleistocene Epoch—previous 1.59 million years before the Holocene.
Wisconsin glacial advance.
Sangamon Interglacial time.
Illinoian glacial advance.
Yarmouth Interglacial time.
Kansan glacial advance.
Aftonian Interglacial time.
Nebraskan glacial advance.
Beginning of Quaternary Period—1.6 million years ago.

The most recent meltback of the ice occurred approximately 10,000 years ago; the worldwide initiation of this major retreat defines the beginning of the Holocene Epoch in which we live. There have been *numerous* smaller glacial advances within the Quaternary; at least ten can be recognized within the Pleistocene of North America.

For Greenland and Antarctica, there has been no interglacial relief; they remain buried in glacial ice. Glaciation in these luckless, high-latitude areas began at least 30 million years ago and continues. The earth today has 6 million cubic miles (25 million cubic kilometers) of ice on its surface. This is probably much more than has existed on earth during most of its history. Our earth is still locked within a glacial period.

We happen to live in a warmer interval, but there

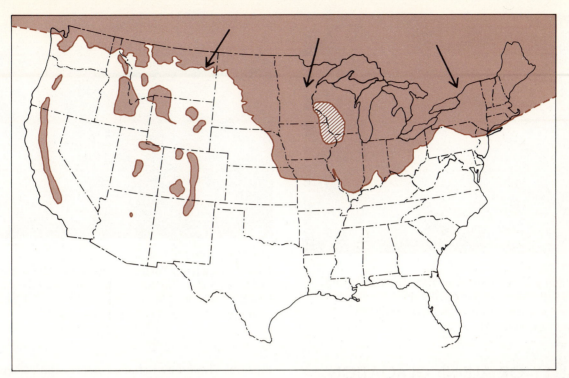

FIGURE 14–10

This map shows the maximum extent and direction (arrows) of glacial advance in the United States during the various periods of ice advance within the Pleistocene. South of the margin of the continental glacier, among the high mountains of the West, valley glaciation and a wide variety of lakes were common during the various ice ages. The diagonally striped area of western Wisconsin, northeastern Iowa, and southeastern Minnesota is often called *the driftless area.* Evidence of glaciation in this area is sparse. (Modified after Dorothy H. Radbruch-Hall and others, 1976, U.S. Geological Survey Map MF 771.)

is no reason to suppose that glaciers will not advance again. As we will see in Chapters Eighteen through Twenty-One, the earth's continents have during many periods been at low latitudes, and have there-fore been totally free of ice. Our current cycle of gla-ciation is, in the long term, an anomaly.

We must be able to recognize the evidence of gla-ciation if we are to interpret past glacial events. A process that is as energetic as glacial erosion leaves behind very distinctive landscapes composed of both erosional and depositional forms.

GLACIAL LANDSCAPES

Comparing glaciers to other erosional agents is in-structive. For instance, glaciers are a far more effec-tive *abrasive* tool than running water. The sole of a glacier may be more rock than ice, so that the base of a glacier acts like very coarse sandpaper, scouring and deeply grooving (Figure 14–11) the rocks over which the glacier glides. Glaciers acquire much of the rock beneath them by wresting away rock beneath it through intense frost wedging, sometimes termed **glacial plucking.**

Glaciers are superb agents of transportation, ca-pable of carrying material of *any size,* from finest rock flour to house-size boulders. At the latitudes in which glaciers accumulate, they are able to engage in intense mechanical weathering through frost wedging. Equally vigorous mass wasting from adja-cent valley walls litters the upper surface of many glaciers with large amounts of rock debris.

Glaciers as depositional agents operate in two modes. As the glacier melts back, much of its load is *dropped* directly from melting ice. Such ice-deposited material is *unstratified and unsorted by size;* it is called **till.** Material deposited from the meltwaters pouring from a glacier is commonly *stratified and somewhat sorted by size;* it is called **stratified drift.** As already noted, the general term for *all* glacially deposited sediment is *drift.*

FIGURE 14–11
Polished and grooved rocks from the Alps. (From Louis Agassiz, 1840, *Etudes sur Les Glaciers,* Pl. 17, courtesy of the History of Science Collections, University of Oklahoma Libraries.)

FIGURE 14–12
Deadman Canyon, Tulare County, California is a typical ∪-shaped valley. The line in color identifies the catenary profile typical of a glaciated mountain valley. (Photograph courtesy of the U.S. Geological Survey.)

VALLEY OR ALPINE GLACIATION

Confined within stream valley walls, every square yard or meter of ice on the sides and the bottom of a glacier has a nearly equal ability to erode. The erosive power of a valley glacier is proportional to the *area* of ice-rock contact.

Because a valley glacier erodes almost equally on its sides and bottom, it changes the typical ∨-shaped, high-elevation valley to a ∪-shaped valley (Figure 14–12). Such an open ∪ shape is *unique to valleys modified by glacial erosion* and consequently serves as an identifying characteristic. Such a shape, called a **catenary profile**, is that assumed by power lines strung between two poles. It is another profile of equilibrium, reflecting uniform distribution of erosional energy along the valley walls and floor.

High-Elevation Landforms

At the highest part of the mountain range, glacial erosion and plucking dramatically roughen the rounded stream divides. Because of intense ice wedging, the cross-sectional profile of formerly rounded stream divides is sharpened into a distinct ∧ shape. Such sharpened stream divides are called **arêtes** (pronounced uh-rehts′) (Figure 14–13); they are a signal that glaciation has affected the very top of the mountain range.

Just beneath the arêtes is a large, natural bowl-shaped ampitheatre, carved into the headwaters area of the stream valley by extreme glacial plucking and

erosion. These bowl-shaped depressions (Figure 14–14) are called **cirques** (pronounced sirks). Though it is not possible to examine fully the base of an active glacier, it appears that cirques are formed by the rotational sliding of ice backed up against the uppermost stream headwall. Jointing also appears to exercise some control over cirque shape as does overall rock structure. In middle- to high-latitude areas of the northern hemisphere, the largest cirques gener-

FIGURE 14–13
This photograph is of an area near Mount Whitney, California. The sharpened former stream divides, known as *arêtes,* are quite prominent in this intensely alpine glaciated region. (Photograph courtesy of the U.S. Geological Survey.)

FIGURE 14–14

High in the Bighorn Mountains of Wyoming, a cirque nestles at the top of Cloud Peak. A remnant of a glacier remains in the uppermost part of the cirque, and a tarn occupies the cirque floor. Notice the large amount of shattered rock talus littering the area; frost wedging in this environment is powerful. (Photograph courtesy of the U.S. Geological Survey.)

ally face northeast—the direction of maximum shading in the summer as well as the direction most protected from the deflating effect of prevailing southwesterly winds.

Where three cirques meet, the arêtes that make up their divides may coalesce to form a **horn,** a generally triangular landform produced by vigorous ice erosion along the headwalls of valley glaciers. The world's best known horn is, of course, the Matterhorn in the Pennine Alps near Zermatt, Switzerland.

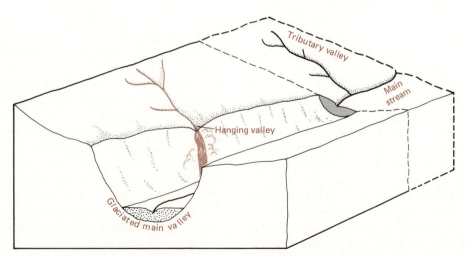

FIGURE 14–15

A ''before and after'' sketch of a hanging valley. The back block, drawn in dashed lines, shows the scene before glaciation; a tributary stream enters the main channel at a common elevation. Glacial erosion, depicted in the front block, has deepened it and changed it to a distinctly ∪-shape. The former tributary valley was little glaciated and so is left high above the main channel, poised as a hanging valley. Its stream flow into the main channel is now a waterfall.

FIGURE 14–16
This close-up view of the granite face of Half Dome,
Yosemite National Park, California, reveals the flat surface
left behind when a valley glacier quarried away the other
part of the dome and left behind a truncated spur. The
rock surface is streaked by stains from lichen growth.
(Photograph courtesy of the U.S. Geological Survey.)

Many cirques are sufficiently deep that they form
lake basins after the glacial meltback; such cirque
lakes are referred to as **tarns** (Figure 14–14).

Middle-Elevation Landforms

Because alpine glaciers exploit preexisting stream
valleys, the valley of the main stream may accumulate very thick snowfields, massive glaciers which
deeply erode the main valley; smaller tributary valleys
will be much less eroded by the thin glaciers that occupy them.

As a result, main valley glaciers will cut down far
more deeply than the glaciers in tributary valleys. After the glaciers melt back in an interglacial period,
the junction between the main and tributary streams
is no longer at a common elevation. The juncture of
the tributary valley with the main valley is now a
hanging valley (Figure 14–15).

The water flow from the tributary valley into the
main stream is now a *waterfall*. Waterfalls may be

FIGURE 14–17
Near Mount Whitney, California this small trough lake
occupies the floor of a vigorously glaciated valley. Notice
the abundant *arêtes* marking the stream divides.
(Photograph courtesy of the U.S. Geological Survey.)

created for other reasons *within a single stream valley*,
but a waterfall occurring where a ∨-shaped tributary
stream valley meets a ∪-shaped main stream valley
almost certainly marks a hanging valley. The extent
of the vertical drop between valleys is a measure of
the differential erosion between the tributary and
master stream.

The great stiffness of a glacier means that it is
much less able to "go around corners" than is a
stream. As glaciers encounter knobs or spurs of rock
projecting out into a sinuous stream valley, they may
simply plane them off along preexisting jointing. The
end result is a **truncated spur.** A stream valley that
has had its marginal spurs truncated by glaciation
not only is widened but is also straightened.

Among the best-known truncated spurs is Half
Dome, in California's Yosemite National Park. This
granite dome was quarried almost exactly in half by
a valley glacier that passed down the Merced River.
The face of the truncated spur is a sheer wall of granite (Figure 14–16), which forms a favored site for advanced technical rock climbers.

It is very common for long, linear lakes to occupy
the floor of a recently glaciated valley. Such lakes
may be notable for their deepness and overall ∪-
shape. Known as **trough lakes,** they are a distinctive and beautiful feature of glaciated valleys (Figure
14–17). Like all lakes, they are temporary features;
the same streams that feed them will fill them with
sediment. Floodplain forms where once there was a
trough lake. Eventually lakes become a part of a
much longer stream profile of equilibrium as the
temporary depression in the profile is filled in.

FIGURE 14–18
The passage of a glacier polished the bedrock surface of the granite in Tuolumne Meadows high in the Yosemite National Park. Lodgepole pines eke out an existence in the shallow soils within the cracks bisecting the surface. Their roots will help tear the granite apart over the centuries. (Photograph courtesy of the U.S. Geological Survey.)

Glacial polish is the surface "finish" on a rock produced by a glacier passing over it, smoothing the surface, and roughly polishing it. Remarkably, more than 10,000 years of high-elevation weathering have caused relatively little of the polished surface to peel off in the example shown in Figure 14–18. Notice also that the lodgepole pines of Tuolumne Meadows are managing to grow in the sparse soil trapped in

FIGURE 14–19
A climber pauses to look at the grooves cut into the Cathedral Veil granite near Half Dome in Yosemite National Park. Imagine the force required to groove solid white granite this deeply! (Photograph courtesy of the U.S. Geological Survey.)

FIGURE 14–20

The Tanana glacier near Eucher Mountain, Alaska glides majestically down its valley. Its surface is streaked with stripes of rock brought to it when tributary glacier after tributary glacier merged their lateral moraines with the main valley glacier. As this glacier moves below snow line, it will dump a large load as morainal material. (Photograph courtesy of the U.S. Geological Survey.)

cracks in the rock. Their roots exert a natural form of mechanical weathering called *root wedging* (Chapter Four).

The rasping of the rocks in the glacier's base also leaves distinctive grooves (Figure 14–19). Such gla-

cial *grooves and scratches* occur on many scales, from hairline scratches on individual cobbles to the gouging of major lake basins like Lake Tahoe on the Nevada–California border. The examples shown in Figure 14–19 are near Half Dome in Yosemite National

FIGURE 14–21

A small glacier on the Three Sisters Mountain in Oregon has reached snow line. Its lateral and terminal moraines make a high ridge of drift. When the glacier melts back, its moraine will mark its point of greatest advance. If the glacier surges down the valley, the moraine will be pushed ahead of it. (Photograph courtesy of the U.S. Geological Survey.)

FIGURE 14–22
This aerial view of Baker's Park, near Silverton, Colorado reveals a well-developed valley train and outwash fan. These features represent the lowest-elevation landforms created by a valley or alpine glacier. Both the valley train and the fan are composed of stratified drift and outwash. Notice the *braided* pattern of the stream choked with glacial debris. (Photograph courtesy of the U.S. Geological Survey.)

FIGURE 14–23
A glacial erratic high up in Yellowstone National Park. The caption for this photograph was furnished by Grove Karl Gilbert, one of the great American geologists of the nineteenth and early twentieth centuries. His caption reads "Perched geologist on perched boulder, both erratics." (Photograph courtesy of the U.S. Geological Survey.)

Park, California. Try to imagine the force required to produce scratches of this magnitude in solid granite!

Moraines are low hills of drift deposited at the margins of a glacier. During the period of active glaciation, morainal material is largely carried at the margins of glaciers. The series of broad stripes of rock debris were brought down as glacier after glacier moved down their tributary valleys and joined to form a single, large majestic glacier (Figure 14–20).

A **lateral moraine** forms along the margin of the glacier where ice contacts the valley wall, creating a natural levee for the glacial mass. A **terminal** or **end moraine** (Figure 14–21) forms at the lowest-elevation end of a glacier, sometimes termed its *snout.* When the glacier melts back, it leaves a system of morainal ridges behind which precisely mark the limits of the glacier's advance. The ridges are created by drift piling up during short periods of rapid advance.

Moraines are an important part of glacial landscapes at the middle and low elevations. They are often marked by bands of vegetation, which find the porous sandy, gravelly morainal material an excellent place in which to grow. Moraines may be composed of ice-deposited till, or less commonly of stratified drift, or of a mixture of the two.

Low-Elevation Landforms

At the lowest elevations glaciated valleys are dominated by features associated with melting, since the glacier has now passed well below snow line. Morainal topography is common, as is the formation of a **valley train** (Figure 14–22), a body of outwash or

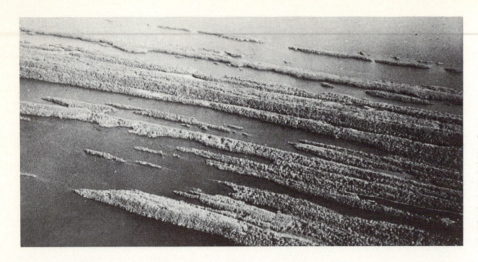

FIGURE 14–24
The fluted topography at the northeast end of Isle Royale in northern Lake Superior reminds us of the erosive power of the glacier that gouged out this huge lake basin and severely grooved the bedrock. (Photograph courtesy of the National Park Service.)

stratified drift deposited by meltwater streams far beyond the terminal moraine. A valley train often grades downward into an **outwash plain,** (Figure 14–22) essentially an *alluvial fan* composed of outwash materials deposited by braided meltwater streams.

Glacial Erratics

A glacial **erratic** (Figure 14–23) is a transported rock, different from the bedrock on which it resides. An erratic is dramatic evidence of the power of glaciers to transport, for erratics have been found tens to hundreds of miles or kilometers from their known bedrock source.

It was finding huge boulders far distant from any probable source that first kindled the curiosity of Venetz, de Charpentier, and other naturalists who roamed the world's high mountains. Since modern glaciers could be seen moving immense boulders downvalley, it required only a leap of imagination to connect the inconsistent placement of erratics with a glacial transportation system. Once that connection was made, erratics became among the most frequently observed lines of evidence for ancient glaciations.

FIGURE 14–25
The scratches on the flat surface of this erratic tell of hard service at the base of a glacier. When such scratches are located on in-place bedrock, they indicate the approximate direction of glacial travel. In North America that travel was generally radial to the Hudson Bay region of central Canada. (Photograph courtesy of the U.S. Geological Survey.)

CONTINENTAL GLACIATION

Continental glaciers spread out across the land surface much as a blob of cold molasses spreads out across a table. They are at once plow, rasp, truck, and architect of lakes.

Pushing eroded soil and rock ahead of their margins, they create numerous complex moraines. Rasping the land surface beneath, they gouge and smooth, creating highly symmmetrical linear landforms. Hauling tons of material, they deposit sheets of drift and till that may cover large parts of a continent. Totally rearranging stream drainage systems, they leave behind an array of lakes ranging in size from ponds to the Great Lakes, which are among the largest bodies of fresh water on earth.

The world's continents are littered with evidence of ancient large lakes that formed beyond the border of continental glaciers in the cool, rainy climates typical of their margins. Such **pluvial lakes** stretched once across several states and Canada; now their dry lake beds form parts of Manitoba, North Dakota, Minnesota, Montana, and Utah, where a small remnant of one pluvial lake is known as the Great Salt Lake.

Continental glaciers are the Goliaths of landscape architects. Overall, they flatten preexisting landscapes by both erosion and deposition, and leave behind gently rolling scenery, and a wide array of distinctive landforms.

Erosional Landforms

Landforms created by continental glacial erosion include grooves and scratches of all sizes. On the largest scale, continental glaciers erode bedrock into linear depressions that become lakes. Such lakes include Pend Oreille Lake in Idaho, Lake Tahoe on the California–Nevada border, Lake Winnipeg in Manitoba (a remnant of pluvial Lake Agassiz), most of the lakes of northern Minnesota and Michigan, the Great Lakes (Figure 14–24), the Finger Lakes of the Adirondacks, and Lake Champlain on the New York–Vermont border.

At a smaller scale, rocks caught between moving ice and bedrock beneath—the agents of abrasion—are severely grooved and scratched by their service (Figure 14–25). Interpretation of scratches on rocks is not always easy; faulting can produce heavily grooved rocks, and heavy earth-moving equipment can intensely striate rocks. The association of grooved and striated rocks with *other* glacial features is, however, quite suggestive. Many erratics and rocks within moraines are strongly abraded.

FIGURE 14–26
This aerial view of rolling morainal topography near Prairie du Sac, Wisconsin is typical of an area of extensive ground moraine. The regolith is drift. The soils are likely to be deep, sandy, and quite rocky. (Photograph courtesy of the U.S. Geological Survey.)

Depositional Landforms

The most general depositional landform produced by continental glaciers is **ground moraine** (Figure 14–26), stratified drift and till left behind on the ground surface over which the glacier has moved. It is commonly bordered by lateral and terminal moraines.

Ground moraine constitutes the soils for large parts of Saskatchewan, Manitoba, Ontario, and the midwestern United States. Since the soil in these areas is composed partly of recently granulated rock and partly of soils stripped from north central Canada, they tend to be fertile, but granular and rocky. Farmers in north central North America are quite familiar with long and laborious efforts to remove very large rocks from new agricultural land; the process is continual for frost heaving brings still more materials up from below.

Eskers (Figure 14–27) are ridges of stratified drift deposited by meltwater streams flowing under the glacier. In map view they are seen as low ridges of stream deposits whose meandering shape is faithful to the physics of stream flow. Excavations into eskers reveal structures in the sand and gravel typical of those in any streambed. The only thing odd about this stream channel is that it was cut *up* into the glacial ice, and therefore *forms a ridge* when the ice retreats.

FIGURE 14–27
A lone jogger runs along the top of an esker near Spalding, Michigan. This sinuous ridge is composed of stratified drift whose textures are reminiscent of ordinary stream deposits. (Photograph courtesy of the U.S. Geological Survey.)

Irish grandmothers used to tell their small grandchildren that eskers had been left by giant "sandmen," who carried leaky bags of sand over their shoulders. They walked over Ireland, sprinkling sand in the eyes of sleepy children, their paths marked by the leaking trails of sand they left behind. Such stories explained what could not be ignored. We have become enlightened since that long ago time, but we have lost a charming story.

Erratics are transported by both continental and valley glaciers. They reflect the power of glaciers to move and deposit materials of any size. Figure 14–28 is of an erratic within Yellowstone National Park,

Wyoming; its size, as compared to the trees growing around it, gives some sense of the power of glacial transportation. The boulder must weigh several hundred tons.

Drumlins are streamlined hills of drift trailing from rock obstructions encountered by the glacier. In map view they have a highly elliptical or teardrop shape. In cross section they are asymmetrical, with the gentle slope facing the direction from which the glacier came (Figure 14–29).

Drumlins often occur in groups of nearly parallel hills. Such drumlin fields are a common part of continental glacial landscapes and are particularly strik-

FIGURE 14–28
An erratic keeps its place within Yellowstone National Park, Wyoming. The size of the erratic, as compared to the trees that surround it, suggests something of the carrying capacity of a continental glacier. (Photograph courtesy of the U.S. Geological Survey.)

FIGURE 14–29
This streamlined hill is a drumlin. Its size is indicated by the trees on its crest. The drumlin is composed of drift plastered against an obstruction overridden by the glacier as it passed from left to right. Drumlins have the same streamlined shape that structures formed by catastrophic floods (see Figure 14-37) and barchan dunes (see Figure 13-21) do. (Photograph courtesy of the U.S. Geological Survey.)

ing when viewed from the air. Drumlins have figured prominently in both American religious and military history. Hill Cumorah, near Palmyra, New York is an important site to those of the Mormon faith, and Bunker Hill, near Concord, Massachusetts is an important part of Revolutionary War history.

As glaciers melt back, they may leave chunks of stagnant ice the size of buildings mired within the thick blankets of deposited drift. As these blocks of ice melt, the drift that once covered them collapses

into the void space created by the melted ice. In this way **kettles,** roughly circular depressions (Figure 14–30, are formed in the ground moraine. In some areas the drift will be peppered with collapse depressions.

In areas of abundant precipitation, the kettle depressions fill with water and are known as **kettle lakes** (Figure 14–31). Many of the smaller roughly circular lakes in north central North America owe their origin to this process.

Moraines form at the margins of continental glaciers, often faithfully recording the oscillating advance and retreat of a glacier as a function of both surging flow and climate changes. Moraines may form ridges many tens of feet or several meters high. These ridges (Figure 14–32) are generally composed of unsorted till, but meltwater streams may also alter them into areas of stratified drift. Within the morainal material are cobbles and boulders that are strongly scratched and grooved, having served their time within the base of a moving ice sheet.

PLUVIAL LAKES

The onset of glaciation is a part of a widespread change in climate. As we have indicated, the region near the margin of the glacier is often an area of coolness and much increased precipitation. The plentiful rainfall collects in low-lying areas as **pluvial**

FIGURE 14–30
The outwash sheet from Hidden glacier in Alaska contained a buried hunk of glacial ice. As it melted beneath the surface, the drift collapsed inward to form an incipient kettle. (Photograph courtesy of the U.S. Geological Survey.)

FIGURE 14–31
These kettles within the ground moraine near Orando, Montana have filled with water. Such kettle lakes form roughly circular ponds and lakes in many glaciated regions. (Photograph courtesy of the U.S. Geological Survey.)

lakes. Two examples of pluvial lakes formed during the Pleistocene will give some idea of how these ephemeral features created much longer-lasting changes in the face of North America.

Lake Bonneville

The Great Salt Lake (Figure 14–33) is the largest lake west of the Mississippi River, so large that the state of Rhode Island could fit within its boundaries. It covers approximately 1800 square miles (4661 square kilometers).

Its predecessor was a pluvial lake ten times its size. Known as Lake Bonneville, it covered almost all of northwest Utah. Its maximum area was 19,000 square miles (50,000 square kilometers) and its maximum depth was up to 1000 feet (300 meters). It was approximately the size of modern Lake Michigan.

Lake Bonneville drained north through Red Rock Pass into the Snake River of Idaho and thence to the ocean. The elevation of the lake's outlet stabilized the lake's surface for thousands of years. Valley glaciers from the Wasatch Range moved down to lake level, producing moraines eroded by wave action of the lake.

The most distinctive landform created by the lake is a series of shoreline features—bars, beaches, deltas, spits, wave-cut cliffs, and shoreline terraces (Fig-

FIGURE 14–32
A terminal moraine, comprising the loose debris pushed ahead of a glacier, marks the lower end or snout of La Perouse glacier in Alaska. The glacier was overrunning the forest on the far left when this photograph was taken in 1889. (Photograph courtesy of the U.S. Geological Survey.)

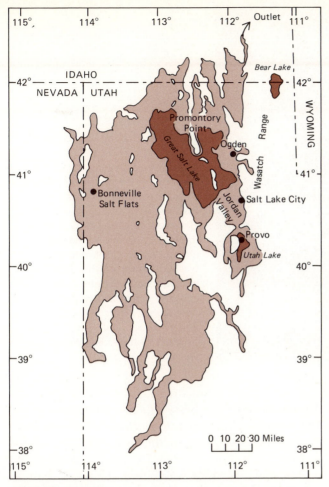

FIGURE 14–33

The Great Salt Lake and Utah Lake form the remnants of the much larger Lake Bonneville, which existed during periods of pluvial climate within the Pleistocene. Lake Bonneville was the size of Lake Michigan and slightly deeper; its surface elevation was controlled by the altitude of its outlet north into Idaho. (Modified from a map by the U.S. Geological Survey.)

ure 14–34). There are several levels of these, each marking a distinctive period of stable lake level in the overall general fluctuation of the lake.

Lake Missoula

The largest lake to form in the Pacific Northwest during the Pleistocene was Lake Missoula. It was formed when ice sheets moving south from Canada invaded northern Washington, Idaho, and Montana, blocking and diverting the Okanogan, San Poil, Columbia, Pend Oreille, Priest, Kootenai, and Clark Fork rivers.

Ice dammed the Clark Fork River near the point where it now empties into Pend Oreille Lake (Figure 14–35).

The water impounded behind the ice dam filled the tributary valleys of the Clark Fork, creating a lake covering 3000 square miles (7800 square kilometers), with a volume half that of modern Lake Michigan. Its surface at the ice dam stood nearly a mile above sea level, at an elevation of 4150 feet (1265 meters), which made it more than twice the depth of Lake Superior. Its old shorelines are faintly visible over large parts of northwestern Montana and northern Idaho.

As water continued to pour into the lake from alpine glaciers in the Bitteroot Range and lobes of glaciers near modern Flathead Lake, the water level began to overflow the top of the ice dam, which held 500 cubic miles (2000 cubic kilometers) of water four-fifths of a mile above sea level. The rapid erosion caused by water overtopping the ice dam set the stage for an extremely energetic event, some time about 18,000 to 20,000 years ago. Within perhaps a day or two, the ice dam was destroyed and the lake drained at a rate unmatched by any flood known to humankind.

The water had only one place to go, and that was down—across the northeastern part of Washington, racing westward toward the Pacific Ocean. Estimating from the landforms that were created, current velocities of the rushing water reached 45 miles per hour (72 kilometers per hour). The maximum rate of flow has been estimated at *386 million cubic feet per second* (11.6 cubic meters per second)—a rate ten times *the combined flow of all the rivers of the world!* The flow rate was 64 times that of the Amazon, the world's largest river, and 1500 times that of the modern Columbia.

The water boiled across Pend Oreille Lake, Rathdrum Prairie, and roared down Spokane Valley. It created giant ripple marks 20 to 30 feet (6 to 9 meters) high, and 2 miles (3 kilometers) long, and it moved boulders the size of three-story buildings, shattering them against one another and creating gigantic percussion marks. As the currents boiled across the lava fields of northeastern Washington, they swept into three natural channels (Figure 14–36). Water scoured the deep loess soils within the channels and carried them all the way to the Pacific.

The three channels, known as the Grand Coulee, Crab Creek, and Cheney–Palouse were deeply eroded into solid basalt. Wild cascades and plunge pools up to 200 feet (61 meters) deep were created. The three torrents, together with dozens of smaller channels crossed and crisscrossed the basaltic area,

stripping soils and basalt and leaving behind some of the strangest scenery in the world. The stripped area is now known as the *Channeled Scabland* of eastern Washington.

To the south near Pasco, Washington, the three floodwaters rejoined and spilled west of modern Walla Walla, Washington into the Columbia River, where they built a massive delta far out into the Pacific. After perhaps a month of torrential flow, streams slowly returned to normal, leaving the scars of a catastrophe beyond our comprehension.

The origin of the scablands scenery, except for its source in Lake Missoula, was worked out by geologist J. Harlen Bretz. He fought for decades a lonely war against the geologic "establishment" in the United States, who *could not* be convinced that such a catastrophic event had ever occurred. Like Wegener who argued for mobile continents at the same time, Bretz could present meticulous empirical evidence, based on lengthy field information, to show that the landscapes were bizarre, *but he could not provide a satisfactory mechanism.*

Bretz could not imagine where all that water had come from—he could show only that it *had.* So he patiently collected still more data from eastern Washington, convinced that data would carry the day. Data did not, but the recognition by others of Lake Missoula and the connection between the emptying of that lake and the Channeled Scablands finally

won acceptance for Bretz's theory. Bretz had simply been working in the wrong place to *find a mechanism;* he had been working in the right place to see the *results.*

The story of Bretz's stubborn refusal to bow to others with more conventional views has two happy endings. He has lived to see his theory accepted, and he has seen it more recently applied to another planet. The surface of Mars (Figure 14–37) has streamlined landforms and chaotic landscapes quite reminiscent of the Channeled Scablands of Washington. How ironic that a concept, bitterly resisted by most of the eminent glacial geologists of the 1930s through the 1950s, should be so readily accepted in the 1980s that it was easily applied to the geology of another world.

SUMMARY

1. Glaciers occupy about 10 percent of the earth's landmass today, largely on the Antarctic and Greenland continents. In the past they have occupied up to a third of the world's continental areas.

2. Glaciation has its onset when northern latitudes experience such limited solar energy in the summer that little snow melts, so that snow accumulates for centuries. For *evidence* of

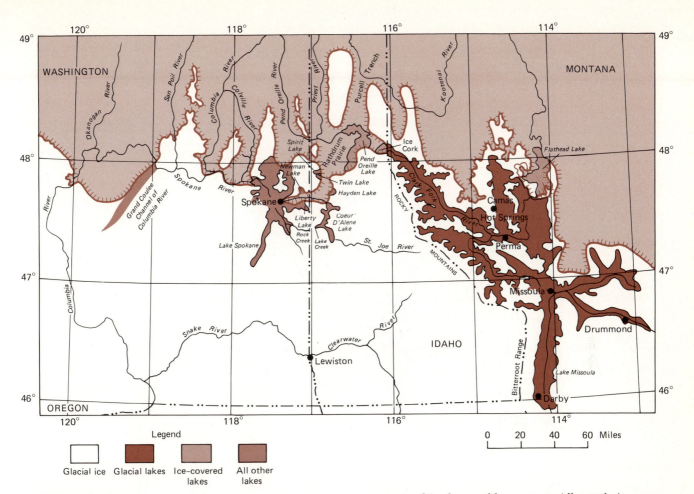

FIGURE 14–35

Glacial Lake Missoula is shown in the darkest color. Its ice "cork" or dam was right at the Idaho - Montana border. Holding half the water of Lake Michigan nearly four-fifths of a mile above sea level, the lake was a reservoir of colossal potential energy. When the lake overtopped its dam and began to rapidly erode it, waters began to spill across the Rathdrum Prairie and Spokane Valley to form what J. Harlan Bretz has called the "Spokane Flood." The flow was ten times that of the world's rivers *combined*. (Map modified from a U.S. Geological Survey map.)

glaciation to occur, there must be a continent in position to accept the accumulating snow.

3. Glaciers are accumulations of snow that flow down stream valleys or across continents under their own weight. As highly viscous erosional agents, they both vigorously erode and transport rocks of all sizes and types. When they move below snow line, they deposit their loads as drift and till in a wide variety of landforms.

4. Glaciers apportion their erosional force to the area of ice–rock contact. They erode by direct abrasion and by plucking materials that have been frost wedged from beneath them.

5. We currently live in the Holocene, a period of some 10,000 years duration in which only the Antarctic and Greenland are experiencing continental glaciation, and only high-elevation, high-latitude mountains have alpine glaciers. The Holocene is an interglacial time, forming the last part of the Pleistocene, which was four times dominated by major mid-latitude glaciation.

6. Glaciers forms lakes of many kinds, including tarns, trough lakes, kettle lakes, linear eroded lakes, and pluvial lakes.

KEY WORDS

arête	crevasse
catenary profile	drift
cirque	drumlin
continental glacier	end moraine

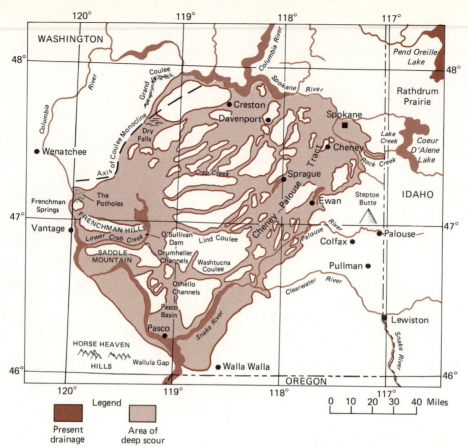

FIGURE 14–36

When the storming floodwaters raged across the Spokane Valley, they split into a northwestern channel, the Grand Coulee, a central channel, the Crab Creek area, and a southeastern channel, the Cheney - Palouse area. Stripping off the overlying loess soils, the torrential floods left numerous smaller channels bare of soil. The waters carved deep channels, leaving bizarre giant ripples, dry falls, and plunge pools, and drove dense basaltic boulders larger than three-story houses far down the channels. Rejoining forces in the Pasco area, the waters stormed on southwest through the Wallula Gap, raising the discharge of the modern Columbia River by 1500 times. The waters finally reached the Pacific Ocean and created a colossal delta 500 miles (805 kilometers) from the burst dam across the Clark Fork River of Idaho. (Map modified from a U.S. Geological Survey map.)

erratic	kettle	terminal moraine	truncated spur
esker	kettle lake	till	valley glacier
glacial plucking	lateral moraine	trough lake	valley train
glacial polish	moraine		
glacier	outwash plain		
ground moraine	Pleistocene Epoch		
hanging valley	pluvial lake		
Holocene Epoch	Quaternary Period		
horn	snow line		
Ice Age	stratified drift		
interglacial epoch	tarn		

EXERCISES

1. The speed of water during the Spokane floods that created the Channeled Scablands has been estimated at 45 miles per hour (72 kilometers per hour). What kind of modern geologic *evidence* would we search for to recover the velocity of a stream that flowed 18,000 years ago?

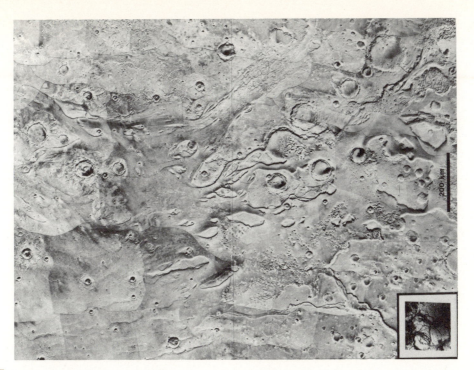

FIGURE 14–37

The outflow channels on Mars are reminiscent of those of northeastern Washington. The channels emerge from the chaotic terrain on the right, flow out onto the plains, creating highly streamlined forms within their channels, and then gradually disappear. Since these outflow channels are geologically quite old, they suggest that Mars in the past had a great deal more fluid water than is detectable today. This could be a sign of long-term climatic changes on Mars. The small inset photograph is a picture of the Channeled Scablands of Washington at the same scale. As you can see, the outflow channels on Mars are of incredible size, with individual channels more than 1200 miles (2000 kilometers) long. (Photomosaic of Viking Orbiter images courtesy of NASA.)

2. In parts of North Dakota near the margin of former Lake Agassiz, there are older river valleys that are several miles or kilometers wide, yet the streams flowing in these valleys are only small creeks. Why should such a small stream occupy a large valley?

3. In parts of Scotland, fields of rich, deep soil are surrounded by fences made of huge boulders laboriously dragged out of the soil. Where did the boulders come from, and how could they have gotten into areas of deep, sandy soil?

4. Along the Mississippi River, the depth of loess diminishes away from the river, yet it is known that loess is deposited by wind, not water. Explain the apparent contradiction.

5. In some mountainous areas the valleys up high in the mountains are ∪ - shaped in cross section, while the same valley lower down in elevation is ∨ - shaped in cross section. How would you explain the differences?

6. As a generalization, the lakes of northern Minnesota are elongate in map view whereas those in southern Minnesota are somewhat circular. There are numerous exceptions, but how could these different shapes be explained?

7. Contrast the ability of wind, water, and ice to sort selectively by size the material they pick up and deposit. What single factor causes the variation in sorting capacity of these three erosional agents?

SUGGESTED READINGS

AGASSIZ, LOUIS, 1840 (translated and edited by A. V. Carozzi, 1967), *Studies on Glaciers*, New York, Harper & Bros, 244 pp.
▶ *Beautifully illustrated and a readable masterpiece.*

BAKER, VICTOR R., 1978, The Spokane Flood controversy and Martian outflow channels, *Science*, v. 202, n. 4374, pp. 1249–1256.

► *Superb review of Channeled Scablands theory and its application elsewhere.*

COVEY, CURT, 1984, The Earth's orbit and the Ice Ages, *Scientific American*, pp. 58–66.
► *Up-to-date review of a critical topic; beautiful illustrations.*

MCINTYRE, A., et al., 1976, The surface of the Ice-Age Earth (CLIMAP Project, *Science*, v. 191, pp. 1131–1137.
► *Brief review of critical data about how it was then.*

PART

4

Earth
Resources

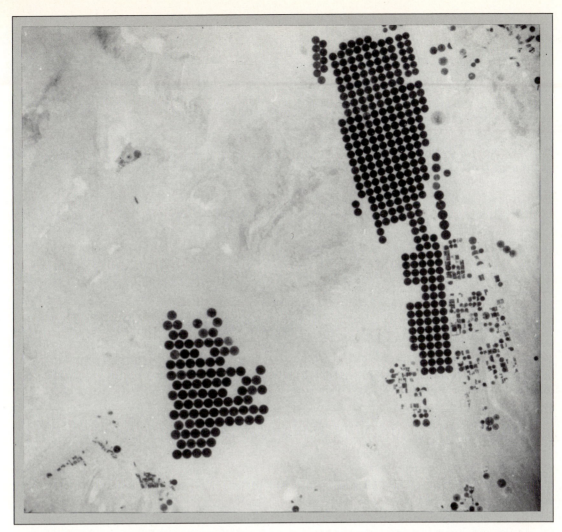

Center pivot irrigation systems stand out as circular oases of green against the desert in northeast Saudi Arabia in the vicinity of Al-Hufuf. Center pivots are a system for pumping water from the ground and distributing it in circles half a mile in diameter from a mobile sprinkler system. Even beneath the Arabian desert there is groundwater, stored for millennia. (Shuttle photography, courtesy of NASA.)

CHAPTER *fifteen*

Groundwater Resources

When the well's dry, we know the worth of water.

BENJAMIN FRANKLIN (1706–1790)

More than two-thirds of our earth is covered with water, yet many people do not have enough. More than half of everything alive is formed from it. Centuries ago it was often the bearer of death, because it was contaminated by typhoid and other devastating plagues. Yet it is the necessity for life, yielding food from once-parched earth. For those who have known severe drought or the agony of unending thirst, its value is greater than gold. Our bodies are composed mostly of water; a change of more than 2 percent in our delicate fluid balance causes death.

Napoleon called it "the greatest necessity of the soldier." Thoreau spoke of it as "the only drink for a wise man." Alas, Coleridge's ancient mariner saw that his vast supply was useless.

The earth's initial supply of water came from volcanism within the earth (see Chapter Eighteen). The water filling the earth's ocean basins is made salty by the addition of chlorine from undersea volcanism and sodium from continental weathering (see Chapter Four). This great reservoir of water is 97 percent of the earth's supply.

The remaining 3 percent is fresh water, evaporated as distilled water vapor from the ocean's surface by solar energy (Figure 15–1) and brought onto the continents by disturbances in atmospheric flow, also caused by solar energy. More than three-quarters of that 3 percent is locked up in glaciers (Chapter Fourteen).

GROUNDWATER AND SURFACE WATER

The remaining fraction of 1 percent of the earth's fresh water is in liquid form. Of that, a small amount forms our rivers and lakes; it is called **surface water.** More than 90 percent of our available fresh liquid water is temporarily stored underground as **groundwater.** The combination of groundwater and surface water totals a little over 1 million cubic miles (4 million cubic kilometers) of liquid fresh water.

SURFACE WATER

Surface water is on its way back to the sea (Figure 15–2). The majority of it, nearly 30,000 cubic miles (120,000 cubic kilometers), is stored in freshwater lakes. The Great Lakes of North America store nearly one-quarter of the earth's surface water, with the large lakes of Asia and Africa each storing another one-quarter. What fills the hundreds of thousands of smaller lakes—and all the world's rivers—is less than one-quarter of the earth's minuscle fraction of surface water.

The stream channels of the world carry approximately 300 cubic miles (1200 cubic kilometers) of water at any one time. This is equal to the amount of water the world uses in four months. Yet, the world's river channels contain only enough water to *maintain* their flow for about *two weeks*. Obviously in

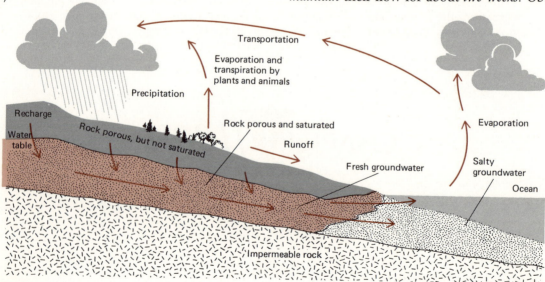

FIGURE 15–1
The components of the hydrologic cycle are *recharge + evaporation + transpiration + runoff ≅ precipitation.* Can you suggest some factors that alter the recharge to runoff ratio? (Diagram modified from U.S. Geological Survey Circular 601-I.)

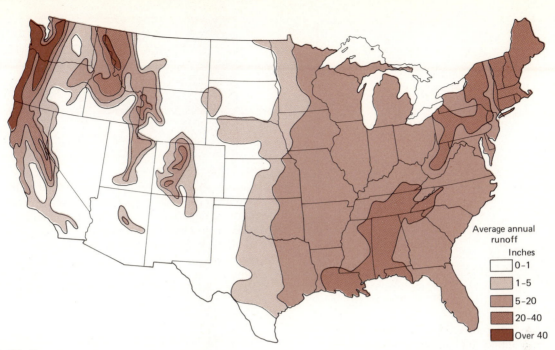

FIGURE 15–2

Average runoff in the coterminous United States. Notice that large portions of the American West have essentially no runoff; drainage is internal, flowing into closed basins. (Modified from U.S. Water Resources Council maps.)

most areas it does not rain every week or two. How do rivers keep flowing year around? Where is all that *extra* water coming from?

GROUNDWATER

Groundwater makes up more than 90 percent of our available fresh water and supplies about one-third of the annual stream flow. It is estimated that there are one million cubic miles (4 million cubic kilometers) of groundwater stored within 2000 feet (610 m) of the earth's surface. *Groundwater* is fresh water stored within the outer part of the earth.

Within the United States there is an estimated 65 quadrillion gallons (246 quadrillion liters) of groundwater within economic reach. Of this amount about 82 billion gallons (310 billion liters) per day are withdrawn from reservoirs to furnish 48 percent of the total population and 95 percent of the rural population with their fresh water. Through time, the dependence on groundwater as the prime source for fresh water is increasing.

Increasing dependence on groundwater has brought an increasing need for expertise in locating, recovering, and preventing contamination of this important resource. *Hydrogeologists* are specialists in this area; they are becoming among the most publicly visible of geologists.

All water flows toward the sea, groundwater being no exception; the rate at which groundwater travels seaward, however, is one-thousandth that of surface water. Groundwater is analogous to a glacier (Chapter Fourteen); pulled by gravity, it moves very, very slowly down to the sea.

Commonly the seaward flow of groundwater is *interrupted* in humid areas by its diversion *into* stream valleys (Figure 15–3). *It is the inflow from groundwater that feeds a stream year around and keeps it running in the absence of rain. Any stream or lake in a humid region* thus possesses *two sources of water:*

1. *Runoff water* (Figure 15–2), which flows directly into it from rain and snow, and,

2. *Groundwater* (Figure 15–3), which flows into it from stream valley walls.

Water Table and Climate

The situation just described is typical of *a humid region* where groundwater trickles *into* a stream and helps maintain its year around flow. In an *arid climate* the situation is reversed; a flowing stream is a

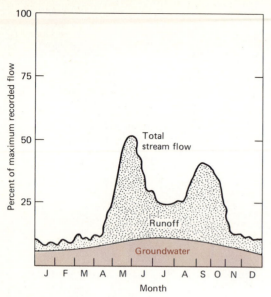

FIGURE 15–3

An example of the sources of stream flow within a humid region. The groundwater contribution remains fairly constant, maintaining a perennial stream.

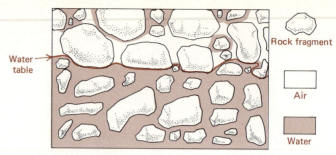

FIGURE 15–4

A much enlarged view of the water table within an extremely permeable sediment. Above the water table, moisture may coat the rock fragments and remain available to plant rootlets. The pore spaces between grains are partially filled with air. Below the water table all pore spaces are totally filled with water; the sediment is saturated, and the water film is continuous, so that any pressure on the water will cause it to percolate.

source of groundwater, in that it feeds runoff water into underground storage basins. Here are the reasons why this occurs.

The boundary between saturated and unsaturated loose rock and soil is the **water table** *(Figure 15–4), a generally planar surface whose elevation is commonly above sea level.* The water table is a familiar concept to anyone who has ever dug a hole and noticed that at some depth (the water table) the soil is so saturated that the hole partly fills with water. The elevation of the surface of water in the hole is the water table at this point.

Above the water table is a zone of unsaturated loose rock and soil, called the **zone of aeration**. The zone extends from the earth's surface downward to the water table. In all but the most arid areas, the zone of aeration contains some moisture, and it is the root zone for many plants. Their roots have the capacity to reach down into the zone of aeration and pull the moisture away from its contact with rock fragments. All of us who have killed a houseplant by overwatering it have discovered that their roots cannot live in the **zone of saturation**. The zone of saturation is the part of the soil beneath the water table that is totally saturated with water.

The water table can be visualized as a surface that separates the zones of aeration and saturation; it is also the surface where groundwater experiences atmospheric pressure. Beneath the water table is a *continuous* mass of water, filling the natural pores and cracks in the rock; the weight of this water causes a rise in pressure within the water mass. *It is this pressure increase plus the force of gravity that drives underground water motion.*

Notice the analogy between the action of groundwater and that of a stream. The mass or weight of water within valley walls, acted on by gravity, "pushes" water down a stream valley. Groundwater, unconfined by valley walls, is "pushed" by pressure differences, with pressure in an unconfined system increasing downward.

STREAMS IN THE HUMID CLIMATE

Stream valleys in a humid area represent a *low-pressure "hole"* in an environment where pressure is otherwise constantly increasing downward. Since streams flow open to the sky, the water at their surface is at atmospheric pressure. The groundwater at the same elevation as the stream valley is *pressured* by the weight of the overlying water (Figure 15–5), and so it *flows toward the stream, a zone of lower pressure.*

Streams, then, are interruptions to the general downward flow of groundwater in a humid climate, where the elevation of the water table roughly parallels the elevation of the ground surface. Streams "rob" groundwater storage and send water to the sea, diverting it from a much more leisurely seaward flow as groundwater. Were it not for the addition of groundwater, the world's permanent or *perennial* streams would run dry in a few weeks (Figure 15–3). They would quickly become intermittent or *ephemeral* streams, much more typical of desert climates.

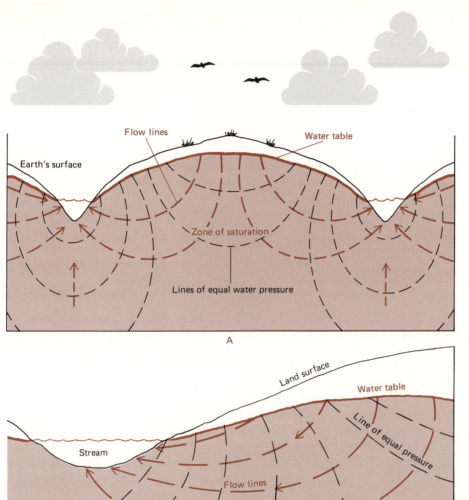

FIGURE 15-5

(A) Groundwater percolates when the recharge from precipitation loads the zone of saturation. The resulting pressure drives water toward low-pressure zones, namely stream valleys. Stream valleys intercept water from groundwater storage and speed its passage to the sea via a set of integrated drainages. (B) The effect of an impermeable layer is to distort flow lines, but the stream remains an interruption to groundwater flow. Any contaminant placed on the land surface may eventually leach both into groundwater storage and into the stream. The profile of the water table nearby is subparallel to the profile of the land surface.

STREAMS IN AN ARID CLIMATE

Arid climates are distinguished by thick zones of aeration and deep water tables—quoting an Arizona water well driller, "It's eighty feet to water and two feet to Hell."

Arid climates are distinguished by the lack of *surface water*, not of groundwater. Even the Sahara, the synonym for aridity, has millions of gallons or liters of water beneath it. Groundwater stored in an arid area may take tens or hundreds of thousands of years to accumulate, but it is there. Unfortunately, it is also deep.

Because the elevation of the water table in arid areas is often at a great depth, *the water table is commonly far below stream valley floors*. As a result, when water is flowing in a desert wash, the valley is a *zone of higher pressure* compared to the zone of aeration surrounding it. The stream "leaks" water *out* of its channel and becomes a *source* of water for the deep zone of saturation (Figure 15-6). It is no wonder that desert streams are ephemeral or intermittent; what little runoff they do receive rapidly leaks into the ground. It is obvious why deserts are areas lacking external drainage (Chapter Thirteen).

Water being *added* to groundwater storage is spoken of as **recharge**, that being removed as *withdrawal*. The most common form of recharge is precipitation; springs and wells are areas of withdrawal. If recharge equals withdrawal, the elevation of the water table remains constant. If recharge exceeds withdrawal, the elevation of the water table rises; at points where the water table intersects the land sur-

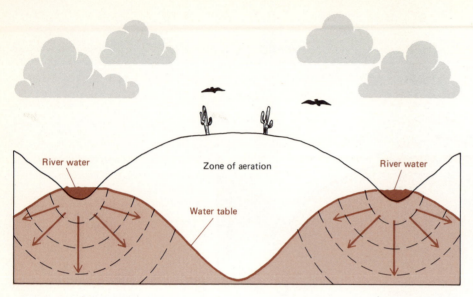

FIGURE 15-6
Compare this diagram to Figure 15-5A. Notice that the water table profile is now the mirror image of the land surface, and streams become *recharge* points instead of *withdrawal* points.

face marshes or swamps develop.

Streams in an arid climate recharge the groundwater table, *raising its elevation near the stream.* Streams in a humid climate withdraw water from the groundwater table, *lowering its elevation near the stream.* Comparing Figures 15–5 and 15–6 indicates that in a humid area, *the elevation of the water table is broadly parallel to the land surface,* while in an arid area, the *elevation of the water table is broadly the mirror image of the land surface.*

Groundwater Motion

As we have seen, water underground moves in response to pressure differences. The movement of water underground is called **percolation,** a term useful in describing the trickling of water forced to find its way through tiny cracks and crevices. The rates of percolation are quite slow by human standards—perhaps a few yards or meters a year—because flow is greatly retarded by friction along the tiny passages through which water must move.

The amount of space, expressed as a percentage of total volume, within fractured rock or loose regolith is its **porosity.** The porosity of earth materials ranges from less than 1 percent to over 30 percent. If a rock is not **porous,** if it does not possess pore space, it cannot contain fluids. If the pore spaces are *interconnected,* the material is **permeable,** capable of *transmitting* fluid. The percentage of interconnected pore space is the rock's or soil's **permeability**. These terms may not apply in areas where groundwater flows through limestone; because limestone is soluble in cold water, groundwater movement in limestones is commonly through open fractures and caverns (see Figure 15–17).

WATER USAGE

All precipitation is part of a single, connected system. Surface water provides recharge to groundwater storage, so that pollution of surface water eventually contaminates groundwater supplies—an increasing problem worldwide. Water abuse and overuse are making headline news, as we learn wise management practices by trial and error.

Our demand for water has doubled in the last 30 years and will double again by the year 2000 (Figure 15–7). Our usage is split among domestic, industrial, and agricultural needs. Each day a total of 2000 gallons (7600 liters) of water is used for the benefit of each citizen of the United States. As we continue to drain this invaluable resource, it is helpful to understand the basics of water demand.

DOMESTIC USE

For domestic and industrial use, the common volume unit is the *gallon,* equivalent to 3.78 liters (Appendix E). The unit of flow is gallons per minute. A gallon a minute (0.063 liter per second) may seem unimpressive, but that equals 1440 gallons a day, enough for two or three homes. A leaking faucet dripping one drop per second wastes 4 gallons per day.

In our homes, an average use is 50 to 150 gallons per person per day. Of that, 20 to 30 percent is direct wastage through leaking pipes, faucets, radiators, and so on. *Domestic use totals approximately 10 percent of water usage in the United States,* with the heaviest

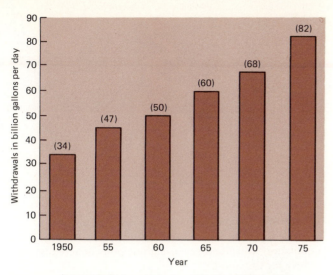

Increase in fresh groundwater utilization, 1950 to 1975

FIGURE 15–7

Groundwater usage in the United States doubles every 20 years, a rate that will bring some southwestern aquifers to total exhaustion within a decade or two.

use being lawn watering. A lawn may require from 30,000 to 200,000 gallons of water per year, depending on size and climate. A fifth of our water is used for flush toilets, the second largest use of water in an average home. The highest per capita consumption generally occurs in the driest, hottest areas of the country where swimming pools, artificial lakes, lawns, and fountains abound.

Water usage directly reflects the habits of American life. Usage sharply peaks between 6:00 and 10:00 P.M., when we eat, wash the dishes, and bathe. Another peak occurs between 6:00 and 9:00 A.M., when we bathe and have breakfast. In hot, dry areas water usage peaks in the afternoon with evaporative air conditioning. Finally, in most American cities there is especially heavy usage on Saturday night; the Saturday night bath is apparently still a tradition.

INDUSTRIAL USE

Of all water used for industrial purposes, 94 percent is used for cooling; much of this water is reused after being cycled through cooling towers. Other typical uses include cleaning and packaging. Few of us appreciate how much water goes into the things we use.

Consider what happens when you grab a 12-ounce (355 milliliter) can of your favorite beverage from the refrigerator. How much water in a 12-ounce can? Almost 17 gallons or 64 liters!

It took 0.05 gallon (0.2 liter) to wash and seal the can, 6 gallons (23 liters) to make the can, and 10.5 gallons (40 liters) to find, mine, grow, fabricate, and transport the raw materials (Figure 15–8) for the can. If the can is not recycled, only 0.5 percent of the total water consumed in producing your canned beverage will have gone to a constructive purpose—quenching your thirst.

Industrial consumption of water accounts for about 6 percent of the total water consumed in the United States. The chief concerns raised by industrial use of water are contamination of groundwater supplies and the elevation of water temperature in streams by repeatedly using its waters for cooling purposes. As electrical power demand increases, so does the need for cooling water for both nuclear- and coal-fired power plants.

AGRICULTURAL USE

Agriculture is our greatest consumer of fresh water. Accounting for a total of *83 percent of the fresh water consumed in America,* irrigated agriculture (Figure 15–9) is an enormous business. It prospers in the Far West, precisely where rainfall is scanty and the sun

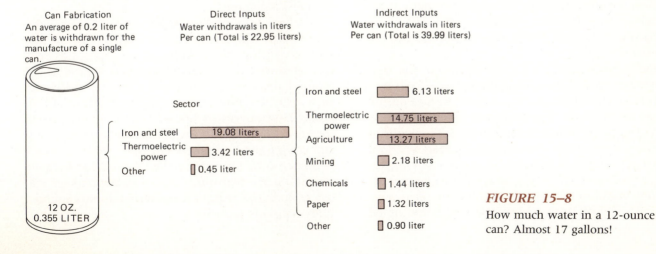

Can Fabrication
An average of 0.2 liter of water is withdrawn for the manufacture of a single can.

Direct Inputs
Water withdrawals in liters Per can (Total is 22.95 liters)

Indirect Inputs
Water withdrawals in liters Per can (Total is 39.99 liters)

Sector

Iron and steel	19.08 liters
Thermoelectric power	3.42 liters
Other	0.45 liter

Iron and steel	6.13 liters
Thermoelectric power	14.75 liters
Agriculture	13.27 liters
Mining	2.18 liters
Chemicals	1.44 liters
Paper	1.32 liters
Other	0.90 liter

12 OZ.
0.355 LITER

FIGURE 15–8

How much water in a 12-ounce can? Almost 17 gallons!

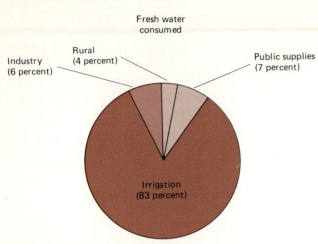

FIGURE 15–9

According to recent data on water *consumption,* the needs of homes and industry pale alongside the needs of irrigated agriculture. Where sunshine is abundant, rainfall is not.

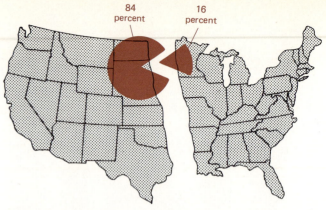

FIGURE 15–10

Fresh water, like oil, is consumed where it is not produced. The arid West, with a fourth of America's population, consumes seven-eighths of its metered water, mostly for irrigation.

shines year around. Irrigation water is commonly measured in *acre-feet,* the amount of water necessary to cover an acre (about the size of a football field) to a depth of one foot. One acre-foot equals 43,560 cubic feet (1233 cubic meters) or nearly 326,000 gallons.

Few realize how much water it takes to grow food. A single egg requires 180 gallons, a bushel of wheat 15,000 gallons, and enough cotton for one shirt or blouse 24,000 gallons. A medium-sized turkey requires 16,000 gallons and a Thanksgiving dinner for eight requires over 40,000 gallons of fresh water. To quote an anonymous poet,

A pint a day is all I say to keep my whistle wet.
But so much more I have to pour before my table's set!

Throughout much of the west, the water used in irrigation is often in excess of dependable recharge. About *one-third* of the 7 million acre-feet of water used in Arizona irrigation is *in excess* of natural recharge. A tank holding one year's excess would be the size of a football field and over 400 miles (640 kilometers) high.

The western states, with one-quarter of the U.S. population, use 84 percent of the nation's fresh water (Figure 15–10), largely because of wastage during irrigation of plants. The newest irrigation system is the center pivot, which delivers water from overhead pipes centered on an irrigation well and moved in a circle (see chapter frontispiece). In one study in Nebraska, *one* center pivot system consumed enough irrigation water in one season to provide a year's supply for a city of 1000, and in one year it consumed

8000 gallons (30,300 liters) of diesel fuel. Spraying water into the arid air caused at least one-half of it to evaporate before it could enter the root zone.

In a study in southern Arizona, it took *six* times as much water to grow an orange as in Israel, where the climate and soil are similar. In Israel, where water is expensive and often obtained by desalinization of seawater, *drip irrigation* systems, which bring the water one drop at a time through small pipes to the root zone, provide water only where the plant needs it.

Icebergs for the Desert?

Because western groundwater supplies are rapidly diminishing, a number of imaginative solutions have been proposed, usually involving large-scale transport of water from areas of abundance to areas of scarcity. The politics of water in the west have always been explosive; the newest scheme is perhaps the most ambitious.

Glaciers, composed of fresh distilled water, contain 77.1 percent of the earth's total supply; 90 percent of which is in the Antarctic ice cap. It has been estimated that the Antarctic ice shelves produce 35 trillion cubic feet (1 trillion cubic meters) of icebergs each year. Recently a few tentative engineering proposals have suggested hooking ocean-going tugs to icebergs ranging up to several hundred million tons, and towing them to places like Saudi Arabia, where estimated costs for water melted from the iceberg are something like two-thirds that of desalinized sea water. A similar proposal for California deserts is under study. Perhaps someday . . .

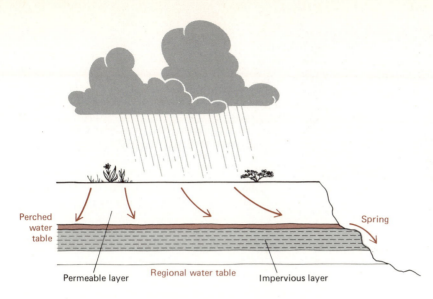

FIGURE 15–11
This sketch indicates one way in which springs are formed. Water percolating downward through a permeable layer encounters an impermeable layer, such as a buried lava flow. The water is now prevented from reaching the deeper regional water table; instead it forms a higher, perched water table. The water travels laterally along the layer and exits at the nearest valley cut through the water table.

WATER WITHDRAWAL

Water for domestic, industrial, and agricultural use comes from any combination of four sources: surface water, springs, conventional wells, and artesian wells. Each source of water has its own unique set of problems and rewards.

SURFACE WATER

The greatest advantage of surface water is its easy availability at low initial cost. It is simply there, and only a pump and piping system are necessary to acquire it. Accessibility, however, comes at a price. Surface water is open to the air, and it therefore collects a wide range of contaminants.

For domestic use, the water must be free of deadly bacteria; the standard treatment is chlorination. A combination of chemical treatments and filtration removes dirt and solid matter of all kinds. The use of activated charcoal and aeration may help remove undesirable odors and volatile chemicals. Still other forms of both biologic and chemical treatment are possible in order to rid surface water of its many contaminants.

SPRING WATER

Spring water is usually a hybrid of surface water and groundwater. **Springs** form when the land surface intersects a water table whose elevation is abnor-

mally high (Figure 15–11).

When the downward movement of recharge water is interrupted by an *impermeable* layer, the water spreads out and flows along the surface of the impervious layer, forming a perched water table. A **perched water table** lies on top of an impervious layer which limits downward recharge into the regional water table.

Depending on the distance traveled underground, spring water may be naturally filtered of much of its solid material, and thus it may emerge as sparkling, clear water. For human consumption, it must normally still be chlorinated to remove bacteria. The most common problem associated with spring water is the incorporation of various soluble chemicals during the water's passage through and along rock surfaces.

In small amounts, some soluble material may be healthful, including such ions as fluorine, calcium, and others. Iron and manganese compounds stain light-colored clothes; sulfur compounds have unpleasant odors; and compounds of arsenic, selenium, and lead may cause acute poisoning. Each spring is individual, its output requiring individualized treatment.

CONVENTIONAL WELLS

A **conventional well** requires pumping to recover water and is a deliberate interruption in groundwater flow beneath the water table (Figure 15–12). In order for water to flow laterally into the well, the layer carrying the water must be permeable. An under-

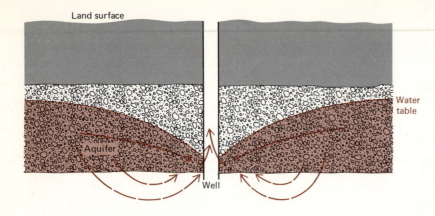

FIGURE 15–12
Cross section of a conventional pumped well. Notice that the well must be drilled far below the water table, as turning on the pump creates a *cone of depression* around the pumping well. This abnormally steep gradient along the surface of the water table causes rapid percolation into the well. The more powerful the pump, the deeper the well must be drilled.

ground layer, capable of serving as a water source, is called an **aquifer.**

An aquifer may consist of a layer of unconsolidated gravel or an area of rock broken by numerous fractures. Or it may be a layer of nearly solid rock, whose slight permeability still allows water to flow

slowly. Regardless of its composition, an aquifer must

1. Contain economic quantities of usable groundwater, and
2. Be sufficiently permeable to allow the water to flow at a usable rate.

Well water may have flowed slowly through many miles or kilometers of rock. Such a lengthy passage *may* clear the water of solid particles and oxidize disease-producing bacteria; consequently some well water *might* be safe to drink without treatment. In areas of contaminated surface water, well water is *far* more suspect.

Most well water is chlorinated as a safety precaution. In all except the most isolated areas, this is simply good, safe practice.

Well water carries two liabilities. First, it is more expensive than surface water to find, to drill for, and to pump (Figure 15–13). The cost of pumping water increases roughly proportional to the depth of the well. Double the depth, and pumping costs approximately double as well. There are economic limits to what is usable water. As the water table declines, so does water quality generally. Another danger to large overdrafts is land subsidence, the collapse of the land surface as the water beneath it is removed.

The *Ogallala aquifer* is an example of an unmanaged heavily pumped aquifer in the central United States. It is a relatively young deposit of limestone and gravel underlying most of Nebraska and parts of western Kansas, Oklahoma, and Texas. Ranging up to 1000 feet (305 meters) in thickness, it originally contained an estimated quadrillion gallons of water, equal to the total volume of Lake Huron. Punctured by thousands of wells, including giant center pivot systems, the Ogallala now yields an *overdraft*, an excess of withdrawal over recharge, nearly equal to the flow of the Colorado River. It is currently estimated

FIGURE 15–13
This is what it was like to drill a water well deep in the piney woods of Georgia in 1903. Although the technology is by now much advanced, the process remains the same. Find an economic source of groundwater, and drill beneath the water table. (Photograph courtesy of the U.S. Geological Survey.)

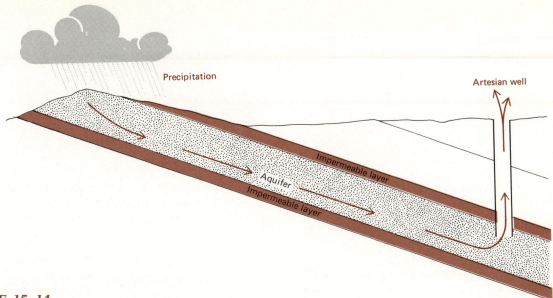

FIGURE 15–14

The components of an artesian system are a tilted aquifer enclosed by impermeable layers and exposure of the aquifer to an area of recharge. Artesian wells puncture the pressurized aquifer and water naturally rises to the land surface. Pressure within the system will be lowered by withdrawals exceeding recharge.

that over 5 million acres (2 million hectares) of irrigated farmland will be bone dry by the year 2020 if nothing is done.

The second liability of well water is that it may contain large amounts of ions dissolved from the mineral material through which it flows. We call water containing abundant dissolved chemicals hard water. Hard water, sometimes defined as water containing more than 500 parts per million or 0.05 percent of dissolved solids, causes scaling and corrosion in pipes, rots out plumbing, and creates real problems in washing things clean. Stains, crusts, and discoloration accompany hard water.

Various types of treatment to rid the water of undesirable dissolved minerals are available, but the costs of treating large volumes of water may be high. The dissolved minerals in hard water come largely from slow, partial solution of lava, limestone, or gypsum. The solution of limestone creates a wide variety of unique landscapes, soon to be described.

ARTESIAN WELLS

An **artesian well** is one in which the water rises to the surface under natural pressure. Artesian wells provide the benefits of well water without the liabilities of pumping costs.

Artesian wells occur wherever the aquifer is naturally pressurized. Most commonly the aquifer layer is tilted (Figure 15–14), and recharge water enters the aquifer above the area of withdrawal. If the aquifer is surrounded by impermeable rocks, water sinking deeper into the sloping aquifer is pressurized by the mass of water above it; the deeper the aquifer sinks, the greater the pressure on the water. The aquifer, in effect, is a tilted pipe, with water added at the high end and recovered at a lower elevation.

If too many wells puncture the aquifer, pressure throughout the whole system drops, much like trying to pump water through a leaky pipe. In well-managed artesian basins, the number and size of wells are strictly regulated. In unmanaged systems, pressures fall as everyone tries to get what they can, while they can.

WATER PROBLEMS

Our problems with water fall into two interrelated categories, its contamination and its overuse. Water become contaminated when careless placement of disposal areas allows any combination of inorganic and organic chemicals, human and animal wastes, radioactive materials, and bacterially contaminated substances to leach downward and enter groundwa-

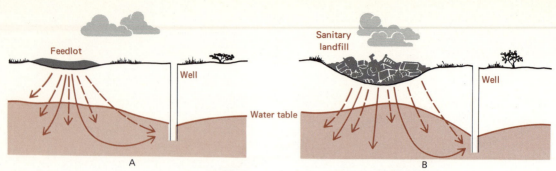

FIGURE 15–15

These two sketches illustrate the concept of groundwater as an open system. Leakage from an animal feedlot or a sanitary landfill may provide fluid contaminants which enter the water table. If a well samples the contaminated flow, the water will be polluted and become both a short-term and a long-term hazard to human health.

ter storage. With overuse of the water, quality is debased, pumping costs rise, the surface settles and collapses, and eventually a resource critical to life is exhausted.

GROUNDWATER CONTAMINATION

Most problems with groundwater contamination stem from failure to understand one very simple idea—*anything placed on the surface will eventually enter groundwater storage* (Figure 15–15). Once contaminants are in the groundwater, they are commonly both difficult and expensive to remove.

The sources of pollutants range from outhouses, to feedlots, to garbage dumps, to pipelines and service stations with leaking gasoline tanks, to runoff from heavily fertilized farms, to industrial waste dumps and abandoned mines. In one startling example a village in England discovered it was drinking water contaminated by a seventeenth-century graveyard for plague victims.

Contaminated wells are being shut down all over North America. When the Environmental Protection Agency tested a thousand public water supply wells, it found 29 percent of them contaminated to some degree. Twenty percent of the 65,000 community water systems in this country fail to meet even the *minimum* standards of the Safe Drinking Water Act. Over 60 percent of tested rural wells—serving 39 million people—were unsafe because of bacterial or chemical contamination.

It appears that groundwater contamination will become the most important environmental concern of the next several decades. As a recent report from the U.S. Geological Survey put it, "[there's] no quick fix in sight."

GROUNDWATER OVERUSE

The city of Las Vegas, Nevada is well known as a place to rest and recreate. Lush green golf courses, trees, and manicured lawns carpet part of this desert oasis. The city uses 24 billion gallons (91 million liters) of water each year—*all* of it from underground sources.

The water table beneath Las Vegas has fallen between 150 to 200 feet (45 to 61 meters), making water so difficult and expensive to pump that the city has now started to pump water from the Colorado River. With no changes, experts estimate that Las Vegas could be out of water by 2020.

As a result of excessive use of groundwater, 3-foot (1 meter) cracks have opened up in concrete walls and pavement near the famed Las Vegas Strip. Such *subsidence cracks* reflect sediment compaction as water, which once helped support the load of the materials above it, is pumped away. As the sustaining underground water table falls, so does the land above it. Not everyone realizes that the weight of a city is carried in part by the strength of the soil or rock beneath it and in part by pressure within the pores of the aquifers.

The specter of sinking cities is, unfortunately, not new. Venice, constructed in the fifth century A.D. on wooden pilings in a muddy lagoon in the Adriatic Sea, has been essentially floating within the mire for centuries. Then in the 1950s great quantities of groundwater began to be pumped from the nearby mainland to meet industrial needs, and the rate of subsidence increased tenfold, accompanied by a rapid decrease in water pressure in the artesian wells.

Other sinking cities include New Orleans, Long Beach, California, London, Bangkok, Houston and

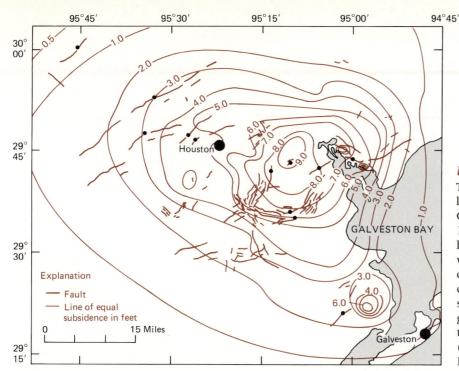

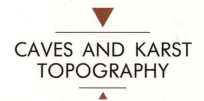

FIGURE 15–16
This map shows the approximate land subsidence in the Houston and Galveston metropolitan area from 1906 to 1978. The land subsidence has caused numerous small faults within the area to become active, damaging house foundations, disturbing utility lines, and shattering streets. Progressive restriction of groundwater pumping has slowed the problem within the 1980s. (Modified from National Science Foundation reports.)

Galveston, Texas (Figure 15–16), Shanghai, Tokyo, and Mexico City. The earthquakes of September 1985 were especially devastating to Mexico City, because it is built on a loosely consolidated lake bed. Much of the damage coincided with areas of highest groundwater withdrawal.

Houston and Galveston are the third largest metropolitan area in the country, and they rely principally on groundwater pumped from unconsolidated clays and sands beneath the cities. Water levels had declined as much as 330 feet (100 meters) before the 1970s, when regulations slowed pumping rates. The collapse of the land surface has created millions of dollars in damages to homes and utility systems; it has also left broad areas of southeastern Houston much more exposed to tidal flooding during hurricanes.

Although subsidence and subsidence cracks are the most visible effects of overuse, in the long run the hidden risks may be more important. Water in desert areas has been accumulating for millennia. If it is sucked dry in a few decades, cities that contain billions of dollars of property may become worthless. It has happened before.

Standing on the arid plains of northern India is Fatehpur Sikri, a beautiful city designed as the capital for the sixteenth-century mogul emperor Akbar the Great. It is a city of graceful, airy palaces and ornate public buildings. Its beauty is little changed by

four centuries, and it is a lovely place to visit.

Crowds are not a problem. The city was abandoned just 15 years after it was completed, when it exhausted the local groundwater supply. It stands as a monument to bad planning, an example reminding us that "Nature to be commanded must be obeyed."

▼

CAVES AND KARST TOPOGRAPHY

▲

All who have explored caves have wondered about their beauty and how they formed. This chapter describes those formed in soluble rocks as well as the landscape produced when the land surface is pockmarked by numerous collapse basins produced by collapse of cave roofs. Such an area may be one where essentially all water flow occurs underground along open fractures and cavern systems.

CAVES

Caves are natural underground chambers with an opening to the surface. There are about 17,000 known caves in the United States. First used for shelter from wild animals and the weather, caves now provide us with beauty and fascinating examples of

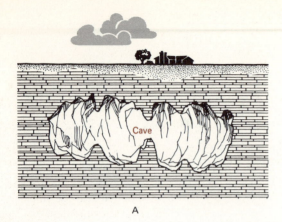

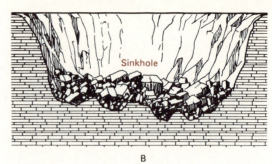

FIGURE 15–17

Before and after cross sections of an area with limestone bedrock in a humid region. A cave first develops as acidic rain slowly dissolves bedrock along major joints and fractures. As the ceiling thins from the continual dissolution, a catastrophic ceiling collapse will form the sinkhole or sink.

the work of underground water. Caves may be formed by any of five processes.

1. *Lava tubes* are tunnels formed in lava when the upper surface freezes while the lava beneath continues to flow, leaving behind an open tunnel or cave within a lava flow. Sometimes these have ice within them and are termed *ice caves*.

2. *Glacial caves* are tunnels formed at the snout of a glacier by meltwater. These drainage tunnels may be the sites of *eskers* (see Chapter Fourteen).

3. Waves battering at the shoreline of oceans or lakes create *sea caves*. They are testimony to the power of wave action in dissolving, abrading, and air-hammering rock at wave level.

4. Wind abrasion in deserts forms *eolian caves*. Through continued collapse of rock they may become quite large.

5. *Solution caves* are formed wherever there is a combination of slightly soluble rocks and slightly acidic water percolating through them. We will describe only solution caves, and those formed by the solution of limestone in particular.

Slightly soluble rocks acted on by percolating groundwater can create both caves of underground beauty and also cause the surface to collapse. In areas of abundant limestone bedrock, caves form when limestone or other soluble rocks dissolve (Figure 15–17A). If the roof of a solution cave collapses, it leaves a **sinkhole**, a large, roughly circular depression (Figure 15–17B). In areas underlain by limestone, having moderate to heavy rain and declining water tables, sinkholes may become a menace.

How Solution Caves Form

Rainwater acquires a small content of carbon dioxide during its fall through the atmosphere, so that most rain is slightly acidic (see Chapter Four). As rainwater passes through the soil, it may pick up still more carbon dioxide from decaying vegetation, which makes it still more acidic. The combination of carbon dioxide and water forms *carbonic acid*, a weak but abundant acid. Other acids, including humic acids, may also form when rainwater passes through vegetation such as decaying pine needles.

As the acid water trickles through limestone or limy sedimentary rocks, it slowly dissolves the calcite of which the rock is composed or cemented, forming small cavities and passageways. As they become large enough for human beings to enter, we call these cavities *caves*. This is the *first* of *two* stages in cave formation.

Most of this early stage in cave formation occurs slightly below the water table, so that there is a continuous flow of slightly acidic water to dissolve the rock and carry away the materials in solution. As nearby stream valleys deepen through time, and the water table slowly declines in altitude, the caves are left behind in the zone of aeration.

The second stage of cave formation, called cave *decoration* then begins. Water filled with carbon dioxide trickles through the zone of aeration acquiring dissolved limestone. Once the trickling water enters the cave, the carbon dioxide gas within the water escapes, much as it does from a carbonated beverage when you open the can. As the water loses carbon dioxide, it loses its capacity to hold dissolved lime-

FIGURE 15–18
Dripstone in Higenbothem's Cave, near Glasgow Junction, Kentucky in 1925. The column and curtain are composed of limestone precipitated by dripping from the ceiling and buildup from the cave floor. (Photograph courtesy of the U.S. Geological Survey.)

stone, and the dissolved limestone is precipitated as **dripstone,** the name given to limestone created by deposition within a cave.

Dripstone (Figure 15–18) is added to the floor, ceiling, and walls of a cave in an enormous array of intricate ornamental forms. *Stalactites* and *stalagmaites*, pillars that hang down from the ceiling and build up from the floor, respectively, are the most common, but the number of ornamental structures formed in caves is legion. The study of the origin and evolution of caves is called *speleology*, (pronounced spee'-lee-ahl-oh-gee). It is a science combining geology and hydrogeology with biology and anthropology.

KARST LANDFORMS

Landforms produced when underground water dissolves massive bedrock have been named **karst** landforms. The word *karst* comes from the Slovenian *krš*, the name of a limestone region of Slovenia along the far western Dalmatian coast of Yugoslavia.

The best-known karstic landforms in the United States are the circular collapse structures commonly called *sinkholes* or *sinks*. On a worldwide basis a wide range of karstic landforms are known, including the tower karsts of China (Figure 15–19), where mountains of solid rock rise abruptly out of flat paddy fields, and the monotonous karstic plains south of the Great Victoria Desert in southernmost Australia.

All karstic landforms arise because some rocks are soluble in natural waters. Specialized instances exist. A few of the sinkholes in south Texas may result from the removal of sulfur by underground mining, and some sinkholes have been formed by the dissolution of underground salt (Figure 15–20). In the majority of cases, *the rock removed is limestone, and the rate of removal is directly proportional to stream runoff.*

In areas where sinkholes (sometimes named *swallow holes* or *dolines*) are common, the land surface is generally rolling, with numerous depressions dotting the landscape. In areas where the water table is high, many sinkholes form lakes whose surfaces are level with the local water table elevation. Most of the myriad lakes of central Florida are of this origin.

In areas of karstic formations there is commonly *no surface drainage*. Rather, water falling on the surface rapidly trickles down crevices along joint systems and flows underground. Underground drainage systems are common and well developed in many karstic areas. Whatever surface drainage there is generally flows into sinkholes.

Sinkholes form by catastrophic collapse of a cavern roof, and are most common in areas where some unconsolidated material overlies the limestone. Lowering of the water table within the unconsolidated material may trigger roof collapse, as the unconsolidated material first compacts and then falls into the cavern beneath it. Sometimes the ground first begins to sag noticeably, forming a shallow depression. More commonly, the rumble of a passing truck causes the already unstable roof to fall.

Sinkholes range in width from several feet to a few hundred yards or meters across; their depth is quite variable. Most sinks are approximately funnel-shaped. As an example, a sinkhole collapse near Boling, Texas in 1983 left a depression 300 feet (90 meters) across and 700 feet (213 meters) deep; it opened up in a few seconds as a pickup truck drove past. Rapid subsidence had been noted in the last few weeks before the collapse.

Little can be done about the obvious hazards of sinkholes forming, except to locate potential sinkholes before starting construction. There are several methods of locating void space underground. Sophis-

ticated instruments are able to recognize the deficiency of mass when air, instead of rock, is beneath our feet.

SUMMARY

1. In humid areas groundwater flow is interrupted by rivers, lake basins, and wells; perched flow is interrupted by springs. The water table, the upper boundary of saturated permeable material, roughly parallels the shape of the land surface. Groundwater flow maintains streams as permanent watercourses.

2. In arid areas groundwater flows out of intermittent streams to deep water tables. Basin recharge causes stream flow to be intermittent and leads to internal drainage. The water table in an arid environment is broadly the reverse or mirror image of the land surface.

3. Water is used for domestic, industrial, and agricultural purposes. The dominant use of fresh water is for agricultural irrigation. Groundwater and surface water systems almost always interconnect, so that contamination of ground-

water commonly results from thoughtless placement of materials on the ground surface.

4. The percentage of groundwater used as a function of dependable supply is largely a function of climate; only in the arid southwestern United States is more water being pumped than is recharged (Figure 15–21). Such usage increases pumping costs, slowly collapses the land surface, creates subsidence cracks, damages building foundations, roads, and utilities, and eventually exhausts the water resource. When recharge exceeds withdrawal, the elevation of the water table rises; if it reaches the surface a marsh or swamp forms.

5. Caves form for the most part below the water table and are enlarged and ornamented above it. The sudden collapse of the roof of a cave forms a sinkhole, a common karst landform. Karst regions commonly have extensive limestone bedrock beneath them and little or no surface drainage. All drainage is accommodated by subsurface river flow along a network of enlarged solution channels. Heavy runoff dissolving limestone at a fast rate fills a region with karst landforms.

FIGURE 15–20
An example of subsidence from extreme southwestern Kansas in an area once crossed by the Santa Fe Trail. This collapse, caused by the dissolving of underground salt by groundwater, occurred overnight in 1859. (Photograph courtesy of the U.S. Geological Survey.)

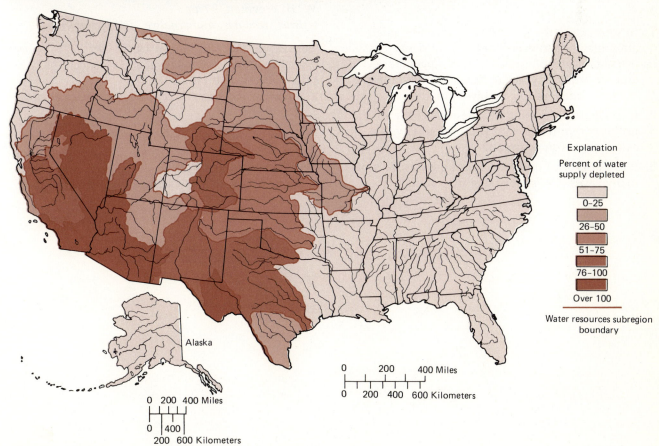

Explanation

Percent of water supply depleted

0–25
26–50
51–75
76–100
Over 100

Water resources subregion boundary

Alaska

0   200   400 Miles
0   400
200   600 Kilometers

0   200   400 Miles
0   200   400   600 Kilometers

FIGURE 15–21
The map patterns illustrate the estimated percentage of water withdrawn compared to water recharged. Estimated depletion is less than 50 percent of supply east of the 95th meridian, but it exceeds 75 percent of supply in the Southwest. In some parts of the Southwest more than 100 percent of the available supply is being withdrawn, which causes groundwater mining. (Modified from a map by the U.S. Geological Survey.)

KEY WORDS

aquifer	permeable
artesian well	porosity
cave	porous
conventional well	recharge
dripstone	sinkhole
groundwater	spring
hard water	surface water
karst	water table
perched water table	withdrawal
percolation	zone of aeration
permeability	zone of saturation

EXERCISES

1. In some areas the groundwater contains small amounts of dissolved fluoride, a nutrient essential to dental health. Check Appendix A and see whether you can figure out the probable mineral source of the dissolved fluoride.

2. It has been observed that limestone caves form upslope from rivers. Can you reason why this association is so common?

3. It has also been observed that the lowest part of many caves is ∨-shaped and may contain a running stream. What sequence of events might cause this to happen?

4. Chlorine has been shown to form carcinogenic (cancer-causing) compounds when injected into water already contaminated with certain organic compounds. Ozone has been suggested as an alternative; how could the safety of ozone be compared with that of chlorine?

5. What is the per capita consumption of water in your campus living unit? How does that compare with the national average of 75 gallons per person per day?

6. Through the ages spring water has often been assumed to have magical or medicinal purposes. Can you suggest some reasons why?

SUGGESTED READINGS

LEOPOLD, LUNA B., 1974, *Water: A Primer*, New York, W. H. Freeman, 97 pp.
► *Brief, readable, authoritative, and recommended.*

MOORE, G. W., AND SULLIVAN, G. N., 1978, *Speleology—The Study of Caves, 2nd ed.*, Teaneck, N.J., Zephyrus Press, 150 pp.
► *Advanced discussion of cave origins, caving, and cave decorations.*

MURRAY, C. R., AND REEVES, E. B., 1986, *Estimated Use of Water in the United States*, U.S. Geological Survey Circular 965, 39 pp.
► *A review of how Americans use and abuse water.*

This image taken by the DMSP Air Force weather satellite at midnight is of the eastern half of the United States. It graphically portrays the relation between electrical energy use and population distribution. (U.S. Air Force photo, distributed by NOAA.)

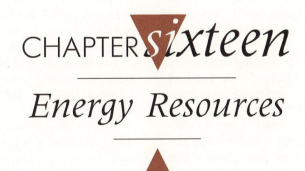

CHAPTER *sixteen*

Energy Resources

Energy is the *capacity* to do work. It appears in many forms, each convertible—at a price—to another form. The energy of a boulder high on a hillside or of water trapped behind a dam is *potential energy,* or the energy of position. The energy of rolling boulders or moving water is *kinetic energy,* or the energy of motion. The energy of motion comes in many forms: electricity is the flow of electrons; nuclear energy is the flow of other subatomic particles. Heat is the movement of molecules.

Spinach is solar energy, stored by *photosynthesis* as chemical energy, largely in the form of complex carbohydrates. If I eat the spinach, my body converts the carbohydrates back to heat *and* to a rich array of compounds to furnish energy to my cells. When I pump my bicycle, muscles contracting and extending in response to electrical stimuli convert the "spinach" into motion, and if my bicycle generator is on—back to light. We humans however are puny resources of physical energy; our maximum output will just keep one 100-watt light bulb burning (Figure 16–1).

Energy is also independence. Energy-rich societies do not need to enslave human muscle to form wealth for the idle few. A world of abundant energy could be a world without toil; energy is the great multiplier of human creativity.

One way to measure the output of nations is to assess the value of a nation's gross national product (GNP), a measure of the value of all goods and services produced by a nation. When we look at the output produced *by one individual* and compare it with the energy consumed *by each individual,* the result is a rising straight line (Figure 16–2). *Increasing the per capita consumption of energy magnifies the value produced by each individual.*

A former director of the U.S. Geological Survey suggested that the life-style of a nation is the product of total raw materials, energy, and food consumed multiplied by education and ingenuity, with this whole product divided by the number of people who share in it. In this sense machines that consume energy and manufacture products are multipliers of our ingenuity and creativity. How, then, do North Americans use energy in the 1980s, and how is this demand measured?

FIGURE 16–1
What is the energy equivalent of 1 "personpower"? We can keep one 100-watt light bulb burning at a sustained pace. For obvious reasons, power is usually expressed in terms of horsepower.

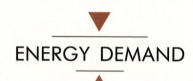

ENERGY DEMAND

In 1985 North America consumed 79 quadrillion Btus of energy. Not a very informative statement, is it? What does that impossible number mean? A quadrillion is a million billions, so 79 quadrillion is 79,000,000,000,000,000. But what is a Btu?

MEASURING ENERGY

We can measure energy by recognizing that the *loss* in energy equals the *work done.* A common measurement of energy is the **erg** (Figure 16–3), the work done when a force of 1 dyne acts through a distance of 1 centimeter. Within the field of energy studies, a more commonly used unit of energy is the B̲ritish T̲hermal U̲nit, or **BTU.** One Btu equals approximately 10^{10} ergs or 10,000,000,000 ergs.

A *Btu* is the amount of heat required to raise the temperature of 1 pound (454 grams) of pure water by 1° Fahrenheit (approximately $\frac{5}{9}$° Celsius). A Btu is also the approximate output of one wooden match. Given the enormity of energy consumption, it is easy to see why energy demand is expressed in quadril-

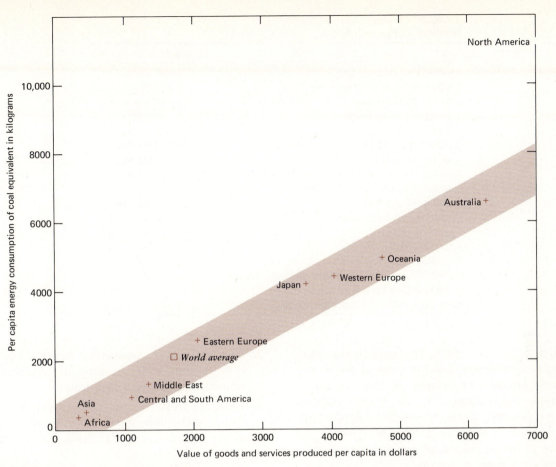

FIGURE 16–2

The per capita energy consumption for selected world areas is compared to the value of goods and services produced. The shaded area includes most of the world with the exception of North America. (Data taken from United Nations Statistical Yearbook; used with permission.)

lion Btus, or more simply as *quads*. Table 16–1 expresses some approximate energy conversion factors.

A very different way of measuring energy is to look at the cost - benefit ratios for familiar fuels. A 5-gallon (19 liter) can of gasoline provides the energy equivalent of a person-year, that is, the hard labor of one person for one year, but it costs much less than the value of an hour of our labor to buy it. A kilowatt-hour equals a week of heavy labor; we pay for it with 30 seconds of effort. Energy for Americans has been cheap, and therefore it is easily wasted.

ENERGY WASTE

The sources energy waste are all around us. A shift from standard to subcompact cars for half of our driving needs would save more fuel than the Alaska pipeline delivers. A ton of freight shipped by truck takes six times the energy as one shipped by rail. A jumbo jet uses a third less fuel *per passenger* than a regular jet. An urban bus uses a fourth the fuel of a standard automobile *per person* to move people around town.

An electric light bulb produces 95 percent waste heat and only 5 percent light, whereas fluorescents convert 45 to 55 percent of their energy to usable light. The pilot light on a gas range consumes one-third of all the fuel the appliance uses; 10 percent of all the natural gas consumed in the United States is used to keep pilot lights burning, although pilotless ignition systems have long been available. One-quarter of space heating is lost through poorly insulated roofs, and another one-sixth is lost through leaky windows.

The generation of electricity requires the conversion of coal, oil, natural gas, or the heat from nuclear

TABLE 16–1
Nominal Conversion Factors

Item	Approximate Equivalents
1 cubic foot natural gas	1000 Btu
1 kilowatt-hour of electricity	3400 Btu
1 ton of refrigeration	12,000 Btu
1 pound of bituminous coal	12,500 Btu
1 therm	100,000 Btu
Average house furnace used for 1 hour	104,000 Btu
1 gallon of gasoline	125,000 Btu
Average annual solar energy captured on average home roof in Arizona	140,000 Btu
1 gallon of #2 fuel oil	140,000 Btu
1 barrel of average crude oil	5.8 million Btu
1 cord of wood	20 million Btu
1 ton of coal	25 million Btu
1 trillion cubic feet of natural gas, or 293 billion kilowatt-hours of electricity, or 180 million barrels of crude oil, or 40 million tons of bituminous coal	1 quadrillion Btu

fission to produce electricity. The initial conversion of heat to electricity wastes 67 percent of the energy in the fuel. Another 14 percent, on the average, is wasted along the transmission line, so that *delivered* electricity represents about *one-quarter* of the fuel energy consumed to produce it. The heat *lost* in the conversion of fuel to electric power *exceeds the heating demands of every home in America.*

If we count all the obvious sources of waste, we discover that about *one-sixth* of our total energy demand is waste. One-sixth of 79 quads is 13 quads, or 2.3 *billion* barrels of oil, an amount equal to *half the petroleum consumed by the United States in one year.* No nation on earth has ever squandered energy in such prodigal amounts; *the energy demand in the United States for the next decade will equal that of all the world in all recorded history.* Where does it all go?

COMPONENTS OF DEMAND

As noted in Figure 16–4, the total consumption of energy in the United States has remained relatively constant and even dropped somewhat since the supply and price disruptions of 1979 vividly drove home our overdependence on cheap foreign oil. Four sectors are the dominant energy consumers. Interestingly, the electrical utility industry is both a source and a consumer of energy. These sectors are

1. Transportation Sector. Moving people and commodities.
2. Industrial Sector. Manufacturing, construction, mining, agriculture, fishing, and forestry.

3. Residential and Commercial Sector. Heating, air conditioning, water heating, and cooking in homes, stores, and institutions.
4. Electrical Utility Sector. Conversion of the heat from combustion of petroleum products or coal, or the heat from nuclear fission to electric power.

Transportation Sector

The transportation sector consumes *exclusively* petroleum products, primarily jet fuel, fuel oil, and gasoline. This sector accounts for *one-quarter* of the total energy use. Of the 9 million barrels consumed each day, 70 percent is gasoline. Of the 1.2 million barrels of jet fuel consumed each day, 80 percent is used by the airline industry, and the rest is used in military operations.

Although Americans are driving more, gasoline consumption has stabilized because of a shift to more fuel-efficient, smaller, aerodynamic, lighter cars. The same pattern is repeated in aircraft, as more efficient engines and aerodynamics combine to cut fuel consumption.

Industrial Sector

Approximately 25 percent of all energy is consumed by the industrial sector. This sector uses a wide variety of fuels; petroleum products account for only 30 percent of the energy consumed. After the 1979 fuel crisis, industry invested wisely in conservation measures and developed multiple-fuel facilities. The flexibility of multiple-fuel plants allows many manufacturing industries to switch rapidly to whichever

```
10⁴⁰
10³⁹  ─ Total output of the sun each day
10³⁸
10³⁷
10³⁶
10³⁵
10³⁴
10³³
10³²
10³¹
10³⁰
10²⁹  ─ Daily receipt by earth of solar energy
10²⁸
10²⁷  ─ Annual U.S. energy consumption
10²⁶
10²⁵  ─ Annual U.S. coal production
10²⁴
10²³
10²²
10²¹
10²⁰
10¹⁹
10¹⁸
10¹⁷  ─ One ton of coal
10¹⁶  ─ One ton of TNT
10¹⁵
10¹⁴  ─ One U.S. gallon of gasoline
10¹³  ─ One cubic foot of natural gas or one horsepower
10¹²
10¹¹
10¹⁰  ─ One Btu
10⁹
10⁸
10⁷  ─ One joule
10⁶
10⁵
10⁴
10³
10²
10¹
10⁰  └─ One erg
```

FIGURE 16–3

This chart portrays the magnitude of energy differences on an exponential scale. Each increase is a power of ten, so that the difference in energy between 1 erg and 1 joule is 10^{107} or 10,000,000 times. The difference between an erg and the total solar output per day is 10^{39}.

fuel is least expensive over the short run.

The widespread investment in conservation measures has increased profitability by cutting waste. As America slowly shifts from "smokestack" manufacturing to an information society, we can expect little increase in industrial demand. Greater efficiencies are realized largely by better handling of "waste" heat.

Residential and Commercial Sector

Energy use in the residential and commercial sector is primarily influenced by weather conditions, with the extremes of summer or winter temperatures requiring higher energy expenditures for comfort. The effects of conservation within this sector have been dramatic; use of petroleum products dropped a total of 28 percent over three years following the Arab oil embargo.

This sector uses fuel oil and natural gas as its primary sources of space and water heating, and electricity as its primary source of air conditioning. The use of wood stoves has greatly lessened dependence on other fuels in areas where firewood is plentiful. Greater efficiencies largely follow from increased use of insulation, conversion to more efficient appliances, and greater reliance on solar energy. *This sector accounts for 16 percent of total energy demand, which is met largely by oil, natural gas, and wood.*

Electrical Utilities Sector

The generation of electricity is slowly becoming a greater and greater part of our total energy consumption (Figure 16–5), and for very good reason. At the point of use, electric power is clean, nonpolluting, and easy to use. At the point of generation, the sources of electric power range from nonpolluting waterpower to the combustion of coal, which may be the major cause of acid rain (see Chapter Four). The generation of electricity by converting water to steam is, unfortunately, an inherently inefficient process; two-thirds of the energy consumed generates nothing more than waste heat, which then becomes a source of thermal pollution of streams and lakes.

Coal has proved to be so much less expensive and more dependable than nuclear power that coal-fired power plants are becoming the norm. Oil and natural gas together provide 15 to 20 percent of demand, nuclear energy and hydropower each provide slightly less than 8 percent, and coal provides all the rest.

The major factor affecting the use of electricity is weather, since climate control in hot climates is essentially an all-electric matter. The *generation of electricity consumes about one-third of our entire energy output.*

ENERGY SUPPLY

In the 1990s our sources of energy will be twofold.

1. *Fossil fuels,* which will provide about 90 percent of our needs.
2. Minor and developing energy sources, including solar energy, geothermal energy, wind energy, and others (Figure 16–6).

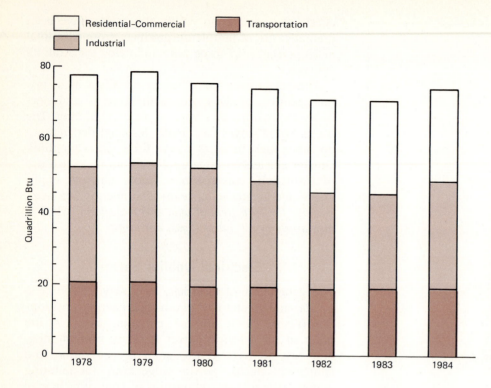

Residential-Commercial ☐ *Transportation* ▨
Industrial ▨

FIGURE 16–4
This graph of the total energy consumption by the United States for the period 1978 to 1984 shows that it has actually declined since the Arab oil embargo; most of this decline has taken place in the industrial sector. (Data and graph from the Department of Energy.)

Fossil fuels are concentrated and highly modified substances derived from ancient plant and animal life. When we burn leaves from the lawn or firewood from the forests, we convert modern *carbohydrates*—chemical compounds of carbon, hydrogen, and oxygen—into a cheery fire in the fireplace. When we burn coal, natural gas, or oil and all its refined products including fuel oil, bunker fuel, jet fuel, propane or butane, or gasoline, we convert *ancient* carbohydrates into heat in a variety of settings. Since what we burn is the residue of *past* life, these fuels are called *fossil fuels*.

Fossil fuels are concentrated solar energy. All animal life depends ultimately on plant life for its existence, and all plant life, with few exceptions, depends on the sun. It is the sun's energy that drives *photosynthesis*, the complex chemical conversion of natural materials into carbohydrates which are stored within green plants. The remains of plants or animals represent concentrations of carbohydrates and fats, lignins, and cellulose—storehouses of energy.

After death, natural processes immediately begin to convert this energy. Scavenging animals and plants may consume the remains and use the energy for sustenance; much more rarely, natural processes may convert the dead into fossil fuels of many types.

A log and a lump of coal differ in only two ways.

1. The coal is hundreds of millions of years older.
2. The coal is a much more concentrated energy source, containing more potential heat energy per unit of weight.

The processes of conversion cause concentration. As Johannes Kepler (1571–1630), German astronomer, reminded us almost four centuries ago, "Nature uses as little as possible of anything." Like nature, the value we place on a fossil fuel depends largely on how concentrated it is.

COAL

Coal is a combustible sedimentary rock composed of carbon and volatile compounds. It is formed from highly altered and compacted plant remains and may also include small amounts of clay, silt particles, nitrogen compounds, and sulfide minerals. The combustion of coal unfortunately converts these materials to fly ash, clinkers, and a variety of oxides of nitrogen and sulfur, which create air pollution, smog, and acid rain.

Coal was first mentioned by Theophrastus (ca. 372–287 B.C.), a Greek botanist, but was little used as long as forests were abundant. Coal has always been a messy fuel to quarry, ship, handle, burn, and clean up after. Its potential for air pollution has long

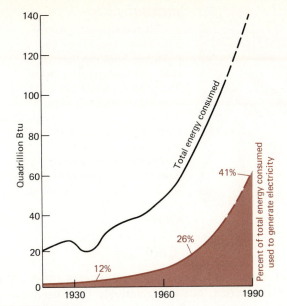

FIGURE 16–5

Notice that from 1920 to 1990 electricity has become an increasing part of the total U.S. energy demand. For the end-user, electricity is clean, convenient, and relatively inexpensive, though the use of nuclear energy is forcing prices up and thereby diminishing demand. (After U.S. Water Resources Council, 1983, *Water for Energy Self-Sufficiency.*)

been obvious (Figure 16–7). In 1306 the king of England decreed death for anyone who burned coal in London while he was there—an early example of environmental activism.

The Origins of Coals

Coal forms in environments where vegetation is lush and waters are stagnant and oxygen-depleted. Natu-

ral environments having these characteristics include marshes, swamps, and some deltaic areas.

Normally, when plants die the vegetable matter is rapidly broken down by bacteria and fungi, so that the materials in the tissues are once again available for uptake. When plant debris falls into stagnant water, however, *anaerobic* bacteria—those that can live in the absence of oxygen—only partially decay the plant material. As the partially degraded plant material is compressed and buried, *peat,* the lowest grade [most poorly concentrated] of coal is formed.

Each successively higher, more compact grade of coal is formed through the rising temperature and pressure applied as more and more plant material or swamp sediment is deposited on top of the peat. The *rank* or *grade* (the terms are synonyms) of coal is a measure of the decrease in moisture and volatiles as the coal is increasingly altered. Highest-rank coals contain the highest percentage of carbon and the lowest percentage of volatiles and moisture.

Prolonged deep burial and the loss of some water will convert peat to a denser, darker, low-rank brownish coal called **lignite,** or "brown coal." Additional weight, time, and dewatering will modify lignite into a still denser and thinner layer (Figure 16–8) of **bituminous coal,** a hard black sedimentary rock which is the standard medium-rank industrial coal. If layers of bituminous coal are crumpled by orogenic forces and mildly metamorphosed, **anthracite,** a metamorphic rock forms. Anthracite, the very highest grade or rank of coal, is glossy and jet black. It is a highly concentrated fuel and commands a premium price; it is used primarily for space heating. Still higher grades of metamorphism will convert anthracite to *graphite,* a mineral composed of pure carbon.

Peat, lignite, bituminous, and anthracite are names

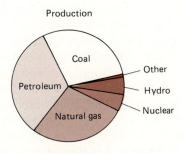

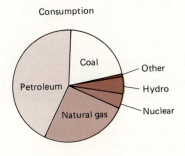

FIGURE 16–6

The consumption of petroleum exceeds our production, which means we must import the difference. The consumption of coal is less than production, which means we must export the surplus. Other energy sources

are consumed in essentially the same quantities as they are produced. Coal, petroleum, and natural gas are *fossil fuels.* (Data from Department of Energy Annual Summary.)

FIGURE 16–7

As this seventeenth-century engraving makes clear, air pollution from the burning of coal is certainly nothing new. The process shown here is the roasting or smelting of ore, which may add large additional amounts of sulfur oxides to the atmosphere. (From G. E. Loehneys, 1679, *Hof Stäats und Regier Kunst,* courtesy of the History of Science Collections, University of Oklahoma Libraries.)

for increasingly higher ranks or grades of coal, which is America's most abundant energy resource. Of the available resource, 1 percent is anthracite, 48 percent is bituminous, 34 percent is subbituminous (a lower grade of bituminous coal), and 17 percent is lignite.

FIGURE 16–8

Here is subbituminous coal in its normal habitat, interlayered with other sedimentary rocks that were deposited in a continental environment. This particular layer has an ominous name; it is called the Big Dirty Coal Bed and was exposed in McCone County, Montana. (Photograph courtesy of the U.S. Geological Survey.)

Coal Use

Coal forms 80 percent of the energy reserves of North America, but it is relatively little used except in the generation of electric power and a few other industrial purposes (Figure 16–6). At one time coal was king, but that was only after the young forest-rich United States began to run out of trees. In 1800 wood supplied 94 percent of our fuel needs; the rest came from waterpower and the muscle power of human beings and animals.

By 1920 coal supplied almost three-quarters (Figure 16–9) of our energy demand. During the 1930s and following, the coal market progressively sagged under the convenience and efficiency of relatively clean-burning, easily transportable, inexpensive oil and natural gas. Now as prices for oil and natural gas have risen, coal is once again claiming a larger part of the industrial and electrical generation markets. The change back to coal has been slowed by the same problems that caused many to abandon coal 50 years ago.

Coal Mining

Coal buried more than 100 yards (90 meters) under the land surface must normally be recovered by *underground shaft mining,* which entails enlisting people into the *most dangerous industrial occupation in the United States.* Because of the medical and insurance costs, plus the ventilation, dewatering, and mine safety costs, underground mining is also quite expensive. A modern underground mine brings only one-eighth as much coal to the surface per hour of human effort as a surface strip mine.

Surface strip mining in the past has had catastrophic effects on the land surface in both eastern and western states, though modern reclamation practices can now limit the damage. The safety and

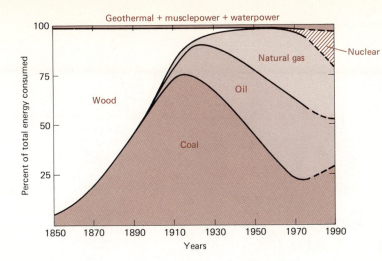

FIGURE 16–9

Trends in American energy use since 1850. Notice that in 1920, coal furnished almost two-thirds of the total energy consumed in the United States, but is currently fulfilling not quite 30 percent of our needs. It has largely been replaced by cheaper, more convenient oil and natural gas. (Data from Bureau of the Census and the Bureau of Mines.)

economy of strip mining must be balanced against the costs of restoring the land, since the costs of land restoration are now built into the costs of recovering the coal.

Strip mining follows a basic mining process requiring six steps.

1. Bulldozers scrape off the top soil and place it in special piles.
2. The rock overlying the coal is drilled and blasted to shatter it.
3. Draglines or power shovels (Figure 16–10) remove the waste rock and stack it in spoil piles.
4. The underlying coal is drilled, blasted, removed by dragline, and loaded onto trucks.
5. The spoil piles are graded back into the void by draglines and bulldozers to recreate a land subsurface.
6. The topsoil is added, and the land surface is graded to natural contours and reseeded with native vegetation, which is maintained until it is successful on its own.

In the past, strip mining in the Appalachian region scraped off the surface of the land, removed the coal, and left things as they happened to lie. From the sulfide minerals exposed by blasting there was intensely acid drainage as well as much intensified soil erosion and destruction of streams and valleys by massive silting. The mined area was simply treated as a resource to be sacrificed. Now with modern reclamation practices damage to the environment is minimal.

In the western United States, where approximately three-quarters of our coal reserves are located (Figure 16–11), strip mining is a logical choice. Most of the coal is near the surface, and its sulfur content is generally lower than that of eastern coals, making it more valuable. But much of the western coal region is an area where water is scarce. Seeding the reclaimed area after mining has sometimes been less than successful. It has also been difficult to preserve the disturbed land surface from flash flooding until the native vegetation is reestablished.

Since many western coal mines are relatively young and modern reclamation practices only 20

FIGURE 16–10

Draglines like this one make removal of coal from the surface a relatively inexpensive. Labor costs are minimal. The environmental costs of stripping coal with draglines remain an ambiguous issue. (Photograph courtesy of the U.S. Geological Survey.)

Coal Fields of the United States

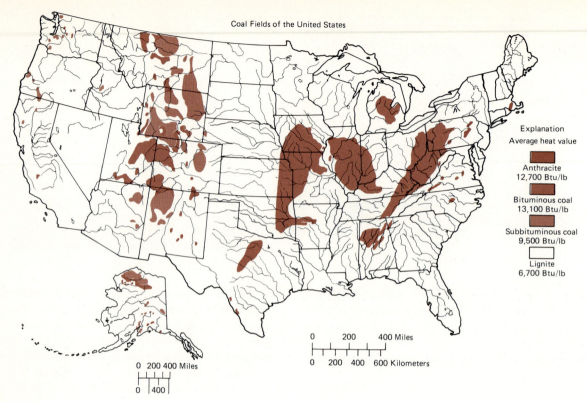

Explanation
Average heat value

Anthracite
12,700 Btu/lb

Bituminous coal
13,100 Btu/lb

Subbituminous coal
9,500 Btu/lb

Lignite
6,700 Btu/lb

0   200   400 Miles

0   200   400   600 Kilometers

0   200 400 Miles

0   400

FIGURE 16–11

The coals of the Appalachians are partly anthracite, whereas bituminous coals are common in the Midwest and in Colorado and Utah. Subbituminous coal is common over the Rocky Mountain area, but lignite is dominant in the northern Great Plains. Notice that the coals of northern Alaska must have formed in a warm, tropical swamp; obviously Alaska has moved quite a way from its location when the coals formed. (Modified from a map by the U.S. Geological Survey.)

years old, no one can yet be certain how all this will turn out. What is certain is that reclamation adds about 10 cents to the cost of a ton of coal, and that progressive coal companies are doing their best. Whether we are sacrificing large areas for a one-time cash "crop" of coal or simply interrupting a current land use before returning the land to equal or better productivity is a continuing debate.

Coal Conversions

One solution to some of the problems associated with the recovery and combustion of coal is to transform the coal to a synthetic natural gas (SNG) through a process termed *gasification*, or to a synthetic liquid or *syncrude*, through the process of *liquefaction*. Either process requires the chemical addition of hydrogen to the coal. Hydrogen is normally obtained by separating it from the water molecules in steam.

The typical ratio of carbon to hydrogen in coal is about 16:1. This ratio is lowered to less than 10:1 to make syncrude, and to about 3:1 for SNG. For the conversion to be economical, the coal must be heated, usually by burning part of it. Both the production of steam and the heating of the coal are a net loss; at the very best the production of SNG or syncrude loses about 30 percent of the energy within the original coal.

The process returns a synthetic gas that is similar to natural gas. This synthetic gas is totally stripped of coal's sulfur and nitrogen pollutants as well as ash. The gasification of coal is *twice* as efficient as converting coal to electric energy.

A large amount of water is required. One-third of the total water needed is used to make the steam for the hydrogen exchange. The remaining two-thirds of the water required is used for cooling; total water demand is very sensitive to cooling needs. The higher the Btu content of the syncrude or SNG, the more heat is wasted, the lower the overall efficiency of the process, and the more water is commonly required for cooling. To produce equivalent amounts of energy, converting coal to SNG or syncrude requires one-fourth the water that the conversion of the same coal to electricity does (Figure 16–12).

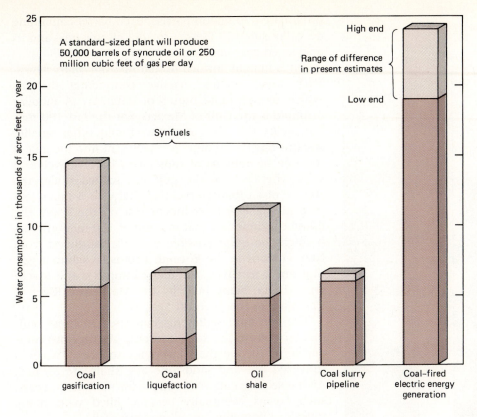

FIGURE 16–12
Estimated water consumption in producing usable synfuel products compared to the water used in a coal slurry pipeline carrying the energy equivalent to that produced in a standard coal-fired generating plant. For the production or transportation of equivalent amounts of energy, synfuels and coal slurry pipelines are two to three times as stingy with water as a coal-fired plant. (Modified from U.S. Geological Survey chart.)

Coal gasification is an old technology. Before World War II almost every eastern city had its own gasworks where coal was converted into a synthetic gas known as "process gas" for lighting and cooking. This type of gas was eventually replaced by natural gas, which had twice the Btu value and was notably cheaper. Now that modern technology can produce a gas equivalent in heat value to natural gas, the issues have largely become economic. SNG may become an important rival to natural gas if the price becomes competitive.

Summary

Coal is a tantalizing fuel because of its abundance; America has centuries of coal supplies within its borders. Its current use is limited to industrial boiler operations and the generation of electricity. It has never made a very satisfactory home heating fuel, and it is useless anywhere in the transportation system. In conventional applications, its mining and combustion raise serious environmental issues which are being addressed. The gasification and liquefaction of coal are old technologies, recently updated; they have some promise of converting America's most abundant fuel into a cleaner, more convenient form. This conversion is, however, costly, with third of its energy content lost largely as waste heat.

PETROLEUM

Through the past eons of geologic time, the tiny carbohydrate- and fat-rich animals and plants of shallow seas have sometimes been preserved by rapid burial under sedimentary layers and modified by temperature and time into a complex mixture of hydrocarbons called **petroleum,** from the Latin term meaning "rock oil."

Petroleum as it is recovered directly by drilling is commonly termed **crude oil.** Crude oil is a mixture of gaseous, solid, and liquid hydrocarbons—chemical compounds of hydrogen and carbon. Crude oil can be refined into tars, asphalts, "pitch," waxes, greases, lubricating oils, motor oils, road oils, fuel oils, kerosene, jet fuels, gasoline, and naptha—to name only a few of the possibilities. Crude oil also may include some fractions so *volatile that they are vapors at room temperature; this material is* **natural gas,** largely composed of *methane,* which is CH_4, a simple hydrocarbon.

Petroleum products are generally much more convenient than coal. Human beings do not have to be sent into dangerous mines to recover them, nor do they have to be shipped in great bulk by freight cars, nor do they have to be stored in cellars and shoveled into furnaces, only to produce a mess of clinkers and ashes. The combustion of petroleum products can

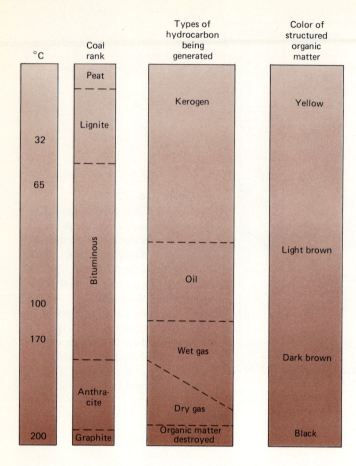

°C	Coal rank	Types of hydrocarbon being generated	Color of structured organic matter
	Peat	Kerogen	Yellow
32	Lignite		
65			
	Bituminous		Light brown
100		Oil	
170		Wet gas	Dark brown
	Anthracite	Dry gas	
200	Graphite	Organic matter destroyed	Black

Energy!

FIGURE 16–13

Comparison of temperature and maturation of coal and petroleum products. Temperature is the dominant factor affecting maturity, but length of burial, type of organic parent material, and type of source rock are other factors. (Data compiled from numerous sources. Concept largely after A. Hood et al., 1975, with the permission of the author and the Bulletin of the American Association of Petroleum Geologists.)

form *smog* (slang for "smoky fog"), but it does not produce fly ash or abundant acid rain. Natural gas is even more convenient and nonpolluting than liquid petroleum. It is an extremely efficient fuel, delivering 97 percent of the energy it contains, and it is essentially nonpolluting. Almost three-quarters of its usual cost is made up of transportation and delivery charges.

Origins of Oil and Gas

Petroleum liquids are often lumped as *oil*, and natural gas is more commonly referred to simply as *gas*. We use these simpler terms as we explore a fascinating question. The known world reserves of oil ap-

proximate 670 billion barrels, each barrel containing 42 gallons or 159 liters. How did all this oil—and its associated gas—form?

Oil is initially formed in marine *sedimentary basins*. Such areas include passive continental margins which receive rapid inputs of sediment. A modern example is the Gulf of Mexico. For the last 100 million years the Mississippi River and other smaller streams have been dumping sediment into the gulf at the rate of millions of tons (see Chapter Twelve) a year. The floor of the gulf has sagged under the weight of all this sediment so that 4 to 5 miles (6.5 to 8 kilometers) of sediment rest on the floor. The filling continues, so that in another 100 million years it might be quite possible to walk southwest from New Orleans to the Yucatan Peninsula without getting one's feet wet. Another example is the Mackenzie Delta, being built by the Mackenzie River of Canada into the Beaufort Sea.

The common sediments deposited in most sedimentary basins are sand and clay. As they are compacted by the weight of overlying sediment, water is squeezed out and sandstone and shale are formed. As the sedimentation continues for millions of years, such basins eventually become filled with many miles or kilometers of sedimentary rock.

SOURCE ROCK

Marine life in the waters above is dominated—in terms of volume—by two types of microscopic life: *phytoplankton*, photosynthetic plant life, and *zooplankton*, animal life. As this material dies, as well as other types of marine life and land life swept out to sea, it settles to the bottom and is incorporated within the rain of sediment on the seafloor. The closer to land, the greater the percentage of continental life included within the sediment. *The buried organic matter is slowly converted within the compacting sediment to mixtures of oil and gas, with the marine life generally becoming oil and continental debris generally forming gas.*

The more organic-rich the sediment and the more rapidly the organic matter is buried so that it can not be scavenged, the greater the volume of oil and gas formed. The *most likely* location for a future oil field is a sedimentary basin that very rapidly fills with organic-rich muds, because shales and certain types of organic-rich limestones compose the **source rock** for oil formation.

Within the source rock the generation of oil and gas is a natural cooking or distillation process. The deeper the source rocks are carried, the hotter they become, both because of the weight of overlying rock and the heat radiating from the earth beneath it. *All*

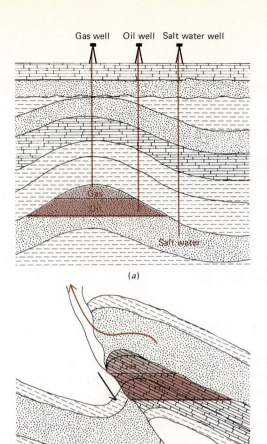

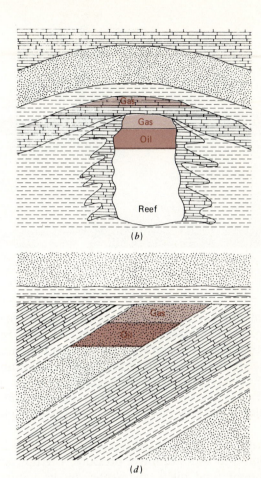

FIGURE 16–14

Gas migrates upward toward zones of diminishing pressure, at the same time oil migrates upward, driven by its buoyancy relative to water-saturated sedimentary rock and diminishing pressure. (A) An anticlinal structure furnishes a trap for oil and gas; a well drilled too far from the center of the structure will produce only saltwater. (B) A buried reef is both the source and the reservoir for oil and gas. (C) A fault provides a seal along part of its area but an avenue of escape for oil and gas traveling along higher layers. (D) An abrupt change in the layering's angle of tilt provides a perfect seal when the lower layer of the upper group of deposits acts as a roof rock. These are only four of several dozen potential combinations which can provide the essentials for an oil field: a source rock, a roof rock, a trap, and a reservoir for oil and gas migrating down pressure gradients.

that is needed to generate oil or gas from the dead organic matter within a source rock is temperature and time.

As the temperature rises, the first material to be generated is a yellow-brownish waxy material termed **kerogen.** As temperature continues to rise (Figure 16–13), the kerogen is progressively converted into lighter, thinner, and progressively more volatile fractions of the original crude oil, analogous to the partial melting of a magma (see Chapter Three). Crude oils are first distilled into a type of natural kerosene, and then a form of gasoline, and finally extremely volatile compounds like naptha, sometimes called "wet gas." At this stage, if heating continues, the "wet gas" is converted into natural gas, or "dry gas."

The whole sequence of generation takes place over temperatures ranging from 150° to 400° Fahrenheit (65° to 200° Celsius) and requires from 5 to 50 million years. The temperatures required mean that the generation of kerogen starts at about 8000 feet (2400 meters); at depths greater than 30,000 feet (9100 meters), temperatures are generally too high for forming oil. At depths greater than 30,000 feet (9100 meters), crude oil is baked into pure carbon, and no fluids or gases remain.

Marine organic matter, largely plankton and algae, forms primarily liquid oil with a little gas; continental organic matter including sedges, leaves, woody fragments, twigs, and bark yield primarily gas, with a little oil.

Accumulation of Oil and Gas

As the oil and gas are being generated in the deeper, hotter parts of the basin, those fluids and gases released from the source rock are under high pressures from the overlying rock. They naturally migrate upward to less pressed, cooler regions. In order to migrate, the intervening rock must be *permeable*, with interconnecting pore spaces (see Chapter Fifteen).

The upward migration of gas is controlled by its expansion into zones of lower pressure. The upward migration of fluid oil is augmented by its natural buoyancy *compared to that of the water-saturated sedimentary rock* it encounters. If nothing stops the upward migration, the gas and oil move all the way to the earth's surface to form bubbling *tar seeps* and *oil seeps*. Without *accumulation*, no commercial quantities of oil or gas can form.

If the upward migration is stopped by an *impermeable* layer or geologic boundary, the oil and gas are *trapped*, and must accumulate. The impermeable layer or geologic boundary is termed a **roof rock** or a **trap**, (Figure 16–14). If oil and gas accumulate *within* a permeable layer, both the pressure of overlying rocks and that from the underlying water force the oil and gas to fill the pore space at the highest level (Figure 16–14), where it remains as a pressured, trapped gaseous fluid. Such a permeable layer, analogous to an aquifer (see Chapter Fifteen), is termed a **reservoir rock.** Reservoir rocks with the greatest permeability tend to be sandstones and some cavernous limestones.

Recovery of Oil and Gas

Approximately two-thirds of all geologists are employed in the search for subsurface reservoirs of oil

and gas. These reservoirs cannot be easily seen, hidden as they are by thousands of feet of opaque rocks. The techniques of oil exploration form a very high level of technology, using extremely sensitive instruments to detect subtle variations in the earth's magnetic field, gravitational attraction, and atmospheric chemistry. A variety of devices ranging from huge trucks that vibrate the outer earth, to air guns, to dynamite are used to send energy waves into the earth; as these energy waves refract and reflect from the layered rock beneath the surface, their pattern reveals the geometries of the layers miles or kilometers beneath the surface (see Figure 1–4).

Highly sophisticated techniques are used to interpret the three-dimensional ancient environments recorded within sedimentary rocks. Entire subfields of very sophisticated physics, technology, and geology are devoted to interpreting information gained from analyzing the rocks on the walls of a well bore. The level of technology surrounding the drilling process, including the recovery of geologic samples and determination of numerous physical properties of the rocks in the well bore, is supported by huge service industries.

Still, drilling for oil and gas in a new area—such wells are called "wildcats"—is largely a "crap shoot" with enormous stakes. Individual wells can cost hundreds of millions of dollars. With all the sophisticated techniques, high-speed computers, and ingenuity, only 1 in 22 wildcat wells finds commercial quantities of oil and gas. As the demand increases, the search for oil and gas extends to increasingly hostile environments whose exploration requires greater costs and higher risks.

Fifty years ago, exploration for oil and gas was easier. Locating oil meant little more than carefully observing surface rocks, surface streams, and oil and

gas seeps, and then drilling the obvious targets (Figure 16–15). When the Spindletop well blew in near Beaumont, Texas, in 1901—gushing 100,000 barrels of oil a day into the air—an anonymous geologist made one of history's worst predictions. He stared at the gusher for a while, then turned to a friend and remarked, ''That's more oil than the whole world will ever use.'' Today the world uses 4 million barrels *an hour*.

GEOGRAPHIC DISTRIBUTION

Oil and natural gas accumulations are anomalies. They are the end product of an extremely chancy chain of events. Oil and gas are only generated in a *limited* temperature region from *certain* kinds of rocks that can function as source rocks *only* when they have received large volumes of little-decayed organic matter. Once formed, the liquids and gases must migrate, but then they must *be trapped in a reservoir rock of sufficient permeability and in sufficient quantity to be commercially worthwhile*. The reserve must then be located through miles or kilometers of intervening rock, and recovered through those same rocks.

The accidents of ancient geography fix the location of economic accumulations. Wherever in the past giant sedimentary basins filled rapidly, cooked to just the right temperature, and released oil and gas to migration, entrapment, and accumulation, the stage was set for an *oil field* (Figure 16–16). *If* the area remained stable and mountain-making did not disrupt the trap, and *if* the reservoir rock is thick and extensive, and *if* the roof rock or trap is large, the oil field may be a giant.

Giant oil fields are those containing more than 500 million barrels of oil. There are approximately 500 of these fields on earth, and they account for 84 percent of the known world reserves of oil and gas. All the giant fields in the western hemisphere make up only 12 percent of the world's known reserves, since most are depleted or are being rapidly depleted (Figure 16–17).

Individual ''supergiant'' fields in the Middle East are as large as the total reserves of the western hemisphere. An area including North Africa, the Middle East, and the west-central part of the Soviet Union warehouses *two-thirds* of the world's known supplies of oil and natural gas. Whoever controls these reserves controls the futures of nations like the United States, Japan, West Germany, Italy, and France, which must import large quantities of petroleum-based energy to keep their highly industrialized economies humming. The accidents of ancient geographies translate into overwhelming economic power in a world where petroleum products are still king.

▼

ENERGY FOR THE FUTURE

▲

A change in the mix of energy sources for the future is inevitable, just as there were numerous changes in the past (see Figure 16–9). Many people have attempted to plot scenarios for our energy futures, but no one can really imagine what innovations in technology could totally change our sources of energy within a few decades. For the next decade or two, our mix will be much as it is today, though increas-

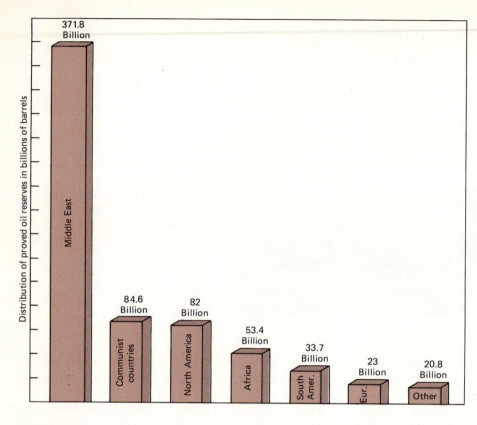

Distribution of proved oil reserves in billions of barrels

371.8 Billion — Middle East

84.6 Billion — Communist countries

82 Billion — North America

53.4 Billion — Africa

33.7 Billion — South Amer.

23 Billion — Eur.

20.8 Billion — Other

FIGURE 16–17

This bar graph displays the known reserves of oil in 1984. The Middle East, a land area about the size of the Great Lakes region, has 55 percent of the oil reserves of the world within its boundaries. (Compiled data from Department of Energy.)

ing prices for fluid oil will probably force an ever greater reliance on coal for all economic sectors except transportation.

But trends are not destinies, unless we fail to manage. Each energy mix brings its own scramble of potentials and pitfalls. *Energy sources are not neutral.* Our choices need to be based on information. A brief review of some alternatives follows.

NUCLEAR ENERGY

Nuclear energy results from atomic fission, which produces heat. One pellet of enriched reactor-grade uranium provides the heat of a ton of coal. Nuclear energy is used solely to generate electricity; it has no other use.

Some have spoken of nuclear energy as a Faustian bargain. (Faust was the mythical hero who sold his soul to the devil in exchange for unlimited power.) Nuclear power seems to hold out the promise of unlimited electric power for the price of a potential nuclear disaster.

After three decades of nuclear experience, the number of casualties attributed to nuclear energy are minimal. In several centuries 18,000 people were killed in coal mines, but fewer than a hundred in all nuclear accidents, even counting the 1986 disaster at

Chernobyl in the USSR. Yet, the risk is still there; an uncontrolled meltdown could still kill thousands and contaminate millions, leading decades later to an epidemic of premature deaths from malignant cancer. Our *perception* of the risks of nuclear power have made it an unwelcome energy source, and disposal of its wastes constitutes an explosive political issue.

No one knows how to decommission an old nuclear power plant. How do we isolate such an area of deadly radioactivity for up to half a million years? It appears that the world reserves of reactor-grade uranium will support only a few more decades of generating of electric power; what do we do when we run out of uranium ore? How do we dispose, safely, of radioactive wastes? What are the possibilities of sabotage and theft of weapons-grade material?

The future of nuclear power looks grim. American companies have long ago given up starting new nuclear power plants, because of the long delays, the legal problems, and a growing recognition that nuclear power is not cost-competitive with electric power produced by the combustion of fossil fuels, nor is it as dependable. For areas stuck with the generation of electricity by nuclear energy, the future is one of climbing power costs, coupled with the nagging small risk, whose vivid portrayal makes it outweigh perceived benefits. In terms of the future, nuclear energy is a technology that has failed.

FIGURE 16–18

Casa Diablo Hot Spring in Mono County, California, a site of geologically recent volcanism, is an example of the geothermal energy available from such western areas. The hot water and steam released are often acidic and highly corrosive. (Photograph courtesy of the U.S. Geological Survey.)

GEOTHERMAL POWER

Geothermal energy is the energy derived from the use of *naturally occurring* hot water and steam (Figure 16–18). Used largely for the generation of electricity and much more rarely for space heating, natural sources of steam and or hot water would seem to be an ideal energy source. Geothermal energy is often portrayed as clean, free, and nonpolluting.

It rarely turns out that way. Natural steam recovered from areas of modern volcanic activity is often loaded with acidic contaminants; without substantial treatment such steam has no industrial or domestic use. The recovery of the steam produces noise levels that require large-scale muffling devices.

Among the promising resources for the future are areas of hot, dry rock at depth. Plans envision pumping water *into* these reservoirs of volcanic heat and recovering the steam or hot water to use to generate electricity. Unfortunately, almost all the areas of suf-

ficiently hot, dry rock are in the west, just where surplus supplies of water are often hard to find.

Geothermal energy currently provides a fraction of 1 percent of our total energy needs, and there seems little reason to think this share will expand. As problems in exploration, drilling, and the handling of corrosive, metal-laden water are solved, geothermal energy may locally provide a larger fraction of energy needs.

OIL SHALE

"Oil shales" are, strictly speaking, not shales, nor do they contain oil. Instead, they are clay-rich deposits within ancient lakes, and the material within them is *kerogen,* the precursor to oil (see Figure 16–13). Within the United States the principal deposits of oil shales are within lake sediments in Wyoming, Colorado, and Utah (Figure 16–19). The kerogens trapped within these deposits could, theoretically, provide *ten times the total oil consumption of the United States to date.* The deposits of just the Piceance Creek Basin would provide the needs of the United States for a century.

So what is stopping their use? Anyone can get oil out of the rock. All one has to do is to mine it, crush it, and heat it, and a mixture of oil and gas leaves the rock. The problem is not obtaining oil but obtaining the oil *at a profit.* In terms of today's oil prices, it simply costs more to recover the oil than it is worth.

The total oil in western oil shales approximates 2 *trillion* barrels, a figure many times larger than the world's known reserves (see Figure 16–17) of liquid oil. If anyone can ever figure out how to recover the kerogen locked in the claystones without expending more heat energy than is recovered, oil shales will become a *major* source of petroleum. Since oil shales are common in many areas around the world, they total an immense resource, which awaits the technology to make them economically attractive.

TAR SANDS

Tar sands are sandstones with pore spaces filled with petroleum that is so thick and viscous that it cannot be recovered by ordinary means. The total world resource is similar to that of oil shales, being about 2 trillion barrels of oil, mostly located in Canada and Venezuela. The tar sands of Alberta, which are the world's largest deposit, contain 12 percent bitumen by weight. Within the United States most of the tar sands are in Utah; the estimated reserve is *equal to* current U.S. reserves of fluid oil (Figure 16–17).

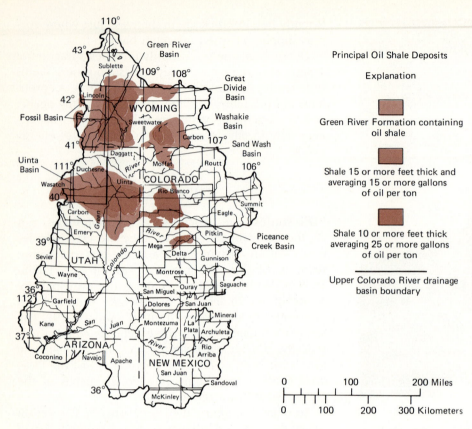

Principal Oil Shale Deposits

Explanation

Green River Formation containing oil shale

Shale 15 or more feet thick and averaging 15 or more gallons of oil per ton

Shale 10 or more feet thick averaging 25 or more gallons of oil per ton

Upper Colorado River drainage basin boundary

FIGURE 16–19

Four zones of oil shale deposits dot the common boundary area between Utah, Wyoming, and Colorado. What the Ute Indians called the "rock that burns" holds the promise of more petroleum products than we have used in all our history—*if* an economic way to recover the oil and gas can be found. Additionally, the quarrying of the necessary quantities of oil shale would make extreme water demands in a generally semiarid area. (Modified from a map by the U.S. Geological Survey.)

Tar sands share the problems affecting oil shales. The stiff bituminous material occurs within highly abrasive sandstones, forming an average of 12 percent of the rock by volume. Recovery normally consists of trenching with huge draglines and scoop shovels, followed by crushing and retorting to drive off the hydrocarbons as fluids and gases. To date, the recovered materials cost more than they bring on the open market. The tar sands await future changes in technology or prices before they shift from being a potential to an on-stream resource.

WIND

Wind has long been used to lift water from wells, and it has been adapted to other forms of work—sailing ships for example—since at least the ninth century A.D. (Figure 16–20), when windmills were first mentioned in Arab literature. Within this century wind has often been used to generate electricity, with battery storage for calm times.

The *promise* of wind is awesome. Estimates place the total amount of energy available from the wind at any one time at *54 times the world's total energy requirements.* The power available from the wind depends on the cube of wind speed, so that areas within the northern Great Plains, the High Plains, and along the New England and Pacific Northwest coastlines have the greatest potential. When wind speed averages less than 8 miles per hour (13 kilometers per hour), the production of power is not economical. For a given wind speed, the power available is proportional to the square of the swept area of the blade; doubling the diameter of the blade increases the available wind power fourfold.

The major advantages of wind power are that it is nonpolluting, free, and limitless. The necessary wind turbines can be up and running in two years, and several decades of experience show that they can be easily tied into an existing utility system for the generation of power. The major problem is wind's intermittent nature. Driven by uneven heating and cooling of the earth's surface, winds in any one locale are quite variable even from minute to minute.

Utility-operated wind turbines in the United States currently generate 4.7 million kilowatt-hours of electricity at costs two to three times the cost of the same electricity generated by conventional plants. The limiting cost is the wind turbines, themselves, which are still very much in the research and development phase.

SOLAR ENERGY

To an astronomer, our sun (Figure 16–21) is a variable star near the outer fringes of the Milky Way. Somewhere near middle age, it is one of more than 10 billion similar stars in our own galaxy.

out over the rising and falling energy demands of an average 24-hour period.

Utilization of Solar Energy

Many homes have been constructed to collect solar heat by means of more efficient windows and skylights and to store it in various combinations of mass, air or water. In the same way, solar collectors on the roof are preheating water for both space heating and use as hot water. Since space and water heating account for about one-quarter of an average fuel bill, the direct use of both passive and active technologies has been responsible for lowering the total energy demand.

The generation of electricity by the sun is accomplished in two ways. In one system, solar energy is focused by mirrors onto a boiler. The resulting steam is run into conventional turbines to generate electricity. Such a system might be particularly useful in hot areas, where intense sunlight imposes a peak demand for electricity for air conditioning just at the same time when a solar boiler system would be most efficient at generating it.

The use of *photovoltaic cells* allows the direct conversion of sunlight to electricity. Technology is currently capable of a 27 percent conversion rate, which means that the cost of electricity generated by photovoltaic cells is not competitive with conventionally generated electric power. The potential is, however, large because efficiencies continue to improve and costs continue to decline. An area of photovoltaic cells about the size of the state of Connecticut or 5 percent of the state of Nevada could power the entire country.

WATERPOWER

A fractional percentage of our total supply of electricity is provided by water-powered turbogenerators. Water backed up behind a dam is allowed to fall through turbines to generate electricity. There seems little reason to assume that hydroelectric power will be much further developed, since few spots of economic potential remain. Hydroelectric power is non-polluting, but the source of its energy is not totally free of concern.

Water backed up into lakes behind dams is wasted by both seepage and evaporation, which renders it more saline. As the process is repeated downstream, the runoff water may become so salty that it is no longer fit for use. The Colorado River, which starts as sparkling clear snowmelt high on the western slopes of the Rockies, arrives at the international

FIGURE 16–20
Wind power has been useful for a long time. Shown in this sixteenth-century woodcut is an artist's view of the use of wind power to drive mining ventilation machinery underground. If the machinery was ever built, mining must have stopped on calm days. (Courtesy of the History of Science Collections, University of Oklahoma Libraries.)

The sun is also a vast raging furnace that fuses hydrogen into helium and produces 70,000 horsepower of energy per square yard (0.8 square meter) of its surface (see Figure 16–3). Our earth, approximately 93 million miles (150 million kilometers) from the sun, intercepts *not more than two-billionths of 1 percent* of the sun's radiant energy, but this intercepted energy is *hundreds of times the total energy we now use.*

In only three days the sun showers the earth with energy equivalent to that of *all* of the earth's fossil fuel reserves. It is no wonder that in an energy-short world we look to the sun as an ultimate resource of nonpolluting, inexhaustible energy.

For all its potential, the sun is also a difficult energy source to utilize. Its energy is diffused over half the earth at any one time, and it is difficult to collect and store. Sunlight is also variable; cloudy days follow sunny days and night follows day. The energy from sunlight must, somehow, be stored and spread

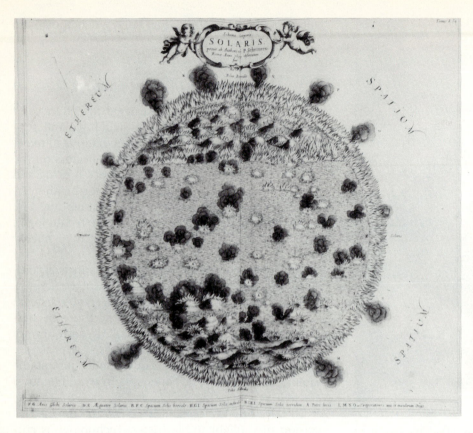

FIGURE 16–21
A view of the sun, drawn not long after the invention of the telescope, shows a fiery body with hills (volcanoes?), flares, and sunspots. (From A. Kircheri, 1665, *Mundus subterraneus*, courtesy of the History of Science Collections, University of Oklahoma Libraries.)

boundary with Mexico so saline that a mammoth desalting plant must intervene in order to allow the United States to meet its treaty obligations for delivery of usable water to Mexico. The lakes behind the dams slowly fill up with sediment; its removal is a generally an unplanned cost in dam construction and maintenance. Removal may be economically impossible.

Although the cost–benefit analysis of dams is a continuing argument, hydroelectric power is often a good bargain for the United States. It is, however, a resource that is essentially fully developed. As our power demands increase, hydroelectric power will become a smaller and smaller fraction of our total power demand.

ENERGY—A PERSPECTIVE

Americans constitute *one-fifteenth* of the world's population, but we use *one-third of its energy*, approximately 21 trillion kilowatt-hours a year. Almost all this energy comes from the combustion of fossil fuels, and this pattern will continue for the next few decades.

As energy costs rise, Americans—accustomed to cheap and abundant energy—are successfully learning to budget their energy. We are painfully learning that with energy as with the rest of life, there is no such thing as a free lunch. We should also recognize that our prodigal consumption of energy brings about an extraordinary productivity on a worldwide basis. From that simple fact flows the affluence Americans take for granted.

With our easy access to abundant energy fading, and others holding the majority of the world's petroleum reserves, we might reasonably ask what the future holds. The answer depends on our vision, not our resources. If we insist on big cars, frigid homes and offices in the summertime, and tropical warmth in the winter, if we remain addicted to petroleum, control over our lives and our fortunes will slowly be handed over to others.

If instead we better insulate our buildings, capture more of the sun, drive smaller cars and drive them less, turn down our thermostats in the winter and raise them in the summer, and vigorously press for renewable technologies, we work toward our own independence. Our future must be less wasteful or we will not have a future.

The challenge is individual; all us can conserve. As Will Rogers said 60 years ago "When we want wood, we chop down a forest. When we want oil, we drill a hole. When we run out, we'll find out how good we really are.''

SUMMARY

1. Energy loss is exchangeable for work completed. All forms of energy are convertible to other forms, but at a loss of total available energy. Energy is the multiplier that magnifies human labor and ingenuity, allowing greater affluence and relieving human toil.

2. Our current sources of energy are almost entirely from fossil fuels, which are concentrated deposits of past solar energy as preserved in solid, liquid, and gaseous materials. Continental materials caught within swamps and marshes are slowly converted into various grades of coal by temperature and time. For North Americans coal is 80 percent of readily available energy.

3. Continental material washed onto deltas and sedimentary basins is slowly converted largely to natural gas; marine organisms are largely converted to liquid crude oil. Both conversions require elevated temperatures, millions of years, and burial to depths of a mile or more.

4. The accumulation of fossil fuels in economic quantities is a rare event at the end of a long chain of circumstances that must all fit together in just the right way. The recovery of fossil fuels depends on a variety of quarrying and drilling operation. Their use elevates levels of atmospheric carbon dioxide, nitrogen and sulfur oxides, smog, and the acidifies rain.

5. Future additions to energy resources include wind power, solar power, and a wide range of technologies that may make alternate forms of petroleum into usable fuels. Because the distribution of fossil fuels around the world is uneven, their import and export constitute a major world business.

KEY WORDS

uanthracite	lignite
bituminous coal	natural gas
Btu	oil shale
coal	petroleum
crude oil	reservoir rock
energy	roof rock
erg	source rock
fossil fuel	tar sand
geothermal energy	trap
kerogen	

EXERCISES

1. Ralph Waldo Emerson once said ''Nature never gives anything away. Everything is sold at a price.'' Compare the price for solar energy to its concentrated forms, the fossil fuels.

2. Today, you probably used the energy equivalent of 7 gallons (26 liters) of oil. Your parents, *when they were your age*, used about one-half as much— a statement that is generally true regardless of age. How was their life-style different than yours?

3. What factors limit the use of photovoltaic cells?

4. Why has the kerogen in the lake beds of the western United States not already been converted into liquid petroleum? What process intervened or what process is missing?

5. Coal contains oxygen chemically bound to carbon and hydrogen to make a *carbohydrate*; oil is a *hydrocarbon*, lacking oxygen. Which of these two classes of fuel could be set on fire underground and continue burning for centuries? Do you know of any examples of this occurrence?

6. Termites release more carbon dioxide into our atmosphere than the combustion of all fossil fuels. The carbon dioxide content of the atmosphere has been increasing since the Industrial Revolution, and the climate has been warming— perhaps as a result of the increasing carbon dioxide. What process *removes* carbon dioxide from the atmosphere? What process buries carbon dioxide as rock?

SUGGESTED READINGS

BALZHISER, R. E., AND YEAGER, K. E., 1987, Coal-fired power plants for the future, *Scientific American*, September, pp. 100–107.
▶ *Review of new combustion technologies; a hopeful look..*

Department of Energy, 1986, *Statistics and Trends of Energy Supply, Demand, and Prices,* Washington, D.C., Superintendent of Public Documents.
▶ *An annual report to Congress; loaded with information.*

Department of Energy, 1987, *Coal Data: A Reference,* Energy Information Administration, Washington, D.C., Superintendent of Public Documents, 96 pp.
▶ *Excellent overview of deposits, reserves, mining, and so forth.*

Federal Energy Administration, 1976, *Comparison of Energy Consumption between West Germany and the United States,* Washington, D.C., Superintendent of Public Documents, F.E.A. Conservation Paper 33A, 104 pp.
▶ *Eye opening, and sobering. A view of the future.*

Office of Technology Assessment, 1984, *Applications of Solar Technology to Today's Energy Needs, Volume II,* Washington, D.C., U. S. Government Printing Office.
▶ *An unbiased look at the future of solar energy.*

PART

5

Geologic
Time
and a
Brief
Earth
History

A contemporary of Nicolaus Steno uses are to argue against the seventeenth-century idea that fossils were simply curiously "ornamented stones," created from the earth like minerals. *Sense,* with the eye of Reason on his chest demonstrates to *"Vain Speculation"*—the ghostly figure on the left—the organic nature of fossils like those within the rock. (From Agostino Scilla, 1670 (1748 ed.), courtesy of the History of Science Collections, University of Oklahoma Libraries.)

CHAPTER *seventeen*

Time and Stratigraphic Principles

time (tīme) *n.* **1.** A nonspatial continuum in which events occur in apparent irreversible succession from the past through the present to the future. **2.** An interval separating two points on this continuum, measured essentially by selecting a regularly recurring event, such as sunrise, and counting the number of its occurrences during the interval; duration.

My dictionary tries, but 27 entries and a quarter-page later, it simply abandons the effort and tells me only that the word time comes from the Old English *tīma* meaning ''time,'' ''season,'' or ''tide.'' From this we learn that our concept of time must have had something to do with seasons or the daily ebbing and flowing of the tide.

HOW DO WE TELL TIME?

At first we marked the passage of time with the rising and setting of the sun, and the slower intervals set apart by the seasons. We could speak of tomorrow, today, or yesterday, of last winter or next spring. Time had a rhythm and a sense of movement. The rising and setting of the sun and moon, the passage of autumn to winter and spring to summer were markers that set apart *intervals*.

The interval from sunrise to sunrise we soon called a *day*. The interval from new moon to new moon became a *month*, a term whose root means *moon*. The advent of a modern calendar replaced the lunar year with its varying number of months with a solar year, the period required for the earth to make one revolution around the sun (Figure 17–1). The time from fall (harvest) equinox to fall (harvest) equinox was defined as a *year*.

The day, the month, and the year all reflect intervals between astronomic events, but the week is without natural origin. The number seven has long had a mystical meaning, perhaps after the seven heavenly bodies recognized by the ancients. Our days are all named after the gods who supposedly ruled them; all are German equivalents of the Latin names except for Saturn's day (Saturday), Sun's day (Sunday) and Moon's day (Monday). In German Mars's day is Tiw's day (Tuesday), Mercury's day is Woden's day (Wednesday), Jupiter's Day is Thor's day (Thursday) and Venus's day is Freya's day (Friday).

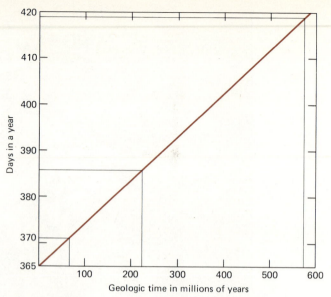

FIGURE 17–1

Paleontologists pointed out that some animals involuntarily record growth lines that may record days, months, or seasons. By counting these lines and comparing them to the evidence of seasons, it is evident that our earth's rotation is slowing down, such that the number of days in a year decreases about 5 days every 100 million years. About 7 billion years from now out earth will slowly grind to a *stop*.

The effect of lunar cycles on our biologic rhythms is obvious. The cycle of the 28-day lunar month is the most obvious in the hormonal cycle of females, but males too have a subtle lunar hormonal cycle. Scientists working through the half-year long polar winters quickly shed their watches and with that lapse into a 24.9-hour cycle of sleeping and waking that parallels moonrise, moonset, and the rise and fall of local tides. Similar rhythms have been observed in the blind. Cut off from environmental cues, the natural biorhythms of our bodies are set to the lunar year of 11 months, not the solar year of 12 months.

Long before we had a watch, we had a clock within us, set at birth to 24.9 hours—the lunar ''day,'' and 28 days, the lunar ''month.'' The source of the rhythmic patterning appears to be the pineal gland, found in animals as diverse as hibernating squirrels and migrating birds. Our biologic clock was the first kind of time we ever knew. Our internal clock is reset for us on a daily basis by the rise and fall of the sun, so that it is forced to accommodate itself to a solar year, keeping us in phase with the seasons.

How do we tell time? We can sense its passage

with *any event that recurs at a constant rate*. The event may be the rise and fall of the sun, the passage from full moon to full moon, or the lapse from harvest to harvest. In a quest to measure its passage with ever-increasing accuracy, we have invented mechanical, digital, and even atomic clocks; their function is still the same—to divide the passage of time into intervals produced by a device having *a constant rate*.

Our discussion omits a question that humans have always asked. When did time begin?

THE LIMITATION OF TIME

The past is not open to direct inspection. We cannot know what color hair Socrates had in the same sense that we know the color of our own hair. We can form conclusions about the past in only one of two ways.

1. Depend on the authority of those who say they know.
2. Evaluate the evidence left behind by past events.

As children we learn by accepting the authority of others, beginning with that of parents and teachers, but we spend the rest of our lives sorting from that imposed authority what we find to be true from experience. But how can we find out when time began?

LESSONS FROM AUTHORITY

The rise of Christianity established for Christians everywhere a focal point in time, much as the Romans had used the legendary founding of Rome as their starting point and the Jews had used the Passover, the building of the Temple, and the reigns of their kings as pivotal dates. In the first half of the sixth century AD.,, the Roman monk Dionysius Exiguus proposed a method of reckoning the Christian era, making the birth of Christ the starting point of modern chronology. The division of all time into B.C. and A.D. quickly spread over the Western world.

Counting backward from the birth of Christ through the genealogies recorded within the Old Testament, it was possible to obtain a date for creation. The Jews set a date that would be equivalent to our 3760 B.C.; others reckoned in other ways. Between 1650 and 1654 Archbishop James Ussher placed the creation at 4004 B.C., a date first proposed by Martin Luther (1483–1546). This date was often inscribed in the King James version of the Bible and furnished a fundamental date for many Christians. Sir John Lightfoot, vice-chancellor of Cambridge University, attempted a still more accurate reckoning; he calculated that creation had taken place during the week of October 18 in 24 B.C., with the creation of Adam at 9:00 A.M. on October 23rd.

The Christian Frame of Reference

Fundamental to accepting the authority of the scriptures was the view that humans had been created at *essentially the same time as the earth*. This placed a single human life in a new perspective.

In an earth a few thousand years old, a human lifetime was a fair proportion of total time. The whole course of history, in a view that lasted through much of the eighteenth century, comprised six cosmic "days" of a 1000 years (one *millennium*) each. Four such "days" had elapsed between the Creation and the birth of Christ. Another 1700 years had passed before the lifetime of eighteenth-century people. Only a few hundred years remained until the allotted 6000 years would be completed, to be followed by the second coming of Christ and the Last Judgment as described in Revelations. This was to be followed by the promised millennium, completing the seven biblical "days" of the earth's total history (see Figure 1–9).

EVIDENCE FROM THE PAST

On an earth whose total time was limited by scriptural authority, evidence from the rocks and the fossils they contained had to be fitted into the prevailing scheme. The role of science was not to challenge the *only* source of truth but to look in the rocks for confirming evidence.

Abraham Gottlob Werner (1749–1817), professor at the Mining Academy in Freiburg, Germany, developed a simplistic way of classifying rocks. He proposed that all over the world rocks had precipitated from a primeval ocean, forming a worldwide, regular, universal sequence. It was a picture of the past that was easy to reconcile with religious authority.

Igneous and metamorphic rocks underlie rocks nearly everywhere, and so they were regarded as the oldest—and were called *Primary* rocks, formed at the Creation. Layered sedimentary rocks overlie the old rocks and were called *Secondary rocks*; they were thought to have been laid down on the Primary rocks during the flood endured by Noah in his ark (Figure 17–2).

Tertiary rocks included all unconsolidated sedi-

FIGURE 17–2
The deposition of Primary rocks was supposed to have taken place during the flood of Noah, followed by Secondary rocks. Notice the dead animals in the ocean, an explanation of the origin of fossils in Secondary rocks. Notice also the bird in the tree returning vegetation to Noah and the ark itself on the right side of the engraving. (From Schuechzero, Johannes Jacob, 1723, *Herbarium Diluvianum Collectum*, (courtesy of the History of Science Collections, University of Oklahoma Libraries.)

mentary rocks lying on top of Secondary rocks and were thought to have been deposited by the last stages of Noah's flood. Some geologists even added a fourth (*Quaternary*) group of rocks, to include volcanic rocks that clearly intruded Tertiary rocks as well as poorly consolidated sedimentary rocks. The fossils in the Quaternary rocks were identical to modern forms of life.

In a world with its history confined within a few thousand years, the assumption that similarity in appearance meant similarity in age made good sense; in a brief time span, like events happen only once. By the end of the eighteenth century, the threefold or fourfold subdivision of rocks based on their appearance was widely used. The rocks of the world were neatly tucked into a 4000- to 5000-year history of a world created, flooded, and little changed since.

The end of the eighteenth century was a time of radical social, governmental, and scientific revolution. That revolution was to destroy the comforting view that mankind had been given dominion over an earth recently created for its sole benefit. In its place, humans slowly unfolded the evidence of an earth with "no vestige of a beginning—no prospect of an end."

FIGURE 17–3
Buffon was probably the most prolific author of the eighteenth century, eventually publishing forty-four volumes of his *Natural History*. Buffon was the first reputable scientist to propose that the earth might be as old as 75 thousand years, pushing time back more than 10 times as far as circumscribed by the Book of Genesis. He also was among the very first persons to consider that animals might evolve, with humans not necessarily created at the same time as earth.

THE EXPANSION OF TIME

The challenge to scriptural authority on the age of the earth came largely from devoutly religious men, many of whom held degrees in both divinity and science. The challenge was mounted from many quarters, as increasingly sophisticated ideas from biology, chemistry, and physics revealed an earth far older than the scriptures could allow.

The earliest challenge came from Georges Louis Leclerc (1707–1788), Comte de Buffon, a French naturalist (Figure 17–3). He was justly famous for his 44-volume *Natural History,* a treatise on the whole of nature written for the general public.

As a part of this work, Buffon (pronounced byoo-fohn') attempted to calculate the time of creation by estimating the rate of cooling of an originally molten earth. He proposed that the earth had begun to cool some 75,000 years earlier, and would probably last another 90,000 years before losing all its internal heat.

His attempt to use an assumed cooling rate to figure earth age was the first of many attempts, all of which were discarded with the discovery of radioactivity at the end of the nineteenth century. But at the time of its publication Buffon's estimate rocked the European world.

The most enduring challenge, however, came from the rocks exposed along the coast of Scotland. The Bible as calendar was destroyed at Siccar Point (Figure 17–4).

UNCONFORMITIES AND TIME

The story of James Hutton (1726–1797), Scottish physician, farmer, and geologist, fills a large portion of Chapter One. It bears only the briefest retelling here. Hutton recognized that uplift, caused by the earth's internal heat, provided a restoring force, countering the losses to the earth's surface by weathering and erosion. Once the earth's heat engine was recognized, it followed that it would go on cycling indefinitely.

Earth time had escaped the limits of scripture. An earth constantly renewed as it decayed could be unimaginably old.

The earth was seen as a great engine, whose purpose was to keep the earth habitable for humankind. In an age intoxicated with the power of the newly discovered steam engine, such an analogy was quite natural. It was no longer necessary to view the earth as created by supernatural forces only an instant ago. On an earth without an engine, everything had to be created as it now was, because nothing could modify it except the forces that wore down its surface. On an earth with an engine, *creation was continuous, not instaneous.*

The recognition of a cyclic earth led Hutton to propose the **principle of uniformity,** which states that the geologic processes of the past were similar to those of today. Using that simplifying assumption, Hutton read the layer after layer of rocks at Siccar Point as sedimentary rocks, deposited at different times on ancient seafloors, turned on end and uplifted, eroded and beveled across their tilted surfaces, and submerged beneath the sea again, with layer after layer of horizontal sediment stacked across

FIGURE 17–4

Siccar Point is a natural exposure of rocks along the coast of extreme southeastern Scotland near Cocksburnpath on the North Sea. The overlying sedimentary rocks we now know are of Devonian age (see Figure 17–17) and were deposited across upturned rocks of Silurian age, intensely folded during the *Caledonian orogeny,* an early Paleozoic mountain-forming event that affected Ireland, Scotland, Scandinavia, Greenland, and the northern Appalachians. (Photograph courtesy of Stan Beus; used with permission.)

Burrows

There are a few rather well-preserved clam fossils in this white, well-sorted sandstone. Thin-bedded laminar bedding, interrupted by burrows. Sand grains somewhat rounded.

Contact between units gradational.

White, well-sorted sandstone. High-angle cross-bedding. Sand grains well rounded, and surface is frosted. One set of tracks observed on bedding surface.

FIGURE 17–5

Modified from a sketch drawn of rocks exposed in a road cut in central Ohio. The rocks are about 450 million years old. After examining the vertical change seen in this column of rocks, see if you can describe what was happening here 4.5 million centuries ago.

the beveled edge (Figure 17–4). To Hutton and his friends, the rocks of Siccar Point were a record of a complete cycle of submergence, deposition, uplift, and erosion, followed by submergence, deposition, uplift, and modern erosion.

Hutton quickly recognized the implications. At the observable rates of earth processes, it would take far more time than anyone could imagine for such a series of cycles to occur, one on top of another. As his friend Playfair phrased it " . . . the mind grew giddy looking into the abyss of time . . . ."

Hutton's ability to reconstruct the past rested on a group of principles for interpreting the *sequence* of events in stratified rocks. The subarea of geology dealing with the reconstruction of past events from stratified rocks is termed *stratigraphy.* At its most elementary level, the science of stratigraphy rests on *nine stratigraphic principles.*

1. Superposition.
2. Original horizontality.
3. Original continuity.
4. Crosscutting relations.
5. Inclusion.
6. Metamorphism.
7. Faunal assemblage.
8. Faunal succession.
9. Correlation.

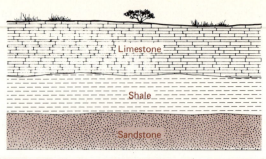

Limestone

Shale

Sandstone

FIGURE 17–6

The principle of *superposition* requires that the sandstone layer had to be deposited before the shale layer, which was deposited before the limestone layer. If orogenic forces were so intense that the layered rocks were turned upside down, then the principle of superposition would no longer apply.

Hutton had applied the first four principles to work out the history of Siccar Point and to propose a world without beginning or end. Others were to use the last two to order events on a worldwide basis and to organize all of the earth's history into discrete chapters, each separated from the other by major events in the history of life.

STRATIGRAPHIC PRINCIPLES

Stratified sedimentary rocks are matchless storytellers. Stacked on top of one another like newspapers, each layer may yield evidence of ancient seas, towering cliffs, teeming reefs in tropical seas, rushing rivers, stagnant, rotting swamps, grinding glaciers, or desert lakes as easily as you and I talk about yesterday. A column of rock (Figure 17–5) chronicles change, the central product of an energetic earth.

Superposition

One of several stratigraphic principles formulated by Nicolaus Steno, the principle of **superposition** states that *in a series of flat—lying sedimentary rocks, the oldest layer is on the bottom, with successively younger layers above* (Figure 17–6). The basic underlying assumption is that the rocks formed by the accumulation of loose sediment, each layer accumulating *on* something older beneath it.

Original Horizontality

A corollary to the principle of superposition is the principle of **original horizontality.** This law states that since sedimentary rocks commonly form by the deposition of loose sediment in water, the layer of

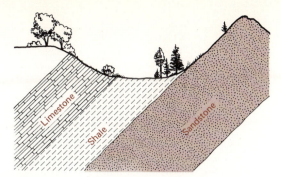

FIGURE 17–7
The principle of *original horizontality* means that these three layers were deposited as horizontal to subhorizontal sheets of sand, mud, and lime particles and then consolidated into hard rocks before tilting. Compare this sketch with Figure 1 –. Does the principle of superposition still apply?

sediment *must have originally formed a horizontal or subhorizontal sheet*. Therefore, if we find tilted layers of sedimentary rock (Figure 17–7), *the tilting had to follow the consolidation of the horizontal sheet of sediment into a horizontal sheet of sedimentary rock.*

Original Continuity

A second corollary to superposition is the principle of **original continuity** (Figure 1–8). Since a layer of sedimentary rock is formed as a *continuous* sheet in three dimensions, parallel layers now separated by erosion (or faulting), must have been continuous *before the erosion or faulting*. The principle of original continuity allows us to "join" layers separated by

FIGURE 17–8
The principle of *original continuity* would argue that the sandstone layers on each side of the canyon wall were at one time a continuous layer, so that stream erosion was much later. Can you describe the total sequence of events that occurred in this area?

valleys many kilometers or miles wide and so more accurately reconstruct the sequence of events. The faulting or valley erosion occurred *after* the continuous layer consolidated.

Crosscutting Relation

Three additional stratigraphic principles apply to work with igneous and metamorphic rocks. The principle of **crosscutting relation** states that an igneous rock *is younger than any other unit it affects*. As in Figure 17–9, the igneous rock is younger than any unit it bakes. Logic insists that a baked rock was in place before it was contacted by magma or lava; therefore the *unit affected is always older*.

Inclusion

The principle of **inclusion** states that *a rock fragment is always older than a rock that surrounds it* (Figure 17–9). As in the sketch, a fragment of foreign rock caught within a cooled magma was obviously torn away from its source by the force of flowing lava. Therefore the solid rock *inclusion* is older than the liquid that surrounds it. Such inclusions are termed **xenoliths,** whose literal translation is "stranger rock."

Xenoliths tell more than their relative age. They also yield evidence of the type of rock that lies beneath an area of volcanism or plutonism. They are gratuitous samples (Figure 17–10). Some extraordinarily violent forms of volcanism may bring xenoliths from as far down as the earth's uppermost mantle.

Metamorphism

The principle of **metamorphism** states that any change in the constitution of a rock *occurred after the consolidation or crystallization of the parent rock*. Thus in Figure 17–9, the contact metamorphism of both the overlying and underlying rock caused by the sill occurred after the formation of the original sedimentary layers and at the same time as the intrusion of the dike. A metamorphic rock always records at least two parts of its history; the existence and character of the parent rock and the timing of the metamorphic change.

THE EQUIVALENCE OF TIME

From the time of the Greeks there was speculation on the meaning of fossils. People as diverse as Aristotle, Leonardo da Vinci, and Nicolaus Steno had de-

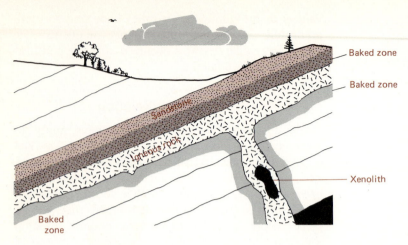

FIGURE 17–9

In this cross section, a fragment or *xenolith* of the lowest rock layer is caught up in the magma, which baked all of the rocks with which it came in contact. Is the igneous rock a sill, dike, or lava flow? What is the oldest event described in this scene? What is the youngest?

duced that fossils were preserved evidence of past life. As we have learned, Nicolaus Steno (see Chapter One) deduced after dissecting the head of a modern shark (see Figure 1–3) that its teeth were *identical* to peculiar fossils called "tongue stones," which were found *within* certain rocks.

The use of fossils to "tell time" and then to establish that distant rocks were of equivalent age was made possible by the last of the nine stratigraphic principles. This last concept was developed independently by two people who could hardly have been more unalike. One was a poor English surveyor with barely a grammar school education; the other was a wealthy French biologist who was educated at the best European universities, became a Baron, and was universally acknowledged as the greatest zoologist of his time.

FIGURE 17–11

The frontispiece from *Strata Identified by Organized Fossils*, shows William "Strata" Smith in 1816. Trained as a surveyor and civil engineer, Smith was interested in the use to which each of the rock layers might be put, and so was concerned with being able to predict where certain valuable layers might be exposed, based on correlation by what he called their "organized fossils." (Courtesy of the History of Science Collections, University of Oklahoma Libraries.)

FIGURE 17–10

This xenolith is a ⅜-inch fragment within volcanic rock on the surface of the moon. Photographed by Apollo 11 astronauts, this photograph displays extremely high resolution. The rock is pitted, probably due to impacts by extremely high velocity micrometeorites. (Image courtesy of NASA.)

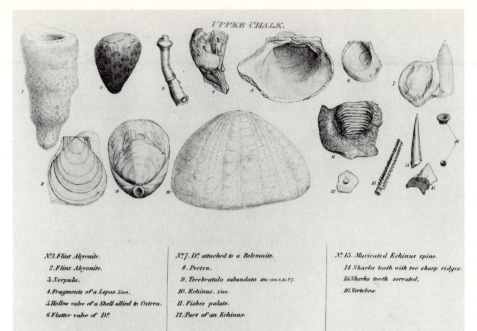

UPPER CHALK.

N.º 1. Flint Alcyonite.	N.º 7. D.º attached to a Belemnite.	N.º 13. Muricated Echinus spine.
2. Flint Alcyonite.	8. Pecten.	14. Sharks tooth with two sharp ridges.
3. Serpula.	9. Terebratula subundata Min Con.s. 11. t. 7.	15. Sharks tooth serrated.
4. Fragments of a Lepas. Linn.	10. Echinus. Linn.	16. Vertebræ.
5. Hollow valve of a Shell allied to Ostrea.	11. Fishes palate.	
6. Flatter valve of D.º	12. Part of an Echinus.	

FIGURE 17–12
Here is an example of a faunal assemblage—in this case a group of fossils unique to a limestone layer termed the "Upper Chalk." Such faunal assemblages were unique to one period of time and one environment—in this case the time when the Upper Chalk limestone was being deposited on the sea floor. (Courtesy of the History of Science Collections, University of Oklahoma Libraries.)

Faunal Assemblages

William Smith (1769–1839), a largely self-taught English surveyor (Figure 17–11), surveyed across the whole of England at a time when England was lacing the land with canals. As the canals were dug, Smith was able to observe the layered rocks or **strata** at excavation sites. He was impressed by the way each *stratum* contained groups or *assemblages* of fossils not found in other strata.

An assemblage of fossils is often called a **fauna** if composed of animal remains or a **flora** if composed of plant remains. Smith had recognized that the assemblage of fossils in each layer or stratum was *unlike* that in the layers above or below it. Smith had recognized the *fact that life changes through time*. Because life changed through time, *each assemblage of life was a unique to an interval of time represented by an individual stratum*, which is a statement of the principle of **faunal assemblage.** Stated another way, the principle says that *similar assemblages of fossil organisms indicate similar ages for the rocks that contain them.*

As Smith put it in 1816 when he began publishing his *Strata Identified by Organised Fossils*, "Each stratum contained organized fossils peculiar to itself. His book included careful line drawings (Figure 17–12) of fossils printed on paper whose color was that of the rock unit that normally contained them. The year before he had published the first geologic map of England, a summation of his years of work as a surveyor and excavator of canals. His only tools were curiosity and a notebook; his only education came from the rocks and the fossils they contained.

Smith helped to establish the scientific foundations of *stratigraphy*. His interest in strata earned him the nickname of "Strata" Smith and the honor of working out the history of layered rock throughout the whole of England. He also helped illuminate the principle of faunal assemblage, recognizing that "the same species of fossils are found in the same strata, even at a wide distance."

Perhaps Smith's greatest contribution was his recognition that the *succession of faunal assemblages matched the vertical succession of rock sequences*—a universal concept now called the principle of *faunal succession*. The **principle of faunal succession** states that *fossil organisms succeed one another in a vertical sequence of sedimentary rocks in a definite and recognizable order, so that each layer of sedimentary rock contains an assemblage of fossils different from that above it or below it.* Because that is observably true, *the relative age of rock layers can be determined from their fossil content.* The principle of superposition now had both a physical and a biologic basis.

Across the English Channel, Baron Georges Léopold Chrétien Frédéric Dagobert Cuvier (1769–1832), French biologist and anatomist, had come to the same conclusion about faunal succession as Smith at essentially the same time, much as Darwin and Wallace were to independently arrive at a causal theory for faunal succession some half century later. Studying the layered rocks around Paris (Figure 17–13), Cuvier also recognized that each layer had its own distinctive fauna, an assemblage unique to an interval of time.

Like Smith, Cuvier recognized the fact that life

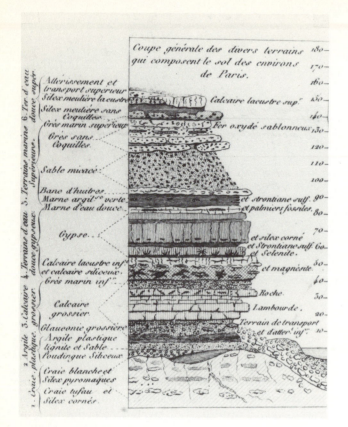

FIGURE 17–13

With this plate from Cuvier, he demonstrated that the vertical succession of sedimentary rocks around Paris also contained a vertical succession of fossils which became progressively more like modern fauna upwards. With the evidence for life change through evolution in front of him, Cuvier chose a very different explanation more in keeping with his devout belief in the truth of the Bible. (From *Essay on the Theory of the Earth* by George Cuvier, Plate 1, courtesy of the History of Science Collections, University of Oklahoma Libraries.)

types change through time, so that *any one assemblage of fossils can only be together during a short interval of time.* The situation is analogous to a devastating volcanic eruption burying your geology class under tons of ash. If both your personal identification and the school records survive, it will be easy for someone to establish later the time of eruption. Your assemblage of individuals—a geology class—is only together during a brief interval of time, a semester or a quarter. Therefore, if you are all found together as "fossils" under the ash, the time of your entombment is apparent.

Cuvier was a fascinating man (Figure 17–14). As probably the most famous biologist in all of Europe, he became perpetual secretary to the Institut National in 1803. He was enormously wealthy, grossly obese, and possessed of a head one-third larger than normal. He was a certifiable genius. He had essen-

FIGURE 17–14

The good baron was a large man, and a virtual dictator in comparative anatomy. His expertise was, however, wide ranging. He, with William "Strata" Smith proposed the principle of faunal succession, the guiding principle for long-distance correlation. (Courtesy of the History of Science Collections, University of Oklahoma Libraries.)

tially memorized the contents of the nearly 20,000 volumes in his personal library.

Cuvier established the fact that life types constantly change through time. He saw clearly that the more ancient the fossil the more unalike modern life it was. Yet, as a devoted believer in the literal words of Genesis, he shrank from the implications of his own discoveries. Instead, he took refuge in an ever more convoluted series of past catastrophes to explain the changes that he saw.

From his own data he could easily have made the leap to some sort of evolutionary theory, as Cuvier's contemporary and antagonist Jean Baptiste Pierre Antoine de Monet Lamarck (1744–1829) had already done in 1809. Lamarck had proposed that species were not fixed, rather that they changed and developed through time; his idea was revolutionary, but he had the wrong mechanism. Lamarck believed that animals changed by passing along to their offspring characteristics they had acquired during the course of their lives. It was relatively easy to prove his mechanism erroneous, but Lamarck did advance

FIGURE 17–15

Plate III, from Volume I of James Hutton's 1795 edition of the *Theory of the Earth.* Entitled "the unconformity with basal conglomerate along the River Jed, south of Edinburgh," this etching depicts a vertical slice into the earth which exposes an angular unconformity between vertical foliation and horizontal limestone strata. The basal conglomerate is sandwiched between the two as a buried residual soil developed on the metamorphic rock. (Courtesy of the History of Science Collections, University of Oklahoma Libraries.)

evolutionary thinking into biology and set the stage for Charles Darwin. As Cuvier lay dying of cholera in 1832, Charles Lyell (1797–1875, Scottish geologist, was already replacing Cuvier's catastrophism with uniformitarianism and writing the textbook that was to profoundly influence Charles Darwin (1809–1882).

Correlation

To *correlate* rocks is to establish time equivalency between them. To show that two rocks now widely separated—perhaps even on separate continents—formed at the same time helps to reconstruct ancient geographies. Without correlation, reconstructing the past from the sequence of rocks exposed in many different areas is like finding 147 unnumbered chapters torn out of an unknown number of different books. Each sequence or chapter tells its own story, but there is no way to relate it to anything else. With correlation, each chapter fits into other chapters from other places, and the whole is a coherent story of events through time.

The principle of **correlation** states that any two rocks exposed in different areas are of the same age *if they contain the same faunal assemblage.* When first proposed, the principle of correlation seemed to offer unlimited power to interpret the relation of rocks from distant areas. We have, however, discovered that the fossil assemblage of an area is *dependent not only on time, but also on environment in which the life forms lived.* Fossils from rocks of the same age that formed partly in the desert and partly on the continental shelf look nothing alike, even though the rocks formed at the same time. Even today, nearly 200 years after William Smith, it is a formidable job to establish time equivalency between continental and marine rocks.

The task of *stratigraphers,* the geologists who interpret past environments by analyzing stratified rock, is made even more difficult when random pages from each "chapter" are missing, and *that is the usual situation.* The record of the past in any one area is rarely complete. The story of the earth's history is riddled with gaps, for erosion erases the record in one area as it adds to it in another.

INTERRUPTIONS IN THE GEOLOGIC RECORD

An **unconformity** is an eroded surface that separates younger strata above it from much older rocks below it. The removal of rock by erosional agents (see Chapters Ten through Fourteen) destroys a detailed record of past events and substitutes in its place a blank. An unconformity is like discovering pages missing in an old diary; you know what dates are missing but not what happened on those dates.

There are three types of unconformities.

1. Angular unconformities,
2. Disconformities, and
3. Nonconformities.

Each of them testifies to a missing record with its own unique combination of geometry and rock type.

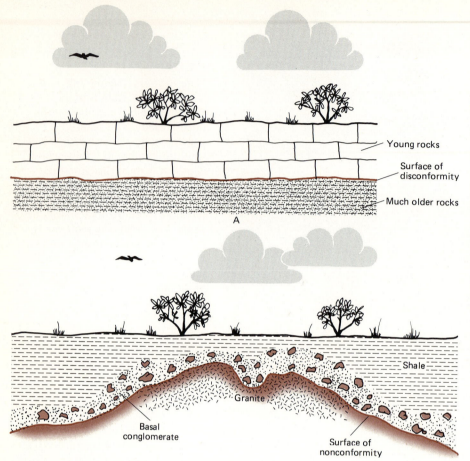

FIGURE 17–16

A *disconformity* (A) is a barely detectable break in sedimentation between two layers which parallel each other. It may be quite difficult to recognize unless fossils within each layer reveal the wide separation in ages above and below the surface of disconformity. A *nonconformity* (B) is an erosional surface between two rocks that *could not have* originally formed together. In the example given, granite and shale can only be contact if granite invades the shale as an intrusive, baking the shale—or—if the granite formed at great depth is uplifted after cooling, and reaches the surface after prolonged erosion, only to be depressed beneath sea level to be buried under mud.

Angular Unconformity

Probably the world's best-known angular unconformity is the one at Siccar Point (Figure 1–10). An **angular unconformity** is an *eroded surface that bevels tilted rocks.*

The beveling of the layers may have been accomplished by mass wasting, running water, glaciers, or wind. The alteration of the eroded surface is a record of prolonged exposure of the underlying rocks to the atmosphere. Soils (see Chapter Four), deposits of lag gravel, or both commonly mark the surface (Figure 17–15).

Disconformity

A **disconformity** is an eroded surface separating two packets of layered rock that are parallel to one another and the surface of disconformity (Figure 17–16A). Since the eroded surface is parallel to the bedding planes of the rocks both above and below it, a disconformity is often difficult to recognize.

Only when someone encounters rocks known to be of dissimilar age in parallel contact are disconformities searched for and sometimes located. A brief period of erosion within rocks of similar age may

leave very, very little evidence that it happened; such a brief interval of erosion is termed a *hiatus.*

Nonconformity

A **nonconformity** is an eroded surface separating two rock types that could not have *formed* in contact with one another. An example would be shale in contact with an underlying granite (Figure 17–16B).

Since granite *forms* tens of miles or kilometers underground, it could not have formed and cooled in natural contact with an ordinary shale. The shale could be baked by a granite intrusion, but it could not have been deposited as a *sediment* across a mass of granite *unless* that granite was first exposed by erosion, then submerged beneath the sea, and then buried beneath the mud that became shale.

THE ORDERING OF TIME

Fossils rapidly became the vocabulary for describing the relative age of rocks, and unconformities became the punctuation marks of this new geologic language. On an earth where there is nowhere a complete vertical column of all rocks of all ages, *fossils furnished the criterion* to link the scattered fragments of one region with the scattered fragments of another—to correlate them into a worldwide *sequence of events*.

The old "rock type" criterion of Werner, with its rigid subdivision of rocks into Primary, Secondary, Tertiary, and Quaternary, slowly fell by the wayside; on an ancient cyclic earth the same events could repeat themselves many times. Therefore, the mineral composition and texture of a rock *could not* specify its relative age.

By the early 1800s geologists knew that the earth was far older than arithmetic juggling of the genealogies within the Bible would allow. At that early time they still had no idea how old it really was, and they had no real framework for ordering the random events they saw into a sequential history. But they had discovered that contrary to the authority of the Bible, the earth's "creation" was continuous, not instantaneous.

By the 1850s geologists had organized historical events within that unlimited time by using the six physical stratigraphic principles (superposition, original horizontality, original continuity, crosscutting relations, inclusion, and metamorphism) to arrange events in correct sequence, and the three biologic stratigraphic principles of faunal assemblage, faunal succession, and correlation to assert that *the irreversible changes in the history of life forms yielded a series of faunal assemblages unique to brief intervals of time*. Using those unique assemblages, geologists could recognize contemporaneous events around the world and fit them into a universally agreed upon ordering system termed *the Geologic Time Scale* (Figure 17–17). It had taken geologists from 1800 to 1850 to establish the framework for sequential time; it took them another fifty years after that to find out how old the rocks really were.

SEQUENTIAL TIME AND THE GEOLOGIC TIME SCALE

Such worldwide recurring events as mountain building, encroachment of the sea, and abrupt changes in forms of life marked off "natural" units of geologic time, even though the *actual* dates of the events were unknown when the geologic time scale was first formed. The **geologic time scale** is a framework that allows the ordering of events by *sequence* or *relative age*.

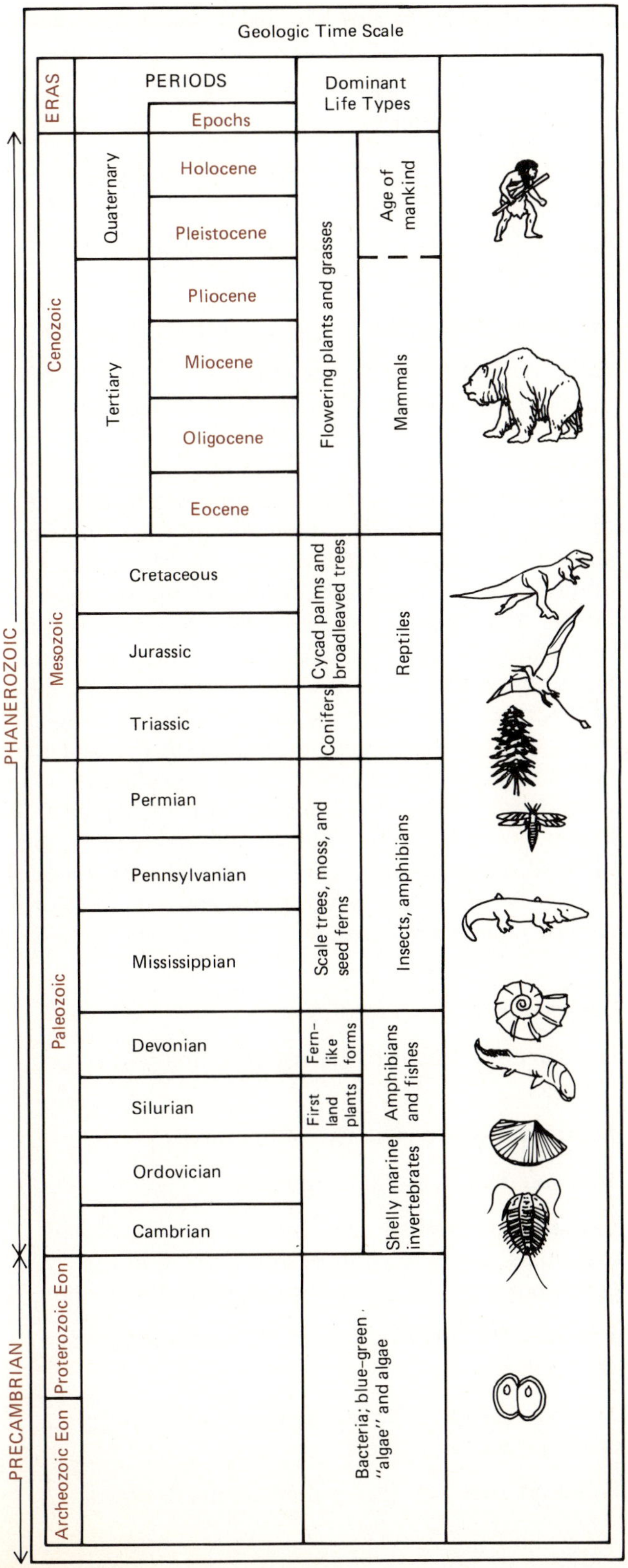

Geologic Time Scale

ERAS	PERIODS	Epochs	Dominant Life Types	
Cenozoic	Quaternary	Holocene	Flowering plants and grasses	Age of mankind
Cenozoic	Quaternary	Pleistocene	Flowering plants and grasses	Age of mankind
Cenozoic	Tertiary	Pliocene	Flowering plants and grasses	Mammals
Cenozoic	Tertiary	Miocene	Flowering plants and grasses	Mammals
Cenozoic	Tertiary	Oligocene	Flowering plants and grasses	Mammals
Cenozoic	Tertiary	Eocene	Flowering plants and grasses	Mammals
Mesozoic	Cretaceous		Cycad palms and broadleaved trees	Reptiles
Mesozoic	Jurassic		Cycad palms and broadleaved trees / Conifers	Reptiles
Mesozoic	Triassic		Conifers	Reptiles
Paleozoic	Permian		Scale trees, moss, and seed ferns	Insects, amphibians
Paleozoic	Pennsylvanian		Scale trees, moss, and seed ferns	Insects, amphibians
Paleozoic	Mississippian		Scale trees, moss, and seed ferns	Insects, amphibians
Paleozoic	Devonian		Fern-like forms	Amphibians and fishes
Paleozoic	Silurian		First land plants	Amphibians and fishes
Paleozoic	Ordovician			Shelly marine invertebrates
Paleozoic	Cambrian			Shelly marine invertebrates
Proterozoic Eon			Bacteria; blue-green "algae" and algae	
Archeozoic Eon			Bacteria; blue-green "algae" and algae	

PHANEROZOIC spans Cenozoic, Mesozoic, and Paleozoic. PRECAMBRIAN spans the Proterozoic Eon and Archeozoic Eon.

FIGURE 17–17

An outline of the geologic time scale. Although the separations among the various intervals of time were originally made on faunal assemblages, the position of many of the boundaries today is also limited by data on physical events.

We are used to sequential time. We talk about "before you were born" or "after the party," recognizing that events occur in a pattern or sequence. The geologic time scale, based primarily on the dominant life types of each time period, is a scale of **sequential time,** sometimes called *relative time.*

Units of Relative Time

All of human history is organized into relative time. We speak of human events as B.C. or A.D., relative to the birth of Christ. Geologists have done the same thing by dividing earth history into two major subdivisions: the **Precambrian,** which includes all time before the development of shelly or skeletonized life, and the **Phanerozoic,** including all time since the Precambrian, and separated from it by the development of skeletal or shell tissue by many life forms.

EONS

There is no universally recognized subdivision of Precambrian time, for the kinds of soft-bodied fossils that are found in the rocks of this interval do not lend themselves to precise subdivisions of time. Using a recent geologic time scale developed in North America, geologists have recognized two **eons**; the older is the **Archean Eon,** and the younger is the **Proterozoic Eon** (Figure 17–17).

ERAS

All of geologic time is further subdivided into **eras.** The eras within the Precambrian are separated from one another both by major events within the history of Precambrian life, major events within Precambrian time, and by the use of radiometric age-dating techniques, to be discussed in the next section of this chapter. Within the Phanerozoic, the eras are broad spans of time (Figure 17–17) having a dominant life type and separated from other eras by *major* events in the history of life. There are nine eras comprising all of earth history.

Phanerozoic Time

Cenozoic Era The term translates as "modern life"; the Era in which we live, dominated by mammals and flowering plants.

Mesozoic Era The Era of "middle life"; dominated by reptiles and broad-leaved trees.

Paleozoic Era The Era of "ancient life"; dominated by marine invertebrates, amphibia, and the first plant life on land.

Precambrian Time

Proterozoic Eon The time period of nucleated cells.

Late Proterozoic Era The Era in which complex soft-bodied, multicellular life developed.

Middle Proterozoic Era The Era of colonial algae; may include the development of nucleated cells.

Early Proterozoic Era The Era defined by the rise of nucleated, single cells.

Archean Eon The time period of nonnucleated cells.

Late Archean Era The Era of cyanobacteria or blue-green "algae".

Middle Archean Era The Era of primitive bacteria.

Early Archean Era Traces of fossils within this Era are doubtful; earliest development of solid rocks

PERIODS

The three eras of Phanerozoic time are further subdivided into **periods,** intervals of time that are shorter than eras and that are distinguished from one another by events in the history of life or physical events. The names of the periods are a curious mixture from the history of geology. The terms include Quaternary and Tertiary, "throwbacks" to the days of Werner. The other names are largely English, since most of the naming was done in the first half of the nineteenth century by English geologists.

In North America 12 periods are recognized; in Europe the Mississippian and Pennsylvanian Periods of North America are combined to make the *Carboniferous* Period. The periods recognized in North America are the following.

Cenozoic Era The Era of mammals.

Quaternary Period The rise of humankind; the Ice Age.

Tertiary Period The rise of mammals.

Mesozoic Era The Era of reptiles.

Cretaceous Period The rise of angiosperms and mammals, the extinction of dinosaurs, and the development of birds mark the end of this period.

Jurassic Period Expansion of the dinosaurs; age of cycads.

Triassic Period Rise of the dinosaurs; development of conifers.

Paleozoic Era The Era of marine invertebrates.

Permian Period The age of ammonoids; the rise of reptiles and conifers.

Pennsylvanian Period The age of cockroaches, horsetails, and scale trees.

Mississippian Period The age of crinoids and blastoids.

Devonian Period The age of fishes and primitive land plants; appearance of amphibians.

Silurian Period The age of eurypterids, earliest land plants, and air-breathing animals.

Ordovician Period The age of graptolites, earliest corals, and earliest vertebrates (armored fishes).

Cambrian Period The age of trilobites and archaeocyathids; development of shells and skeletons.

EPOCHS

The two periods within the Cenozoic Era have been subdivided into named epochs, quite brief periods of time. The original subdivision was determined by Charles Lyell, based on the percentage of modern molluscs found within each group of strata. His method, in hindsight, was marred by the fact that rates of evolution and extinction are both variable, but the usefulness of the subdivisions remains. The epochs are

Quaternary Period

Holocene Epoch The interglacial time in which we live.

Pleistocene Epoch Four periods of major glacial advance.

Tertiary Period

Pliocene Epoch

Miocene Epoch

Oligocene Epoch

Eocene Epoch

Paleocene Epoch

CHRONOLOGIC TIME

With time firmly tied to stratigraphy, and gaps in the chronology of strata in one locale filled by observing stratified rocks of the same age elsewhere, the out-

lines of earth history were drawn by the middle of the nineteenth century. It was clear to all that the deposition of the pile of sedimentary rock that represented even Phanerozoic time must have taken an enormous period to accumulate—but how long was enormous?

WHAT IS A CLOCK?

In order to *measure* time, rather than *order it in sequence,* one needs a device whose *rate* is known, and some way to measure the length of time that rate has continued. If we know the rate of the apparent passage of the sun across the sky, we can measure the changing position of its shadow and create a *sundial.* If we know the rate with which sand falls through a circular opening, we can create an egg timer or an *hourglass.* If we can cause gears, driven by a spring, to move at a constant rate, we can say that it has been 8 hours and 21 minutes since midnight, and we have a *clock,* a device that delineates **chronologic time,** a kind of sequential time which states the *precise interval since* some fixed event happened.

The ability to *measure the interval by counting the number of units set off by a device of constant rate* is what distinguishes chronologic time from sequential time. Thus to speak of yesterday is to speak in sequential time, to speak of 8:21 a.m. is to speak in chronologic time. But what kind of clock could be used to measure the interval since the earth began?

EARLY ATTEMPTS

Both the uniformitarianism of Charles Lyell (Figure 1–11) and the evolution by natural selection proposed by his friend Charles Darwin (1809–1882) (Figure 17–18) *required* that the earth be unimaginably old. Estimates ranged up to many hundreds of millions of years.

Thus, when the physicist William Thomson (1824–1907), later known as Lord Kelvin (Figure 17–19), announced in 1846 that his calculations of the length of time since the primitive molten earth first acquired a solid crust showed the earth to be 100 million years old, and during almost all that time too hot for life to have lived on its surface, geologists were outraged.

In much the same way that the English physicist Sir Harold Jeffreys 80 years later almost destroyed the concept of continental drift with mathematically elegant calculations based on faulty premises (see Chapter Ten), Kelvin, Irish by birth and Scottish in education and residence, severely crippled both Lyellian uniformity and Darwinian evolution with math-

FIGURE 17–18
Charles Darwin, exponent of the evolutionary approach to the origin of species. (Courtesy of the History of Science Collections, University of Oklahoma Libraries.)

FIGURE 17–19
William Thomson, better known as Lord Kelvin, as he appeared in 1897. Operating in the context of the physics of the day he severely limited the concepts of both uniformity and evolution. (Courtesy of the History of Science Collections, University of Oklahoma Libraries.)

ematically elegant calculations based on an erroneous understanding of how the earth released heat. Then as now, it is difficult for empirical constructs to argue with the elegance of mathematics.

As the years went by, continuing calculations by Lord Kelvin kept shortening the time interval, bringing total earth age down to 24 million years. The laws of physics bound geologic explanation as tightly as the scriptures ever had, and both evolution and uniformity seemed in a shambles by the end of the nineteenth century. Darwin came progressively to doubt his own theory. With a now severely restricted time scale, reluctantly agreed to by some geologists, there simply had not been enough time available for the processes of gradual change proposed by both Lyell and Darwin to mold life and landscapes as we see them.

In 1899 three events that were to have profound consequences occurred. John Joly, an Irish chemist, first presented an elaborate calculation of the age of the earth—based on the amount of salt in the world's ocean. The idea had been suggested by Edmund Halley (of Halley's comet fame) in 1693. Joly's *assumption* was that the first oceans were of distilled

water and that all the sodium in them had resulted from weathering of continental rocks in the interim. Joly calculated an earth age of 80 to 90 million years based on the *rate* at which sodium was added to the sea—Joly's "clock." The paper bristled with mathematics, but it rested on enormous uncertainties about the amount of sodium-rich salt that is naturally withdrawn from the ocean by evaporation and the formation of thick layers of rock salt. Like Kelvin, Joly's equations were based on false initial premises, and he soon abandoned using salt in favor of research with the newly discovered radioactivity.

In the same year, Lord Kelvin, by then Britain's leading physicist and a great popular hero for his work with the telegraph, once more presented his ideas on earth age deduced from his "clock," the cooling rate. American geologist Thomas C. Chamberlin (1843–1928), who believed that planets accreted in a cold state and were never molten, pointed out that all the long strings of numbers had "defects of premises that condition the whole process." He then proceeded to take apart, one by one, Kelvin's arguments concerning the cooling of a molten earth. Then he suggested an extremely early version of radioactivity as an energy source that might have operated for "an unknowable period."

Still in the same year, Marie Curie (1867–1934) was laboring to purify radium, one of two radioactive elements she had discovered. It was she who named the recently discovered process by which uranium gave off invisible rays 'radioactivity." She studied the radiations given off by uranium and concluded, along with Ernest Rutherford (1871–1937), that there were three different kinds of radioactive particles and rays (see Chapter Two)—alpha α, beta β, and gamma γ.

With the discovery of radioactivity, Kelvin's *assumption* of an initially molten earth with a fixed *rate* of cooling was destroyed, and the earth gained a "clock" whose intervals were fixed by newly discovered laws of physics. This new clock would not give the earth Hutton's *unlimited time,* but would eventually reveal it to be far older than Lyell and Darwin had dared imagine. A century after Hutton, the age of the earth was open to measurement.

PRINCIPLES OF RADIOACTIVITY

In order to understand how radioactivity provides both a clock and at the same time an internal "heat engine" that drives earth processes, we must first

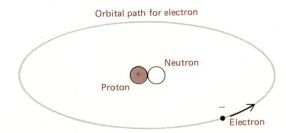

FIGURE 17–20

Schematic sketch of a deuterium atom, whose nucleus consists of an uncharged neutron (open circle) and a proton having a positive charge (Colored circle). The nucleus is orbited by a single electron that has a negative charge. Deuterium is an isotope of hydrogen with an atomic weight or mass number of two; the atomic number is one.

step aside to review some elementary chemistry. A somewhat more extensive review is offered in Appendix B. **Radioactivity** is the spontaneous disintegration of the atomic nucleus by emitting radiation in one of three forms. The mysterious particles and rays discovered by Madame Curie and British physicist Lord Rutherford make a logical place to begin.

Types of Radiation

The *gamma rays* are uncharged electromagnetic waves, identical to the X ray. They pass readily through most materials, having a penetrating power approximately 10,000 times that of alpha particles. The *alpha particle* is a positively charged atomic particle of mass number 4, identical to the nucleus of the helium atom (He^{+2}), consisting of two protons and two neutrons, written ^4H. Alpha (α) particles can pass through only a few inches or centimeters of air; their great size, mass, and positive charge make them easy to stop. The beta (β) particles are tiny negatively charged atomic particles called **electrons**. They are shot from a nucleus when a neutron is split into a proton and an electron. The penetration power of β-particles is approximately 100 times that of alpha particles.

The Atom

Atoms of all elements consist of a **nucleus,** which contains nearly all the mass of the element. The nucleus always includes **protons,** positively charged atomic particles, and commonly includes **neutrons,** electrically neutral particles (Figure 17–20). Surrounding the nucleus is a cloud of orbiting, negatively charged **electrons**.

The number of electrons must equal the number of protons in a stable atom to maintain electrical neutrality. The **atomic number** of an atom specifies *the number of protons or electrons within the atom.* The **atomic weight** or *mass* of an atom is the *average* weight of one atom of the element, usually expressed as its weight in relation to that of one atom of carbon, standardized at 12.

ELEMENTS

All atoms of the same element have the same atomic number. *The number of protons or electrons is what separates the elements from one another.* As an example, the atomic number of potassium is 19; that of calcium is 20. Potassium has one less proton in its nucleus and one less electron surrounding its nucleus than does calcium.

Within a single element the number of *neutrons* in

the nucleus may vary. This fact was recognized early in the twentieth century when precisely determined atomic weights turned out to be nonintegral numbers. The atomic weight of hydrogen is 1.00797, for example; it would be 1.00000 if *all* its atoms had only one proton and no neutrons.

ISOTOPES

In the most common form of hydrogen, a single proton forms the nucleus, which is orbited by a single electron; this form *would* have an atomic weight of 1.00000 when measured. *Adding a neutron* (Figure 17–20) to the nucleus causes the atomic weight to increase to 2, while the atomic number, the number of protons, *remains 1*.

A new form of hydrogen has been created, informally called "heavy hydrogen," and more properly termed *deuterium*. Deuterium is an **isotope,** *a neutron-number variant to the nucleus with a single proton number*. An isotope is one of two or more elements which are almost identical in chemical properties, but differ in atomic weight because the nucleus has a greater or lesser number of neutrons. Add still another neutron and an isotope of even heavier hydrogen is created, one called *tritium*. Tritium has an atomic weight of 3, although its atomic number remains at 1. The element hydrogen has three isotopes consisting of forms with 0, 1, and 2 neutrons added to the nucleus.

Hydrogen gas consists of ordinary hydrogen, written in chemical symbology as H^1, and much smaller amounts of its heavier isotopes, symbolized as H^2 and H^3. Averaging together the large amount of ordinary hydrogen and the tiny amounts of its heavier isotopes that are normally present yields an atomic weight of 1.00797—the average weight of the three isotopes of hydrogen when combined together into hydrogen gas.

Radioactive Decay

No element whose nucleus spontaneously disintegrates can remain as the same element. As it emits particles of itself, it must be *changed into* a different element. The process of change from one radioactive "parent" element to a different "daughter" element is called **radioactive decay.**

HEAT AND CLOCKWORDS

Radioactive decay is the process by which mass is converted to energy, a concept worked out by Einstein in 1905. Part of the energy released by radioactive decay is in the form of *heat*. It was Lord Rutherford who wrote the basic equations governing radioactive decay in 1904. At that time he calculated the amount

of heat released by the tiny amounts of radioactive elements present in most rocks and concluded that "the time during which the Earth has been at a temperature capable of supporting the presence of animal and vegetable life may be very much longer than the estimate made by Lord Kelvin from other data." Only a year later it was shown that the amount of helium produced by the radioactive decay of uranium in uranium minerals was far greater than could have been present had the decay only been going on for the few tens of millions of years allowed as the total age of the earth by Lord Kelvin.

Lord Kelvin, who warmly congratulated Mme Curie for her enormous contributions to the early study of radioactivity, remained unimpressed with it as a source of either heat or measurement of time. He bitterly opposed any concept that radioactive atoms were disintegrating or that the energy released came from within the atom. He died convinced that everyone else was all wrong.

HALF-LIFE

The atomic weight of one isotope of uranium is 238, written ^{238}U. The atomic weight of helium is 4, written ^4He. Uranium-238 decays through a long series of intermediate steps an isotope of lead that has an atomic weight of 206, written ^{206}Pb. Lead-206 can be formed *only* by the radioactive decay of uranium-238, and lead-206 is stable.

The series $^{238}U \rightarrow {}^{205}Pb$ occurs with the loss of both α- and β-particles. During the process of decay, eight nuclei of helium (β-particles), each having a mass number of 4, are emitted so that the following equation explains why ^{206}Pb is formed. $^{238}U - (8 \times 4) = {}^{206}Pb \ \Sigma \ 8{}^4He$. Uranium-238 thus changes into ^{206}Pb in eight steps; in each step four units of atomic weight (the helium particle) are lost: $238 - 32 = 206$.

Ordinary lead consists of a mixture of four isotopes, ^{204}Pb, ^{206}Pb, ^{207}Pb, ^{208}Pb, and has an atomic weight of 207.19. Lead-204 is not formed by radioactive decay; the other three stable isotopes of lead are formed from radioactive decay of different parents with ^{206}Pb forming from the decay of ^{238}U, ^{207}Pb from ^{235}U, and ^{208}Pb from ^{232}Th (thorium).

The time required for the original number of atoms of a radioactive parent to be reduced by one-half is called **half-life.** The half-life of ^{238}U is 4.51 billion years. After this period of time one-half of the nuclei of ^{238}U will have changed to ^{206}Pb (Figure 17–21), so that the ratio between the amount of ^{238}U and the amount of ^{206}Pb is 50:50.

The end of one half-life is the beginning of another, for the *rate* of radioactive decay is an *absolute constant*. After 9.02 billion years (Figure 17–21), *three-quarters* of the uranium nuclei will have decayed into lead; the ratio in abundance between ^{238}U and ^{206}Pb will be 25:75 after two half-lives have passed.

MINERAL CLOCKS

The combined creative efforts of American, British, French, German, and Polish physicists and chemists at the beginning of the twentieth century were responsible for the discoveries that revolutionized our understanding of the earth in the ways already described. We have learned that radioactivity is the source of the earth's internal heat energy, and that radioactive decay can serve as a natural clock.

The *constant rate of radioactive decay now provided the ideal "clock."* In 1907 an American chemist demonstrated that in minerals *of known sequential age, the uranium to lead ratio was progressively different. The older the mineral, the greater the lead-to-uranium ratio.* Radioactive decay, whose rate of change is unaffected by any event, became the ultimate tool for measuring the chronologic age of rocks.

In practice, minerals containing radioactive elements are carefully separated from igneous and metamorphic rocks. After ultra-precise chemical analysis of the abundance of isotopes, the ratio of the radioactive *"parent"* isotope to the stable *"daughter"* isotope is obtained. *This ratio is inversely proportional to the length of time radioactive decay has been occurring.*

Radiometric Dating

Radiometric dating, the process of finding chronologic time from analyzing parent:daughter isotopic ratios, operates from three basic assumptions.

1. The *number* of atoms, n, that decay in one year in relation to the total number of atoms, N, in a unit volume is invariant; the probability of any one atom decaying remains constant. The proportion $n \div N$ is called the **decay constant**, usually symbolized as lambda, λ. As with all attempts to keep chronologic time, a process operating at a constant rate is required to mark off the intervals since an event—in this case crystallization of the mineral. This assumption has been tested many times and so far has always been correct; it also follows from the laws of quantum chemistry.

2. *No atoms of the daughter isotope were present when the mineral formed,* or the amount originally present can be determined.

3. The *mineral or rock has remained a closed system since it was formed.* Nothing has leaked in or out. This third assumption may or may not be correct, but leakage can generally be both predicted and recognized. The leakage of either parent or daughter nuclei into or out of the system is one of the chief causes of inaccuracy in radiometric dates.

In order to cross-check the validity of a single radiometric date, accurate dating techniques normally require that *more than one parent:daughter isotopic pair from the same rock be analyzed,* allowing a comparison between the dates obtained. Fortunately, many radioactive isotopic pairs are known. A few of the parent:daughter pairs are shown in Table 17–1, along with the half-lives of the parent.

CHRONOLOGIC AGE DETERMINATION

A wide variety of techniques are used to provide radiometric or chronologic dates from materials as diverse as bones, hair, shells, cloth, whole rocks, feldspars, micas, amphiboles, zircons, and a range of radioactive minerals. Radiometric dates define the time at which the mineral that contained radioactive

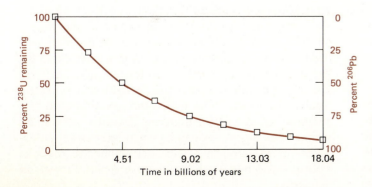

FIGURE 17–21

Decay curve for Uranium-238 to Lead-206. *All decay curves follow this general form.* In EACH successive half-life period, the number of nucleii of the parent atom—in this case Uranium-238—present at the beginning of that half-life is halved.

TABLE 17–1
Radioactive Isotope Pairs

Parent Isotope	Half-life in years	Daughter Isotope	Source of Material
Carbon-14	5730	Nitrogen-14	Bones, shells, wood, charcoal
Uranium-235	707 million	Lead-207	Zircons, uranium minerals
Potassium-40	1.27 billion	Argon-40	Whole volcanic rocks, micas, feldspars, and amphiboles
Rubidium-87	47 billion	Strontium-87	Whole rocks, feldspars, micas
Samarium-147	106 billion	Neodymium-143	Minerals containing rare earths

elements *formed or crystallized.* For this reason radiometric dating is restricted for the most part to young organic matter that once contained carbon-14 and to igneous and metamorphic rocks whose crystallization yields a mineral without any daughter products. *Crystallization resets the radiometric "clock" to zero.*

The dates available from radioactive parent-to-stable daughter ratios are the date of *crystallization* for igneous rocks and of last cooling for metamorphic rocks. Reconstitution of the contained minerals Sets or completely resets the radiometric clock. The new mineral begins with a limited amount of radioactive parent elements within it; the amount diminishes *geometrically* (see Figure 17–21) with time.

Note that the rate of decay is geometric, yet the probability of any one atom decaying remains constant. Therefore, the total number of decaying atoms is higher when the population of radioactive parent atoms is higher. Since the population of radioactive atoms decreases at a geometric rate, the rate of radioactive decay decreases geometrically.

Following is a very brief review of some of the common methods of radiometric dating.

Carbon-14

Carbon-14 (^{14}C) is continuously being created in our atmosphere by cosmic ray bombardment, which causes nitrogen-14 (^{14}N) to absorb a neutron, emit a proton, and thus change to radioactive ^{14}C. Once formed, ^{14}C is incorporated into all living substances. When the living animal or plant dies, the ^{14}C begins to change back to ^{14}N by β-particle emission. We would show this decay as $^{14}C \rightarrow {}^{14}N \Sigma \beta$

Since the half-life of carbon-14 is so short (see Table 17–1), very small samples or samples over 40,000 years old are difficult to date accurately. The chronologic age is determined by directly measuring the radioactivity in the sample, then converting the observed emission rate directly to age. A cyclotron, a circular accelerator for atomic particles, can measure directly the amount of remaining ^{14}C from quite small samples, yielding radiocarbon dates up to 100,000 years. Cyclotrons have also been used to assay accurately the amount of ^{10}Be (beryllium) whose

half-life is 1.5 million years, a half-life which nicely covers events within the last 20 to 30 million years.

Uranium–Lead

Two radioactive isotopes of uranium (^{235}U and ^{238}U) decay, with vastly different half-lives (see Table 17–1) to unique isotopes of lead. Thus, almost any mineral bearing uranium contains *both* radioactive isotopes and *both* stable lead isotopes. Now the mineral *zircon* is generally used; in its crystal structure, isotopes of uranium substitute for zirconium.

Typical radiometric dating techniques obtain the parent:daughter ratio on *both* sets of uranium–lead pairs. If the rock involved has had no leakage of material, both ratios should indicate the same radiometric age (Figure 17–22), and the two ages obtained are said to be *concordant.* The *pair* of ratios plots on Figure 17–22 at a point along the *concordia curve;* different points along this curve represent different ages and their corresponding $^{206}Pb \div {}^{238}U$ and $^{207}Pb \div {}^{235}U$ ratios.

The uranium–lead method is one of the oldest radiometric dating techniques; it is particularly suitable for older rocks that contain the mineral zircon or other uranium-bearing minerals. Young rocks contain little lead of radiogenic origin, so their analysis requires extraordinarily precise spectroscopic methods.

Potassium–Argon

Any argon-40 within minerals that contain potassium must result from the radioactive decay of potassium-40, since argon-40 can form in no other way.

With a half-life of more than 1.2 billion years (see Table 17–1), potassium–argon dating is suitable for the oldest rocks on earth—and for the moon as well. All that is required is that the gastight mineral being analyzed contain some potassium, an abundant element in many crustal minerals. Because very small amounts of argon gas can be easily measured, potassium–argon dating has also been applied to rocks as young as 50,000 years if they are especially rich in potassium.

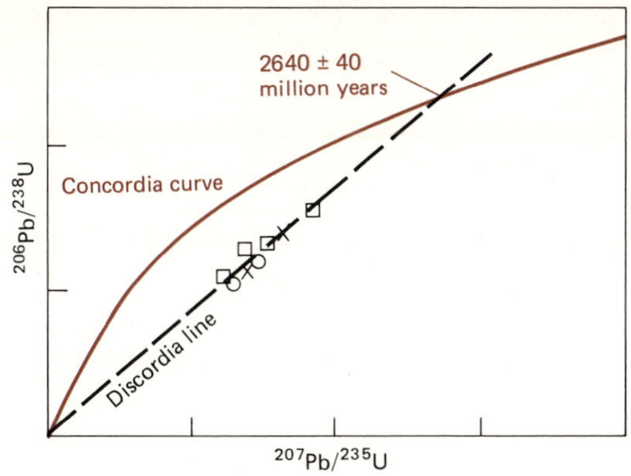

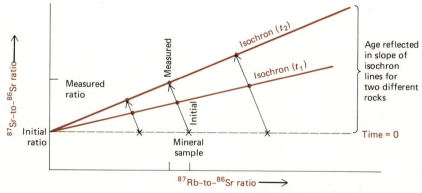

FIGURE 17–22

A typical concordia plot. The colored curve plots the changing ratios of uranium and lead isotopes through time. The concordia curve is plotted so that the amount of radiogenic lead increases upward for Lead-206 and to the right for Lead-207. The dashed or *discordia* line is drawn through the results of chemical analyses of *several different* uranium-bearing minerals from one deposit. The interception of the discordia line with the concordia curve yields the only age at which all of the different uranium minerals could have formed. This is the age of the deposit. (Example after K. R. Ludwig and J. S. Stuckless, 1978, *Contrib. Mineral. Petrol., 65,* pp. 243-254, with permission.)

Argon-40 is an inert (unreactive) gas, and it may diffuse from minerals when they have been heated to a high temperature. The loss of ^{40}Ar yields apparent radiometric ages that are too low. Unless it can be demonstrated that the rocks have never been intensely heated, potassium–argon radiometric ages are usually regarded as *minimum ages*.

Rubidium–Strontium

The rubidium–strontium method analyzes the ^{87}Rb, ^{87}Sr, and ^{86}Sr contents of a number of different minerals from the same rock. Strontium-86 is *not* formed by radioactive decay; it is *nonradiogenic*. At the time of crystallization of each mineral, the ratio of radiogenic ^{87}Sr to nonradiogenic ^{86}Sr was constant throughout the whole rock, because the isotopes were perfectly mixed in the magma as it rose.

With the passage of time, the ratio of radiogenic ^{87}Sr to its parent ^{87}Rb increases, since the radioactive rubidium is decaying to radiogenic strontium. The ratio for each mineral is *different, since each mineral started with a different amount of rubidium.* The ^{87}Rb ÷ ^{86}Sr to ^{87}Sr ÷ ^{86}Sr series of analyses of the different minerals in a rock, when plotted on a graph, define a line called the *ioschron*, meaning equal time; its slope defines the rock age (Figure 17–23). The isochron interception with the ^{87}Sr to ^{86}Sr axis defines the *initial ratio* of radiogenic to nonradiogenic strontium.

Dating Sedimentary Rocks

Conventional radiometric dates of the *minerals* in a sedimentary rock would simply yield the date when their *source* minerals first crystallized in an igneous or metamorphic rock; such a date would be useless, since the minerals in sedimentary rocks come from many sources, and we want the date *when the sedimentary rock formed.*

There is one mineral—*glauconite*—that forms *during* sedimentary consolidation and is rich in potas-

FIGURE 17–23

Rubidium-87 decays to strontium-87; both parent and daughter isotopes are shown here as a ratio to strontium-86, *not* produced by decay and hence constant. At the time of crystallization of a magma ($t=0$), the ratio of ^{87}Sr to ^{86}Sr in all the minerals was a constant value, which depended on the *source* of the rock. The plots of the isotope ratios of three minerals are shown; they establish a line whose *slope* is a function of the length of time the ^{87}Rb to ^{87}Sr decay has been going on. The interceptions of the sloping line with the ^{87}Sr to ^{86}Sr axis establishes the *initial ratio*, which indicates the probable rock source. At a still later time (t_2), the measured ratios will be still higher, since more time has passed and more ^{87}Sr will have been produced. The slope of of the line is greater the older the rock is.

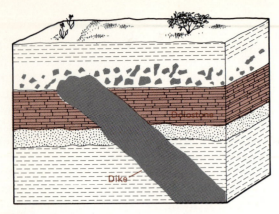

FIGURE 17–24

This hypothetical section illustrates one way in which igneous rocks can be used to provide an approximate date for a sedimentary rock. After the dike intruded the colored layer, it was then eroded (note fragments) as the upper layer of sedimentary rock was being deposited. If fossils in both layers indicated they were of similar age, and Rb-Sr dating of the dike yielded a date of 220 million years, the colored layer must be slightly older than 220 m.y., and the upper layer slightly younger than 220 m.y. Continued comparisons of the *same* layers elsewhere may allow their age to be ever more precisely bracketed. Once that occurs, then the fossil assemblages each layer contains will also have a chronologic time value.

sium. But glauconite commonly retains argon gas poorly, so it cannot usually be used for radiometric dating. Sedimentary rocks may, however, be *indirectly* dated by determining the date of the igneous rocks that intrude them, overlie them, or underlie them (Figure 17–24).

Fission Track Dating

When uranium-238 decays, the particles released in radioactive decay (*fission*) damage the crystalline or glassy material through which they pass. An example of this is *smoky quartz*, a mineral that owes its beautiful smoky brown color to radiation damage to its crystal lattice.

Examining a polished surface of a rock after it has been etched with acid reveals radiation damage, which is exposed as a series of pits across the mineral surface; these lines are called *fission tracks*. Using a microscope, we can count the number of tracks per unit of area; the track density is a direct function of the age of the material and the ^{238}U content, which must be measured.

The fission track method, unlike other radiometric methods, does not measure the *products* of decay—instead it measures the *effects* of radiation damage. Depending mostly on the amount of ^{238}U originally in the specimen, the ages obtainable from fission track dating range from a few years to many billions

of years. Materials that have been dated include crystals from many rocks, amber, ash, obsidian, and pottery.

OTHER TECHNIQUES

There are a wide variety of other chronologic dating techniques. None of them depends on radioactive decay. Instead each depends on observation or on unique *rate* processes. Some examples follow.

Tree Ring Dating

The science of tree ring dating, called *dendrochronology*, depends on an old idea—that of counting the number of rings in the trunk of a tree to learn the tree's age, and a new idea, that the variation in width between annual rings is a function of climate. Since climate changes from year to year, it "writes" on each tree a *unique* record for the life of the tree.

Comparing the annual patterns recorded in one long-lived tree with the corresponding patterns in another of a slightly different age gives zones of record overlap; the *total* record of climate back to nearly 8000 B.C. has been obtained. By checking the pattern of tree rings from any sufficiently large piece of a very old tree, dead tree, or even a burned dead trees against the "master record" extending back not quite 10,000 years, we can date when a tree was killed quite accurately. Dendrochronology has been especially useful in archaeology and in dating quite recent volcanic events that leave charred logs buried under ash or lava.

Amino Acid Racemization

It has recently been noted that after death some of the amino acids in an animal's bone *slowly* shift their chemical structure from one form to another *at a steady rate*. The form to which the amino acids shift is unknown in living things.

Since the *rate* of change appears to be constant, examining the extent of change allows a type of age dating called *amino acid racemization*, (pronounced race'-meh-zay-shun). Such dates have generally correlated well with carbon-14 dates on the same material. This method can use relatively small bone samples to yield dates within reasonable limits of error for materials formed within the last 100,000 years.

Obsidian Hydration

A new technique, useful for dating materials worked by human beings within the last few thousand years, is *obsidian hydration*. Obsidian is a volcanic glass

(Chapter Three) with pronounced *conchoidal fractures* (see Figure 2–15). Because it can be worked to a sharp edge by coalescing fractures, it was often used to form projectile points.

The technique depends on measuring the length of time since a fresh surface was cut on obsidian by noting the thickness of the *hydrated* film that slowly develops on the buried glass. The rate of hydration is proportional to the square root of time. The longer the surface has been exposed, the thicker the *hydration rind*, an altered zone where water has has worked its way inward from the surface.

THE MEASUREMENT OF THE GEOLOGIC TIME SCALE

With rocks divided largely on the basis of fossil content, the geologic time scale was sequential (Figure 17–17) but it still lacked any sense of the age difference between eras, periods, and epochs. The early attempts to estimate, using salt and the earth's outward heat flow, had suggested total earth age within a few tens of millions of years.

The rapid application of radiometric methods within the first half of this century brought the recognition that our earth was *billions of years old*. The earth's time scale had been shriveled by Genesis, expanded to infinity by Hutton, shrunk to unbelievably brief periods by Kelvin, and then expanded to *46 million centuries* (4.6 billion years), the currently accepted age for the probable solidification of planet Earth.

With the application of radiometric methods, the boundaries of each *era, period, and epoch* were established in chronologic time. The advent of paleomagnetism (Chapter Ten) made available an additional dating technique; suitable rocks formed within the last 180 million years may also be dated using this technique.

THE PRECAMBRIAN

Rocks formed within the Precambrian turned out to include *seven-eighths* of all of earth history. Beginning 4600 million years ago and ending 570 million years ago (Table 17–2), the events of this 4-billion-year interval are very briefly outlined in Chapter Eighteen.

Within this vast span of time, the sparse fossils are largely unicellular and generally unsuitable for correlation, although fossil algae have served for this purpose in some areas. More generally correlation

must depend on radiometric dates, since 87 percent of Precambrian rock is of igneous or metamorphic origin. There are no generally agreed on subdivisions for the Precambrian because there are no fossils suitable for correlation.

Life types include a wide range of bacteria and blue-green "algae"; both groups are composed of single cells lacking a nucleus. The development of nucleated cells, the oxygenation of the primitive atmosphere through photosynthesis, and the formation of the earth, followed by the origin and development of continents to two-thirds of their modern size, are the major events of Precambrian time.

PALEOZOIC ERA

The Paleozoic Era comprises almost 350 million years of earth history, less than a tenth of Precambrian time. The rich collection of well-skeletonized fossils in many Paleozoic rocks have enabled them to be subdivided into seven periods and many smaller subdivisions. The major physical event of this era is the collision of the Americas with all other continents to form Pangea. The formation of the Appalachian Mountains records this supercollision. The events of the Paleozoic Era are briefly outlined in Chapter Nineteen.

Life in early Paleozoic rocks is a rich treasure of *invertebrate* (lacking a backbone) marine animals plus a few vertebrate fish. In about the middle of the era, both plant and animal life began to colonize the continents for the first time. The evolution of all life for the first 3 billion years of earth history took place wholly within the oceans.

MESOZOIC ERA

The Mesozoic Era comprises approximately 160 million years. The dominant physical event that ended this era was the separation of Pangea into continental fragments, including the separation of the Americas from Eurasia and Africa across the widening Atlantic Ocean. The events of the Mesozoic Era are briefly outlined in Chapter Twenty.

The fossil life recorded in Mesozoic rocks is rich and varied, and it includes a wide range of reptilian fauna. Our imagination is always caught by the dinosaurs, who both flourished during this era and perished at the extreme end of it in one of the major episodes of extinction ever recorded on earth. The causes of this extinction event are still controversial. Some dinosaurs had taken to the air as flying dinosaurs, which evolved into birds. Dinosaurs also took to the sea, and on the land they developed into some of the largest creatures ever to live.

TABLE 17–2
Major Intervals of Geologic Time

Interval	Duration in Millions of Years	Major Events
Cenozoic Era	0–65	Rise of Rocky Mountains; dominance of mammals and flowering plants.
Mesozoic Era	65–225	Rise and decline of reptiles; dinosaur extinction; breakup of Pangea.
Paleozoic Era	225–570	Formation of Pangea and Appalachian Mountains; first skeletons, first continental animal and plant life.
Precambrian	570–4600	Formation of solid earth and primitive atmosphere, origin and development of unicellular life; atmospheric modification coincident with cell nucleation.

CENOZOIC ERA

The Cenozoic Era comprises the last 65 million years, slightly more than the last 1 percent of all time. The excellent preservation of the fossils and the preservation of physical events as landforms has given us a record so rich in detail that the Cenozoic record can be separated into fine subdivisions. The events of the Cenozoic Era are briefly outlined in Chapter Twenty-One.

The Cenozoic is the age of mammals, including a unique upright bipedal mammal called *Homo sapiens* whose presence has had enormous impact on the ecosystems of the earth. It is also a period of overall temperature decline and major continental glaciation on the Antarctic continent by the Miocene Epoch, and extending into central north America at the end of the Pliocene Epoch 1.8 million years ago. For North America the Cenozoic begins with the formation of the Rocky Mountains, which continue to form today.

Reviewed in only four chapters, the story of the earth may seem like an old-time movie. The actors will seem to move on and off their earthly stage at a dead run, while the sea roars in and out, and mountains pop up in the air like blisters on a cheap tire. The pace of change seems rapid only because of our perspective (Figure 17–25). If we were to live for 100 million years, the skidding of continent into continent, the rise and the fall of the sea, and the prosperity and death of the dinosaurs would hardly be noticed.

Even chronologic time, then, is relative.

SUMMARY

1. Our knowledge of the passage of time must have first depended on astronomic events. Our bodies are "set" to lunar time, but they are "reset" each day by the rising and setting of the sun to solar time. Reading from the authority of the scriptures, human beings and earth began about 6000 years ago.

2. Hutton recognized that there must be a heat engine within the earth providing an uplifting force to counter the more obvious forces that wore down the earth's surface. Once it was firmly established that the earth had a restoring force, then creation was regarded as a continuous, not an instantaneous event, and Hutton saw a world with no evidence of a beginning or an ending.

3. Six physical stratigraphic principles have served to order events in time, and three biologic stratigraphic principles have served to allow the ordering of worldwide events into a common framework called the *geologic time scale*. The development of this scale depended on the development of means to *order* events in *sequential time* by faunal succession.

4. The search for processes of uniform *rate* that would allow the *measurement* of time ended in the early part of this century with the development of *radiometric time*. This time depends on the decay constant exhibited by a wide range of radioactive elements, whose parent elements decay at a geometric rate through half-life intervals to stable daughter elements. The radioactive age is given by the parent:daughter isotopic ratio.

5. The subdivision of geologic time into Precambrian and Phanerozoic intervals, which were further subdivided into progressively smaller intervals called *eons, eras, periods, and epochs*, was

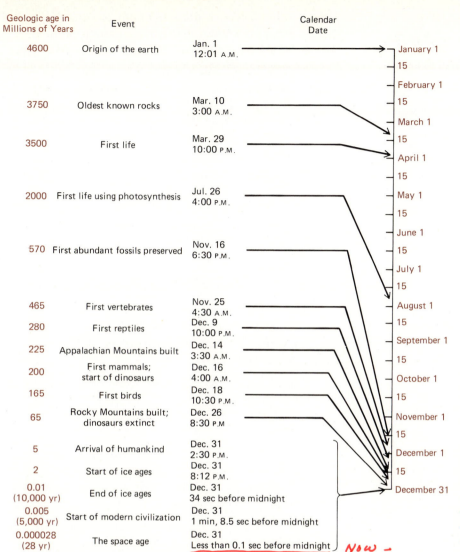

Geologic age in Millions of Years	Event		Calendar Date
4600	Origin of the earth	Jan. 1 12:01 A.M.	January 1
			15
			February 1
			15
3750	Oldest known rocks	Mar. 10 3:00 A.M.	March 1
			15
3500	First life	Mar. 29 10:00 P.M.	April 1
			15
2000	First life using photosynthesis	Jul. 26 4:00 P.M.	May 1
			15
			June 1
570	First abundant fossils preserved	Nov. 16 6:30 P.M.	15
			July 1
			15
465	First vertebrates	Nov. 25 4:30 A.M.	August 1
280	First reptiles	Dec. 9 10:00 P.M.	15
225	Appalachian Mountains built	Dec. 14 3:30 A.M.	September 1
200	First mammals; start of dinosaurs	Dec. 16 4:00 A.M.	15
165	First birds	Dec. 18 10:30 P.M.	October 1
65	Rocky Mountains built; dinosaurs extinct	Dec. 26 8:30 P.M	15
			November 1
			15
5	Arrival of humankind	Dec. 31 2:30 P.M.	December 1
2	Start of ice ages	Dec. 31 8:12 P.M.	15
0.01 (10,000 yr)	End of ice ages	Dec. 31 34 sec before midnight	December 31
0.005 (5,000 yr)	Start of modern civilization	Dec. 31 1 min, 8.5 sec before midnight	
0.000028 (28 yr)	The space age	Dec. 31 Less than 0.1 sec before midnight	

Now →

FIGURE 17–25

Imagine that all of earth history was could be compressed into the events within a single year. Humans would walk upright on the afternoon of New Year's Eve and begin using stone tools after dusk, while our life would occupy less than the last tenth of a second of the year.

originally accomplished on the basis of recognized changes in the fossil assemblages in the strata of each sequence. Later recognition of the role of physical events and dating of each break in time by radiometric means much improved the usefulness of the geologic time scale.

6. Our earth is 46 million centuries old, its internal energy is fueled by radioactive decay, which provides both the energy to make changes and the clock to measure the intervals between these changes. The recognition of immensity of geologic time is the greatest intellectual gift offered by the study of geology.

KEY WORDS

angular unconformity

Archean Eon

atomic number

atomic weight

Cambrian Period

Cenozoic Era

chronologic time

correlation, principle of

Cretaceous Period

crosscutting relation, principle of

decay constant

Devonian Period

disconformity

electron

eon

era

fauna

faunal assemblage, principle of

faunal succession, principle of

geologic time scale

half-life

inclusion, principle of

isotope

Jurassic Period

Mesozoic Era

metamorphism, principle of

Mississippian Period

neutron

nonconformity	proton
nucleus	Quaternary Period
Ordovician Period	radioactive decay
original continuity, principle of	radioactivity
	radiometric dating
original horizontality, principle of	sequential time
Paleozoic Era	Silurian Period
Period	strata
Pennsylvanian Period	superposition, principle of
	Tertiary Period
Permian Period	Triassic Period
Phanerozoic	time
Precambrian	unconformity
Proterozoic eon	uniformity, principle of
	xenolith

EXERCISES

1. Certain types of fossils are useless for correlation because they are restricted to a particular rock type rather than a specific time interval. Why would a particular group of life-forms be unique to a specific sedimentary type of rock?

2. Suppose you found in the same rock a mixture of marine and continental fossils. What would you conclude?

3. If the mythical element *kryptonite* had a half-life of 10 minutes, of the 10,000 nuclei of kryptonite originally present, how many would remain at the end of an hour?

4. Why would very, very small fossils be of particular value in correlation of rocks at depth that were penetrated by two oil wells?

5. Suppose that in a rock of unknown age the ratio of ^{238}U to ^{206}Pb is 75:25. How old is the rock in millions of years? In what era would this rock be placed? What kinds of fossilized life might be expected to be present in the rock?

6. Scientists believe that the length of the day is increasing by 0.002 second every century (2 seconds every 100,000 years) as the earth's rotational speed is slowed by tidal friction and other forces. If the rate of day lengthening has been constant, what was the length of the day when the earth formed 4600 million years ago?

7. In a remarkable recent study, a man blind since birth was shown to have a sleeping–waking cycle of 24.9 hours, the length of the lunar day. His sleep onset was "remarkably coincident . . . with a local low tide." What would you make of this?

8. Uranium-235 is now 0.72 percent of natural uranium. How could we compute the percentage of natural uranium at the time of earth formation? (It can be done, and you may want to do it. The answer is 17 percent.)

SUGGESTED READINGS

BERRY, W. B. N., 1968, *Growth of a Prehistoric Time Scale,* San Francisco, W. H. Freeman, 158 pp.
▶ *Excellent review of the history of the development of the geologic time scale.*

CALDER, N., 1983, *Timescale,* New York, The Viking Press, 288 pp.
▶ *Popularized science at its best; all will learn from this approach.*

GOULD, S. J., 1987, *Time's Arrow; Time's Cycle,* Cambridge, Mass. Harvard University Press, 221 pp.
▶ *Wonderful scholarship on our conflicting views toward time; engaging reading.*

GOULD, S. J., 1983, False premise, good science, *Natural History,* October, pp. 20–26.
▶ *America's most literate geologist describes Kelvin's attempt to destroy an entire profession not his own with a flawed understanding (in hindsight!) of the earth's heat source. See also his paper on tidal braking of earth's rotation in the April 1979 issue of the same magazine.*

HITCH, CHARLES J., 1982, Dendrochronology and serendipity, *American Scientist,* v. 70, pp. 300–305.
▶ *The story of one man's determination to make tree-ring dating work.*

MORRIS, RICHARD, 1985, *Time's Arrows,* New York, Simon & Schuster, 240 pp.
▶ *An intriguing look at scientific attitudes toward time. Wetherill, George W., 1982, Dating Very Old Objects, Natural History, September, pp. 14–20*
▶ *A lucid discussion of the principles and techniques of radiometric dating.*

Much about geology has changed since 1890 when this photograph was taken of geologists at work in Yellowstone National Park. The discoveries outlined in this chapter have depended on some of the most sophisticated scientific equipment available. They have also depended on mules, backpacks, and boot leather—commodities that are still much used by geologists. (Photograph courtesy of the U.S. Geological Survey.)

CHAPTER *eighteen*

The Precambrian

Repetition is the only form of permanence that nature can achieve.
                                    GEORGE SANTAYANA (1863–1952)

Life is an offensive, directed against the repetitious mechanisms of the universe.
                                    ALFRED NORTH WHITEHEAD (1861–1947)

Precambrian time spans about 4 billion years, or seven-eighths of all earth history. The restless recycling of rocks that characterizes planet Earth has largely converted Precambrian sedimentary rocks to igneous and metamorphic rocks; in fact, 87 percent of all Precambrian rocks are igneous and metamorphic. Because there is no agreed on subdivision of Precambrian time, this chapter separates this interval of time only into the Archean and Proterozoic eons. At this level of review, finer subdivisions that have been proposed are not useful.

The crystalline Precambrian rocks furnish us with numerous opportunities for radiometric dating; they also help establish a Precambrian chronologic time scale (see Figure 18–1). These ancient crystalline

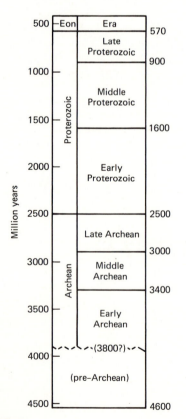

FIGURE 18–1
Subdivisions of Precambrian time, as commonly used in the United States.

rocks contain most of the earth's mineral wealth, including almost all its gold, silver, chromium, and iron.

Precambrian sedimentary rocks are uncommon, but when found and studied carefully, they yield information about the accumulation of the earth's primitive atmosphere and ocean. These old rocks also yield evidence of a resource that is, as far as it is now known, unique to planet Earth—life. The history of life began in the early reaches of Precambrian time.

THE OLDEST EVENTS

Much as human history merges into prehistory and anthropology, geologic studies of the oldest events merge into *astrophysics*, the science that deals with the origin of stars and planets. Both astrophysical and geologic knowledge set limits to what are reasonable models for the origin of earth.

Any model for the origin of the earth must be able to form an earth that is at the correct distance from the sun and neighboring planets, and it must provide for earth's orbital path, its overall density, its diminishing spin rate, and its spin direction. The earth, once formed, must have an overall chemical composition like that of modern earth and must provide a process that will separate the earth into shells of inwardly increasing density. The chemical distinctiveness of the earth must also be accounted for—any model of the earth must answer questions about why only the earth is left with abundant water and an oxygen-rich atmosphere. Any model for the origin of the earth must also provide the correct amount of parent radioactive elements to create the earth's internal heat engine, which drives the earth's continuing evolution.

This chapter presents one particular model for the origin and differentiation of the earth into its modern state. This model satisfies the requirements just stated, but it is not the *only* model that does. The model presented is simply a reasonable model for how the earth *might* have formed and changed. Dealing with the available information about the earliest earth, we can easily separate even the oldest events into two stages.

1. A *primitive earth*, formed by the collision of particles which accreted as a solid, cool planet of homogeneous chemistry. At this stage earth had a thin, primitive atmosphere which lacked oxygen. From 4.6 billion years ago, often written as 4.6 Ga (from the Latin *giga anna*, meaning

billion years [ago]), to sometime before 3.8 Ga, events left a record on other planets and moons, but no terrestrial record exists because the earth is so dynamic that it has totally modified its earliest rocks. This interval is called the *pre-Archean* or *Hadean*. The name Hadean, from the Greek meaning "beneath the earth," is given to this period of earth's history for which there is no known *rock* record. It was followed by

2. An *early or Archean earth*, produced when rapid heat buildup precipitated a differentiation of the earth into core, mantle, and crust. This event was followed by an enormous outpouring of volatile materials from within the now-molten earth, which strongly modified the primitive atmosphere and produced the earth's ocean. These events left a partial record on earth reflected in Archean rocks; Archean is from a Greek word meaning "ancient."

THE PRIMITIVE PRE-ARCHEAN EARTH

It is certain that the earth as it initially formed was never molten throughout; if it had been, it would have lost water locked up in the crystal lattices of its minerals, and with it other elements like nitrogen and carbon. Fortunately for the upright human creatures who were to evolve 46 million centuries later, water was retained within the minerals of the primitive earth. Inert gases—those that could not chemically combine into minerals—almost entirely escaped from the primitive earth. Their abundance on earth is a tiny fraction of their abundance in the total universe.

Most models of the earliest earth suggest that it formed by the accretion of planetary particles into a solid, cool, homogeneous planet. The average composition of the material of primitive earth was rich in oxides and silicates of iron and magnesium—much like the overall composition of stony and chondritic meteorites.

About 80 percent of all meteorites that fall to earth are chondritic. Broadly, chondritic meteorites are composed of iron and magnesium silicates and iron-nickel alloys; in form, they are rounded aggregations of olivine embedded in a fine-grained matrix of iron–nickel alloys, pyroxenes, and olivine. Chondrites, like all meteorites, appear to be fragments ejected from extremely primitive planets during planetary bombardment as well as fragments formed when primitive bodies exploded or collided during an early phase in the formation of our solar system. As such, they are samples of what primitive planetary matter may have looked like, and it is from their accreted mass that the Hadean earth was formed.

Differentiation of the Primitive Earth

As the earth grew in size, attracting planet-forming chondritic particles by its gravitational pull, the processes of *collision, gravitational contraction, and radioactive decay* caused planetary temperatures to rise. How could the growing amount of heat be released?

Solids can transmit heat only by **conduction,** the transfer of heat from a hotter to a cooler solid. Rocky materials are very poor conductors of heat, which is why many building materials like gypsum, brick, asbestos, mica, and stone are used for insulation. As heat was trapped within the earth by its insulating properties, temperatures began to rise. More heat was being produced than could flow away by conduction. As temperatures rose, the earth, in part, began to melt.

A

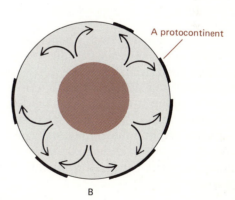

A protocontinent

B

FIGURE 18–2

(A) The primitive pre-Archean or Hadean earth is heated above the melting point of iron by *conduction*. The molten iron begins falling inward, displacing less dense material which moves upward. (B) After the melting and density differentiation event is complete, the earth has a core and a viscous, hot mantle which releases heat to the surface by *convection*. A few continental granitic masses are shown on the surface as dark lines. The floating continents are driven by mantle convection in a form of plate tectonics; the Archean Eon has begun.

Among the materials with a low melting point is metallic, elemental iron, and it undoubtedly melted early on. Being dense, the iron as it melted began to sink toward the earth's interior to form its core (Figure 18–2). As the molten iron sank, large amounts of gravitational energy were released, heating the earth and melting it still further, forming the mantle and core.

THE EARLY EARTH

The melting of the earliest earth destroyed its primitive homogeneous state. It had become a *differentiated* planet, one whose components had separated into concentric shells, the core and the mantle, of decreasing density outward. As a result of differentiation, the earth began to release heat by convection.

Planet Earth had acquired its dynamic nature. It had gained the power to continuously change itself, using the heat brought to its surface by rising plumes of hot, low-density, fluids coming from the mantle, which, in turn, is heated by the convecting outer core (see Chapter Nine). **Convection**, a heat transfer process occurring only in fluids, transfers heat from less dense, hot, rising fluids to denser, colder, sinking fluids.

Convection is a far more effective mechanism for heat transfer than conduction, so that temperatures within the earth no longer continued to rise. As discussed in Chapter Ten, we believe that it is the continuing convection in the mantle that drives both the horizontal and vertical motion of plates and cools the earth.

In a largely molten earth, materials arranged themselves in zones of graduated density, the heavy nickel – iron materials sinking to the central core, magnesium and iron silicates rising to form the mantle, and trace amounts of aluminum, leftover silicon, sodium, potassium, calcium, and magnesium forming fragments of primitive continental crust (Figure 18–2) on the surface of the rapidly spinning sphere. Still lighter materials vaporized to modify the primitive atmosphere.

ORIGIN AND MODIFICATION OF THE PRIMITIVE ATMOSPHERE

We have briefly mentioned that the earliest atmosphere was probably thin, and totally lacking in oxygen. It is probable that it was dominated by outgassed methane and ammonia, somewhat similar to the atmosphere of Jupiter, the only planet sufficiently massive to retain its original atmosphere. Theory suggests that the intense volcanism that must have been common on the early earth catastrophically modified the earth's earliest atmosphere, adding abundant carbon dioxide and water. Such an atmosphere would have been mildly reducing, *lacking oxygen*.

Radiometric dates from meteorites, thought to be impact-ejected fragments of a primitive, perhaps exploded planet, cluster around 4.6 Ga , a date consistent with the oldest rocks returned from our moon. *This date marks the origin of the primitive earth.* The current abundance ratios of lead-206 and lead-207 are also consistent with *the earth's becoming a solid by accretion 4.6 Ga.* Although *no rocks of this age have yet been found on earth,* it is probable that the first continental rocks (Figure 18–1) interacted with the earth's modified atmosphere.

The oldest rock formations are in southwestern Greenland, near Godthaab. Termed the Isua Series, these are largely marine sedimentary rocks, strongly metamorphosed by burial to pressures typical of 25 to 40 miles (15 to 25 km) depth at temperatures of 1020° Fahrenheit (550° Celsius). Analyses of radiogenic pairs date their recrystallization near 3.8 Ga. Both uranium – lead and samarium – neodymium methods yield comparable dates. *This date marks the beginning of the early Archean earth, one which furnishes a direct geologic record.* Still older *minerals* are known; detrital grains of zircon separated from rocks in the Jack Hills area of western Australia yield uranium – lead ages near 4200 million years ago, or 4200 Ma (from the Latin *mega anna* meaning million years).

The most ancient Isua rocks are dark metavolcanics, sandstones, and chemical metasediments, including small exposures of a very interesting rock called *Banded Iron Formation*, which will be discussed later. The chemistry of these rocks is consistent with an atmosphere composed of molecular nitrogen, abundant carbon dioxide, water vapor, and small amounts of oxygen. As we have noted, the same melting event that gave the earth its density stratification and its convective regime also outgassed enormous quantities carbon dioxide and water vapor, modifying the primitive highly reduced atmosphere into a much less reduced atmosphere, but one still lacking in free oxygen.

An equilibrium had developed between the melted interior and the developing atmosphere, but the volatiles outgassed from the earth were largely trapped—except for the lightest gases, hydrogen and helium—by the earth's gravitational field and retained. Thus the earth acquired both a modified atmosphere and its oceans. Its atmosphere differed from the present atmosphere largely in being oxygen-poor, containing less than 1 percent oxygen.

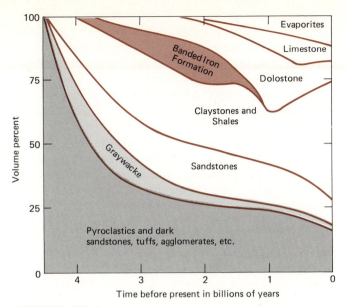

FIGURE 18–3

Relative proportions of the major sedimentary rock groups through time. (Adapted from R. B. Hargraves, 1976, *Science, 193*, pp. 363–371, copyright © 1976 by the American Association for the Advancement of Science, which was adapted with permission from F. T. MacKenzie, 1975, in J. P. Riley and J. Skirrow (eds.), *Chemical Oceanography*, Vol. 1, p. 309, copyright 1975, by Academic Press, which was derived from data from Alexander B. Ronov, 1964, *Geokimiya, 8*, p. 715. Published with permission from all cited sources.)

Evidence for an early oxygen-deficient atmosphere comes largely from two distinctive characteristics of the Archean weathering environment.

1. Iron was transported to the oceans in its soluble form as part of the dissolved load of streams. In the presence of oxygen, soluble iron compounds rapidly oxidize into insoluble iron compounds. Therefore the soluble iron in streams must have been in contact with an oxygen-deficient atmosphere.

2. The mineral *uraninite*, essentially uranium dioxide, was deposited in some early Proterozoic rocks. It, too, cannot survive in a weathering environment when oxygen is abundant.

ORIGIN OF ARCHEAN OCEAN

Nearly three-quarters of the earth's present surface is water, home to all life for the first 3 billion years or more of earth history. As we have indicated, this immense volume of water was probably formed during the catastrophic loss of volatiles from within the earth that so profoundly modified the atmosphere.

We know that more than three-quarters of the gas produced by modern volcanoes is steam, and that volcanoes are today the places where the earth vents its internal volatiles. Again, it is logical to assume that the early earth was abundantly volcanic. Evidence indicates that the early earth was hot and losing a very large amount of heat, and it is very clear that the oldest rocks were largely dark rocks of volcanic origin (Figure 18–3).

It appears that the earliest ocean had a composition similar to that of the modern ocean, though it may well have had more magnesium and less sodium. In rocks 2 billions years old there are casts of gypsum and halite crystals, minerals found in abundance dissolved in today's oceans. The sequence of minerals within the oldest evaporite deposits is also consistent with an ocean similar in composition to the modern ocean. Limestones were forming by 2 Ga (Figure 18–3), preceded by more than a billion years of their more magnesia-rich variety, *dolostones* (see Chapter Five).

A significant difference between modern carbonate sediments and Precambrian carbonates is that Precambrian limestones have consistently higher dolomite to calcite ratios, reflecting the much higher magnesium content of the Precambrian ocean. There are several possible explanations for this, all having to do with an important fact of the entire Precambrian. *Following differentiation, the earth lost heat at rates perhaps three to five times that of the modern earth.* What has been the thermal history of planet Earth?

PRECAMBRIAN TEMPERATURE

All lines of evidence point to the same conclusion: *the earth has been cooling since the early Archean.* All the kinetic energy acquired during the earliest phase of bombardment by particles, all the gravitational energy released by the settling of denser material to the core, and all the heat produced from the radioactive decay of the much larger initial quantities of uranium-235, uranium-238, thorium-232, and potassium-40 would have given the early Archean earth a much higher temperature than it has today (Figure 18–4).

How ironic that Kelvin was partially correct (see Chapter Seventeen). Our earth is cooling, though never from as hot a temperature as he had predicted; yet it cools by convection more rapidly than the conductive, solid earth he had assumed. Still, it is far more efficient at maintaining a high near-surface temperature gradient than he recognized, since it continues to generate its own radioactive heat.

Heat coming to the earth's surface today exceeds

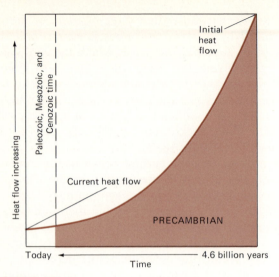

FIGURE 18–4

This curve gives the change in heat flow through time. The earth's heat sources were the energy from meteoritic bombardment in pre-Archean time, gravitational contraction, and radioactive decay. At the beginning of the Archean these combined sources must have produced about five times the modern heat flow.

that produced solely by radioactive decay, which may account for more than half of the current heat loss. This implies that the earth is losing not only its radioactive heat but also the energy stored in the earth by its formative processes. It appears that planet Earth has had its temperature peak and is now slowly cooling down.

Several lines of evidence point in this direction. The simplest one is direct chemical computation. If we take the amounts of daughter elements within the earth today and determine when their parent radioactive atoms were first sorted into crystals, it becomes evident that the early earth had very large amounts of radioactive elements. The heat output from those large volumes of intensely radioactive elements should have been quite large.

Another line of evidence is more indirect, but intriguing. A type of lava called **komatiite** (pronounced koh-mat'-ee-ite) is known from Archean basaltic lavas. With one exception, it has not erupted since Archean time. Komatiite lavas contain up to 30 percent magnesium, a figure many times that of ordinary basaltic lavas, and they appear, on the basis of recent laboratory experiments, to form only at depths of up to 125 miles (200 kilometers), well within the mantle, at temperatures up to 2000° C and under pressures up to 2 million pounds per square inch (14 billion pascals).

The temperature necessary to generate komatiite

magma at this depth is 500° hotter than temperatures of lavas observed today. Since pressure increases with depth in the earth, and melting temperature increases with pressure, the Archean generation of komatiite lavas is strong evidence that melting in the Archean took place at *much* greater depths than it does today.

There have been few komatiite eruptions since the Archean because the depth at which melting takes place within our cooling earth *has progressively moved toward the earth's surface*. When at some time in the distant future the depth of melting reaches the earth's surface, all igneous activity and plate tectonics will stop, and the earth will become dynamically as dead as our moon. Our earth is, however, alive with energy, and with one other resource singularly confined to earth—life.

PRECURSORS TO LIFE

The first evidence of life is found within Precambrian rocks. Even the oldest of all rocks—those from southwestern Greenland—yield microscopic objects that have been interpreted by some as bacteria or at least "fossil-like." Analysis of the isotopes of carbon within graphite from the Isua rocks has also suggested that it was of organic origin. From slightly younger rocks there is still more evidence of bacterial life. At some time in Archean history, the evidence for simple unicellular fossil life becomes indisputable. The only argument is exactly when.

Given the undoubted existence of unicellular life in Archean oceans, a difficult question remains. How did it get there?

There are various answers to that question. A sampling of proposals for the origin of are listed. Each has its own adherents. As we search for an understanding for what processes led from nonlife to life, questions of meaning are immediately entangled.

▶ *Divine creation*. Calling on a miraculous, divine creation lies outside the bounds of science, since it is untestable. Genesis provides two conflicting outlines of creation. Creation stories are common to all cultures.

▶ *Meteoritic infall*. Testing of meteorites has revealed all five of the critical amino acids in place. Perhaps life came to earth from elsewhere in space and landed in a fertile environment. This proposal transfers the problem of creation elsewhere, but does not solve it.

▶ *Clay catalyst*. Perhaps the structures of clays provided the foundation plan and chemical "push" for the organization of natural amino acids into replicating molecules. Numerous experiments are highly suggestive.

▶ *Atmospheric lightning*. Numerous laboratory experiments demonstrate that lightning striking into a primitive atmosphere bathed in ultraviolet rays produces simple amino acids, precursors to life. The earliest Archean environments would have provided all the necessary conditions.

▶ *Deep-sea vents*. The combination of hot water, sulfur compounds, metals, and clays found in modern ocean floor vents makes them an ideal place for the formation and modification of the amino acids essential to life.

The jump from amino acid molecules to ever more complex organic molecular chains able to reproduce themselves is an *enormous* jump. No one has yet demonstrated what the intervening stages between complex organic proteins and living, replicating proteins would be like. It remains a formidable piece of biochemistry and the subject of continuing research.

Although scientists cannot be certain yet how life was first formed, they can say what the first life had to be like in order to survive on the early earth. It had to be quite *small* so that it exposed a maximum of its surface area to the sea to maintain its internal temperature. It had to be *anaerobic*, capable of living without oxygen. It had to be *heterotrophic*, deriving its nutrition from the consumption of other amino acids, just as we do today. It had to *live in the oceans*, because no living matter could have survived on the continents, which were bathed in intense ultraviolet radiation. The chemistry of life is still essentially the chemistry of seawater plus carbon.

▼ PRECAMBRIAN LIFE
▲

The earliest Precambrian life-forms are **prokaryotes** (pronounced pro-carry'-oats), a group of single-celled microscopic organisms distinguished *from all other forms of life* by the absence of a cell nucleus (Figure 18–5). The word *prokaryote* literally translates as "prenuclear." Most biologists consider the split between prokaryotes and *all* other life-forms, whose cells have nuclei, as *as the fundamental subdivision of life*. This subdivision is even more significant than the older distinction between animal or plant. Prokaryotes form the kingdom *Monera* in biologic classifications (Appendix F).

THE PROKARYOTES

The prokaryotes include all bacteria and a specialized life-form known either as **blue-green algae** or as **cyanobacteria.** Prokaryotes are so unlike any other living thing that at their simple level of cellular organization they are neither plants nor animals—or—they are both. They are among the simplest life-forms known on earth, and they survive to this day.

We know bacteria primarily because some of them make us sick. Blue-green algae are the greenish scum on ponds. All blue-green algae are not blue-green; a reddish version gives the waters of the Red Sea its name. Blue-green algae are not properly algae, either—they are more correctly classified as *cyanobacteria*. Both terms are used interchangeably within this text, since this is current practice.

Cyanobacteria are an extremely diverse and sturdy group. One large group lives in hot springs whose temperatures may approach boiling (Figure 18–6). Others enter into a mutual arrangement with fungi and form the plant group we call *lichens*.

Prokaryotic cells have their genetic material largely diffused through their cells, and they are sexless. Prokaryotes lack chromosomes, major genetic bodies seen in all other cells; chlorophyll, if present, is diffused throughout the cell. All prokaryotes reproduce by *binary fission* (Figure 18–5). The genetic cell material—one chromosome—is divided equally, and each new daughter cell is *an exact copy* of the original. Reproduction is totally asexual since there is no mingling of genetic material from two different parents. Because offspring are exact copies or clones of the single parent cell, little change from generation to generation is possible.

All cycanobacteria and many other modern kinds of bacteria are **photosynthetic,** a name given to organisms that can provide their own food by the re-

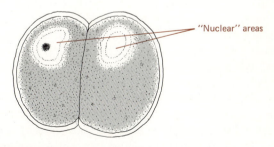

FIGURE 18–5
Sketch of idealized prokaryotic cell undergoing binary fission. The lighter-colored portions of each cell are the very poorly defined nuclear areas devoid of a cell wall. In fission the nuclear material is split equally between the two offspring (compare with Figure 18–9.)

FIGURE 18–6

Calcareous algae floating in a hot spring in Yellowstone National Park. Such algae reveal the enormous adaptability of cyanobacteria or blue-green algae. Their survival in an early earth was assured by sunlight but threatened by the waste gas they gave off—oxygen. (Photograph courtesy of the U.S. Geological Survey.)

action of sun with the chlorophyll in their cells as they take in carbon dioxide and give off oxygen and water vapor. Thus cyanobacteria and many other bacteria are **autotrophic**, creatures that provide their own food. It appears, however, that photosynthesis was a "trick" that took some time to accomplish. The oldest life type was probably **chemosynthetic**, deriving its energy from the direct conversion of chemical compounds. These early organisms could not endure exposure to oxygen, but lived by combining hydrogen gas and carbon dioxide, producing water and methane as waste.

First Life

Methane-producing bacteria, the **methanogens,** are the likeliest candidate for the very first living thing. They are so unique that some biologists would insist they are not really "true" bacteria at all and should instead form a sixth kingdom of life. They are often classified simply as *archaeobacteria*, that is, *ancient bacteria*. On fragmentary evidence, methanogens or sulfur-fixing kindred appear to be the most primitive of all prokaryotic life, though the poor preservation provided by extremely old fossiliferous rocks will not allow them to be distinguished from other bacterial groups.

The methanogens are *anaerobic*, unable to live in the presence of oxygen. Their requirements for growth were perfectly met by the early Archean at-

mosphere, rich in hydrogen and carbon dioxide and depleted in oxygen. One subgroup lives by oxidizing elemental iron and hydrogen to produce hydrogen gas; they are the only creatures known which use metallic iron as an energy source for growth.

Modern methanogens are a minor fragment of the tree of life, driven into "hiding" by the flood of oxygen released into the atmosphere by their photosynthetic progeny. They are found in a few hot springs and in stagnant mud everywhere; their days of dominance are billions of years past.

Bacterial life has been observed in rocks whose age is 3.5 billion years in two different areas of the earth; one area is in northwest Australia, the other is in South Africa. The material from western Australia comes from the Warrawoona group and includes stromatolites, wavy laminations in limestone produced by lime-secreting blue-green algae, and a variety of filamentous bacteria in chert. The South African example is from the Swaziland rocks and includes only spheroidal microfossils.

As mentioned earlier, fragmentary evidence of "fossil-like" carbon-rich structures have been detected in the oldest whole rocks yet found on earth; we may very tentatively conclude that there was unicellular life as soon as the earth cooled sufficiently to allow rocks to preserve it. Let us look at some of the oldest *undoubted* unicellular fossils (Figure 18–7).

Chert is an *extremely* fine-grained or microcrystalline rock composed wholly of silicon dioxide. Rela-

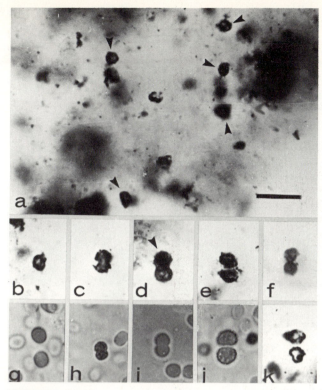

FIGURE 18–7

All photomicrographs are at the same scale (about 800x at the scale of reproduction). Arrows in part *a* point to microfossils in the chert. Parts *b* through *e* show stages in cell division as observed in fossils in the chert. Parts *f* and *k* are fissioned cells. Parts *g* through *j* are examples of binary fission in *modern*prokaryotes. (From A. Knoll and Elso S. Barghoorn, 1977, *Science, 198,* Fig. 1, pp. 396–398, copyright © 1977, by the American Association for the Advancement of Science. Used with permission of Elso S. Barghoorn.)

tively common in sedimentary sequences, it provides an ideal medium for preserving the details of microscopic life. In the cherts of the Swartkoppie Formation at the base of the Fig Tree Group of South Africa are tiny microspheroids (Figure 18–7) that are similar to photosynthetic prokaryotes. They have the following characteristics.

1. They are within the size range of modern prokaryotes.

2. They have a size distribution similar to that of modern prokaryotes.

3. Their forms are like those of modern and fossil prokaryotes.

4. They yield carbon isotope ratios like those of photosynthetic organisms.

5. About one-quarter of the microfossils have been "caught" in several stages of fission cell division.

On the basis of the Swartkoppie and other areas, the existence of prokaryotic life is well documented by 3.5 Ga; its existence before that is probable but not proven.

Life for photosynthetic prokaryotes must have been easy. There were no predators, and the sun guaranteed an unending supply of food. The atmosphere's reservoirs of carbon dioxide, however, were being steadily depleted both by the formation of limestone and by photosynthetic prokaryotes who took in carbon dioxide and exhaled oxygen and water, just as photosynthetic organisms do today. As the earth's atmosphere began receiving larger and larger quantities of oxygen from photosynthesis, the atmosphere began its *second* modification.

Its primitive atmosphere, largely composed of methane and ammonia in pre-Archean time, was first modified in early Archean time by outgassing of huge volumes of water and carbon dioxide by volcanoes. This event oxidized the ammonia into molecular nitrogen and the methane into carbon dioxide and water vapor. The second modification was the continuing introduction of oxygen into the atmosphere in mid-Archean time; the consequences were far reaching.

Oxygen and Life

The oxygenation of the earth's atmosphere was the result of photosynthetic life. *Oxygen both arises from life and supports many forms of life.* What is much less well known is that oxygen *is poisonous to all life.*

The toxicity of oxygen is due to its great chemical reactivity. We use this property to our benefit when we apply hydrogen peroxide, an oxygen source, as an antiseptic for minor wounds. It is destructive to cellular material with which it comes in direct contact.

We defend ourselves against the reactivity of oxygen, while simultaneously taking advantage of it. Our entire cardiovascular system is an elaborate defense system that allows us to use indirectly the enormous energy source of oxygen metabolism while defending us from its direct reactivity. Oxygen-based respiration releases ten times the energy that anaerobic fermentation does. Fermentation is the name given to a group of chemical reactions that split complex organic compounds into simple substances; an example is the anaerobic conversion of sugar to carbon dioxide and alcohol by yeast, a process fundamental to the brewing industry.

In spite of our elaborate defense mechanisms, low levels of oxygen damage to cells do continue to occur. Many biologists believe that aging is little more than cumulative oxygen damage to our cells.

The photosynthetic algae – cyanobacteria had begun a *unidirectional major change in the composition of the earth's atmosphere*. The algae were to become more abundant, and some were preserved in the reef-like fossil masses of the **stromatolites**. The stromatolites (Figure 18–8) formed when the respiration of algae caused lime muds to be deposited around their tissues. Mounds of stromatolites became increasingly common, and are in rocks from 2.8, 2.5, and 2.0 Ga. Among these younger rocks the Gunflint Chert from Ontario, Canada has been dated at 2.0 Ga; it contains marvelously preserved fossils of blue-green algae. In the Gunflint rocks algal colonies are indisputably plentiful and photosynthetic.

As colonies of algae became more abundant, they created "air pollution" on a massive scale, steadily altering the atmosphere by increasing its content of oxygen—a gas poisonous to the creature that was making it. Since there were no animals around to breathe it in and convert it back to carbon dioxide, another way to defend against an increasing problem had to be found.

FIGURE 18–8

Stromatolites such as this one, shown here in a polished cross section to illustrate the details of layering, are formed when algal respiration causes lime muds to precipitate in layers on the algal mats. Stromatolites are common fossils from 3.5 Ga forward. (Photograph courtesy of the U.S. Geological Survey.)

THE EUKARYOTES

The formation of the *eukaryotic* cell appears to have been the response to the rising oxygen level of the atmosphere. Though no one can be certain why such changes take place, the rise of eukaryotic cells occurred as oxygen levels accelerated above 1 percent, so the timing is suggestive.

Eukaryotes are cells that have a nucleus (Figure 18–9). The word *eukaryote* means "truly nucleated." The development of a nucleus pulls the critical genetic and reproductive material into the center of the cell and isolates it behind a "second cell wall," giving the vital contents much more protection against potential oxygen damage.

All living matter except bacteria and blue-green algae is composed of one or more eukaryotic cells. In many ways all life is miraculously alike. Life-forms as different as people and mushrooms have similar nucleic acids and proteins that are composed of the same 20 amino acids. We share similar enzyme systems and methods of obtaining energy from food.

The eukaryotes represent a major advancement over the prokaryotes. The prokaryotes, a genetically conservative group, are locked into reproduction by binary fission; only an occasional "accident" will allow any change at all. Bacteria and blue-green algae that formed as fossils billions of years ago are indis-

tinguishable from many living species. For change to be possible, the asexual reproduction of the prokaryotes had to give way to the sexual reproduction *charactertistic of eukaryotes*. For all of us who live in a society preoccupied by sex, it may come as a jolt to discover that sexual reproduction, and all the potential for change that it creates, was probably "invented" by the first eukaryote—the green algae—at least 1600 to 900 million years ago.

Single prokaryotic cells in that long ago time began to incorporate other highly specialized cells in a mutually beneficial arrangement. Mitochondria probably resulted when an oxygen-using predator invaded an anaerobic victim, which then developed a tolerance for the attacker. Plastids, the structures that form food from water and sunlight, might also have been incorporated, like Jonah in the whale. Cilia and flagella, which allow nucleated cells to move, may have resulted from the merger of whipping spirochetes with other bacteria. Eukaryotes seem to suggest that the biologic world is not all cutthroat; sometimes unlikely partners triumph.

The end product of this change, really a series of successful alliances spanning at least a billion years, was the eukaryotic cell. This life-form was capable of *mitotic* cell division, an advanced form of asexual reproduction, which may be the forerunner of *meiosis*,

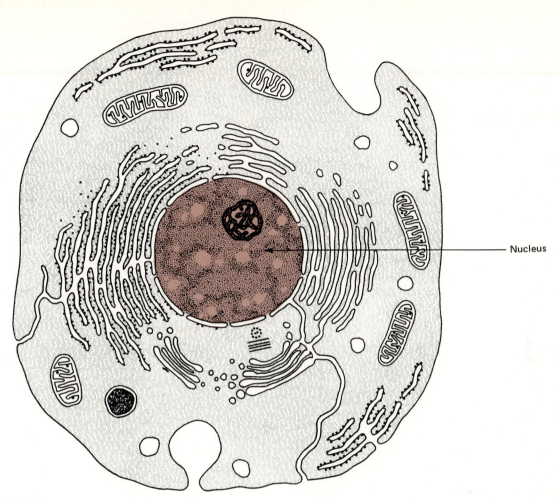

Nucleus

FIGURE 18–9

Sketch of an idealized eukaryotic cell. The colored central area is a well-defined nucleus with a surrounding nuclear membrane. Within the nucleus are chromosomes and other genetic material. The outer cell wall surrounds a variety of organelles, highly specialized cellular components. Compare with Figure 18–5.

a highly specialized division of the genetic material that alternates with sexual fertilization. Sexual reproduction, with its fertilization and meiosis, allows *random* chromosomal pairing, making each offspring different than its parent.

The development of the eukaryotic cell was a major milestone in the history of life. It allowed for mitosis and meiosis, which allowed for genetic variation, which led to the great diversity of life-forms we accept as normal today. Without the development of eukaryotic cells, the earth would still be populated with only two life-forms—bacteria and blue-green algae. And where would that leave us?

When Did Eukaryotes Arise?

Because the development of the eukaryotic cell is the key to the development of all the forms of life that followed, both biologists and geologists would obviously like to document the timing of the event. But, unfortunately, nature has played a nasty trick on us.

As prokaryotic cells die and degrade, they create central areas of darker dead organic matter that are indistinguishable from the nucleus of a fossilized eukaryote. We simply are left with no clear way to distinguish eukaryotic from prokaryotic life within the Archean or most of the Proterozoic. That the change occurred is certain; that we will ever be able to put our finger on a time line and say "There is the break" is very uncertain.

Our best estimate is that eukaryotic life was thriving by 1.6 Ga, based on a variety of subtle biologic and chemical evidence. Beyond this best estimate, we just do not know. There is, however, still another way to "skin that cat." We can look for metazoans.

METAZOAN LIFE

The classification **metazoa** includes all life consisting of more than one cell. *All metazoan life consists of eukaryotic cells.* Therefore, if we can locate the oldest metazoan fossil, we can be certain that it is formed of eukaryotic cells, and that the earliest eukaryotic cells had formed before the advent of this metazoan.

The oldest metazoan candidate is the filamentous *green algae*, generally regarded as the group from which all more complex plants and animals developed. Some types of modern green algae are unicellular eukaryotes that reproduce asexually by mitosis and sexually by a primitive form of meiosis. Others, equally unicellular, live in colonies of hundreds to thousands of cells, many of which have become specialized for reproduction, feeding, excretion, and the like. Individual cells living in colonial form are so dependent on one another that they will die if removed from the colony.

Like so much of nature, the change from unicellular to metazoan life is gradual, and it becomes difficult to recognize what we could even mean by the first metazoan. We cannot escape complexity by attempting to define it. If we can accept the 1.6-billion-year-old fossils as metazoic green algae, we have found the earliest proven eukaryote. If we cannot, we must look to younger rocks, where the evidence of a metazoan experiment is truly overwhelming.

The Ediacaran Fauna

In every continent except Antarctica, a fauna of metazoans is well preserved in rocks yielding ages of from 670 to 570 Ma. The first discovery came from the Ediacaran Ranges of South Australia, and so metazoic fauna of this age are commonly called **Ediacaran fauna.** The fossil material is of entirely soft-bodied animals, including a number unlike anything alive today. Of those with some modern affinities, we can recognize creatures that *look like* flatworms, echinoderms, sea pens, and jellyfish. They most assuredly are not, in fact, any of these.

These curious, generally flat creatures are obviously metazoans, and they include animals that are remarkably complex, having sophisticated organ systems and three-layered body walls. Clearly the flourishing of the Ediacaran fauna marks the end of the extremely limited diversity of Archean and all but the latest Proterozoic fauna. For three billion years, life had been all bacteria and blue-green algae.

This trend to greater diversity will continue across the Precambrian – Paleozoic boundary, with almost explosive diversification of life within the earliest Pa-

leozoic (Figure 18–10). But life will continue on *without* the Ediacaran fauna. With a few exceptions, this group of soft-bodied, shallow-water invertebrates, some up to a yard or a meter long, did not survive into the Paleozoic. They were an experiment that failed. No one knows why.

A major extinction, the loss of the Ediacaran fauna, marks the boundary between the Precambrian and all later life. Such widespread extinctions, cutting abruptly across many classifications of life, are called **mass extinctions;** *these events mark the boundaries between all the Phanerozoic eras.* The extinction at the end of the Paleozoic wiped out more than half of all the families of marine invertebrates. Another one at the end of the Mesozoic wiped out nearly a third of all kinds of marine life and extinguished the most popular of all past beasts, the dinosaurs, together with their reptilian kin that flew in the air, and swam in the sea.

We will be discussing mass extinctions again. For now, the Precambrian ends with a death knell for the Ediacaran fauna. Within the youngest Proterozoic sedimentary rocks there are fragments of worm tubes and other trace fossils of more rounded life types that would soon explode into numerous major groups of Paleozoic invertebrates. Still found in these same rocks are the conservative bacteria and blue-green algae that had for so long been "an offensive, directed against the repetitious mechanisms of the universe."

▼
PRECAMBRIAN PHYSICAL EVENTS

▲

The geologic record starts with the Archean, extending from the formation of the oldest rocks at least 3.8 Ga to the beginning of the Proterozoic Eon at 2.5 Ga. The oldest known rocks are the Isua Series in Western Greenland, but Archean rocks are known from Africa, northern Scandinavia, eastern Siberia, northeastern South America, western Australia, Minnesota, Wyoming, and large areas of central and northern Canada (Figure 18–11). Archean rocks form 10 to 15 percent of all exposed continental crust, and approximately half of all continental crust.

The Archean was the time during which the first continents formed and grew in both volume and area. Perhaps as much as 50 percent of the North American continent was formed by 2.5 Ga, at the end of the Archean. The origin of Archean rocks is quite controversial, reflecting two contradictory lines of evidence.

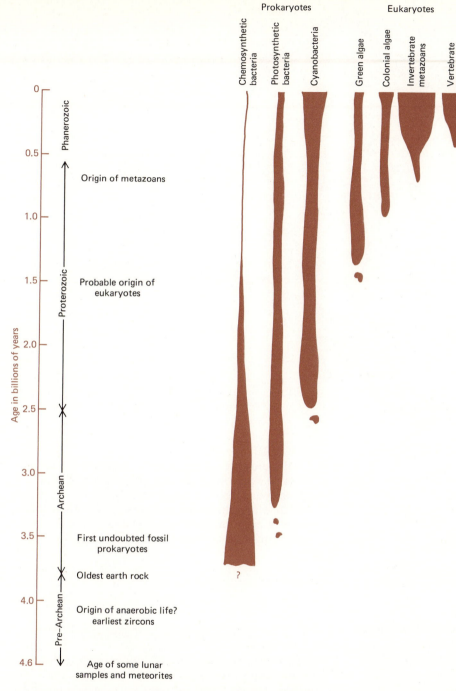

Prokaryotes   Eukaryotes

Chemosynthetic bacteria · Photosynthetic bacteria · Cyanobacteria · Green algae · Colonial algae · Invertebrate metazoans · Vertebrate metazoans

Age in billions of years

Phanerozoic

0

Origin of metazoans

0.5

1.0

Proterozoic

Probable origin of eukaryotes

1.5

2.0

2.5

Archean

3.0

3.5   First undoubted fossil prokaryotes

Oldest earth rock

4.0   Origin of anaerobic life? earliest zircons

Pre-Archean

4.6   Age of some lunar samples and meteorites

FIGURE 18–10
The width of bars is roughly proportional to abundance. The chemosynthetic bacteria went into serious decline as the oxygen level rose. The age of eukaryotic development is poorly constrained.

▶ It is quite probable that heat flow from the earth's interior to the surface was from two to five times that at present (Figure 18–4). The origin of komatiitic lavas has already been cited as one example of the much greater depth of melting in Archean times.

▶ Yet, the metamorphic rocks at Isua, in common with other Archean rocks, reveal a *thermal gradient*—the rate of temperature increase into the earth—similar to that of the modern earth.

Some of the fundamental questions these contradictory lines of evidence raise may be easily stated.

1. How does the evidence of normal heat flow in continental rocks square with the evidence of much increased heat flow throughout the Archean earth?

2. What processes operated to form and expand continents in a regime of high heat flow?

3. Are Archean rocks like other rocks?

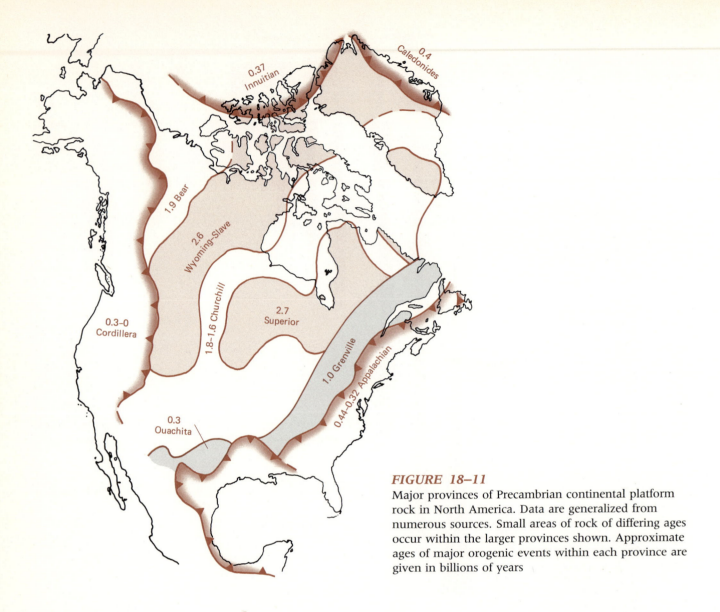

FIGURE 18–11
Major provinces of Precambrian continental platform rock in North America. Data are generalized from numerous sources. Small areas of rock of differing ages occur within the larger provinces shown. Approximate ages of major orogenic events within each province are given in billions of years

We will take each of these questions in turn, starting with the last and easiest question.

ARCHEAN ROCKS

Archean rocks form a part of the earth's great **shields** or **cratons,** areas of very old, seismically stable igneous and metamorphic rocks that underlie essentially all continental rocks (see Chapter Ten). Three kinds of rock associations are common in the Archean; each is either much less common or unknown in younger rocks. These associations are (1) terranes of high-grade granite gneiss, (2) Banded Iron Formation, and (3) low-grade greenstone belts (Figure 18–12).

Granite Gneiss Terranes

Large areas of light-colored gneisses dominate the surface exposures of many Archean terranes. The gneisses are typically strongly foliated and have undergone various amounts of partial melting, with numerous granitic veins penetrating and paralleling the foliation. The overall composition is decidedly intermediate in chemistry, similar to the volcanic equivalents called *andesite* (see Chapter Three). The terranes of granite gneiss appear to be the oldest of all the rocks, and they are typical continental crust.

Associated with these gneisses are amphibolites, metamorphosed equivalents of basalt, as well as marbles, quartzites, and schists—many of which are rich in graphite whose biogenic origin is in dispute.

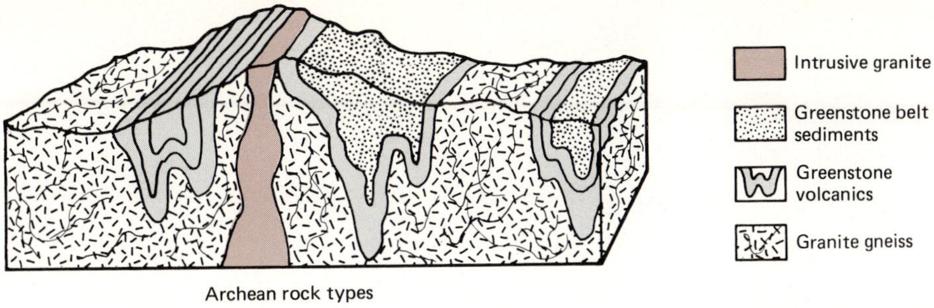

Archean rock types

■ Intrusive granite

▒ Greenstone belt
sediments

W Greenstone
volcanics

X Granite gneiss

FIGURE 18–12

A cross section of a typical Archean shield shows the relations of greenstone belts to the intermediate gneisses and granitic gneisses that enclose them and the granites that often intrude them.

Included among the sediments within these gneisses is the unusual rock type called Banded Iron Formation (Figure 18–13).

Banded Iron Formation

Found worldwide, Banded Iron Formation (BIF) is a rock composed of hematite, magnetite, and various forms of finely crystalline silica. It is an extremely common rock within Archean and Early Proterozoic time (Figure 18–3), but it was no longer formed after 1.6 Ga.

Its origin is also controversial, though there is general agreement that the rock formed on shallow-water tidal flats under some rather specialized chemical conditions. Today, silica tends to be transported in alkaline environments, whereas iron is transported in strongly acidic surroundings. How then did iron and silica combine in such massive volumes?

The Precambrian ocean may well have been saturated in silica, perhaps from submarine volcanism; the ocean was also anaerobic, an environment which would have kept iron in solution. As photosynthetic bacteria continued to release oxygen, the iron was rendered insoluble and precipitated in a silica-saturated environment. As all the soluble iron was finally oxidized, deposition of BIF terminated, and iron no longer gobbled oxygen as fast as it was emitted into the atmosphere. With iron no longer available to consume oxygen, the amount of oxygen in the atmosphere began to increase rapidly.

The range of origins proposed for this rather strange rock suggests the difficulty in trying to suggest a reasonable environment of deposition for a rock that has not been formed in 1600 million years. Among the origins suggested are the following.

▶ *Biologic.* Iron and silica were precipitated from seawater by biologic activities.

▶ *Continental weathering.* The iron was released in its soluble form by weathering in an anaerobic atmosphere and precipitated in the oceans as the environment became oxygenated by photosynthetic organisms.

▶ *Evaporitic.* Iron and silica were evaporated from seawater.

▶ *Hydrothermal.* Hot springs on the seafloor enriched the seawater with iron, which was then precipitated.

▶ *Upwelling.* Iron and silica were precipitated on the seafloor by the upwelling of chemically enriched seawater.

The earliest occurrence of (BIF) is at Isua, western Greenland where its age of 3.8 billion years places it among the oldest Archean rocks. Since BIF was laid down when the amount of oxygen within the atmosphere rose above some very low level, the presence of (BIF) among the oldest rocks casts real doubt on models of an early Archean atmosphere that totally lacked oxygen.

FIGURE 18–13

Collected near Ely, Minnesota, this specimen of Banded Iron Formation is composed of alternating layers of dark magnetite-rich and lighter hematite-rich rocks in a matrix of finely crystalline silica. Around the world rocks of this type are unique to the Archean Eon and the Early Proterozoic Era.

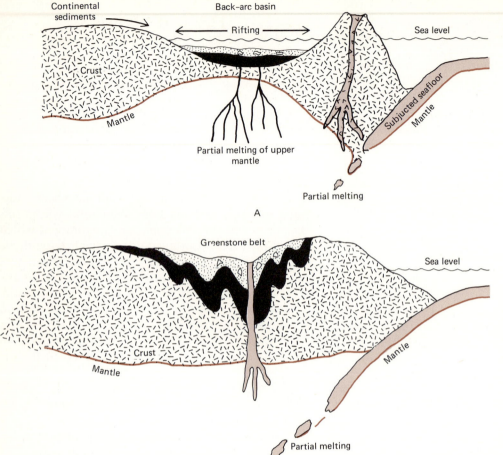

FIGURE 18–14

These sketches suggest one of the modes of origin for greenstone belts. Having formed as a back-arc basin deposit of basalt, volcanic sediment, and continental sediment sandwich, the collision of the arc with an active plate margin converts it into a high-grade gneiss and greenstone belt couplet.

The *last* occurrence of BIF is within 1.6-billion-year-old rocks, an age that coincides with the *first* formation of *red beds*. These rocks are sandstones which are colored a deep reddish orange by the iron oxides they contain. Their development coincides with the oxygenation of the last of the earth's reduced iron by an oxygen-rich atmosphere, so that iron in its reduced, soluble state became unknown.

Greenstone Belts

Rocks of the greenstone belt include a variety of dark-colored volcanic and sedimentary rocks partially derived from volcanics. Their name comes from the pronounced dark-green color of many rocks within this belt.

The belts occur as irregularly shaped, elongate slivers, embedded within the granite gneisses. Most of the rocks are strongly folded, with the degree of metamorphism ranging from weak to moderate. Most of the belts are discontinuous ribbons surrounded by very large areas of monotonous light-colored granitic gneisses. The areas of greenstone

range in size up to 600 by 150 miles (1000 by 250 kilometers), with the boundary between the greenstone masses often being unclear.

Most greenstones can be subdivided into an upper, predominantly sedimentary group—with deeper-water sedimentary rocks at the base and shallow-water sedimentary rocks at the top—and a lower largely volcanic group. The volcanic units commonly include komatiites, ultramafic basalts, and basalt intermixed with chert and BIF.

THE ORIGIN OF CONTINENTS

Continents today increase in size by the irreversible chemical differentiation of the upper mantle into oceanic crust, followed within a few hundred million years by the crust's consumption along island arcs and active continental margins or magmatic arcs (see Chapter Ten). An example is the western edge of South America where the Andes grow by the addition of the lower-density, lower melting point (see Figure 10–25) materials from the Nazca plate. Volume is added by both plutonism and volcanism.

It appears that as far back as the record can be read, the earth has had continents that are relatively thick. The metamorphism of the 3.8-billion-year-old Isua rocks took place at depths and temperatures typical of lower continental crust. It appears that the ancient Archean granite gneisses are the equivalents of the materials currently being added to the west coast of South America, and that the greenstones of that period were formed from the metamorphism of the volcanics and sediments of back-arc basins like the Sea of Japan (Figure 18–14). *Just as today's continents are enlarged by consumption of seafloor, they apparently were so enlarged beginning in the earliest Archean.*

Plate tectonics seems also to have been operating to cool the earth in the earliest Archean, though perhaps at different rates and geometries than are characteristic of a much cooler modern earth. By no means would all geologists agree that there was Precambrian, much less Archean plate tectonics; for a decidedly different view, see the work of R. B. Hargraves referenced at the end of this chapter. He cites the enormous difficulties of keeping the continents from remelting as rapidly as they form on an earth with perhaps five times its modern heat flow.

HEAT FLOW AND ARCHEAN PLATE TECTONICS

Those favoring Archean plate tectonics argue that the greater heat flow must have brought much higher rates of plate motion and seafloor spreading. Spreading rates proposed would create 6 to 8 square miles (15 to 20 square kilometers) of new seafloor each year—up to six times the present rate. Such amounts of seafloor could have been generated by more rapid spreading or by numerous smaller oceanic plates having many more miles or kilometers of spreading axes. Either way, the rapid production of oceanic seafloor on an extremely hot earth meant that the seafloors were rapidly consumed at continental margins, and that the continents grew rapidly in size.

On an earth with high heat flow, and therefore a steep geothermal gradient, the asthenosphere would have been far more fluid than it is today, which would have made horizontal motion of plates much easier. The "transit time"—the time required between the generation of seafloor at a spreading center and its partial melting beneath a continent—would have been much shorter at higher spreading rates. The slab subducted would have been much hotter and therefore less dense. However, the pres-

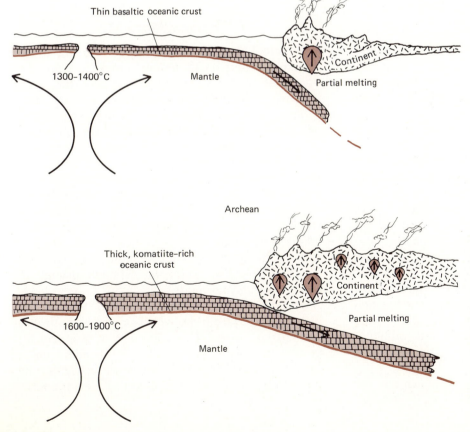

FIGURE 18–15

The difference between Archean and modern plate tectonic systems is primarily the much greater heat flow from a much hotter Archean mantle. The oceanic crust may have been thicker and the angle of the subduction zone affecting komatiite lavas flatter, leading to wide zones of continental accretion and magmatism during Precambrian time.

ence of komatiitic lavas must have increased the density of seafloor basalts and enhanced subduction.

The angle of dip or tilt of the subduction zone would have been shallow, leading to much wider zones of continental compression and volcanism (Figure 18–15) than are common now. Our narrow, linear orogenic belts reflect a greater dip of the colder subducting slabs which move much more slowly.

As discussed in Chapter Ten, the inability to subduct continental crust is one of the fundamental facts of the earth. The transfer of material from the upper mantle to the continents is both irreversible and permanent. Since the Archean, the continents have been rifted into fragments many times and sent "piggybacking" across the earth's surface to collide with other continental masses to form mountain chains. Archean continental rock has been largely recycled by weathering into sedimentary rocks (see Figure 5–13), but they, too, are returned to the continent as they collide with its margin.

Thus continents can grow only in size and depth through time. Estimates are that the area of the earth's surface occupied by rocks older than 3.0 billion years is probably between 5 to 10 percent. By the end of the Archean at 2.5 Ga, *at least half* of the current surface area of the continents had been formed.

PROTEROZOIC ROCKS

The Proterozoic Eon extends from 2.5 to 0.570 Ga (570 million years ago), a time that includes half of all earth history since the pre-Archean. It was a time of continued continental expansion (see Figure 18–11), with North America growing to 80 to 85 percent of its total area by the end of the Proterozoic.

Proterozoic rocks show some distinct differences compared to their Archean predecessors. The greenstone belts strongly infolded into high-grade intermediate gneisses and granitic gneisses are unique to the Archean. Banded Iron Formation continued to be deposited through the Early Proterozoic Era to about 1.6 Ga, making up 15 percent of Early Proterozoic sedimentary rock on a worldwide basis.

As indicated earlier, when the sedimentary rocks of the the Banded Iron Formation were no longer deposited, hematite-rich reddish sandstones, the *red beds*, became abundant. Their development is taken to indicate an increased supply of oxygen, for when oxygen was scarce, weathering made iron soluble in its reduced *ferrous* form, and the *soluble* iron was transferred to the seas along river systems.

Today, essentially no soluble iron enters the world's ocean. It is instead oxidized to its *insoluble*

form and remains behind on the continents, where it primarily serves as a pigment, staining rocks various shades of red, orange, yellow, purple, or brown. Thus the end of BIF deposition 1.6 Ga appears to have signaled a switch to oxidation of iron on the continents.

The common rocks of the Proterozoic Eon are gently dipping, thick packages of sedimentary and volcanic rocks which were deposited in broad downwarps within the crust. The sedimentary rocks are largely stable platform sediments like quartz-rich sandstones, conglomerates, algal limestones, and shales. Volcanic rocks shifted away from the komatiitic magnesium-rich basalts typical of the Archean to more "normal" olivine-deficient platform basalts, as well as potassium-rich andesites and rhyolites. The shift to accumulating extensive sequences of platform sedimentary and volcanic rocks marked a fundamental shift in geologic process; *it is this shift that sets off the Proterozoic from the Archean.*

Proterozoic rocks may have formed along passive margins and intracratonic basins within a single, giant supercontinent. Paleomagnetic evidence from several areas suggests this, as do attempts to rematch the edges of Proterozoic mobile belts. The combination of thick, abundant platform sedimentary rocks in sites later deformed into mountains invites a discussion of the history of passive margins.

Passive Margins

Continental margins facing rift oceans are passive margins where sedimentary rock accumulates. Sometimes called a *geosyncline* or a **geocline**, a passive margin includes the continental shelf and rise.

The sediments deposited on the modern continental shelves are characteristically clean quartz-rich sandstones, shales, and limestones. Comprising about 15 percent of the world's total surface area (see Chapter Nine), the continental shelves are broad aprons of sediment accumulation stretching from shoreline to a prominent break in slope at depths averaging 600 yards (550 meters). The wedge of sediments is generally coarser-grained nearest the shore and progressively finer-grained offshore, with carbonates being the dominant sediment beyond the reach of sands and muds.

The continental shelves spread the load of sediments received from the world's rivers by both wave and current action. This collection of sediment, deposited in *shallow, sunlit* water, forms about half of the total geosyncline. The shoreward section is called the *miogeosyncline* or more simply the **miogeocline** (Figure 18–16).

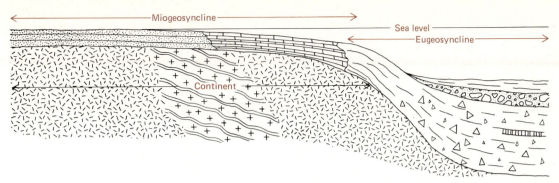

FIGURE 18–16

The miogeosyncline or miogeocline encompasses the continental shelf; the eugeosyncline or eugeocline includes all materials seaward of the miogeocline. The continent is composed primarily of eugeoclinal material welded to it plus volcanic and plutonic rocks injected into it and extruded onto it. The folding, metamorphism, and plutonism of geoclinal material making up the continent *predates* the *formation* of the modern geocline shown here by many millions of years.

Seaward from the miogeocline is a different domain of sediment accumulation, known to oceanographers as the continental rise and to geologists as the *eugeosyncline* or **eugeocline,** the deeper-water part of the geosyncline. The eugeocline stretches seaward for many miles or kilometers, grading gradually into the deep ocean floor (Figure 18-16).

Eugeoclinal sediments include **turbidites,** accumulations of coarse sand abruptly deposited on the prevailing muddy sediments by sedimentary avalanches from the upper rise. Earthquakes may kick off such avalanches, better known as *turbidity currents* because they flow downslope as dense mixtures of sediment and water at high velocities. Mixed with the turbidites are typical deep-water clay beds, dark, thin limestones, and cherts. Fossils are sparse and include primarily siliceous marine animals.

The Wilson Cycle

J. Tuzo Wilson, an early researcher in plate tectonics, proposed that over several hundreds of millions of years as seafloor spreading reverses direction, a geoclinal area evolves from being a passive margin at the edge of a spreading ocean to being an active consumptive margin. The **Wilson cycle** outlines the complete sequence of events, which may be thought of as divided (Figure 18–17) into four parts.

1. Initial rifting by the spreading center.
2. Development of a passive margin.
3. Reversal of the spreading direction and development of an active margin.
4. Closure and collision.

As a miogeoclinal – eugeoclinal couplet moves from stage 2 to stage 3, the seafloor, which has been passively carrying the continent, detaches from it and begins to be subducted beneath it. Its lighter-weight fractions begin to melt, and andesitic volcanoes begin to erupt along the magmatic margin (Figure 18–18). The miogeoclinal material is thrust toward the continent and is crumpled into open asymmetric folds, which die out toward the continent. The shoreline, the base level for streams that have been emptying into the sea, begins to rise, reversing the flow of streams and sending their flow back into the continental interior.

The eugeoclinal materials, much closer to the onrushing seafloor, are intensely crumpled, compressed, sliced, and metamorphosed at high pressure. Converted into an **ophiolite suite,** a group of metamorphic rocks characterized by an association of high-pressure rocks and minerals (such as blueschists) mixed with cherty deep-sea sedimentary rocks, the former eugeoclinal rocks are *forced* onto the continent. The pressure literally *welds* the eugeoclinal rocks onto the continent from which they earlier came as sediments. As closure and collision are completed, the geocline, which has been compressed and thrust deeper, begins its rise as a mountain range.

As the former active margin becomes a passive margin once again, the miogeoclinal rocks of the second cycle accumulate across the eroded eugeoclinal rocks of the first cycle. The continents always recover all the sediment washed from their heights. Their miogeocline is thrust back onto the continent, and their eugeocline is compressed and welded back onto the continent from which it came.

As the Wilson cycle indicates, *continents add to*

Rift initiation

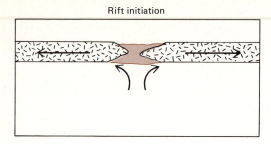

Passive margin

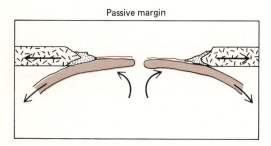

Active margin

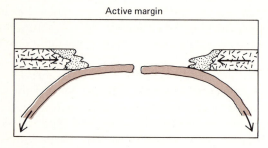

Closure and collision

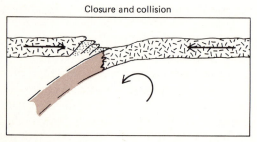

FIGURE 18–17
The Wilson cycle probably operated more swiftly and with much smaller continental landmasses than are present now. The slowing of heat flow that has been going on since the earth first completely melted limits the maximum, but not the minimum, size of ocean basins that may be formed by rifting and spreading.

their own area through time by welding on successive geoclines and by *adding new plutonic and volcanic material from the consumption of the seafloor beneath either magmatic or volcanic arcs.* From each cycle the continents emerge enlarged, their boundaries extended into the ocean basins that feed them.

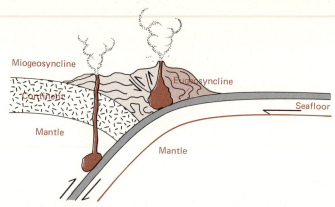

FIGURE 18–18
During a collision of an ocean basin with a continent, the miogeoclinal rocks are squeezed and faulted upward and inward toward the continental interior, forming fold mountains like the Appalachians. The compression and accretion of the eugeoclinal sediments to the continental margin may create extreme pressures and metamorphose them, forming ophiolite and blueschist (see Chapter Six) zones.

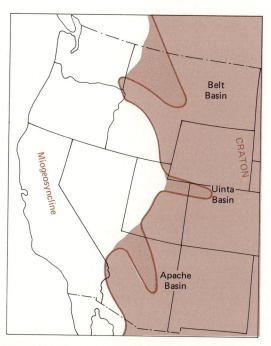

FIGURE 18–19
A very generalized reconstruction of the probable western shoreline of the United States in Late Proterozoic time, about 800 Ma. The continental platform was interrupted by three major cratonic basins in which thick muddy and limy sediments were deposited during late Precambrian time. (After Jack E. Harrison et al., 1974, U.S. Geological Survey Professional Paper 866.)

Proterozoic Examples

The Wopmay geocline of far northwestern Canada is one of the best studied of all Proterozoic orogenic belts. The sedimentary rocks within the Wopmay include a normal range of geoclinal sediments. They were disturbed four times within 80 million years by the repetitive opening and closing of an ocean basin against the edge of the Wopmay margin. These "high-frequency" Wilson cycles centered around a period of about 1.9 Ga.

This example of remarkably "short-lived" oceans suggests, but does not prove, that the rate of seafloor spreading in the Proterozoic was still considerably faster than modern rates. It also suggests that the amount of heat flow was still high, facilitating both the opening and the closure of small, newly rifted basins. Other Proterozoic examples include the circum-Ungava and Coronation geoclines of Early Proterozoic age in Canada and the much later (1.0 Ga) Grenville belt (Figure 18–11). At about the same time (1400 to 900 Ma) intracratonic troughs stretching from Canada south to Arizona opened and received deposits of silt and coarser sand that ranged up to 10 miles (16 kilometers) in thickness (Figure 18–19).

Rocks formed in these old basins now have many different names, and they form scenery as diverse as Glacier National Park, the Uinta Mountains, and a lower part of the walls of the Grand Canyon. They record sedimentation in an environment like that of our modern Gulf of Mexico over a period of hundreds of millions of years. By the end of Precambrian time, the western shoreline of the United States roughly paralleled the western boundaries of Utah, Idaho, and Arizona (Figure 18–19).

Among the rocks of the Late Proterozoic Era, those formed between 900 and 570 Ma, there is abundant evidence of extensive glaciation. In this period glacial deposits, often called *tillites*, occurred widely and have been found in Australia, southern Siberia, China, western North America, South Africa, Brazil, and Scandinavia. They provide an interesting clue to the decisive mass extinctions that closed out the biologic history of Precambrian time.

PRECAMBRIAN ROCKS AND MINERAL WEALTH

Precambrian rocks underlie most of North America (Figure 18–11), but they are largely hidden under overlying Phanerozoic rocks. Within the United States (Figure 18–20), the Adirondacks of New York State are an area where a complex uplift has removed the overlying Phanerozoic cover. The Blue Ridge zone in the Appalachians also exposes both Precambrian and early Paleozoic igneous and metamorphic rocks. Most of northern Michigan and Wisconsin have exposures of older Precambrian rocks, largely buried under drift.

Within the central United States Precambrian rocks are exposed in the St. Francois mountains of southeast Missouri, the Arbuckle and Wichita mountains of southern Oklahoma, the Llano Uplift of central Texas, and the Black Hills of South Dakota. Within the western United States exposures of Precambrian rocks are common within the Rocky Mountains, the Uinta Mountains of Utah, and the Grand Canyon of Arizona, and along an extensive belt of isolated ranges that make up southern Arizona, southeastern California, and Nevada.

PRECAMBRIAN MINERAL DEPOSITS

The Precambrian rocks are the greatest storehouse of mineral wealth in the world. More than 80 percent of the world's economic deposits of minerals are mined from Precambrian host rock. Iron is the dominant resource, mined throughout the world from Banded Iron Formation and allied rocks. The lucky accident of ancient chemistry that created BIF left a reserve so abundant that it forms the basis of industrial technology for the whole world. Raw iron costs us less per pound than dog food; each of us uses a little less than than a ton of iron and steel every year. Our needs are largely met from the BIF and allied rocks of the Lake Superior region, whose resources are adequate to meet our needs for many centuries. Since iron and steel account for over 95 percent of all metals used by humankind, and since over 90 percent of the world's iron comes from Precambrian rocks, the importance of these ancient sources cannot be overemphasized.

The dwindling reserves of uranium ore in the United States come from young Phanerozoic sedimentary rocks, but most of the world's supply comes from igneous and metamorphic rocks that contain the mineral sometimes called *pitchblende*, or more properly *uraninite*. These deposits occur in several regions of Canada, the USSR, and Africa. There is little of this material left; it is estimated that worldwide reserves will be depleted within fewer than 50 years.

Most of the world's need for gold is filled by reusing old gold, since no one is careless with this valuable mineral. The gold in your dental filling may

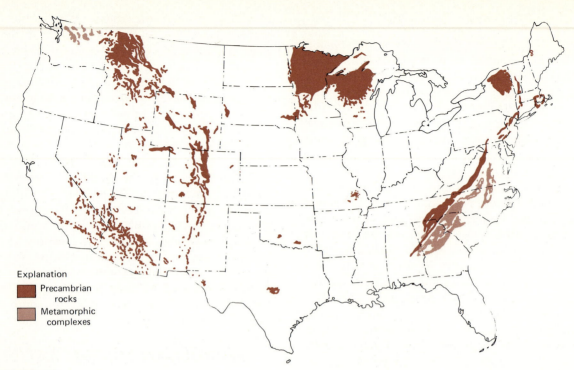

FIGURE 18–20

Surface exposures of largely Precambrian rocks. Also shown are metamorphic complexes that may include some rocks of Precambrian age. (From P. B. King, 1976, U.S. Geological Survey Professional Paper 902.)

once have been a piece of jewelry or a coin. The world demand for new gold is met from deep mines in South Africa, where gold occurs in Precambrian gravel deposits that may also contain diamonds and uranium. Within the United States, one source of Precambrian gold is the Homestake Mine in South Dakota, where gold occurs in Precambrian metamorphic rock.

Nickel is used primarily as an alloy with steel, creating a noncorrosive alloyed metal that has great strength at high temperature. The majority of nickel comes from mineral segregations in ultramafic gabbro bodies, largely of Precambrian age. *One* mine in Sudbury, northern Ontario, furnishes about one-third of the world's total nickel demand from deposits in a Precambrian gabbroic body. This deposit is particularly interesting as it was formed owing to the impact of an ancient nickel-rich meteorite from outer space.

SUMMARY

1. The Precambrian includes seven-eighths of all geologic time; newer discoveries may continue to push the date of its beginning still farther back in time (Figure 18–21). Our earth began probably as an accretion of meteoritic particles; swept from space, they formed a planet of relatively uniform magnesium - iron - silicate composition. Heated by its own gravitational attraction, radioactive decay of elements, and the continuing bombardment of meteorites, the earth began to melt, allowing the densest material to form the core. The sinking iron further accelerated heating, creating a largely molten planet and complete differentiation of the rapidly rotating planet into concentric shells of less density outward. A protocrust was formed, and continuing volcanism released large amounts of volatiles, which modified the primitive atmosphere of methane and ammonia into one containing abundant water vapor, hydrogen, and carbon dioxide. Much of the water vapor precipitated to form the earth's ocean, as the pre-Archean or Hadean phase of the earth's formation was completed.

2. The Archean was marked by the development of marine chemosynthetic organisms as soon as the oceans cooled below the boiling point; their ancestors continue to live in hot springs and stagnant muds today. The origin of methanogens

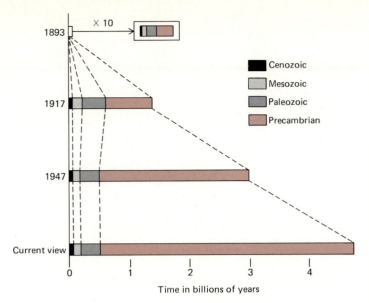

FIGURE 18–21
The advent of radiometric dating quickly overturned the 1893 estimate that earth history extended back only 55 million years. Once radiometric dating established the age of Phanerozoic rocks, further changes came almost exclusively from finding older and older rocks. The process continues, lengthening the span of time assigned to the Precambrian. (After H. L. James, 1970, *American Journal of Science*, (Bradley Volume), Vol. 258-A, pp. 104–114, with permission.)

remains a mystery, but as they developed into photosynthetic bacteria and blue-green algae, these groups of prokaryotic unicellular life began to oxygenate the atmosphere.

3. The biologic record of the Proterozoic is marked by the development of first the eukaryotic cell and then the eukaryotic green algae, capable of sexual reproduction. The algae adopted a colonial life-style which merged into an interdependent life-style indistinguishable from multicellular or metazoic life. The first major expansion of metazoa was to form the Ediacaran fauna, life-forms that became extinct at the end of the Proterozoic Eon. The prokaryotes survived this first mass extinction, as did elements of rounded metazoa whose progeny were to flourish within the earliest Paleozoic Era.

4. Archean events were dominated by the growth of continents to perhaps half of their current size. In a regime of high heat flow, such growth was accomplished by consuming komatiitic lavas along both magmatic and oceanic island arcs. Continental shields of Archean rock are made up of three distinctive rock suites: (1) high-grade granitic gneisses of intermediate composition, (2) infolded low-grade greenstone belts, and (3) Banded Iron Formation, a common Archean rock formed as oxygenation of the atmosphere rendered soluble iron into its insoluble form.

5. When the Proterozoic Eon began at 2.5 Ga, there was a marked shift from the gneissic and greenstone terranes of the Archean to platform sedimentation within both intracratonic and marginal geoclines. Orogenesis broadly followed the Wilson cycle, with repeated episodes of ocean basin opening and destruction. The continental cratons grew through accretions of geoclines and additions along both magmatic and island arcs. In latest Proterozoic time a series of intracratonic rifts developed along what is now Montana, Utah, and Arizona. The sediments that filled them mark the close of Precambrian time. Among Late Proterozoic sedimentary rocks are a large series of tillites, which are glacially derived sedimentary rocks.

KEY WORDS

autotrophic

blue-green algae (cyanobacteria)

chemosynthetic

conduction

convection

craton

cyanobacteria (blue-green algae)

Ediacaran fauna

eugeocline

eukaryote

geocline

komatiite

mass extinction

metazoa

methanogen

miogeocline

photosynthetic

prokaryote

ophiolite suite

shield

stromatolite

turbidite

Wilson cycle

EXERCISES

1. Referring to Figure 5–13, why is it so difficult to find unmetamorphosed Archean rock?

2. What are the limits to the lateral extent of continents?

3. For North America, radiometric dates on cratonic rock show a roughly circular pattern "younging" outward from an Archean center. How could this pattern be explained?

4. The earliest known fossil life-forms are spherical. What are the advantages of being spherical to a tiny life-form in the ocean?

5. During the earliest Precambrian, the year might well have had over a thousand days (see Figure 17–1). On such a rapidly spinning earth the moon would have been much closer to the earth. How would this have affected ocean tides? How would the tidal effect affect tidal life?

6. In order for clay to be available to form a "template" for first life, what process must first occur?

7. What are at least two separate reasons why Precambrian rocks hold essentially all the earth's mineral wealth?

8. We now know there was life on earth as soon as the oceans formed. Does this fact tell you anything about the chances of finding life in other solar systems?

SUGGESTED READINGS

ARNDT, NICHOLAS T., 1983, Role of a thin, komatiite-rich oceanic crust in the Archean plate-tectonic process, *Geology*, v. 11, pp. 372 - 375.
▶ *One explanation of plate tectonics in a high heat flow regime.*

CLEMMEY, HARRY, AND BADHAM, NICK, 1982, Oxygen in the Precambrian atmosphere: An evaluation of the geological evidence, *Geology*, v. 10, pp. 141 - 146.
▶ *Challenges orthodoxy; proposes that the Archean atmosphere was fully oxygenated.*

GLIKSON, A. Y., 1979, The missing Precambrian crust, *Geology*, v. 7, pp. 449 - 454.
▶ *Another challenge to orthodoxy; proposes an expanding earth to account for "missing" Proterozoic crust.*

GOULD, STEPHEN JAY, 1984, The Ediacaran experiment, *Natural History*, February, pp. 14 - 20.
▶ *Elegant discussion of the Ediacara and the results of mass extinctions.*

HARGRAVES, R.B., 1976, Precambrian geologic history, *Science*, v. 193, no. 4251, pp. 363 - 371.
▶ *Vigorously champions an alternative view to Precambrian plate tectonics.*

HOFFMAN, PAUL F., AND BOWRING, SAMUEL A., 1984, Short-lived 1.9 Ga continental margin and its destruction, Wopmay orogen, northwest Canada, *Geology*, v. 12, pp. 68 - 72.
▶ *A brief review of one of the best documented Proterozoic orogenic belts in the world. A reminder that all theories ultimately depend on testing by the field geologist.*

MARGULIS, LYNN, 1982, *Early Life*, Boston, Science Books International, 160 pp.
▶ *Written for the layperson, this is an absolute treasure by a major authority in the field. A find!*

REYMER, ARTHUR, AND SCHUBERT, GERALD, 1986, Rapid growth of some major segments of continental crust, *Geology*, v. 14, pp. 299 - 302.
▶ *Points out the astonishing difference in crustal growth rates of the Precambrian and the Phanerozoic and suggests some cogent reasons for them.*

SAGAN, DORION AND MARGULIS, LYNN, 1987, Bacterial bedfellows, *Natural History*, March, pp. 73 -85.
▶ *Wonderful story of the role of luck and shrewd observation in science; presents a compelling tale of the origin of eukaryotic organelles.*

SCHOPF, J. WILLIAM (ed.), 1983, *Earth's Earliest Biosphere*, Princeton, N.J., Princeton University Press, 643 pp.
▶ *Exhaustive summary of Precambrian life; fine overall reference, somewhat technical.*

SIMONSON, BRUCE M., 1985, Sedimentological constraints on the origins of Precambrian iron formations, *Geological Society of America Bulletin*, v. 96, pp. 244 - 252.
▶ *Thorough discussion of the conflicting origins presented for this abundant but exotic rock type.*

WINDLEY, BRIAN F., 1984, *The Evolving Continents*, New York, John Wiley & Sons, 399 pp.
▶ *Another terrific reference on the development of continents through time, with special emphasis on Precambrian as well as elegant summaries of Phanerozoic examples; highly recommended.*

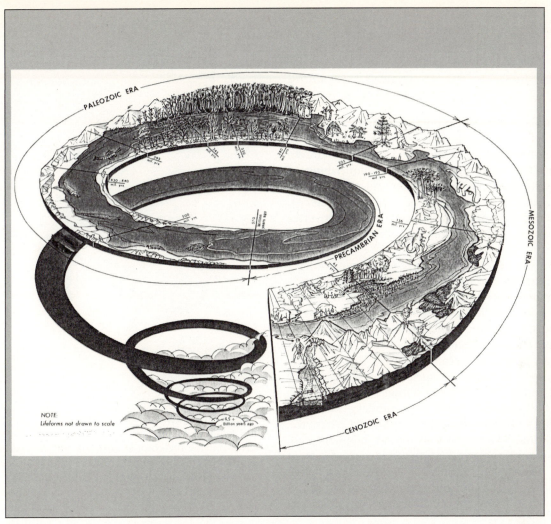

This spiral depicts in highly schematic form the parade of events and life within the past 570 million years. (Diagram courtesy of the U.S. Geological Survey.)

CHAPTER *nineteen*

The Paleozoic Era

Knowledge consists in understanding the evidence that establishes the fact, not in the belief that it is a fact.

CHARLES T. SPRADLING

The boundary between events of the Precambrian—all time before the Cambrian—and the Cambrian Period, the oldest period of the Paleozoic Era, is drawn based on the rapid appearance and flourishing of metazoic life about 570 Ma. After the mass extinction of the Ediacaran fauna (see Chapter Eighteen), between 570 and 500 Ma, essentially every invertebrate phylum now known was established in the fossil record.

The mechanisms that triggered this radiation of species remains one of the major unanswered questions both in the study of ancient life and in evolutionary theory. Both the beginning and the ending of the Paleozoic Era are marked by mass extinctions; at the termination of the Paleozoic *at least four-fifths* of all Paleozoic species became extinct. The Paleozoic Era—literally the age of ''ancient life''—began 570 Ma and ended 245 Ma (see Chapter frontispiece).

The Precambrian-Cambrian boundary has now been placed at a specific locality in Siberia where the exposure and fossils allow a precise judgment call. The boundary is placed at the first occurrence of what is often called *Tommotian fauna*, a distinctive group of shelly tubular and conical fossils similar in appearance to worm tubes and simple conical gastropods (snails).

Just beneath these fossils are those of the Ediacaran fauna, sometimes placed in the *Eocambrian,* an informal term for the last few hundred million years of the Proterozoic Eon. Whatever the terminology, this chapter deals with the rich history of the physical events that shaped the North American continent and the life that inhabited its marginal seas, which partially flooded the continent many times. Near the midpoint of this era, life colonized the continents, which had been barren for 3.5 billion years.

As we examine younger and younger rocks, the details of geologic history become progressively clearer, because the record is less disturbed by erosion, metamorphism, and tectonism. Each succeeding era covers a briefer and briefer time span. The Paleozoic Era includes only 7 percent of geologic time, the Mesozoic 4 percent, and the Cenozoic a little over 1 percent. Compared to the Precambrian, the periods of time covered by this chapter and the next two are trivial, but their physical and biologic events set the stage on which we play out our lives.

Details of the record in Paleozoic rocks are occa-

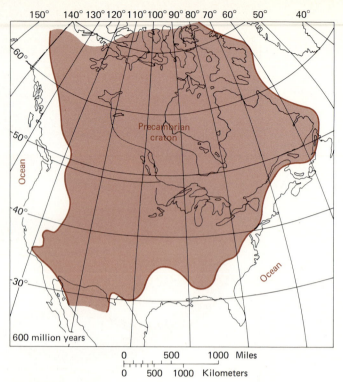

FIGURE 19–1

This diagram depicts North America 600 Ma, just before the beginning of the Paleozoic Era. All the continental crust outside the colored area has been added to the North American continent since Precambrian time. (After P. B. King, 1976, U.S. Geological Survey Professional Paper 902.)

sionally given to illustrate *how* geologic history can be interpreted (see the opening chapter quotation). As Bertrand Russell (1872–1970), English mathematician and philosopher, pointed out, ''A fact in science is not a mere fact, but an instance.''

PHYSICAL EVENTS

At the beginning of the Paleozoic, Precambrian rocks of diverse ages (see Figure 18–11) made up the North American craton or shield (Figure 19–1). From the Eocambrian through the Cambrian, the rocks were exposed to erosion, and deposition continued in seas marginal to the craton. The western continental margin stretched from British Columbia south an unknown distance into Mexico (Figure 19–2); the eastern boundary roughly paralleled the present site of the Appalachian mountains.

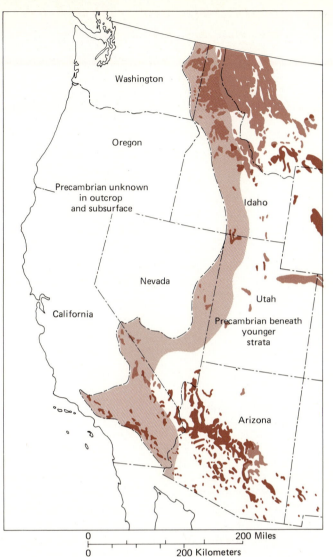

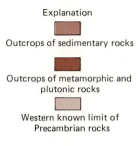

FIGURE 19–2

This diagram depicts the nature of the western continental boundary at the end of Precambrian time based on the present-day distribution of surface exposures of Precambrian rocks plus knowledge gained from the penetration of Precambrian rocks in deep wells. (After P. B. King, 1976, U.S. Geological Survey Professional Paper 902.)

LATE PROTEROZOIC-CAMBRIAN CONTINENTAL RIFTING

Interspersed within the Eocambrian sedimentary rocks are layers and pods of volcanic, largely basaltic rock. Along the southern margin, in what is now southwestern Oklahoma, intracratonic rift basins opened, much like those along the western shoreline (see Figure 18–19) that developed in Late Proterozoic time. Called **aulacogens,** their development together with the production of volumes of basalt signaled an episode of continental rifting and breakup. Today much the same sequence of events along the East African Rift Zone (see Chapter Ten) and the Red Sea signals the breakup of the African and Arabian plates.

This worldwide episode of reorganizing cratons was coincident with widespread glaciation. The breakup of what has been called "Proterozoic Pangea" created within the Late Proterozoic and Early Paleozoic the distinctive sequence of rocks that form in the earliest stages of continental rifting. These include long, linear belts of basaltic volcanics. The southern Oklahoma aulacogen was the site of both gabbroic and granite intrusions during the Early Paleozoic, accompanied by an abnormally thick sequence of coarse sedimentary rocks. The plutonic association is well exposed in the Wichita Mountains of southwestern Oklahoma, along with a sedimentary assemblage formed by deposition into the aulacogen during the Cambrian Period.

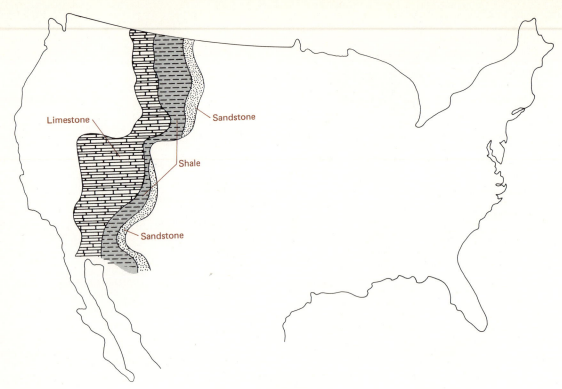

FIGURE 19–3
A sketch map of the distribution of major rock types whose age is 540 million years in the western United States. (Data compiled and generalized from several sources.)

THE WESTERN CONTINENTAL MARGIN

As an example of the evolution of a passive margin, let us look at the sequence of rocks forming in one instant of time, 540 Ma. The map in Figure 19–3 is drawn by plotting the distribution of rock types of the same geologic age; such a map summarizes a considerable amount of individual investigative effort. As you can see, the broad patterns revealed are remarkably simple: sandy sediment grades eastward into an area lacking deposits of this age and westward into shale and still farther offshore into marine limestones. Patterns of deposition on modern continental shelves or miogeoclines are similar, with the coarsest sediment along the shoreline, with muds offshore grading progressively seaward into limy sediment.

The kinds of Cambrian fossils preserved in these rocks are consistent with the western shoreline of North America at this time being a continental shelf whose shoreline trended almost due north-south, in terms of present geography. Paleomagnetic evidence (see Chapter Ten) shows that during Cambrian time the shoreline would have stretched east-west, with the equator crossing the Great Lakes and the Gulf of

Mexico (see Figure 10–3). What is now the eastern part of North America would have been the southern half, with the present western half being the northern half. Climate would have been tropical, since the geographic position of Cambrian North America was bisected by the equator.

If we draw a diagrammatic east-west cross section across the western United States (Figure 19–4), based on what is exposed in deep canyons and numerous oil wells, we observe that the sandstone at the old shoreline is quite thin, and it is deposited directly on the granitic and metamorphic rocks of the craton. Farther west the sandstone grades into thicker shaly and then limy units. Such a cross section would be nearly *identical* to one drawn today south from Baton Rouge, Louisiana into the Gulf of Mexico. As James Hutton reminded us long ago, the present is one key to understanding the past. Look at what a study of modern environments tells us about rock sequences in the past.

The Facies Concept

Notice that for any one brief interval of time, three *different* rock types, sandstone, shale, and limestone, *are being deposited simultaneously*. Therefore any one

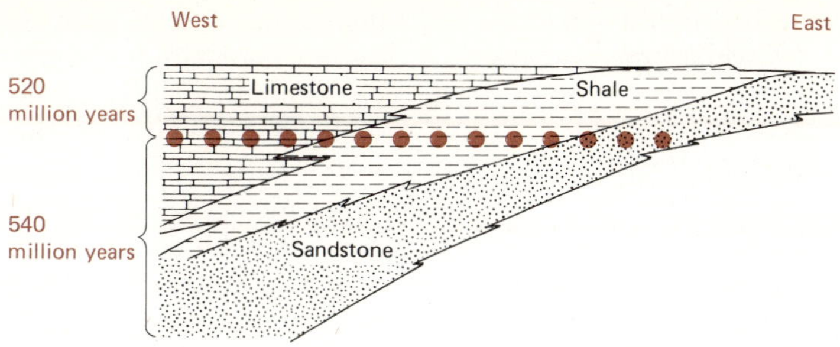

FIGURE 19–4

A highly diagrammatic cross section of the western United States showing changes in thickness and type of rock for two broad time intervals. Deposition during the 540- to 520-Ma interval was essentially continuous; the dates given are averages. The series of colored dots designates a distinctive index fossil which allows a fairly precise separation of time intervals. During this, and any other time interval, sand was being deposited near the shore, mud farther offshore, and limy material still farther offshore. *One single rock type is not the same age everywhere but becomes progressively younger toward the continental platform or craton.*

rock type is not the same age everywhere; each rock unit *becomes progressively younger toward the craton.*

The recognition that rocks of differing appearance could be of the same age was one of the milestones of the history of geology. William "Strata" Smith and Baron Cuvier suggested independently that although unlike *rock* types or sequences could reappear through time, the sequence of *fossil* assemblages through time was irreversible (see Chapter Seventeen). Fossil forms appear in a determinable order through time, and once extinct, they never reappear. Geologists **correlate** two separate rock units by determining their mutual time relations with either radiometric or faunal means.

Geologists have coined the term **facies** (from the Latin *facies*, meaning shape or form) for *the overall aspect and common characteristics of a unit that distinguish it from other units.* Defined originally by the Swiss geologist Armand Gressly in 1838, the term has been widely accepted and applied to a variety of different aspects of natural units. At least three different *kinds* of facies can be described.

▶ **Lithofacies.** Rock units sharing a common composition, such as a *granite gneiss facies* within Archean rocks, or a common environment of formation, such as a *greenschist facies* of metamorphic rocks, or the *red bed facies* of sandstones.

▶ **Biofacies.** A group of modern or fossil organisms having a common environment, such as a *back-reef facies,* or a common form of fossil life, such as a *shelly facies,* a group of rocks which contain bivalve fossils.

▶ **Stratigraphic facies.** Aspects of rocks which are *all the same age,* such as the *shaly facies* of Middle

Cambrian rocks or the *stromatolitic facies* of Eocambrian rocks.

Notice that during the geologic time indicated by the colored dots in Figure 19–4, there are sandy facies, shaly facies, and limy facies simultaneously being deposited side by side. The stratigraphic facies occur in a *horizontal plane,* yet each sequence of lithofacies—sand, shale, lime—occurs in a *vertical array,* superimposed on one another (Figure 19–4).

Johannes Walther (1860–1937) was a German geologist who often found modern processes the most satisfying explanations of ancient phenomena, another statement of uniformity. He recognized the interrelatedness of horizontal stratigraphic and vertical lithologic facies, and in 1894, he established what he termed the **principle of succession of facies:** *only facies that can be observed beside each other at the present time will be elsewhere found directly superimposed.*

He believed that understanding modern environments would always provide the clues for interpreting ancient facies, just as knowledge of modern organisms provides the "architectural plans" for the reconstruction of extinct species. Walther's principle leads us to expect that certain facies will show only limited kinds of transitions to other facies. Because along modern continental shelves we observe that sandstones pass seaward into clay-rich muds and then into limy sediment, we should expect limy muds to pass vertically (upward or downward) into clay-rich muds. Based on the sequence of modern sediments, we should expect there to be *recurring patterns* of sedimentary rocks in a vertical column, depending on the *environments* in which those rocks were formed. Unlike fossil assemblages, a limited number of vertical sequences of lithofacies *will* recur, with the exact sequence controlled by the characteristics of the environment in which the sediment formed.

THE EARLY PALEOZOIC TRANSGRESSION

To demonstrate Walther's principle in action, we now make a somewhat more elaborate cross section across the western United States for the whole of Eocambrian and Cambrian time, from approximately 700 to 500 Ma (Figure 19–5). Having done this, we can then describe *the sequence of past events*—the area's history. Notice that the oldest sedimentary rocks are confined to the marginal sea, and that progressively younger rocks encroach farther and farther east onto the craton.

The rocks sketched in Figure 19–5 and the fossils they contain tell us that throughout this time, sea level steadily rose against the land and flooded the margin of the craton. *The story is essentially the same all around the North American craton.* Progressively younger and younger Paleozoic rocks are located inward toward the center of the Precambrian craton. Geologists call such an encroachment of the sea onto the land a marine **transgression.**

Since there is a general relation between distance offshore and grain size, when there is a transgression, any one point is progressively farther offshore and receives progressively finer and finer sediment. A vertical column of marine rocks that *becomes progressively finer-grained upward* is a classic expression of Walther's principle of succession of facies—*such a sequence recurs wherever there is marine transgression across a miogeocline.*

If the sea level falls, the sea retreats or regresses from the land, an event called a marine **regression.** A regression is marked at any one point by an increasingly shallow sea and by a column of marine rocks that *becomes progressively coarser grained upward.*

When there is a regression, layers of recently deposited sedimentary rock are always uncovered and exposed to erosion (Figure 19–6). As erosion ensues, the record of deposition is progressively lost, producing an *unconformity* (defined in Chapter Seventeen). The amount of time or the erosional interval represented by the gap or unconformity *progressively increases toward the shore,* since this is where the first exposure occurs at the beginning of regression. As the regression continues, the vertical sequence is marine rocks with an overlying unconformity overlain in turn by continental rocks. Regressive rock sequences are often truncated by the forces of erosion, whereas transgressive deposits tend to be preserved.

THE CRATONIC SEQUENCES

The record of marine and continental deposition of sedimentary rocks on the craton is riddled with unconformities. Estimates are that up to 80 percent of the rocks deposited on the craton have been eroded, their remnants deposited in the flanking marginal seas around the craton. The sea may transgress and largely cover the craton, as in Late Ordovician time when the seas essentially covered North America, but their regression will leave thin sedimentary rocks exposed to numerous erosional agents.

Paleozoic rocks on the craton give us only occasional glimpses of the totality of Paleozoic history. Cratonic sedimentary rocks form thin packages bounded above and below by unconformities. The cratonic rock record—*thin, discontinuous, and riddled with unconformities*—is fundamentally different from that recorded in the marginal geoclines at the same time. The record left behind in the sedimentary rocks formed along the passive margins is essentially continuous.

In recognition of the uniqueness of the record retained in the sedimentary rocks of the craton, L. L. Sloss, an American geologist, has proposed that cratonic rocks provide a *sixfold* division of Phanerozoic time. Each packet of sedimentary rocks is termed a **sequence,** a unit of sedimentary rocks resulting from one *major* transgression and regression (Figure 19–7).

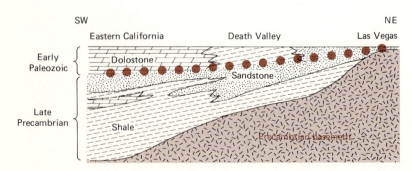

FIGURE 19–5

A geologic cross section from Las Vegas, Nevada west to eastern California. The colored dots define the strata whose fossils mark the boundary between Precambrian and Paleozoic rocks. The shoreline must have been just east of Las Vegas, Nevada about 570 Ma. (After P. B. King, 1976, U.S. Geological Survey Professional Paper 902.)

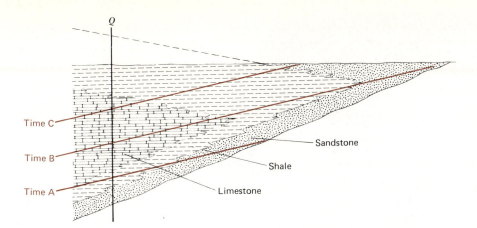

FIGURE 19–6
A diagrammatic transgressive
sequence of rocks followed by a
regressive sequence, all shown in
cross section. The colored lines
indicate planes of *equivalent time*.
Time B is the time of maximum
transgression. Notice that at locality
Q the vertical sequence moving
upward is at first finer and then
coarser, marking the shift from
transgressive to regressive phases.
The entire sequence is bounded by
an *unconformity*, because the upper
part of the regressive rock series has
been eroded away.

TWO CRATONIC EXAMPLES

Often called the **stable interior,** the area of cratonic
rock between the marine marginal basins is little affected by seismic events. The stable interior includes
all of central Canada and all of the central United
States between the Appalachian and Rocky mountains.

The surface of the craton is not a totally flat, fea-
tureless sheet. By slow, gentle movements its surface
is formed into a series of gentle rises and basins. One
well-developed rise is the *Transcontinental Arch*,
which stretches from the Great Lakes region southwest to Nevada. Among the best-studied basins is the
Michigan Basin, a roughly circular depression centered on what is now the state of Michigan. A brief
review of its history will provide an example of the
kind of tectonic activity characteristic of stable interiors.

The Michigan Basin

In Middle Silurian time (ca. 420 Ma) a circular area
centered on northeastern Michigan began to subside
rapidly, dropping 3300 feet (1 kilometer) in 10 million years, for an average subsidence rate of 3 feet (1
meter) per 10,000 years. The cause of such rapid
subsidence is unknown; something must have upset
the isostatic balance of this part of the stable interior,
causing a broad areas to undergo very gentle depression. A slow change in elevation over a broad area is
known as *epeirogeny*, as mentioned in Chapter Nine.
As epeirogenic movement continued, reefs began to
form around the margins of this shallow basin.

Like their modern counterparts, these ancient
reefs were largely composed of limestone skeletons
secreted by small animals that lived within the shallow, warm, sunlit water (Figure 19–8). Reefs de-

velop because agitated, well-oxygenated, sunlit water is ideal for the production of microscopic life,
which is the food base for all other life. As diverse
animal and algal communities develop, the growth

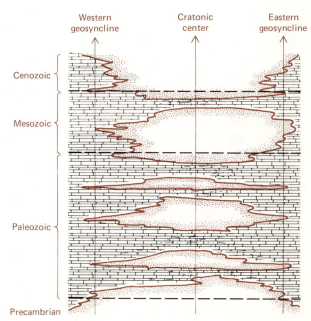

FIGURE 19–7
The cratonic record is best understood by considering it
as a thin series of *sequences*, thin units of sedimentary
rocks bounded above and below by unconformities.
These unconformities divide the sequences and produce
gaps in the historical record. The extent of the gap
increases toward the center of the craton and dies out
into the margin geoclines, where sedimentation may be
essentially continuous. (Adapted from L. L. Sloss, 1963,
with permission and with the permission of the
Geological Society of America, Inc. Copyright 1963.)

FIGURE 19–8
An example of reef material from a 370-million-year-old reef in central Arizona. Life types captured as fossils include sponges, corals, and stromatoporoids. (Photograph by Curt Teichert, U.S. Geological Survey.)

of skeletal organisms is so great that the reefs grow upward from the seafloor. Sponges, stromatoporoids, algae, and corals, as well as the calcareous skeletons of snails, clams, and sea lilies, contributed to the reef mass marginal to the Michigan Basin.

As the basin subsided and the depth of water increased, reef-building animals that needed light to live built their limestone homes higher. A substantial mound of reef rock built up to near sea level. In time the reefs became substantial fringing barriers along the margin of the basin. The reef rock encircling the Michigan Basin is the same age as the limestone that forms the lip for Niagara Falls and the surface rock for part of the upper midwestern United States.

The patchwork of fringing reefs restricted the flow of seawater into the interior of the basin. Michigan was near the Silurian equator, so the climate was hot; any seawater spilling in through the fringing reefs and entering the restricted shallow basin rapidly evaporated (Figure 19–9). As evaporation continued the water within the restricted, shallow basin became more and more saline, and deposits of gypsum (Appendix A), anhydrite, and rock salt progressively formed in the basin center.

Mining of these massive deposits of basin evaporites furnishes rock salt for table and industrial use, potassium-rich rock for fertilizers, and gypsum for the building industry. If you are in the Midwest, the salt in your shaker has been waiting about 420 million years for you to recycle it. Not only are reef eva-

porites useful, but the reef rock itself, rich in lifeforms and highly permeable, may be the host for important petroleum accumulations.

The association of evaporitic sedimentary layers with fringing reefs is a common one, predicted by Walther's principle of faunal succession. Within the Devonian time period the same association formed near Alberta, Canada. It recurs in the Permian rocks of west Texas and southeastern New Mexico.

Eustatic Cycles

Near the end of the Paleozoic Era, during the period European geologists call the Carboniferous and Permian periods and American geologists the Mississippian, Pennsylvanian, and Permian periods (360 to 250 Ma), the eastern and central craton was covered by more than 50 repetitive, highly cyclic synchronous sequences of marine and nonmarine strata, including extensive coal deposits. Similar rock sequences synchronously deposited were common on stable platforms worldwide.

Such rock sequences are formed by **eustatic** cycles, worldwide changes in sea level that produce unique sets of rock units formed by repetitive transgressive-regressive cycles. What processes could cause small changes in sea level to occur over all the continents at the same time?

The answer is still controversial, but two processes may be involved.

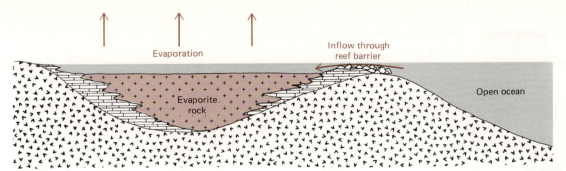

FIGURE 19–9

This cross section could be from any highly restricted basin undergoing evaporation, such as the modern Persian Gulf or the Silurian Michigan Basin. Seawater enters the basin through a series of fringing barrier reefs.

As water evaporates, what remains becomes denser and sinks. Evaporation of the trapped water continues, and evaporite minerals that grade laterally into reef limestones are deposited around the basin.

1. Change in the volume of the ocean basins, or
2. Change in the volume of water in the basins.

One recent proposal is that past worldwide cyclic changes in sea level are best accounted for by episodic volcanism along the world's oceanic ridge system. As the ridges were inflated with magma (Chapter 10), they decreased the volume of the "container," and sea level rose. Then as the ridge cooled and contracted, the ocean basins gained volume, and water came back off the land. Often termed *pulsation tectonics*, the idea receives strong support from the study of the correlation of periods of high spreading rates with periods during which the seas made the greatest advances onto the land.

Others have suggested that the cyclic rise and fall of the sea reflected *Milankovitch cycles* (see Chapter Fourteen). The onset of glaciation lowered sea level as it tied up more water, and the warming trend caused a rise in sea level. Since there was widespread glaciation on the southern Gondwana continents during the Pennsylvanian and Permian periods (see Chapter Ten), the correspondence between the timing of each individual cycle (estimated at a few hundred thousand years) and the cycles of glacial advance and retreat is quite suggestive.

Whatever the causes, these cyclic deposits, formed in a tropical climate approximately 300 Ma, contain a major coal resource for both the United States and Europe. The eroded remnants of rocks of this age stretch from Rhode Island to Oklahoma (Figure 19–10) and provide an important component of the industrial energy for the heartland of the United States.

As we have learned, cratonic rocks are naturally divided into six *sequences* (see Figure 19–7), four of which are of largely Paleozoic age. We began this chapter by describing the oldest sequence whose transgression began in latest Precambrian and continued through Late Cambrian, by which time the sea had covered all of North America except the Hudson Bay area of Canada. The second sequence culminated in a marine transgression that covered almost all of North America, *except for* an area that includes much of what was to become the Appalachian Mountains.

In this area the eastern passive margin was beginning by Middle Ordovician time (about 470 Ma) to shift from the passive to an active phase within the Wilson cycle (see Chapter Eighteen). The Atlantic Ocean was closing as onrushing Europe and Africa headed westward (actually northward in terms of Ordovician geography) toward a series of collisions with North America that were to raise the Appalachians into the sky.

THE APPALACHIAN RECORD

The southern Appalachians reveal the geometry of a simple, moderately deformed geocline, with the strongly to gently folded and faulted miogeocline forming the Valley and Ridge and Appalachian Plateau areas on the continental side of the range, and the strongly metamorphosed and plutonized eugeocline forming the Piedmont and Great Smoky Mountain areas on the seaward side. The northern Appalachians reveal a much more complex series of events, dominated by plutonism, metamorphism, and the formation of ophiolite suites accompanied by repeated accretionary episodes.

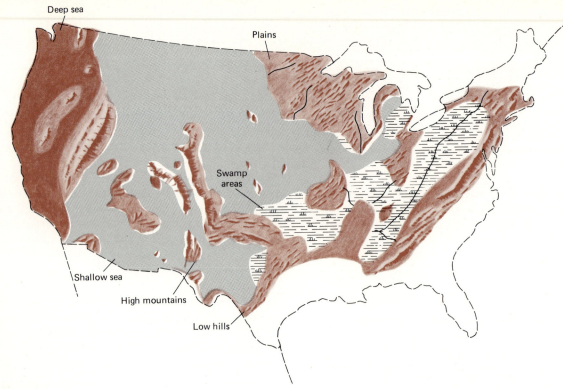

FIGURE 19–10

A reconstruction of the geography of the United States 300 Ma. In areas shown as swamps in the central and eastern United States are major coal deposits today. The coal deposits form about one-quarter of total United States reserves. (Map from the U.S. Geological Survey.)

The story of the sequence of events in the Appalachians is associated with a worldwide Wilson cycle. The formation of the fold belt is closely correlated with the opening of what has been called the *Iapetus* Ocean between "Europe" and "Africa" and the "Americas" in Early Cambrian time (570 Ma), a part of the general rifting that separated the Proterozoic supercontinent into smaller fragments. Based on faunal, paleomagnetic, and ophiolitic suite evidence, the Iapetus Ocean basin was open and growing in size through much of Silurian time (about 570 to 410 Ma), but it began closing within the Silurian and was closed by Middle Devonian (380 Ma) to Mississippian (340 Ma) time.

The major fold belts that resulted are called the Caledonides belt (pronounced Cal'-eh-dohn-eye-dees) in Greenland and northwest Europe (*Caledonia* is the ancient Roman name for Scotland); the Taconian and Allegheny belts in the Appalachians in eastern North America; the Ouachita belt (pronounced wash'-it-tahs) in Oklahoma and Arkansas; and the Marathon belt in Texas. The deformation that resulted can be separated into five distinct **orogenies** (see Chapter Nine), periods of mountain formation. All these directly or indirectly affected North America.

1. **Taconian.** A Late Ordovician (480 to 450 Ma) orogeny affecting the northern Appalachians along an island arc complex.
2. **Caledonian.** Ordovician to Lower Devonian (460 to 400 Ma) orogeny affecting Britain, Scandinavia, and eastern Greenland. The locus of this mountain-forming event migrated to eastern Canada by the Middle Devonian.
3. **Acadian.** Devonian (400 to 350 Ma) orogeny affecting the northern Appalachians. This represented a a continuation of the Caledonian orogeny in North America and was caused by the collision of northern North America with Europe. The northern Iapetus Ocean basin closed during this time.
4. **Alleghany.** Pennsylvanian and Permian (320 to 260 Ma) orogeny affecting primarily the southern Appalachians, marking the final

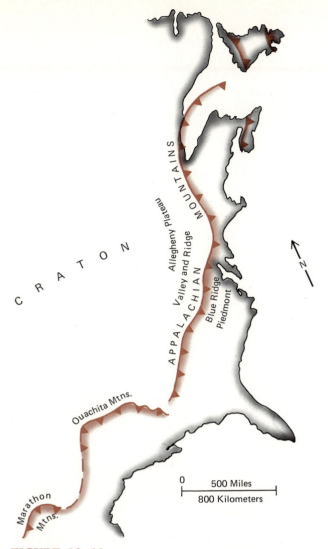

FIGURE 19–11

This sketch map shows a reconstruction of the relation between the Appalachian geocline, which stretched well north into Canada and joined the Caledonian belt in eastern Greenland, and the southern geoclines, which were deformed into the Ouachita and Marathon mountains. Landward from the geoclines is the craton, composed of Precambrian rocks of many different ages (see Figure 18–11).

collision of North America with Africa. Once called the *Appalachian orogeny*, this event completed the formation of the Appalachians.

5. **Ouachita.** Pennsylvanian and Early Permian (320 to 270 Ma) orogeny which affected what is now the southern passive margin for the first time. This orogeny culminated in the formation of the Ouachita and Marathon mountains as the

Ouachita–Marathon geocline collided with Colombia and Venezuela.

Late Proterozoic to Late Ordovician

The history of the Appalachians begins with Late Proterozoic extension and rifting of the margins of the North American craton. In Canada, the rifts were filled with a 4-mile-thick (6 kilometers) sequence of coarse continental sediments shed by rising Precambrian landmasses. These sedimentary rocks are exposed on the Avalon Peninsula of Newfoundland. A similar sequence in the Great Smoky Mountains of North Carolina and Tennessee formed across the Cambrian-Precambrian boundary. Basaltic volcanism created pods of lava and flows along rifted zones.

The Cambrian and Early Ordovician periods are dominated by marine transgression and by the formation of localized offshore island arcs; almost from their inception, episodes of localized subduction occurred along these arcs. The onset of the Taconian orogeny in the Northern Appalachians was marked by the first accretionary event in the north, in which slices of the Iapetus oceanic floor and materials from the offshore island arcs were caught between the colliding miogeocline and the Avalon rocks. The minimum foreshortening (loss in width across the deformed belt) caused by this collision is 100 percent. To the north the rising Taconic landmass began shedding coarse continental deposits, a sequence collectively called the *Queenston delta*.

Devonian to Pennsylvanian

Following continued deposition of coarse reddish sediments, the Acadian orogeny took place in the Middle Devonian (380 Ma) as Canada obliquely collided with Europe. The unstable point of collision migrated generally southwesterly down the northern and central Appalachians. The Acadian was a full-blown continental collision, creating mountains reaching Himalayan proportions. Large deposits of coarse, feldspar-rich reddish rocks, called the *Catskill delta complex*, were shed from the newly elevated mountains. Like the older Queenston rocks, these deposits became coarser-grained and thicker toward their mountainous source to the east.

The Acadian orogeny generated large volumes of granite and intense metamorphism, forming plutonic domes in the northeast, and extensive domains of regionally metamorphosed rocks. The angular unconformities between strongly tilted 350-million-year-old rock and slightly younger undeformed rock allow a rather precise determination of the timing of the main pulse of the Acadian orogeny—the most severe orogeny affecting the northern Appalachians.

Pennsylvanian to Permian

The continuing rise of the Appalachians had reversed easterly drainages. Rivers that had flowed into the Iapetus Ocean now drained into the continental interior, creating extensive swamps in which Late Paleozoic vegetation was slowly converted to coals. The locus of deformation shifted to the south as the Alleghany orogeny began when Africa and North America collided.

The Alleghany orogeny caused the miogeocline to be pushed westward, moderately folded, and thrust-faulted. One of the main thrusts, the Blue Ridge (Figure 19–11), is estimated to have a minimum westward displacement of 90 miles (150 kilometers). Regional metamorphism and plutonism occurred in the Avalon area once again. The features associated with continent-continent collision mark the embrace of Africa and the eastern seaboard of North America.

Still farther south the collision of northern South America with southern North America deformed a passive margin undisturbed since its rifting in Late Proterozoic time. The onset of subduction is marked in Mississippian time (340 Ma) by turbidity flows into a deep-water basin associated with volcanism to the south. The formation of the island arc implies that subduction was underway by Middle Mississippian. The arc filled in during the Pennsylvanian Period; later strong horizontal compression directed northwestward forced the geocline-arc system up onto the craton. By Early Permian time, horizontal strata overlay strongly tilted sediments, marking the termination of 220 million years of intermittent unrest along the eastern and southern margins of North America.

PANGEA

During the Pennsylvanian Period, western Europe south of the Caledonides collided with Africa, creating the *Hercynian orogeny,* whose deformed rocks can be recognized over most of western Europe, northwestern Africa, and southernmost Britain. Europe collided with Asia to form the Ural Mountains within the Permian Period in what is called the *Uralian* orogeny.

A worldwide episode of collision marked the end of the Paleozoic Era. By the end of the Paleozoic all the world's continents became locked together into a single supercontinent, **Pangea.** Wegener (see Chapter Ten) had been right after all; his mechanism for continental motion has been shown to be faulty, but

his reconstruction of the past was basically correct.

The formation of Pangea, with the consequent formation of the Caledonides-Appalachian range—a continuous mountain range stretching from Texas to Oklahoma, Alabama, Newfoundland, Greenland, Scotland, and Scandinavia—was only one of the results. The Hercynian ranges of western Europe and northwestern Africa and the Urals of Europe and Siberia were also created by the collisions coalescing cratonic fragments into one supercontinental mass.

The formation of Pangea also had profound effects on the history of life, for the creation of a supercontinent radically changes the distribution of environments and the kinds of life that can inhabit them. Coincident with the formation of Pangea was one of the greatest of all mass extinctions. The dying out apparently cut randomly across a wide array of life-forms; the offspring of its survivors made the Mesozoic a wholly different world from the Paleozoic.

It is to this Paleozoic world our studies now turn. The Paleozoic seas teemed with marine invertebrates and primitive vertebrates as well as the conservative prokaryotes from the distant past. By the earliest Devonian, the continents, barren since their origin, hosted amphibia, insects, and a wide array of land plants. There is even some fragmentary evidence even of Ordovician land-dwelling life.

PALEOZOIC LIFE

To step from the limited world of Precambrian life, with its bacteria, cyanobacteria, and green algae, to the Paleozoic world was to enter one teeming with diversity. Within the first 10 million years—one-thirtieth of all Paleozoic time—an explosive radiation of marine animals without backbones had filled the Paleozoic seas with a wide array of life-forms.

The reasons for this sudden gain in diversity remain a mystery, though the explosive development of new life-forms often follows mass extinctions, a phenomenon seen at the close of the Precambrian. Part of the reason we *see* such diversity is because earliest Paleozoic life rapidly developed both external and internal *skeletons,* a type of specialized (and more easily preserved) organ system totally unknown in all the previous 3.5 billion years of Precambrian life.

THE FIRST SKELETONS

The development of skeletons is the major biologic event separating Paleozoic from Precambrian life. Skeletons offer many advantages. External skeletons provide a defense against predators and a covering against which muscles can work and from which internal organs can hang. Internal skeletons allow body size to increase by providing a strong framework to support body weight. Skeletons also provide a flexible, jointed frame which makes sophisticated forms of motion possible.

Increase in body size carries its own reward. Larger size is closely tied to longevity, productivity, survival through prolonged starvation, defense against predators, and effective maintenance of body temperature. Increases in size within Precambrian life had been possible only by

▶ Adopting a colonial life-style, with progressive specialization of cells.

▶ Adopting a very flat body form (the *Ediacaran* fauna) to keep cellular tissues close to the surrounding water.

But with all the advantages possession of a skeleton confers, it has one major drawback. A skeleton is a heavy weight that must be moved around at the expense of the organism's energy.

Preservation

From the perspective of **paleontologists,** geologists who study the preserved evidence of past life, skeletons confer another advantage. A skeleton *dramatically* improves the odds that some evidence of the organism will be left behind after death.

Because of the efficiency of scavenging organisms, as well as the agents of weathering and erosion, the odds that any one boneless organism will be preserved are essentially nil. These odds are improved if a hard skeleton is available and the creature is *rapidly* buried—in tar, quicksand, ice, tree resin, volcanic ash, or more commonly a continuous rain of sediment onto the seafloor. Rapid burial denies access by scavengers to the remains and slows the decay process. After death the soft parts eventually decay, but a variety of processes can act to preserve the skeleton.

Those processes include

1. **Permineralization.** Deposition of hard minerals in the pore space within skeletal tissue; this is a common method for the preservation of large bones.

2. **Replacement.** Conversion of organic tissue to hard minerals by molecule-by-molecule replacement; an example is petrified wood.

3. **Molding.** The imprint or impression of the organism into mud or other soft material; many leaf fossils are preserved in this way.

4. **Casting.** The natural filling of an impression, recreating the three-dimensional animal; casts are a common mode of preservation for shells.

5. **Carbonization.** The creation through partial decay of a thin film of carbon along bedding planes; the thin film preserves a detailed record of living tissues.

6. **Mummification.** Drying the organism through burial in tar seeps or in extremely arid environments produces a mummified replica of the organism.

7. **Resin Traps.** Trapping and enclosure of small insects in the sticky, pitchy, or resinous materials on some trees leaves an exquisitely preserved fossil.

8. **Pyritization.** The preservation of tissue by pyrite replacement may occur in highly stagnant watery environments without much oxygen.

FIGURE 19–12

These are cross sections of one of the earliest Paleozoic fossils. Known as *archaeocyathids*, these fossils are several inches or centimeters long and are of uncertain affinity. Their cone shape with a hollow cavity is very typical, reminiscent of ancient sponges—a literal translation of their name. (Photograph courtesy of the U.S. Geological Survey.)

9. **Freezing.** The "quick-freezing" of animals after they have fallen through the surface of an icy lake; the preservation may leave tissue in a little-altered state. A few Pleistocene animals were "quick-frozen."

EARLY PALEOZOIC LIFE

The fossil life-forms found within the Cambrian, Ordovician, and Silurian periods (from 570 to 410 Ma) were exclusively marine and largely **invertebrate.** Invertebrates are "the great spineless majority"; today their representatives include about 97 percent of the some 20 million living species.

Two of the episodes during which the seas advanced across the continent to the maximum extent occurred during the three periods that form the Early Paleozoic. Consequently, the majority of Early Paleozoic life developed in shallow seas that covered most of the craton and its largely passive margins.

The diversity of life is amazing. An incomplete list of early Paleozoic life-forms would include *archaeocyathids, trilobites, brachiopods, gastropods, pelecypods, corals, stromatoporoids, ostracods, bryozoa, graptolites, echinoids,* and *eurypterids.* The Paleozoic world was full of life.

Archaeocyathids

Among the earliest of all Paleozoic life-forms is the *archaeocyathid* (Figure 19–12), a cone-shaped or vase-shaped creature whose skeleton was composed of calcium carbonate. The name literally translates as "ancient sponge"; archaeocyathids are indeed similar in appearance to primitive sponges. Further study of these organisms indicates that they are so totally unlike any other life-form that they should be placed in a *phylum* of their own. A phylum is the primary subdivision of *kingdom,* the largest taxonomic unit of life (see Appendix F).

Trilobites

Trilobites (Figure 19-13) belong to a very large group of abundant marine *arthropods,* a phylum of complex animals making up over 80 percent of all animals living today. Arthropods, which have a horny, segmented, external skeleton with jointed limbs, include all insects, lobsters, crabs, shrimps, barnacles, spiders, millipedes, and centipedes (Appendix F).

Trilobites were unique to the Paleozoic Era; they are also that era's most characteristic fossil. Consequently wherever they are found they are considered

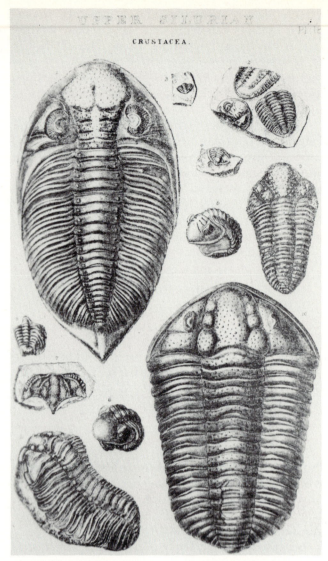

FIGURE 19–13

These are trilobites, taken from 410-million-year-old rock (Late Silurian) by the geologist who named the period. The head has two distinct eye sockets, the body is composed of numerous segments, and the tail of this particular species is very poorly developed. (From Roderick Murchison, 1872, *Silurian System,* courtesy of the History of Science Collections, University of Oklahoma Libraries.)

an *index fossil* for Paleozoic rocks. An **index fossil** is a life-form which underwent rapid evolutionary changes, occurred over a widespread area, and was unique to a single interval of time. Because of the trilobites' rapid rate of evolution, elaborate markings, and widespread occurrence, individuals make excellent index fossils for the periods within the Paleozoic Era.

Trilobites gain their name from their characteristic separation into three lobes: head, body, and tail.

FIGURE 19–14
Here are different views of *Leptocoelia*, a brachiopod that was fossilized 390 Ma. Its shells are intensely wrinkled, probably to provide extra strength. The lower and upper shells are of different size but are symmetrical about a plane *perpendicular* to the hinge line where the shells join. (Photograph courtesy of the U.S. Geological Survey.)

They have an exterior skeleton somewhat like that of the modern horseshoe crab. It is developed from *chitin,* a horny organic compound similar to the material that composes our fingernails and the external skeletons of many insects. The trilobites were seafloor scavengers, eating decaying materials there, much as crabs do today.

Because they lived on the dark seafloor, trilobites developed the most sophisticated eyes ever preserved. The lens elements of trilobite eyes were made of single crystals of calcite and came in compound and simpler forms, both of which had *doublet lenses.* The doublet lens so highly corrects for various optical defects that it was not discovered by human beings until the seventeenth century. Trilobites accomplished this optical miracle 5.7 million centuries earlier. They created natural lenses of such optical perfection that they have yet to be duplicated in the natural world.

Brachiopods

The *brachiopods* are a phylum of shelled animals that at first glance can be mistaken for clams (Figure 19–14). They are often called "lampshells," because the open shell is shaped something like old Roman oil lamps. Their range extends from the Lower Cambrian to the present; they still inhabit the seafloor today, though in vastly reduced numbers and variety.

More than 30,000 species of fossil brachiopods are known; they are among the most abundant Paleozoic fossils. Brachiopods differ from their "lookalike," the clam, in the organization of their body. The plane of mirror symmetry (see Chapter Two) of brachiopods is *perpendicular* to the line along which the shell hinges; the plane of mirror symmetry of a clam is *parallel* to this line (compare Figures 19–14 and 19–15).

Pelecypods

Pelecypods are one of the molluscs, a phylum of lifeforms second only to the arthropods in abundance. The molluscs include all edible shellfish, *gastropods,* including snails, *chitons, scaphopods* or tusk shells, *cephalopods* including squids and octopuses, and *monoplacophorans,* an almost extinct group of primitive limpetlike molluscs with a cap-shaped shell. The clams or *pelecypods* live along the floors of both oceans and freshwater lakes. Symmetrical in a plane which parallels their hinge line (Figure 19–15), clams are a common component of many faunal assemblages laid down throughout the Phanerozoic.

Gastropods

Gastropods range from the Upper Cambrian to the present, occupying marine, freshwater, and airbreathing environments. Characterized by an asym-

FIGURE 19–15

Pelecypod shells are themselves asymmetrical, but each shell is the *mirror image* of the other. The plane of symmetry *parallels* the hinge line where the two shells join. Compare with Figure 19–14. (Photograph courtesy of the U.S. Geological Survey.)

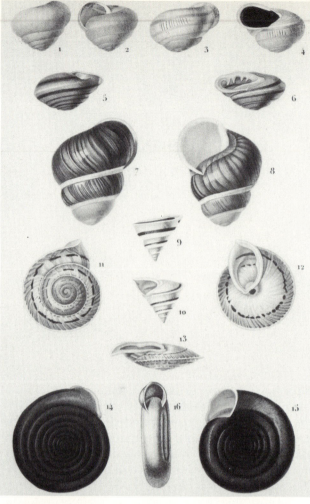

FIGURE 19–16

Here are various kinds of coiled gastropods. Some are coiled in cones; others coil within a single plane. Coiling is the result of displacement of the body of the animal from its foot, which is extended out of the shell opening. Such displacement offers advantages in streamlining and water flow. (From A. D'Orbigny, 1849, courtesy of the History of Science Collections, University of Oklahoma Libraries.)

metric calcium carbonate shell which is often spiraled and closed at the apex (Figure 19–16), the gastropod "carries its house on its back" wherever it goes.

Eurypterids

Among the most unique life-forms of the Middle and Late Paleozoic were the "sea scorpions," or *eurypterids* (Figure 19–17). Ranging from the Ordovician to the Permian, they inhabited shallow brackish or freshwater lagoons. Looking very much like the modern land scorpion, eurypterids must have been fearsome predators. Though successful during most of the Paleozoic, they failed to survive it—fortunately for swimmers. Imagine a scorpionlike animal 9 feet (3 meters) long swimming toward you! Eurypterids were closely related to modern scorpions, spiders, ticks, and mites.

Graptolites

Among the typical fossils found in black shales of the Early to Middle Paleozoic are the *graptolites*, small colonial marine organisms of uncertain biologic classification. The organism had a chitinous external skeleton; it dangled down from a surface float along a series of branches which formed a colony.

Graptolites have been excellent index fossils wherever found, since they were distributed globally, evolved rapidly, and are commonly preserved. They are a particularly abundant Ordovician index fossil.

FIGURE 19–18
Graptolites, shown here in a sketch of the surface of a rock slab, are very important index fossils for the early Paleozoic, they were especially abundant in the Ordovician. This species is *Didymograptus murchisonii*. (From R. Murchison, 1872, *Silurian System*, courtesy of the History of Science Collections, University of Oklahoma Libraries.)

to be some form of ancient, undecipherable writing left within the rock by forces, spirits, or agents unknown. Modern practice places graptolites within their own phylum, emphasizing their distinctiveness.

Crinoids

The most important class of Paleozoic echinoderms, the *crinoids* have a five-sided radial symmetry; the body usually consists of a central disk with plumelike arms. The crinoids belong to the same phylum as starfish. Ranging from the Ordovician to the Holocene, the crinoids are commonly called "sea lilies." On the seafloor in modern reefs, sea lilies are a small but beautiful part of the marine environment. Attached by a long stalk (Figure 19–19) to the seafloor, crinoids wave their arms in the water to bring food to themselves. The crinoid stalk or stem is a rather common fossil, usually found as circular fragments. The head and arm are much less commonly preserved.

Crinoids became so prolific by Middle Paleozoic times that their fragments sometimes make up entire rock layers covering hundreds of square miles or kilometers. A single limestone layer in Iowa is estimated to contain over 230 trillion individuals.

FIGURE 19–17
This sea scorpion or eurypterid is a small one, about 9 inches (23 centimeters) long. It inhabited brackish water about 430 Ma. (Photograph courtesy of the U.S. Geological Survey.)

Ranging from the Middle Cambrian to the Pennsylvanian periods, they are preserved as carbon films on the bedding plane surface (Figure 19–18). Their name translates literally as "writing on stone," for when they were first discovered they were thought

LATE PALEOZOIC LIFE

The latter part of the Paleozoic Era includes the Devonian, Mississippian, Pennsylvanian, and Permian periods, an interval of time that stretches from 410

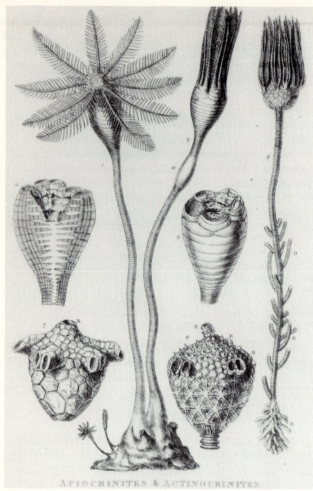

APIOCRINITES & ACTINOCRINITES

FIGURE 19–19

This view of the complete animal, rarely fossilized in its entirety, suggests why fixed crinoids are called "sea lilies." The jointed stems are a fairly common fossil find, but the budlike head and appendages are commonly not well preserved. Some crinoids similar to this live on in the seas today. (From Rev. William Buckland, 1837, courtesy of the History of Science Collections, University of Oklahoma Libraries.)

to 245 Ma. During this time the marine life-forms already mentioned continued their proliferation and were joined by still others.

This latter part of the Paleozoic opens with the *first colonization of the continents* about 410 Ma; until then they had remained lifeless and barren. With a wholly new series of environmental niches to conquer, continental life-forms rapidly expanded, moving in 160 million years from awkward amphibians to tough little reptiles and from primitive rootless, leafless plants to large trees inhabiting Late Paleozoic swamps.

We will first review a group of marine life, mem-

bers of which had their ancestry within the earlier Paleozoic but who reached their peak of abundance in the second half of the era. We will then watch the transition from life-forms in the sea to habitation of the land.

Foraminifera

More commonly called "forams," the *foraminifera* are an abundant group of single-celled animals with calcareous or siliceous external skeletons. They are superb index fossils, stretching from the Cambrian to the Holocene, but they became extremely abundant in the second half of the Paleozoic. Their granulated skeletons may make up entire limestone layers, and their kindred still form limy oozes in the deep ocean floor of today.

These tiny animals, whose name translates as "pore bearers," are studied under the microscope (Figure 19–20). They are ideal fossils for geologic research within the petroleum industry, for their small size allows them to be recovered whole within the rock chips brought up from the drilling bit. One very large group has shells the size and shape of wheat grains; they look for all the world like wheat grains entombed in the rock.

Radiolaria

Another group of microscopic animals are the *radiolarians* (Figure 19–21). They form their shells out of silica, creating delicate frameworks of opal. The skeletons are in a dazzling array of geometric shapes, which are revealed as extraordinarily beautiful under the microscope. The radiolaria have been fossil constituents of siliceous rocks since the beginning of Paleozoic time, and their progeny inhabit the seas today.

Ostracods

The *ostracods* are another group of small animals, about the size and shape of a small (Figure 19-22) lima bean. They are members of the Crustacea, a varied subphylum of arthropods (see Appendix F) which also includes barnacles, crabs, shrimps, and lobsters. Present as fossils since the Late Cambrian, ostracods are excellent index fossils, and they form important parts of many later Paleozoic faunal assemblages.

Bryozoans

Bryozoans are colonial, minute forms sometimes called "moss animals." They have nothing to do

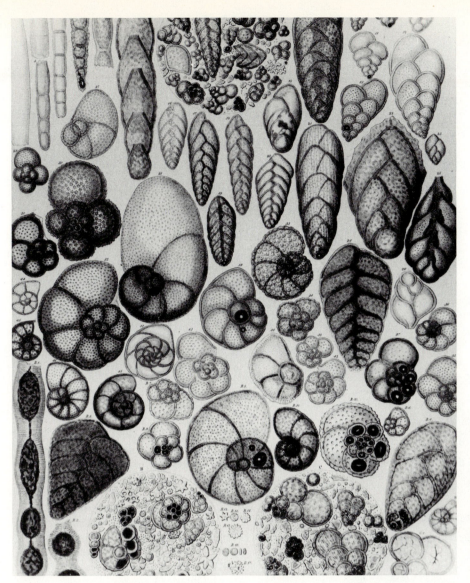

FIGURE 19-20
Here are quite a variety of "forams," a type of single-celled animal that furnishes important index fossils. Their tiny shells may be either siliceous or calcareous. (From Christian G. Ehrenberg, 1854, *Mikrogeologie das Erden und Felsen*, courtesy of the History of Science Collections, University of Oklahoma Libraries.)

with moss, however. These tiny animals build colonies of calcareous material (Figure 19–23) that resemble rough twigs, branches, or filamentous tubes. The colonial structures are fine index fossils for the Ordovician through the Holocene because they are particularly sensitive to environmental changes. Studied largely at the microscopic scale, members of the phylum Bryozoa are relatively common fossils.

Echinoids

The *echinoids*, the largest group of echinoderms, a diverse phylum that includes starfish, blastoids, and crinoids, are animals with a bun-shaped exterior skeleton (Figure 19–24) composed of generally united calcareous plates which are polygonal and arranged in fivefold radial symmetry. Probably the best

known of the echinoids is the "sand dollar," but the echinoids are an extremely diverse group.

Characteristically spiny, many echinoids live out their lives attached to the sea bottom or they may creep very slowly along. Their spines not only supply additional strength and a means of locomotion, but also give predators a mouthful of painful reminders to leave echinoids alone.

Stromatoporoids

The *stromatoporoids* are an important group of highly generalized (Figure 19–25) calcareous sponges ranging from the Ordovician to the Holocene. They were an extremely important group of reef-building animals in the Middle Paleozoic. Along with the corals, stromatoporoidal fragments and stromatoporoid reef

FIGURE 19–21
Radiolaria, like forams, are a group of unicellular animals. They live within a delicate symmetrical framework of opaline silica formed into numerous geometric shapes. Under the microscope, these fossils are intricately beautiful. (From Christian G. Ehrenberg, 1854, courtesy of the History of Science Collections, University of Oklahoma Libraries.)

colonies may form huge moundlike structures within extensive marine limestone units.

Corals

The *corals* are coelenterates (see Appendix F) which have a characteristic three-layered wall surrounding a body cavity. The corals belong to the same class as the sea anemones and have adopted both solitary and colonial life-styles. Solitary corals (Figure 19–26) are often cup-shaped; they live out their lives attached to the seafloor, straining food through their digestive systems. The tiny colonial coral animals build extensive, usually calcareous, external skeletal colonies within which they live. These colonies may branch (Figure 19–27) or radiate, or they may develop as tight polygonal clusters.

Corals were an abundant part of Paleozoic fauna; they range from the Ordovician to the Holocene. As in the past, corals continue in their role as reefbuilders; they are responsible for the great barrier and fringing reefs in shallow marine environments worldwide.

Fish

Fish are members of the phylum Chordata, a highly organized group of complex life-forms distinguished by having a protected central nervous system with a spinal cord. They are **vertebrates,** animals with spinal columns somewhat like those of human beings.

Fish stretch back into the Late Cambrian (510 Ma) and are a dominant force in the world's modern

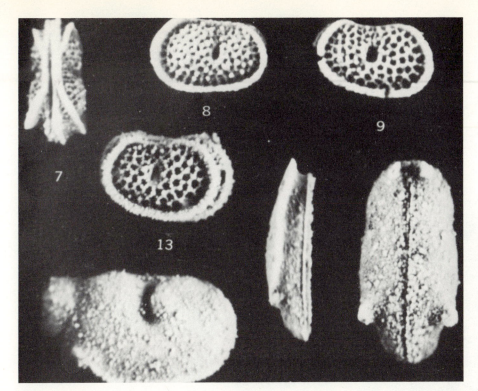

FIGURE 19–22
Ostracods are bean-shaped creatures inhabiting shells that are about one-eighth inch (one-third centimeter) long. These are *Chironiptrum,* a fossil ostracod from rocks whose age is Early Devonian (390 Ma). (Photograph courtesy of the U.S. Geological Survey.)

oceans. The earliest fish had an external skeleton and commonly lacked paired fins and jaws. Their internal skeletons were cartilaginous. These bizarre, heavily armored fishlike creatures are called the *Agnatha,* meaning jawless. With their heavy external skeletons they must have been poor swimmers and were probably filter or deposit feeders living on the seafloor. The modern lamprey belongs to this group.

Whatever their other shortcomings, the agnathans were the first life to have a spinal column. The central nervous system, with its enhanced capacity for coordination of nerve impulses over a still larger and longer body, would be the key to the development of life on land.

The development of a jaw gave rise to the *placoderms* (meaning armored skins), a type of Devonian fish (Figure 19–28) with paired fins and either bony or cartilaginous jaws. Growing to 10 yards (ca. 9 me-

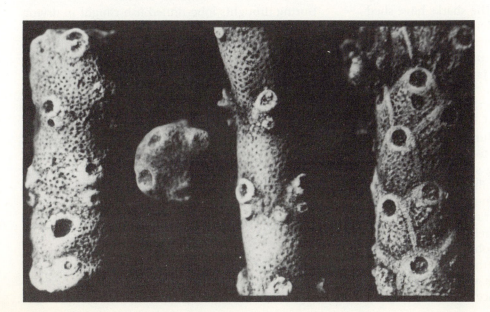

FIGURE 19–23
These are the calcareous "houses" within which minute animals called *bryozoans* once lived as colonial life-forms. They are from 55-million-year-old rock. (Photograph courtesy of the U.S. Geological Survey.)

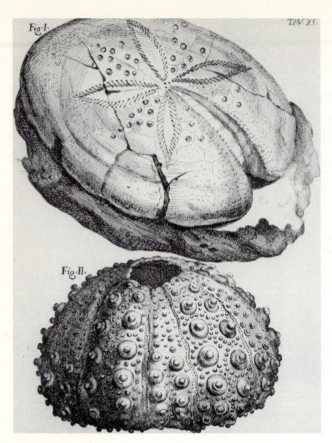

FIGURE 19–24
The echinoids are a group of spiny-skinned marine animals with fivefold symmetry. The upper figure is of a type called the "sand dollar." (From A. Scilla, 1748, *Vain Speculation. . .*, courtesy of the History of Science Collections, University of Oklahoma Libraries.)

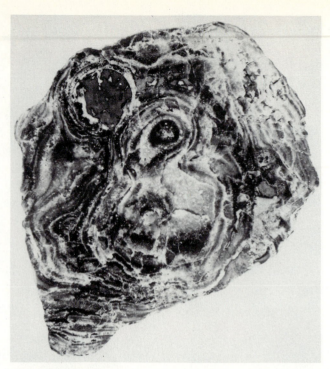

FIGURE 19–25
This polished section of *Stromatopora*, a typical stromatoporoid, illustrates the group's highly generalized form. The stromatoporoids are a poorly understood group of Early and Middle Paleozoic reef builders of uncertain biologic kinship. (Photograph courtesy of the U.S. Geological Survey.)

ters) in length, they must have been ferocious hunters, but they failed to survive the Devonian, giving way instead to sharks 360 Ma. The sharks have shed their cumbersome external skeleton, but they have retained their cartilaginous internal skeletons. With paired appendages, a powerful musculature, a streamlined form, and efficient jaws, sharks are the perfect killing machine.

The Devonian was a time for the explosive radiation of fishes. Both sharks and bony fishes (those having a bony skeleton) arose during this time, with most modern fishes belonging to the last category. Bony fish number 40,000 living species and include essentially all fishes that are familiar to us.

Bony fishes are subdivided into two groups. The larger group consists of the ray-finned fish; the smaller group contains only six relict species of lungfish and lobe-finned fish. The lobe-finned fish, long known from fossil remains, was thought to be extinct until in 1939 a living specimen was caught off the east coast of South Africa. Since then other speci-

mens of this living fossil, the *coelacanth* (pronounced seal'-oh-canth), have been caught. As the name suggests, this odd fish has sets of fins that are highly modified into lobes of muscular flesh.

During droughts lobe-finned fish caught in drying lakes and tidal pools might well have used these leglike fins to pull themselves across mud flats to another pond or stream, much like the modern "walking catfish." Walking catfish are able to wiggle out of drying ponds and waddle distances of several miles or kilometers on their fins in search of a water-filled depression; they are particularly well known in Florida. Any animal that could manage to survive on land would have a whole new environment open to it, because there was no other animal competition. Although land plants had appeared by 400 Ma, there was nothing to eat them—yet.

Amphibians

Lobe-finned fish gave rise during the Devonian to a successful experiment—the Amphibia. Ranging from the Devonian to the Holocene, amphibians are con-

FIGURE 19–26
These two views of *Caninia*, a solitary or horn coral, illustrate the kind of calcareous "house" the coralline animal leaves behind after death. (Photograph courtesy of the U.S. Geological Survey.)

sidered the most primitive four-legged vertebrates. Having lungs, internal nostrils, and two sets of paired bony limbs, these animals are still dependent on a watery environment because their eggs must hatch and the larval stages develop only in water. The only amphibians surviving today are the salamanders, frogs, toads, and the little-known caecilians.

The earliest amphibians were, not surprisingly, quite fishlike, having descended directly from the lobe-finned fish. They probably spent much of their time in water, leaving its safety only occasionally to exploit the empty environmental niche awaiting them.

Living on land presented three major new challenges.

▶ Fully supporting the body's own weight, which in water had been largely supported in water by its buoyancy.

▶ Directly breathing air, unfiltered through a gill system.

▶ Avoiding drying out.

By selecting amphibian fossils of progressively younger age, we can watch the early amphibians move their eyes from the sides to the top of their head, develop lungs from gill sacs, and greatly strengthen their skeletal systems. Even with all these improvements, their dependence on water for egg and larval development greatly limited the environmental niches that could be exploited by amphibians. It still does.

Reptiles

The answer to the problem of the amphibian egg was the development of an entirely new kind of creature, well-adapted to life on land. *Reptiles*, which emerged in the latest Paleozoic, are born from an **amniotic egg,** similar to a chicken egg. The embryo is safely enclosed in its own "water," is surrounded by enough food to get by, and is enclosed and protected by a permeable shell. The amniotic egg evolved in stages from the amphibian egg, with each succeeding stage providing better security for a land-dwelling organism. Reptiles slowly evolved through the rest of the Paleozoic Era into Mesozoic creatures with much stronger skeletons. These new creatures walked on legs placed directly under their bodies and they had better lungs—two adaptations needed for a completely terrestrial life.

Within the Mesozoic reptiles dominated the amphibians from which they had come, as well as everything else. Present-day reptiles include turtles, lizards, snakes, alligators, and crocodiles. Within the Permian lived groups of reptiles known as *cotylosaurs* and *pelycosaurs*, an odd group often called the "finback" reptiles. These cold-blooded reptiles had large vertical fins sticking straight up into the air from their backbones, giving the impression of a lizard with sails. The elaborate fins may have been a mech-

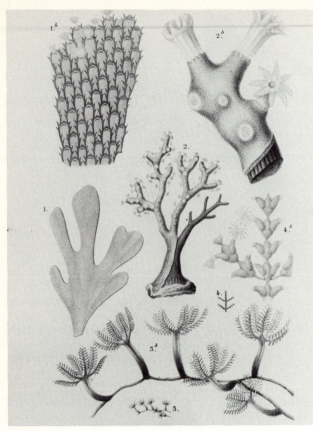

FIGURE 19–27

These sketches give several views of colonial coral. The upper right sketch shows the living animals within the calcareous tubes of the colony. The other sketches show various common shapes for coral colonies; the one in the center is often termed "staghorn coral." (From D'Orbigny, A., 1849, *Dictionaire Universelle. . .*, courtesy of the History of Science Collections, University of Oklahoma Libraries.)

anism for heat exchange in the hot, dry Permian climate. Still other reptiles returned to the sea, including a group called the *mosasaurs*; these became briefly successful as air-breathing marine reptiles.

EARLY LAND PLANTS

The earliest plants, by some classification schemes, were photosynthetic bacteria. Though the "plantness" of blue-green algae (cyanobacteria) might be disputed, there is little question that eukaryotic green algae were and are aquatic plants. Another type of aquatic plant is the *diatom*, a microscopic single-celled plant whose death leaves behind opaline silica networks of incredible variety and beauty (Figure 19–29). Diatoms differ from radiolaria in that they are plants.

Natural accumulations of diatom networks form what are called *diatomaceous earth*, often used as a filter in swimming pools, as an absorbent, and as an abrasive in kitchen cleaning powders. Diatoms span the Phanerozoic and are found in both fresh and marine waters today.

Moving marine plants onto land required eight major readjustments for survival.

1. Withstanding wide fluctuations in temperature, moisture level, and sunlight.
2. Providing internal fluids, while no longer being bathed in fluid.
3. Reproducing in an environment nearly devoid of water.
4. Respiring in a medium of air rather than in water.
5. Preventing excessive evaporation during periods of strong wind or high temperatures.
6. Developing systems that withdraw moisture and nutrients tightly held by mineral fragments within the soil.
7. Developing other systems to transport moisture up to the plant's top against the rising pressure of its own weight.
8. Maintaining the plant's structure erect in high winds.

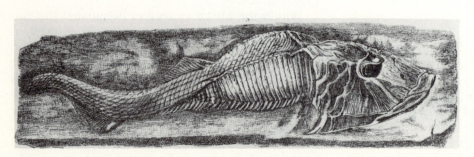

FIGURE 19–28

This is *Cephalaspis Lyelli, Agassiz,* a primitive freshwater fish from the Old Red Sandstone of Britain. Although odd-looking by today's standards, this was a classic Devonian fish. (From R. Murchison, 1872, courtesy of the History of Science Collections, University of Oklahoma Libraries.)

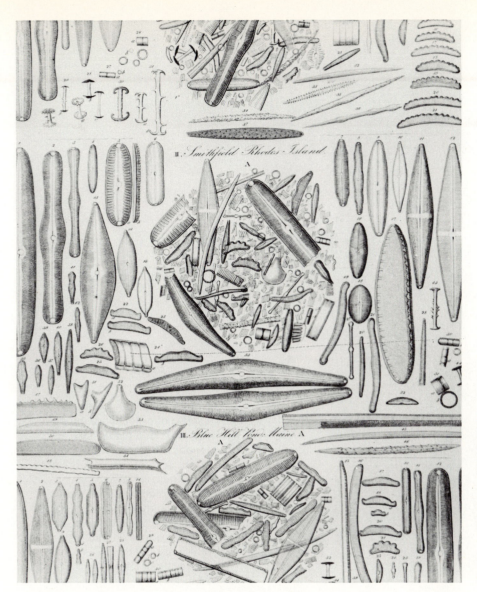

II. *Smithfield Rhode Island*

III. *Blue Hill Bay Maine*

FIGURE 19–29
Diatoms are minute plants whose opaline skeletons make spectacular microscopic fossils, as well as scrubbing powders and water filters. Shown here are three different sets of diatoms from ponds in New England. (From Christian G. Ehrenberg, 1854, courtesy of the History of Science Collections, University of Oklahoma Libraries.)

Plants met this daunting list of challenges by developing root systems to hold them fast and extract moisture and nutrients from minerals, internal vascular plumbing systems or vessels to transport and distribute the moisture and maintain erectness in soft-skinned stems, tough exterior skins to minimize evaporation and maintain erectness, and complex reproductive strategies; the earliest one was spreading spores by wind. *Spores* are an asexual, usually single-celled, reproductive apparatus produced by nonflowering plants like fungi, mosses, and ferns. They are an extremely durable cell to cast into the wind.

As an example of an strategy intermediate between a fully upright and erect plant and an alga surrounded by water, consider moss. Lacking an efficient system for fluid transport, moss must remain small and can live and reproduce only where it is moist. The mosses and their kindred the liverworts have left a fragmentary fossil record since the Devonian.

The movement of life onto land had to await the development of an *ozone shield*. Ozone is a form of molecular oxygen made up of three combined atoms and written O^3. The ozone level in our atmosphere is largely created by solar energy modifying molecular oxygen (O^2) into ozone; lightning has a similar effect. The pleasant smell we associate with lightning and thunderstorms is that of ozone. Without ozone in our atmosphere, the full intensity of ultraviolet radiation from the sun would strike the earth and destroy all life. As the earliest marine prokaryotes, safely hidden from damaging ultraviolet rays by sea-

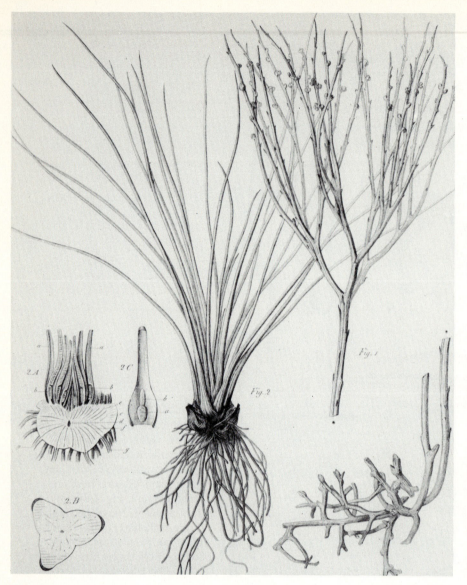

FIGURE 19–30
This is *Psilotum triquetrum*, a living example and a distant relative of psilophytes, the most primitive land plants. Like psilophytes, this plant lacks true leaves or roots, having instead only branched stems and inverted stems beneath the ground. (From A. Brongniart, 1857, *Histoire des Vegetaux Fossiles*, courtesy of the History of Science Collections, University of Oklahoma Libraries.)

water, began to generate oxygen, solar energy combined oxygen atoms to form ozone.

During the long course of increasing oxygenation of the atmosphere, the ozone level also increased, reaching its peak when the oxygen level was at 10 percent of its present level, about 420 Ma—a time *coincident with the development of the first land life,* though there is some fragmentary evidence of worm-like creatures in fossil soils which date back to 448 Ma. The development of diverse land life apparently had to await the formation of enough ozone to form an effective ultraviolet shield.

As the oxygen level has continued to climb to its current value, which is more than one million times its initial level, we must note that the ozone content of the atmosphere has decreased by 20 percent of its

highest (Silurian) value; the accumulating oxygen limits the penetration of ultraviolet light, decreasing the rate at which ozone forms. During post-Silurian time all life has learned to adjust to some ultraviolet radiation. The tanning of our skin is one way the body defends itself from immediate damage, though the strong correlation between prolonged exposure to the sun and skin cancer suggests that the damage from ultraviolet may simply be cumulative, taking years to develop.

Psilophytes

The oldest land plants date from the Late Silurian (410 Ma) and are known as *psilophytes.* The ancestors of the psilophytes are unknown, but they probably

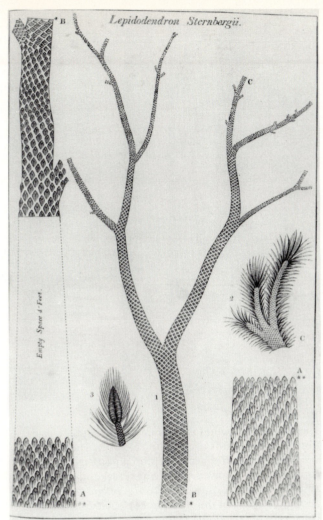

Lepidodendron Sternbergii.

Empty Space 4 Feet.

Fossil Tree found prostrate in a Coal Mine at Swina in Bohemia.

FIGURE 19–31

This is a sketch of *Lepidodendron sternbergii,* a scale tree found in a coal mine. Scale trees are a common constituent of coals, since they grew in Late Paleozoic coal swamps in both Europe and North America. (From Rev. William Buckland, 1837, *Geology and Mineralogy. . .,* courtesy of the History of Science Collections, University of Oklahoma Libraries.)

include the filamentous green algae. Psilophytes have simple branching stems which lack both leaves (Figure 19–30) and true roots. The branching stems carry out photosynthesis; the branching stems underground have primitive root hairs which enable plants to stay in place and draw water and nutrition from the soil. Psilophytes reproduce by the generation of spores.

The early psilophytes were confined to the Silurian and the Devonian, but a few remote relatives (Figure 19–30) share their characteristics. The psilo-

phytes differentiated into four important groups which became dominant Late Paleozoic plant types.

▶ *Lycopsids* or scale trees.
▶ *Pteridosperms* or seed ferns.
▶ *Sphenopsids* or scouring rushes.
▶ *Pteropsids* or ferns.

Lycopsids

The scale trees are represented today by club mosses and the ground pine, but once they were the lofty giants of Late Paleozoic coal swamps. Rising to 100 feet (30 meters) in height (Figure 19–31), scale trees had true roots and leaves. They dominated the Late Paleozoic in both abundance and size. The leaves grew *directly out of the trunk,* and were arranged in spiral rows. When the leaves were shed, the trunk was left with leaf scars that looked something like the scales of a reptile skin—hence the term "scale trees." The lycopods have extended from the Devonian to the Holocene.

Pteridosperms

Often confused with their kindred the "true" or "spore-bearing" ferns, the seed ferns that arose in the Devonian were the earliest group of plants to yield true seeds. A seed is, in effect, a miniature plant nurtured in the plant's ovary, the female reproductive organ, and then cast out onto the ground. It survives, encapsulated by a food store with a permeable covering which rapidly splits after being damp for a while.

The lowly ferns of today are the puny offspring of these Late Paleozoic giants, which often reached 40 feet (12 meters) in height. The seed ferns are the probable source of all Mesozoic seed-bearing plants, but they themselves became extinct in the Jurassic.

One of the seed ferns, *Glossopteris* (Figure 19–32), is an important fossil found only on southern continents that underwent a major Permian glaciation; at the same time the northern continents yielded only lycopsid flora. Wegener used this evidence (see Chapter Ten) to support his arguments for continental drift.

Sphenopsids

"Horsetail" or "scouring rush" grows along wet ridges in many localities; the plants are true "living fossils," for their ancestors were abundant in Pennsylvanian coal swamps. They have a simple, strongly

FIGURE 19–32

Photograph of the leaf of the seed fern *Glossopteris*. These ferns were the most primitive seed-bearing gymnosperms, which were to dominate the Mesozoic. (Photograph courtesy of the U.S. Geological Survey.)

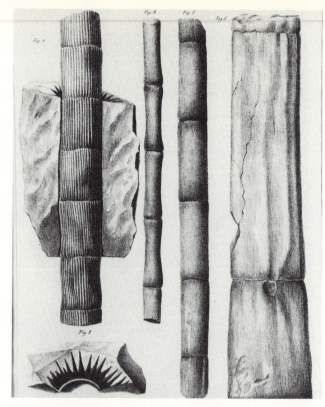

FIGURE 19–33

These sketches are of fossils of *Calamites*, a sphenopsid from the coal forests of the Late Paleozoic. The joints are very prominent; the stems are strongly striated and contain some dissolved silica, which makes them slightly abrasive. (From A. Brogniart, 1857, *Histoire des Vegetaux Fossiles*, courtesy of the History of Science Collections, University of Oklahoma Libraries.)

jointed single stem with spores carried in capsules at the tips of the stem. The plant stems, which preferentially incorporate finely divided silica, can be crushed and used to scour cookware on a campout—hence the name scouring rush.

The fossil representatives of these plants, such as *Calamites*, (Figure 19–33) grew to heights of 40 feet (12 meters) or more and were as big around as washtubs. The modern equivalents may reach a height of a large dog, and they snap off easily along their jointed stems when disturbed. The sphenopsids have extended from the Devonian to the Holocene.

Pteropsids

The pteropsids or "true ferns" grew to heights of 60 to 70 feet (18 to 21 meters) in the Late Paleozoic coal swamps. Like all true ferns, the pteropsids reproduce by the occasional growth of spore sacs underneath the leaves. These spore sacs burst open at certain times, releasing spores into the wind on the reproductive gamble that a few of the thousands will find a receptive locale. Since spores are extremely durable, they may survive decades and then burst into life when conditions are right.

Ferns are, of course, abundant today, and all ferns bear spores. The range for pteropsids has been from the Devonian to the Holocene.

LATER LAND PLANTS

Along with their ancestors the seed ferns, several of the latest Paleozoic land plants reproduced by sexual processes, producing seeds. The most primitive seed-bearing plants are collectively called *gymnosperms*, meaning "naked seeds." Examples of trees whose seeds are exposed include modern conifers such as pine and spruce trees. Of the gymnosperms that arose from the pteridosperms, three groups were important in latest Paleozoic time. They were

1. **Cycads.** A group of trees of generally palmlike appearance, having long unbranched trunks with crowns of palmlike leaves; they lacked flowers. One surviving member of this group is the sago palm, which bears fruit at the top of its trunk.

2. **Cycadeoids.** A group of trees whose range was from the Permian to the Mesozoic. They were similar in appearance to cycads, except that cycadeoids had primitive flowering centers and were probably the ancestors to all flowering plants.

FIGURE 19–34
These are leaves from a modern ginkgo tree. Ginkgoes are "living fossils," little changed since the Permian. They are extremely tolerant of adverse environmental conditions. (Specimen courtesy of Velma A. Barnes.)

3. **Conifers.** A group of trees bearing seeds in cones and having evergreen foliage of bladelike or needlelike leaves. This group includes the modern pine, yew, redwoods, spruce, and hemlock.

The most primitive conifers stretch back to the Mississippian and include the Cordaites, a primitive treelike plant intermediate between conifers and seed ferns; this lofty plant became extinct in the Permian. Modern cycads are commonly separate male and female plants. Commonly separated into its own division of the gymnosperms is the *ginkgo*, (pronounced gink'-oh) tree, which is represented in modern times by a single species (Figure 19–34). The distinctive fan-shaped leaves of the ginkgo can be found as fossils all the way back to Permian time. Ginkgoes continue today as a conservative, tough tree, capable of adapting to difficult environmental conditions after 250 million years of doing so.

INSECTS

No discussion of Paleozoic life would be complete without mentioning the *insects*, six-legged arthropods having flexible external chitinous skeletons and a body subdivided into three parts; they are commonly winged. The insects are the most numerous arthropods and may be the most numerous life-form on earth, with an estimated total mass exceeding that of humankind.

About 12,000 species have been described as fossils, ranging from the Silurian to the Holocene.

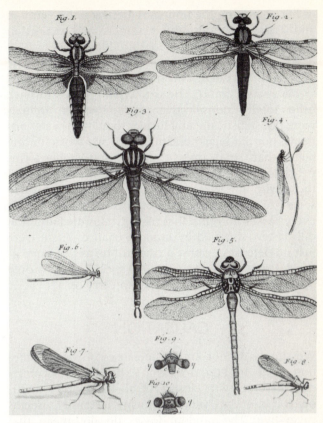

FIGURE 19–35
Dragonflies like this are preserved in detail in the fine-grained sediments from Late Paleozoic swamps. With wingspans reaching over 2 feet (0.6 meter), they would have been agile creatures. Flying insects stand a very poor chance of preservation, which makes these exquisite specimens remarkable. (From M. DeReamur, 1717, *Memoirs pour Servir à l'Histoire des Insects, Vol. 6,* courtesy of the History of Science Collections, University of Oklahoma Libraries.)

Among these insects are many of our old "buga-boos," including mosquitoes, cockroaches, wasps, grasshoppers, locusts, and houseflies. The group also includes the dragonflies (Figure 19–35). As an indication of how durable cockroaches really are, the Pennsylvanian Period (320 to 280 Ma) is often called the "Age of Cockroaches."

THE PERMO-TRIASSIC EXTINCTION

The Paleozoic Era closed with the formation of Pangea and one of the greatest extinctions in the history of life. Spanning the boundary between the Permian and Triassic Periods, the Permo-Triassic extinction event occurred within 10 to 20 million years.

Approximately half of all the families of marine organisms perished, and from *85 to 95 percent of all species became extinct*. The extinction appears to have been totally nonselective. Shallow-water animals were not statistically any luckier than deep-water animals, nor did bigger creatures fare any worse or any better than small ones. Complex and simple forms perished at similar rates. Creatures with lungs and those with gills were equally affected. In short, there seems to have been no protection. Rather there is every reason to think that most groups died and that only a few survived by random chance. There were just no defenses against a catastrophe as pervasive as the destruction of up to 200,000 species, leaving only 5 to 15 percent of all life types to carry on.

CAUSATION

We simply do not know the causes of mass extinction. That the Permo-Triassic extinction was coincident in time with closure of all the world's continents into a supercontinent certainly seems suggestive. The random losses do not mean that there were no determinable causes; rather such seeming randomness may reflect a highly complex web of environmental interactions. Mass extinctions are events contingent on a cascading web of consequences that follow from seemingly trivial differences along the way. As Darwin said, "A grain in the balance may determine who should live and who should die."

Stability is the norm for complex natural systems evolved over millions of years; the shifts are rare, rapid, and episodic. However we choose to look at it, the processes whose results seem random to us left only a few thousand species to carry on with life. Those who survived to enter the Mesozoic may have been no better adapted, no more fit, than those who did not. At this level of extinction, we do not see the "survival of the fittest." The remnants of Paleozoic life were the lucky survivors who set out to build the Mesozoic world.

SUMMARY

1. The Paleozoic Era encompasses all time from 570 to 245 Ma. At its boundary with the Precambrian the Ediacaran Fauna became extinct, skeletonized marine invertebrates began to flourish, and the Proterozoic supercontinent broke up along a series of rifts as the Iapetus ocean basin opened. At its boundary with the Mesozoic Era four-fifths of all marine invertebrate species and many amphibia and reptiles perished as the Iapetus ocean basin was closed.

2. Paleozoic rocks on the North American craton are separated into sequences, thin packages of sedimentary rocks characteristic of shallow marine environments, by disconformities whose time value increases toward the cratonic center. The western geocline remained passive throughout the Paleozoic. The southern one was subducted and deformed from the Pennsylvanian through the Permian because of the collision of North America and South America; this collision resulted in foreshortening, thrusting, folding, and metamorphism of the passive marginal sedimentary rocks that formed the Ouachita-Marathon mountain belt.

3. The eastern geocline was first deformed into an island arc complex in Late Ordovician time, creating the Taconic Highlands within the northern Appalachian geocline. As the collision between North America and Europe continued in Devonian time, the Caledonian orogeny gave way to the Acadian orogeny in the northern Appalachians. The collision between Africa and North America followed in Pennsylvanian and Permian time, creating the Alleghanian orogeny, which completed the formation of the Appalachian Mountains, whose eugeocline forms the Piedmont and Blue Ridge areas and whose miogeocline forms the Valley and Ridge and Allegheny Plateau areas in the central and southern Appalachians.

4. Epeirogeny on the craton formed broad domes and basins, including the Michigan Basin, a circular downwarp created in the Silurian. This basin became filled with a deep section of evaporitic rocks surrounded by fringing reefs. Highly cyclic eustatic sedimentation is recorded over much of the eastern craton in the last third of the Paleozoic. Such worldwide synchronous sedimentation may have been caused by a decrease in the volume of the ocean basin through volcanism in spreading centers, or by a decrease in the amount of water through glaciation, perhaps driven by Milankovitch cycles.

5. Paleozoic animal life included a wide array of marine invertebrates, of whom the trilobite is probably the most distinctively Paleozoic fossil.

Vertebrate fossils began with early Paleozoic agnathids, but they rapidly move toward more modern cartilaginous and bony fishes. The development of lobe-finned fishes allowed vertebrate life to move onto land in the Devonian, leading to the development of amphibians, and, in turn, their development into reptiles.

6. Paleozoic plant life included the survivors from the Precambrian; the green algae from this earlier time were the probable ancestor of all plant life to come. The earliest land plants were psilophytes, followed rapidly by the development of scale trees, "horsetail trees," ferns and seed ferns, and later by the development of cycads and conifers.

7. As much as 80 to 95 percent of all marine species perished in the Permo-Triassic mass extinction. This event also decimated the fishes, amphibians, reptiles, and almost all plant life except the gymnosperms, bacteria, and algae.

8. The Paleozoic Era closed with that mass extinction, and consolidation of all the earth's landmass into Pangea. The North American continent was now bisected by the Permian equator, which ran southwest from Newfoundland to Utah.

KEY WORDS

Acadian orogeny	orogeny
Alleghany orogeny	Ouachita orogeny
amniotic egg	paleontologist
aulacogen	Pangea
biofacies	regression
Caledonian orogeny	stable interior
correlate	sequence
epeirogeny	stratigraphic facies
eustatic	succession of facies, principle of
facies	
index fossil	Taconian orogeny
invertebrate	transgression
lithofacies	unconformity
	vertebrate

EXERCISES

1. The shell thickness of modern oysters that live in the wave-washed zone is proportional to wave height or tidal energy. Noting that early Paleozoic oysters have extremely thick shells, what can we conclude about early Paleozoic tidal height? Would this square with our knowledge that the Cambrian day lasted about 21 hours and the Cambrian year had 408 days, which means that our moon was somewhat closer to earth?

2. Would Walther's principle of succession of facies apply to evaporitic facies in the center of the Michigan Basin? Why or why not?

3. What are the *probable* causes of worldwide eustatic disconformities? Why might disconformities be worldwide but angular unconformities be confined to a limited area?

4. It has been said that all cells of every living thing are themselves strictly aquatic organisms. What could this mean?

5. We note that trilobites had exquisitely developed eyes with very high light-gathering power. What does this tell us about their normal environment?

6. What are the advantages of a seed over a spore?

SUGGESTED READINGS

GOULD, STEPHEN JAY, 1986, A short way to big ends, *Natural History*, January, pp. 18–28.
▶ *Fascinating discussion of how we think about time and organic complexity.*

GRAHAM, LINDA E., 1985, The origin of the life cycle of plants, *American Scientist*, v. 73, pp. 178–186.
▶ *Elegant discussion of the a recent model for the colonization of latest Silurian continents by aquatic plants.*

KING, PHILIP B., AND BEIKMAN, ELLEN M., 1976, *The Paleozoic and Mesozoic Rocks: A Discussion to Accompany the Geologic Map of the United States.* Washington, D.C., U. S. Government Printing Office, 76 pp. (U.S. Geological Survey Professional Paper 903).
▶ *Thorough review of the overall picture. Fine bibliography.*

MACURDA, D. BRADFORD, JR., AND MEYER, DAVID L., 1983, Sea lilies and feather stars, *American Scientist*, v. 71, pp. 354–365.
▶ *Modern crinoids, movable and immobile, illustrated in beautiful color.*

McMenamin, Mark A. S., 1987, The emergence of animals, *Scientific American* April, pp. 94–102.
► *Review of Eocambrian and Ediacaran fauna and their environments.*

Milne, David, et al., 1985, *The Evolution of Complex and Higher Organisms*, NASA Special Paper 478, 193 pp.
► *Sweeping review of the history and origin of complex life.*

Morris, Simon C., 1987, The search for the Precambrian-Cambrian boundary, *American Scientist*, vol. 75, pp. 157–167.
► *Modern review of the boundary problem, and how it was resolved.*

Oliver, Jack, 1980, Exploring the basement of the North American continent, *American Scientist*, v. 68, pp. 676–683
► *An outline of how we know what we know about the Appalachian past in three dimensions. The project de-scribed is ongoing—and yielding exciting new information every year.*

Patrusky, Ben, 1987, Mass extinctions; the biological side, *Mosaic*, vol. 17, no. 4, pp. 2–13.
► *Modern scholarship on the causes of mass extinctions.*

Poag, C. Wylie, and Ward, Lauck W., 1987, Cenozoic unconformities and depositional supersequences of North Atlantic continental margins: testing the Vail model, *Geology*, v. 15, pp. 159–162.
► *An example of how scientists test the theory of synchronous sedimentary packages; technical, short, and fascinating.*

Rudwick, Martin J. S., 1986, *The Great Devonian Controversy: The Shaping of Scientific Knowledge among Gentlemanly Specialists*, Chicago, University of Chicago Press, 494 pp.
► *An extraordinary piece of scholarship about how science really works.*

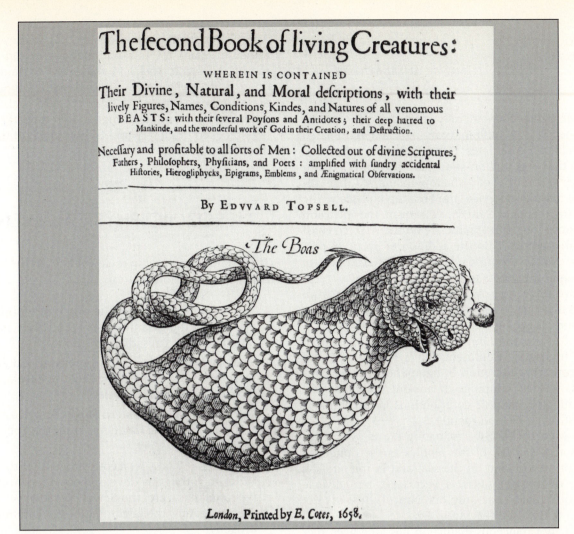

The second Book of living Creatures:

WHEREIN IS CONTAINED

Their Divine, Natural, and Moral descriptions, with their lively Figures, Names, Conditions, Kindes, and Natures of all venomous BEASTS: with their several Poyfons and Antidotes; their deep hatred to Mankinde, and the wonderful work of God in their Creation, and Destruction.

Necessary and profitable to all sorts of Men : Collected out of divine Scriptures, Fathers, Philofophers, Physicians, and Poets : amplified with sundry accidental Hiftories, Hierogliphycks, Epigrams, Emblems, and Ænigmatical Obfervations.

By EDVVARD TOPSELL.

The Boas

London, Printed by E. Cotes, 1658.

Three centuries later we know that boa constrictors do not swallow children and that boas originated late in the Mesozoic Era as highly modified lizards. Those 300 years record the shift from dependence on authority to dependence on experience and reason. (From Edward Topsel, 1658, *The History of Four Footed Beasts and Serpents,* courtesy of the History of Science Collections, University of Oklahoma Libraries.)

CHAPTER *twenty*

The Mesozoic Era

Land is immortal, for it harbors the mysteries of creation.

ANWAR AL-SADAT (1918–1981)

The Mesozoic Era—literally the era of "middle life"—began 225 million years ago and terminated 160 million years later (see chapter frontispiece). Within that 160-million-year span there was a major advance and retreat of a shallow inland sea that created a cratonic sequence, the breakup and scattering of the fragments of Pangea, volcanism and intrusion of plutons along a magmatic arc on the western coast of North America, and the addition of up to 25 percent of North America's landmass by accretion of continental fragments to its western margin.

As we have already learned, the quotation from Egypt's martyred leader that opens this chapter is not quite correct geologically; land does harbor the mysteries of its creation, but in the geologic time scale it is not immortal. Continents have beginnings, and they grow from accretion of fragments of other continents, welded scrapings of ancient seafloors, and magmatic and volcanic contributions from continental crust, seafloor, and mantle.

The Mesozoic is the fascinating era in which we see the rise and fall of the reptiles as the dominant life-form on the land, in the sea, and in the air. The demise of the dinosaurs, a very large and diverse group of reptiles, was only one part of the mass extinction that closed the Mesozoic.

The Late Cretaceous extinction has long been recognized as a major crisis in the history of life, and scientists have developed several models to explain why it happened. Among the most intriguing of these is a model suggesting that the Late Cretaceous extinction was coincident with the impact upon the earth of a giant meteorite. As we will see, the evidence for an impact event at the Cretaceous - Tertiary boundary is convincing, but the explanations for *how* this impact causes the observed patterns of extinction are much less convincing.

The Paleozoic had begun with the fragmentation of the Pangean supercontinent, accompanied by an enormous increase in the diversity of life. It ended with closure of the Iapetus Ocean basin; the formation of the Marathon–Ouachita–Appalachian–Caledonian mountain chain; the consolidation of Pangea; and a precipitous decrease in the diversity of life through mass extinction. The Mesozoic began with the fragmentation of Pangea and an increase in the diversity of life; it ended with another mass extinction which cut randomly across all life-forms.

Mesozoic rocks are widely exposed over North America, particularly in the West, where they are admired in many of our national parks and monuments. Mesozoic rocks in the West contain three-quarters of the reserves of coal within the United States. Like eastern coals, these western coals formed when rising mountain ranges—this time the Rockies—blocked drainages and turned much of the West into swampland.

▼

MESOZOIC PHYSICAL EVENTS

▲

This chapter will briefly review a few events in North American Mesozoic history.

▶ The breakup of Pangea, the origin of Florida, the opening of the Atlantic Ocean, and the Cassiar-Sonoman and Palisadian orogenies.

▶ Widespread Cretaceous transgression and eustatic worldwide cycles, paralleled by the development of widespread coal deposits.

▶ The growth of far western North America by accretion of exotic fragments from the Pacific Basin.

▶ The Jurassic and Cretaceous orogeny in the Far West, including the development of an Andean-type magmatic arc, the Sevier-Nevadan orogenies, and the beginning of the Laramide orogeny.

THE BREAKUP OF PANGEA

As noted in Chapter Ten, the Red Sea is forming as continental rocks are extending and thinning owing to an underlying spreading center creating elongate rift basins. At the same time there is basaltic volcanism, seismic activity, and rapid infilling of the elongate basins with locally derived sediment. As spreading continues, basaltic ocean floor forms at the center of the rift, and an elongate, narrow ocean basin is taking shape. Is the present one key to the past?

The Palisadian Event

During the Triassic, continentally derived sediments in eastern North America were trapped in fault-bounded rifted basins from Georgia to Newfoundland (Figure 20–1). These sediments originated in the high Appalachians to the west. The resulting reddish sedimentary rocks form a discontinuous record

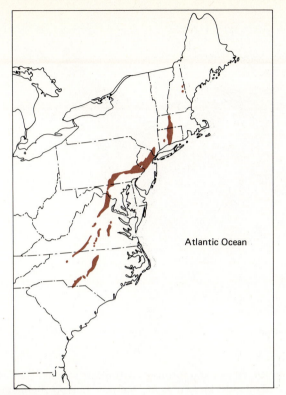

FIGURE 20–1

The colored areas are Triassic sedimentary rocks exposed in fault-bounded rift zones in the eastern United States. Jurassic rifts are just off the shoreline. (From P. B. King and H. M. Beikman, 1976, U.S. Geological Survey Professional Paper 903.)

of the stretching and rifting of the eastern margin of North America.

Within these generally reddish sediments are numerous *diabasic* (fine-grained *diorite*; see Figure 3–8) and basaltic dikes and flows, as well as the footprints of small dinosaurs. Many of these downfaulted troughs were themselves tilted and penetrated by numerous basaltic sills. The Palisades of the Hudson River, a series of cliffs, are the eroded edge of a thick sill that intruded into these sediments about 195 Ma.

The sequence of events just described, which is typical of continental margin rifting, has been termed the **Palisadian event,** a Triassic and Jurassic episode. The thick wedge-shaped conglomerates from rising fault-block mountains were deposited as alluvial fans, which were, in turn, tilted and intruded by ultramafic sills with the development of numerous unconformities. The unrest continued with development of similar rifted offshore basins along what is now seafloor. Plutonic bodies, several of which were of ultramafic composition, were also formed from the Jurassic to the Cretaceous.

Fossils in the earliest marine rocks shed from the eastern continental margin date the complete separation of Europe from North America at 180 Ma. Radiometric dates from the oldest ocean floor basalts along the eastern coast are consistent with this date. The opening by spreading of the Atlantic Basin meant that the Pacific Basin began to close.

The Cassiar–Sonoman Orogeny

Subduction along the western margin occurred as North America began to override the Pacific Ocean floor. This convergence along the shoreline created the **Cassiar–Sonoman orogeny.** The development of thrust faults, intense folding, abundant volcanism, the development of minor plutons, and the creation of numerous unconformities were all a part of this orogenic event of Early Triassic time.

It was initiated by the westerly motion of the North American plate. The subduction of eugeoclinal sedimentary rocks and volcanics formed distinctive blueschist associations. The associated ophiolitic suites extend from central British Columbia to Oregon, and are juxtaposed with basaltic volcanics and ultramafic plutonic and volcanic rock. These units record the subduction of Pacific oceanic lithosphere and a western island arc system.

The Origin of Florida

When the continent split, most of the rifting occurred along the line where the continental fragments first joined. Such a "weld" would continue to be a weak area, subject to reopening as the crust was placed in tension.

One old join, however, failed to yield, and a part of Africa remained where it was to become part of North America. This piece of what once was ancient Africa is called Florida. The **suture,** the line along which the continental fragments are joined, has recently been located in southeast Georgia. The oldest Floridian rock is part of the African shield; along with some younger rocks, it records the history of western Africa, not eastern North America. From the point of the old African–North American suture in south Georgia northward, Europe and North America separated along the line on which Iapetus had closed.

THE CRETACEOUS TRANSGRESSION AND CYCLIC SEDIMENTATION

On a worldwide basis, Cretaceous rocks overlap older rocks, suggesting a large marine transgression.

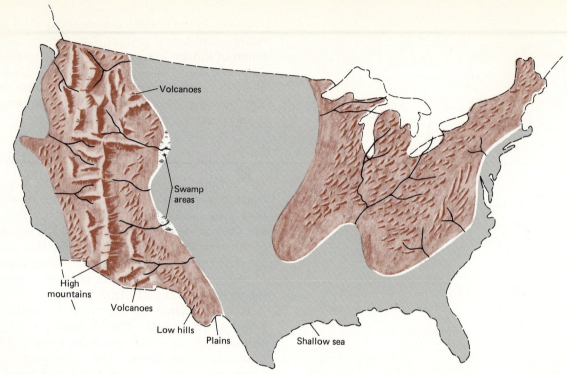

FIGURE 20–2

A sketch map of the general geography of the United States in Early Cretaceous time—100 Ma. The shallow sea was at its maximum transgression. To the west the highlands were shedding immense amounts of sediment onto the craton and forming coals along subsiding areas. The highlands, known as the Mesocordilleran Highlands, were an active sediment source throughout the Jurassic and Cretaceous periods. (Map from U.S. Geological Survey.)

By the middle of the Cretaceous only the Midwest, central Canada, and a long island arc stretching from Alaska through the Yukon, British Columbia, Idaho and south all the way to Baja California remained above the transgressive Cretaceous ocean. From the Northwest Territories south through Alberta, Montana, North Dakota, South Dakota, Colorado, Utah, New Mexico, Oklahoma, and Texas a Cretaceous seaway (Figure 20–2) was the site of one of the most unusual series of cyclic sequences in geologic history.

The north–south shoreline shifted continuously during the whole of the Cretaceous. Subsidence outstripped sedimentation, resulting in the deposition of quite thick sections of Cretaceous rock in the Far West. It was there that the ocean was deepest. Along the main extent of the seaway, the oscillating transgression and regression of the sea created over a hundred cyclic sedimentation units; these can be correlated worldwide, often to within half a million years.

The eustatic control of these worldwide cycle is obvious. The rates of transgression and regression can be measured because there are some 137 unique fossil zones and frequent thin layers of altered volcanic ash whose minerals yield radiometric dates (see Chapter Seventeen). The cyclic sequences themselves are quite symmetrical, with both regressive and transgressive sequences equally preserved by rapid subsidence.

Within these cycles are immense coal deposits, approximately two-thirds of the coal reserves in the United States. Preserved largely as regressional deposits, both low-grade lignite and medium-grade bituminous coals occur (see Chapter Sixteen), with lignitic coal confined to the Dakotas and Montana (Figure 20-3).

SUSPECT TERRANES AND MESOZOIC ACCRETION

It has long been known that by the end of the Paleozoic the edge of the North American craton ran roughly north–south from central Nevada to eastern British Columbia (Figure 20–4). Since that time, the western edge of the continent has grown by the addition of nearly 25 percent of its total land area in a belt nearly 400 miles wide (640 kilometers), shown in white on Figure 20–4.

This accreted crust, which includes essentially all of Alaska, was added by the collision of crustal frag-

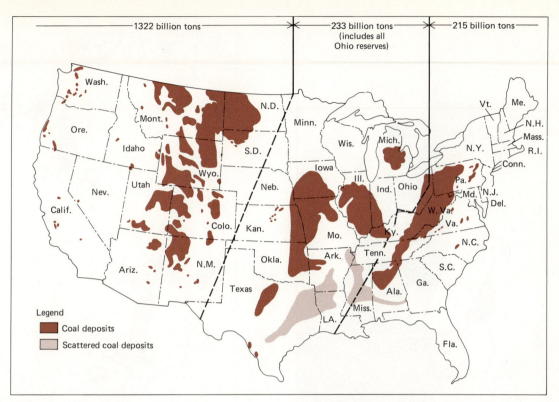

FIGURE 20–3

Major coal reserves in the United States. Western coal, entirely of Mesozoic age, accounts for three-quarters of the total reserve. (Map from Environmental Protection Agency, 1977.)

ments whose origin can be broadly traced by paleomagnetic studies (see Chapter Ten) to somewhere within the Pacific Basin. The individual fragments, which have been called *microplates,* range from tens to thousands of miles or kilometers in size. Virtually the entire Pacific coast from Alaska to Baja California originated by the "docking" and accretion to the continental margin of these microplates. The process of continental growth by the accretion of microplates is referred to as *collage tectonics* or **microplate tectonics.** The western example was hardly the first; the accretion of several microplates figures prominently in detailed histories of the origin of the Appalachians.

Using paleomagnetic and fossil information, we can speculate on probable sources of the fragments, but it is not possible to be certain of their point of origin. Many of them are fragments of oceanic crust, others are clearly fragments of other continents, and a few appear to be hybrids of continental and oceanic crust. Because their geographic origin is in doubt, these microplates are termed **suspect terranes.**

It is clear that many of these suspect terranes wan-

dered thousands of miles across the Pacific Ocean before "docking" with the growing North American continent, much as India moved north for perhaps 3800 miles (6000 kilometers) before colliding with Asia to form the Himalayas (see Chapters Ten and Twenty-One). Nearly 200 of these suspect terranes have been identified by observing striking discontinuities in the geologic history, age, and paleolatitude of rock units that are now sutured together side by side.

Each terrane is a completely separate geologic entity, made up of a sequence of rocks and rock types wholly different from those of its neighbors. Where exposures will allow, the contact between terranes is commonly extremely sharp and most often forms a thrust fault (see Chapter Nine). Each terrane is totally bounded by faults that isolate it from neighboring terranes. These terranes originated in wholly different environments and latitudes and reveal very different geologic histories. Their paleomagnetic and fossil ages indicate that the episode of accretion of suspect terranes to form the geologic patchwork that is North America west of Nevada lasted from 200 to 50 Ma.

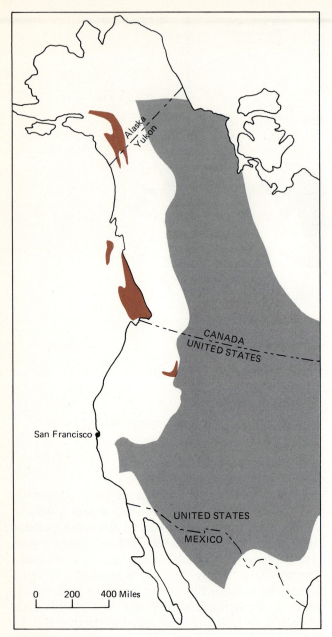

FIGURE 20–4
This sketch map shows the margin of the Paleozoic craton in a gray stipple; the white area includes the area added by Mesozoic suspect terrane accretion. One of the most prominent terranes, *Wrangellia*, is shown separately in color. In the white area there are nearly 200 similar suspect terranes. (Map courtesy of the U.S. Geological Survey.)

The *Wrangellia terrane* (Figure 20–4), which extends from southern Alaska to eastern Oregon, is a good example of a well-traveled terrane. It is composed of Late Paleozoic volcanic rocks and thin marine Permian sedimentary rocks which are overlain

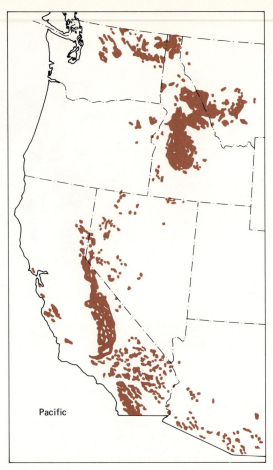

FIGURE 20–5
Jurassic and Cretaceous intrusive bodies in the western United States. (From P. B. King and H. M. Beikman, 1976, U.S. Geological Survey Professional Paper 903.)

by a thick sequence of basalt; after these units formed, they subsided beneath the seafloor, with carbonate rocks and then deep-water settlings deposited on them. It appears from paleomagnetic evidence that Wrangellia then moved northward some 60 degrees of latitude, the equivalent of 3600 miles, or one-sixth of the circumference of the earth, before it docked with North America about 100 Ma. Since that time it has been further fragmented along faults that have carried fragments from Oregon to Alaska.

OROGENIC DEVELOPMENT OF THE WEST

A series of orogenies spans the whole of Jurassic and Cretaceous time, and they continued well into the Cenozoic Era, affecting the whole of western North America from the modern Rocky Mountains to the Pacific shore. Although it is possible to separate each

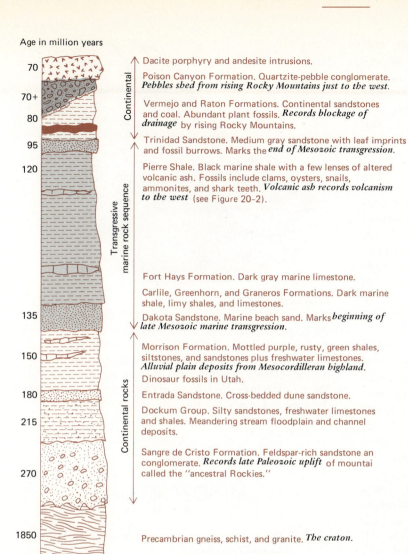

Age in million years

70 — Dacite porphyry and andesite intrusions.

70+ — Poison Canyon Formation. Quartzite-pebble conglomerate. *Pebbles shed from rising Rocky Mountains just to the west.*

80 — Vermejo and Raton Formations. Continental sandstones and coal. Abundant plant fossils. *Records blockage of drainage* by rising Rocky Mountains.

95 — Trinidad Sandstone. Medium gray sandstone with leaf imprints and fossil burrows. Marks the *end of Mesozoic transgression.*

120 — Pierre Shale. Black marine shale with a few lenses of altered volcanic ash. Fossils include clams, oysters, snails, ammonites, and shark teeth. *Volcanic ash records volcanism to the west* (see Figure 20–2).

Fort Hays Formation. Dark gray marine limestone.

Carlile, Greenhorn, and Graneros Formations. Dark marine shale, limy shales, and limestones.

135 — Dakota Sandstone. Marine beach sand. Marks *beginning of late Mesozoic marine transgression.*

150 — Morrison Formation. Mottled purple, rusty, green shales, siltstones, and sandstones plus freshwater limestones. *Alluvial plain deposits from Mesocordilleran highland.* Dinosaur fossils in Utah.

180 — Entrada Sandstone. Cross-bedded dune sandstone.

215 — Dockum Group. Silty sandstones, freshwater limestones and shales. Meandering stream floodplain and channel deposits.

270 — Sangre de Cristo Formation. Feldspar-rich sandstone an conglomerate. *Records late Paleozoic uplift* of mountai called the "ancestral Rockies."

1850 — Precambrian gneiss, schist, and granite. *The craton.*

Continental

Transgressive marine rock sequence

Continental rocks

FIGURE 20–6
Schematic column of sedimentary rocks exposed within the Philmont Scout Ranch region of north-central New Mexico. Within these rocks is the story of the origin of the Rocky Mountains. Events within the Cenozoic Era are omitted. The major events are shown in color, the evidence for them in black. (Simplified from G. D. Robinson et al., 1964, U.S. Geological Survey Professional Paper 505.)

of these events, it may be more helpful to think of a single orogenic pulse that began in middle to late Jurassic time (170 Ma) in the Far West and moved in a generally easterly direction for 100 million years.

The earliest event, spanning the Permo–Triassic boundary, has already been described as the Cassiar and Sonoman orogenies of Canada and the western United States. By the beginning of Triassic time (240 Ma), an island arc was fully developed along the west coast of North America. In the Early Triassic the *Sonoma terrane* was thrust eastward onto the continental margin, an event reminiscent of the Taconian orogeny in the northern Appalachians (see Chapter Nineteen), but with the direction of thrusting reversed.

By Late Triassic, the arc complex was replaced by a foredeep, and an Andean-type (see Chapter Ten) magmatic arc consumptive margin was being cre-

ated. By Late Jurassic time (150 Ma), widespread deformation was occurring over the Far West, including both thrusting and plutonism. This Jurassic episode is termed the **Nevadan orogeny**, recognizable by strongly deformed Jurassic and older rocks overlain by little-deformed Cretaceous rocks. Jurassic plutons (Figure 20–5) were also formed at this time. Similar plutons must underlie Andean volcanoes today.

The Sevier and Laramide Orogenies

A later event, the **Sevier orogeny** (pronounced severe), occurred in the Late Cretaceous. This orogeny is recognized by strongly folded Jurassic and Early to Middle Cretaceous rocks in the Cordilleran region, plutonism, eastward-directed thrusting, and initiation of the Cordilleran uplift. Huge wedges of coarse conglomeratic rocks, which become thicker and coarser closer to the rising mountains, were deposited.

The transition from island arc collision in the Far West to uplift along the western Cordillera, that mountainous area from the Pacific margin to the eastern face of the Rockies, *reflects a wholly new way to form* strongly deformed uplifted areas. As we have seen, the Appalachians, Ouachitas, and Marathons formed because of the progressive collision of Europe, Africa, and South America with North America. The Late Mesozoic and Early Cenozoic deformation across the whole of the Cordilleran did not result from continent–continent collision, but rather from continent–ocean interaction owing to *high rates of oceanic plate consumption along a gently dipping Benioff zone* (see Chapter Ten).

One of the end results was the production of very large volumes of plutonic rock; from Mexico to British Columbia the Late Mesozoic was a time for the formation of batholiths. Many of the radiometric dates range from 75 to 100 Ma (Figure 20–5). These massive plutons provide evidence that very large volumes of subcrustal rocks melted. This melting occurred during periods of fast plate motion as the oceanic ridges inflated to maximum volume, creating the rapid Middle Cretaceous marine transgression just described. Thus, periods of abundant intrusion at spreading centers created high spreading rates, which in turn created abundant plutonism under low-angle consumptive margins. These plutons must at one time have had volcanic chains above them; the ash is interbedded with the transgressive sequences. Erosion long ago removed the Late Mesozoic volcanoes, but not the masses of now-crystalline magma that once fed them.

The Mesozoic Era closed with the initiation of what is termed the **Laramide orogeny**, the series of orogenic events that was to form the Rocky Mountains in Cenozoic time (Table 20–1). Along a zone extending from northwestern Mexico to northwestern Canada, the Cretaceous and older sedimentary rocks (Figure 20–6) that had been accumulating in the western geocline were thrust eastward onto the craton. The thrusting was intricate and reflected total shortening and telescoping of the crust in the range of 50 to 150 percent. The thrust faults in some cases carried rocks more than 100 miles (160 kilometers) from their source. Among the rocks thrusted to the east were parts of the suspect terranes that had docked and accreted along the far western shoreline. The events that were to culminate in the Rocky Mountains in Cenozoic time had been set into motion.

FIGURE 20–7

This view of the modern chambered nautilus illustrates the geometry of a coiled cephalopod. The creature lives in the last compartment and uses its tentacles to wave food into its mouth. Previously occupied chambers are walled off, except for a tube that runs through the whole shell. The nautilus uses this tube to maintain the air pressure in the shell to provide just the right amount of buoyancy.

MESOZOIC LIFE

At both the beginning and the end of the Mesozoic there was widespread extinction of preexisting faunas. The Mesozoic Era began with an earth impoverished of half of all Paleozoic families. Among the amphibians, 75 percent had disappeared; the reptiles had been decimated by 80 percent. Extinction had come to very diverse groups of marine invertebrates. Essentially all the Paleozoic corals and crinoids and all the blastoids had disappeared. The largest and most successful groups of foraminifera and brachiopods had exited, as had the eurypterids and the trilobites. Among the cephalopods, 90 percent of the coiled ammonites had left the stage. The straight-shelled groups continued the decline that had begun in the Ordovician. There had been sizable reductions among both the bryozoa and the other molluscs. The Early Mesozoic was inhabited by a few survivors of the greatest dying in the entire history of life.

INVERTEBRATES

As the Mesozoic began, the stony or lime-secreting corals that had been so abundant in Paleozoic seas

TABLE 20–1
Major North American Orogenic Events

Name	Affected Area	Age Range	Geologic Evidence	Origin
Taconic	Northern Appalachians	Ordovician - Silurian	Thrust sheets; ultramafic intrusions, conglomerate wedges, granites	Island arc complex collision
Acadian	Northern Appalachians	Devonian	Conglomerate wedges, granites, metamorphism, unconformities, folds	Collision of North America with Europe
Appalachian	Eastern North America	Pennsylvanian	Conglomerate wedges, granites, metamorphism, folding, thrusting, unconformities	Collision of North America with Africa
Ouachita	Arkansas, Oklahoma, Texas	Pennsylvanian - Permian	Folding, thrusting, low-grade metamorphism	Collision of North America with South America
Cassiar - Sonoman	Western cordillera	Early Triassic	Volcanism, thrusts, folding, minor plutonism, ophiolites	Island-arc complex collision
Palisades	East coast	Triassic - Jurassic	Block faulting, ultramafic and basaltic sills, rifting, clastic wedges in rift basins	Rifting of Pangea
Sevier - Nevadan	Cordillera	Late Jurassic - Middle Cretaceous	batholiths, volcanism, thrusts, folding, unconformities	Island-arc collision shifting slowly to simple subduction and evolution of magmatic arc
Laramide	Rocky Mountains	Late Cretaceous	Massive batholithic emplacement, folding, vertical tectonics	Low-angle Andean-type rapid subduction
		Early Tertiary	Conglomerate wedges, metamorphism, volcanism, thrusts, block faulting	Low-angle oceanic subduction
Cascade	Northern Cordillera	Middle Cenozoic	Volcanism, metamorphism, thrust faulting, folding	Island-arc complex subduction
Basin and Range	All of Cordillera	Late Cenozoic	Block faulting, basin filling, volcanism	Crustal doming and extension over spreading center

exited. In the Middle Triassic they were replaced by a new group of reef-building forms, the *scleractinian* corals. Other reef-building forms included the clams and oysters, which became so supremely abundant that many Late Mesozoic limestones are little more than deposits of oyster shells.

The oceans, however, were dominated by the *cephalopods*, a group of molluscs which had inhabited the Early Paleozoic seas as squidlike creatures living in the last chamber of their long, straight, conical shells. These *orthoceracone cephalopods* were successively replaced by coiled cephalopods, whose squidlike inhabitants again lived in the outermost chamber of the spiral shell (Figure 20–7).

The intersection of the individual cell walls with the spiral shell creates lines of varying complexity, called *sutures*. The simplest suture pattern is one of straight lines, as in the modern nautilus (Figure 20–

FIGURE 20–8
A variety of coiled cephalopods. The patterns on the exteriors of most of the shells are caused by the intersection of numerous interior walls with the exterior of the shell. The intricate folding was perhaps an adaptation for greater strength. (From G. W. Leibniz, 1749, *Protogaea*, courtesy of the History of Science Collections, University of Oklahoma Libraries.)

7). Increasingly complex patterns evolved throughout the Mesozoic (Figure 20–8).

One group, known as the *ammonites*, had become nearly extinct by the end of the Triassic. After the Triassic they then flowered into a coiled, complexly sutured group which dominated the sea life of the Jurassic and Cretaceous periods. Because of the rapid change in their suture patterns as well as their abundance and their rapid rates of evolution, ammonites make ideal *index fossils*; their study allows very precise discrimination of marine fauna on a worldwide basis. Some of these creatures approached a yard (about 1 meter) in diameter. Several persons are needed to pick up their fossilized shells.

Another unusual fossil unique to the Mesozoic is the *belemnite*, an animal with a conical *internal* skeleton (Figure 20–9) belonging to the belemnoid family; this group is related to modern squids. Their abundant fossil skeletons have often been called "fossil cigars."

The variety of cephalopods in the Mesozoic shows that life belonged to the swift and agile, for cephalopods were capable of a kind of jet propulsion and could turn rapidly. They could also emit an inklike substance in defense and use their tentacles and beak in offense; they were successful predators. They had a complex brain and nervous system, a cartilaginous skeleton, and highly sophisticated eyes, much like those of human beings.

On land, freshwater and air-breathing gastropods were abundant, as were clams, ostracods, and freshwater sponges. The Mesozoic was a time of increased diversity and abundance, including in particular the development of the bees and insects, the dominant land invertebrates.

The rise of insects was related in time to the development of a new type of plant, the **angiosperm**. Angiosperms are plants that are able to reproduce sexually by means of an ovary enclosed in a flower, the visual "bait" that attracts insects who inadvertently pollinate the ovary while indulging in the tasty nectar and pollen. The angiosperms lose some of their pollen as food for the bees and butterflies, but in return they are fertilized, which causes them to produce seeds, often enclosed in fruit, from which they reproduce themselves. During the Mesozoic, this mutually beneficial relation developed between insects and the flowering and fruit-bearing plants; it continues to this day.

PLANTS

The Paleozoic plant world had been made up of psilopsids, scale trees, horsetails, ferns, seed ferns, ginkgoes, cycads, cordaites, and conifers. Of this diverse group a few remnants of the psilopsids, scale trees, and horsetails survive today. The seed ferns and cordaites died out. By the middle of the Mesozoic only the ginkgoes, cycads, and conifers remained. The cycads and conifers are classic *gymnosperms*, the *naked-seed* plants, which remained the dominant elements

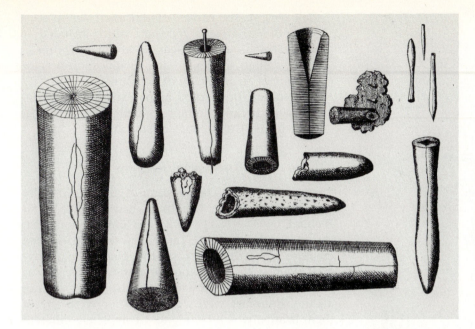

FIGURE 20–9
The fossil remains of *belemnites*, a variety of cephalopod that in life surrounded the internal skeletons shown here. It is obvious why the remains are sometimes termed "fossil cigars." (From G. W. Leibniz, 1749, *Protogaea*, courtesy of the History of Science Collections, University of Oklahoma Libraries.)

of the Mesozoic flora until almost the end of the era.

By the end of the Mesozoic Era *angiosperms*, the flowering and fruiting plants with *enclosed seeds*, became dominant and remain so today. The reproductive structures in gymnosperms are cones with naked seeds (Figure 20–10); those in angiosperms are enclosed in flowers. Because flowers are quite showy and are largely in the colors most easily seen by insects, these creatures visit and pollinate the flower's ovaries, as we have just indicated. Once pollinated, the flower develops seeds within its ovary, which then swells into a fruit. The fruit protects the seeds from drying and helps disperse them. When animals eat the fruit, they release the seeds unharmed but now coated with the rich natural fertilizer of their excrement.

Angiosperms comprise a tremendous variety of plants that are quite familiar to us—willows, grasses, roses, corn, oaks, tomatoes, beans, wheat, lilies, and dandelions. The explosive growth in the abundance and variety of angiosperms at the end of the Mesozoic has always been a puzzle to **paleobotanists**, biologists and geologists who study ancient plant life. No doubt the growth of angiosperms was assisted by the rise in the abundance of pollinating insects.

The Mesozoic Era began with enormous regression of the sea, as the continents coalesced and Permian glaciation of the southern continents became widespread. The vigorous continental climates were injurious to many plants. The major marine transgression of the Cretaceous Period had the reverse effect. Climates became universally mild, and continental life prospered under ideal conditions.

VERTEBRATES OTHER THAN REPTILES

In the Mesozoic seas, which included the growing Atlantic and the shrinking Pacific, fish prospered. Both cartilaginous fish and those with bony skeletons entered a golden age and filled essentially every marine environment available to them. The lungfish and lobe-finned fish remained as unimportant, sparse members of the marine community, as they do today. The Late Paleozoic decline of the amphibians continued; today only the frogs and salamanders remain as inconspicuous survivors. The line of descent from fish to reptile was to stretch from lobe-finned fish through the amphibians to the first awkward, waddling reptile. From that inauspicious start, reptiles began their climb to preeminence.

REPTILES

The Mesozoic is commonly called the "Age of Reptiles." They dominated the Mesozoic land, sea, and air as no other animal group had before them. These vertebrate animals with amniotic eggs had four, two, or, if snakes, no legs. Characteristically cold-blooded, they must live in climates where the average temperature is near their desirable body temperature; they raise their temperature by lying in the sun or diminish it by finding shade or water. Reptilian skin is horny, usually consisting of folded, overlapping scales.

Early Mesozoic reptiles can be broadly subdivided into two groups, each of which went in wholly different evolutionary directions.

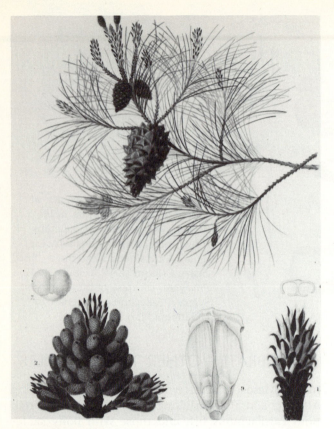

FIGURE 20–10

The reproductive structures of *gymnosperms*, plants which do not enclose their seeds, are cones. These are the male and female cones of a common evergreen. (Courtesy of the History of Science Collections, University of Oklahoma Libraries.)

► **Therapsids.** The mammal-like therapsids originated in the Late Permian from the *pelycosaurs* or "sail-back reptiles." Unlike the sprawling lizards, they walked on all four legs, which were directly under their streamlined bodies. Although regarded as reptiles, they *may* have been warm-blooded and had hairy skin. Their teeth, skeletal arrangement, and the pattern of the blood vessels in their fossilized skeleton were also very much like those of modern mammals. At the end of the Triassic, one branch, the *theriodonts*, led directly to the mammals.

► **Thecodonts.** Ancestors of the reptiles known as dinosaurs, plesiosaurs, pterosaurs and birds, thecodonts were unique to the Triassic. Unlike the therapsids, the *thecodonts* ran solely on their two hind feet. Because the weight of their bodies had to be supported from their hips, their pelvic structures were extensively modified to support

their weight. The earliest modification was to a group of *lizard-hipped dinosaurs*, which included both carnivorous (meat-eating) and herbivorous (plant-eating) species. A later modification was to *bird-hipped dinosaurs*, all of which were herbivorous. Both lizard-hipped and bird-hipped dinosaurs became dominant reptiles in the Jurassic and Cretaceous periods.

Terrestrial Reptiles

There is no formal classification of life that includes the word "dinosaur," a term coined in 1842 by Sir Richard Owen (1804–1892), an English zoologist, for an *Iguanodon*, a large Late Cretaceous bird-hipped creature which was the first of its kind to be discovered (Figure 20–11). Dinosaur translates literally as "terrible lizard." Both words are distortions, for not all the dinosaurs were "terrible," nor were any of them lizards, which belong to a completely different group of reptiles.

As already mentioned, dinosaurs are extinct members of two large reptilian groups (see Appendix F), the *saurischians*, which were the lizard-hipped dinosaurs, and the *ornithischians*, the bird-hipped dinosaurs. Each of these groups can be further subdivided.

Saurischians   Lizard-hipped dinosaurs.
   Theropods   Carnivorous, mostly two-legged dinosaurs. Late Triassic through Cretaceous.
   Sauropods   Herbivorous, largely four-legged dinosaurs, including the largest land animals of all time. Late Triassic through Cretaceous.

Ornithischians   Bird-hipped herbivorous dinosaurs.
   Ornithopods   Primitive, largely duck-billed dinosaurs. Late Triassic through Cretaceous.
   Stegosaurs   Armored dinosaurs with plates, spikes, and armored tails. Jurassic to lower Cretaceous.
   Ankylosaurs   Heavily armored dinosaurs with thick, bony plates encasing the tail, back, and body sides. Late Jurassic to Cretaceous.
   Ceratopsians   Horned dinosaurs with heavily armored head and neck. Cretaceous.

THEROPODS

Therapods include a diverse group of carnivores that remained largely bipedal, walking on their hind legs only, and that balanced themselves with their long tails. Within the Triassic, theropods remained rather small, but some members of this group grew to enormous size in the Jurassic and Cretaceous and must

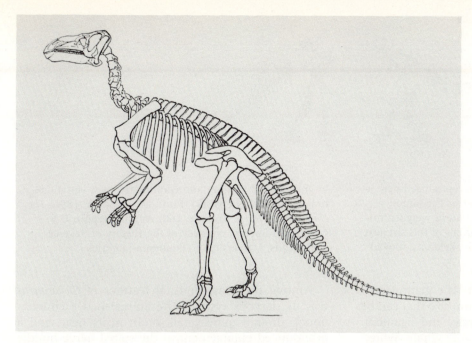

FIGURE 20–11
This is the skeleton of *Iguanodon*, an ornithopod that was the first dinosaur ever discovered. This reproduction is about one-eightieth the natural size of the adult dinosaur. Its discovery prompted thoughts that it was a ''terrible lizard.'' Like all ornithopods, it was a herbivore; its build suggests it may have been fleet afoot. (From O. C. Marsh, 1896, courtesy of the History of Science Collections, University of Oklahoma Libraries.)

have been fearsome predators. Best known of the therapods is *Tyrannosaurus,* which reached up to 50 feet (15 meters) in length, and stood perhaps 20 feet (6 meters) high. The skull was huge in proportion to body size and contained a vicious set of curved, serrated teeth which were the size of bananas (Figure 20–12).

SAUROPODS

When we think of dinosaurs, we usually think of the *sauropods*, some of which became the largest land animals that have ever lived. Animals such as *Brontosaurus,* (Figure 20–13), now properly called *Apatosaurus,* or *Brachiosaurus,* reached lengths of 60 to 90 feet (18 to 30 meters) and may have weighed as

FIGURE 20–12
The fossilized head of *Tyrannosaurus rex* suggests how well adapted this creature was for a carnivorous life-style. The teeth in this 100-million-year-old specimen remain awesome. (Photograph by W. T. Lee, U.S. Geological Survey.)

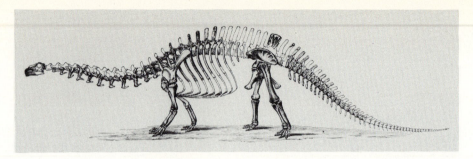

FIGURE 20–13

The skeleton of *Brontosaurus excelsus* as reproduced here is about one two-hundredths its natural size. The largest land animal that has ever lived, *Brontosaurus* weighed in at up to 80 tons (72 metric tons). Animals of this size test the limits of what can be supported on skeletons, moved by muscle and tendon, and coordinated by a vertebrate brain and nerve network. (From O. C. Marsh, 1896, *The Dinosaurs of North America*, 16th Annual Report, U.S. Geological Survey, courtesy of the History of Science Collections, University of Oklahoma Libraries.)

much as 80 tons (72 metric tons). The giant sauropods probably reached the maximum size that can be supported on land by bone, muscle, and ligament. The largest living creature is the blue whale, whose 100-ton (90-metric ton) body mass must be largely supported by water.

The sauropods became wholly quadripedal, needing all four legs and very broad cushioned feet to hold up their enormous weight. The skeleton was extremely dense and heavy, with huge vertebrae and pelvis but with a smallish skull.

ORNITHOPODS

The most primitive of all bird-hipped dinosaurs, the *ornithopods*, were small- to medium-sized and were both bipedal and quadripedal. They were plant eaters whose teeth and jaws were especially modified for the crushing of large amounts of green plant material. The *Iguanodon* (Figure 20–11), the first dinosaur ever to be described, was an ornithopod.

STEGOSAURS

Like all the bird-hipped dinosaurs, the stegosaurs were herbivores, both smaller and lighter than their meat-eating theropod kindred. Stegosaurs must have looked like tasty meals for *Tyrannosaurus*, but they had two sets of defensive mechanisms. With their lighter and more efficient skeleton, they were likely fleet afoot and might have been able to run from harm. When backed into a corner or ambushed, the stegosaur could defend itself with a wicked spiked tail (Figure 20–14), and massive triangular bony plates protected its spine. The spiked tail would have been an awesome weapon, and the bony plates an unpleasant mouthful. In addition to providing protection, the spinal plates probably also served as heat regulators, serving the same function as a car radiator; the plates were full of blood vessels.

Among the most unusual features of *Stegosaurus* was its nervous system. The animal had a conventional, though quite small, brain in its head and a pronounced enlargement of the spinal nerve bundle in its tail to form a sacral plexus twenty times the size of its brain. Some have suggested that this unique animal had "two sets of brains," with that in the rear controlling locomotion and the defensive use of the tail. As Bert Taylor, a columnist for the Chicago *Tribune*, wrote in 1912,

Observe by these remains, the creature had two sets of brains. . . no problem bothered him a bit, for he made both head and tail of it. . . He thought twice before he spoke, and thus had no judgement to revoke. . . .

Apparently "two sets of brains" were still not enough, for the stegosaurans were the first large category of dinosaurs to become extinct.

ANKYLOSAURS

Ankylosaurs were armored quadripedal bird-hipped dinosaurs with plated, spiny armor covering their tail, back, and body sides. The tail terminated in a huge bony mass, which must have been used as a bludgeon or club. Once they were down on their belly, ankylosaurs must have looked like an armored pillbox and were probably just as dangerous to attack. Ranging up to 20 feet (6 meters) long, ankylosaurs were inoffensive plant eaters.

CERATOPSIANS

Medium to small quadripedal dinosaurs famous for their horns, the ceratopsians had a skull uniquely modified to make up a fourth of the total animal length. The enlarged skull underlay a heavily armored head and neck, defended by a series of awesome pointed spines (Figure 20–15), similar to those

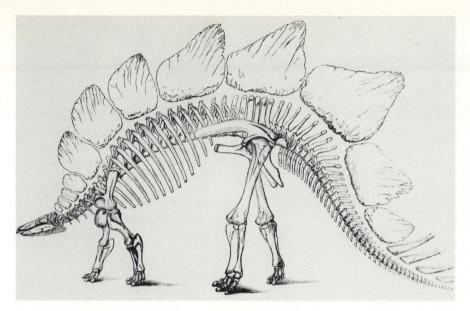

FIGURE 20–14

Stegosaurus represents one of many different kinds of defensive adaptations by prey. The spiny tail must have been an awesome weapon. The bony triangular plates along its spine served as defensive armor and possibly absorbed and dissipated heat. (From O. C. Marsh, 1896, *The Dinosaurs of North America*, courtesy of the History of Science Collections, University of Oklahoma Libraries.)

of a rhinoceros. Like the rhino, ceratopsians were probably herd animals. With their powerful neck muscles, heavy bodies, huge skulls, and multiple pointed horns, the ceratopsians were formidable foes when face on. Any meat eater that attacked one of these was in for a long afternoon.

Marine Reptiles

One of the surprises to those who study evolutionary trends is that they are seldom unidirectional. Although the benefits of marine life had been abandoned by the amphibian ancestors of the reptiles, several groups of reptiles returned to the sea with lungs, eggs, and legs. Lungs were used by creatures who needed to come up for air, legs were modified back into flippers, and instead of hatching from eggs, the young were born live.

There were three main lineages of marine reptiles.

▶ **Icthyosaurs.** The most fishlike of marine reptiles, icthyosaurs had well-developed fins, streamlined bodies, powerful tails, and elongated jaws supporting numerous pointed teeth (Figure 20–16). Middle Triassic through Cretaceous.

▶ **Plesiosaurs.** Turtlelike in body plan, but without the shell and with a long, graceful neck (Figure 20–17), plesiosaurs paddled through the water catching fish in their jaws. They reached lengths of 40 feet (12 meters). Jurassic through Cretaceous.

▶ **Mosasaurs.** Essentially marine lizards, mosasaurs reached 30 feet (9 meters) in length, with powerful tails and paddlelike limbs. Their pointed and curved teeth were well adapted for grasping prey (Figure 20–18). Cretaceous only.

All these swimming reptiles, so obviously well adapted to marine existence, disappeared at the end of the Mesozoic Era. Only their modern relatives, the crocodiles, turtles, and some of the snakes, remain adapted to a partly aquatic existence.

Airborne Reptiles

Flight imparts a freedom from the limitations of geography that constrain all other living things, but the

FIGURE 20–15

The third horn is hidden in this view of the massive skull of *Triceratops,* one of the last of the ceratopsian dinosaurs. The frill is the massive bony "collar" covering the neck. The neck muscles must have been huge to support this weight. (From O. C. Marsh, 1896, *The Dinosaurs of North America*, courtesy of the History of Science Collections, University of Oklahoma Libraries.)

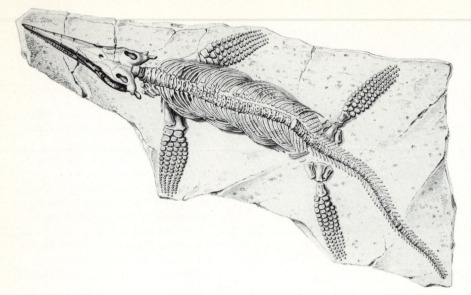

FIGURE 20–16
Notice the streamlined form, long snout, and bony flippers. The *icthyosaurs* were remarkably fishlike in form and must have been superb swimmers. The bony flippers are a readaptation of legs, a superb example of convergent evolution. (From Alcide D'Orbigny, 1849, *Dictionaire Universelle...*, courtesy of the History of Science Collections, University of Oklahoma Libraries.)

demands of flight are formidable. To fly, an animal must overcome gravity with a combination of muscle and strong though lightweight hollow bones. Powerful "arm" muscles are needed to flap wings, which are generally extended arms or fingers, and rigid breastbones and vertebrae are required for the muscles to work against. Flying also requires keen vision, a delicate sense of balance, and superb coordination; landing requires retractable, muscular feet.

The enormous energy needed for active flight re-

quires that all flying creatures be warm-blooded and covered with an insulating membranous skin; insulating structures may include hair, as in bats, or feathers, as in true birds.

The problems of flight have been solved in several different ways. The insects solved the challenge with thin, rapidly beating wings. Bats, a mammalian entry, developed sonar and soaring flight. Among the reptiles, the *pterosaurs*, literally "winged lizards," were the *only* solution. They filled the air during the

FIGURE 20–17
The *plesiosaur* was a graceful animal which must have been adept at using its long neck to dart at and grab unsuspecting fish. Everything about the skeleton shows that the animal had readapted well to a marine environment from a turtlelike ancestry. (Photograph courtesy of the U.S. Geological Survey.)

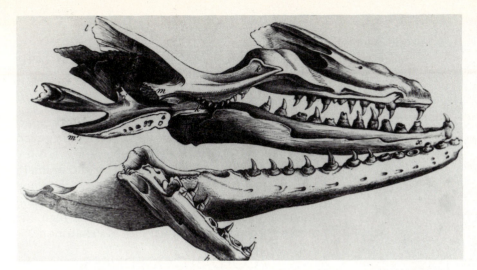

FIGURE 20–18
When these jaws were first found
in a chalk quarry in 1770, they
created a sensation. No one could
imagine that a 10-foot (3-meter)
"lizard" had ever lived, much less
have been marine. The animal
was named a *mosasaur*. Its teeth
clearly suggest that it was adapted
to a life of predation. (From
Reverend William Buckland,
1837, *Geology and Mineralogy
Considered with Reference to Natural
Theology*, Vol. II, courtesy of the
History of Science Collections,
University of Oklahoma
Libraries.)

Jurassic through the Cretaceous periods and may have had hairy skin, unlike all other reptiles. The true birds, creatures with feathers, hollow bones, and powerful flight muscles, descended from dinosaurs and originated in the Jurassic; they have filled the skies ever since.

THE PTEROSAURS

Pterosaurs are animals with one "finger," the fourth digit, enormously extended in length and covered with wing membranes, much like the modern bat. The best-known group are the *pterodactyls*, whose name literally means "wing-fingers" (Figure 20–19). They had batlike leathery wings, powerful necks, and pelicanlike jaws full of teeth. They must have been fierce competitors for food.

Through time the pterodactyls increased in size. The largest flying creature ever known was a pterosaur discovered in Texas with a probable wing span of 51 feet (15 meters), larger than many modern

fighter aircraft. It may have occupied the same environmental niche as the modern vulture, which eats dead meat. Wind tunnel tests indicate that if such a creature faced into even a light breeze with its wings extended, it would have risen into the air like a kite and soared away. Excellent preservation of a few specimens has revealed the presence of hair, normally an insulating material found only on warm-blooded creatures. This finding is consistent with the energy demands of pterodactyl flight, duplicated in 1986 by a working model.

AVES—THE TRUE BIRDS

The birds belong to a group of vertebrates wholly separated from the reptiles; the group name is *Aves*. Birds are a large subdivision of the vertebrates having feathers, warm-bloodedness, scaly feet, and forelimbs developed into true wings.

The oldest bird skeleton ever found is that of *Archaeopteryx*, a Jurassic ancestor to all other birds. The pigeon-sized *Archaeopteryx* is known from only three extremely well-preserved skeletons in fine-grained Jurassic limestone (Figure 20–20). If it had not been for the excellent preservation which records the presence of feathers as well as swiveling wrist bones and three claws, the fossil creature would have quickly been identified as a reptile, perhaps a pterodactyl. It has a thecodontlike skull and teeth, long legs, theropodlike feet, and highly modified forelimbs covered with feathers, which must have functioned as wings.

By a lucky accident of preservation, we have in *Archaeopteryx* a missing link—a true intermediate between the reptiles and birds. The skeleton is broadly reptilian, but its delicacy and geometry are birdlike. Because the creature had well-developed feathers,

FIGURE 20–19

The *pterodactyl* may have been a superb glider. The completion of a working model and wind tunnel tests show that its membranous wings were capable of great endurance in flight. Though replaced by the birds, the *pterodactyls* were the first creatures larger than insects to take to the air. (Photograph courtesy of the U.S. Geological Survey.)

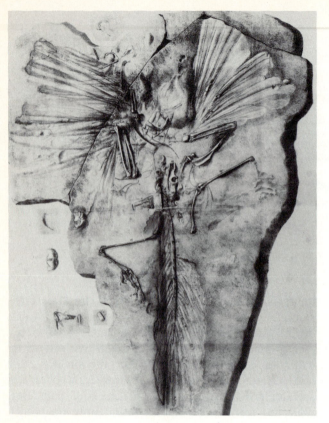

FIGURE 20–20

This drawing is by Richard Owen, a brilliant British zoologist of the last century who never accepted his friend Darwin's mechanism for evolution. The creature is *Archaeopteryx*. Notice the impression of primary wing feathers on the two forelimbs and the well-preserved feathers on the jointed tail. The whole animal was about the size of the modern crow and is about as close to an intermediary between reptiles and birds as paleontologists will ever get. (From Richard Owen, 1863, *On the Archaeopteryx of Von Meyer*, courtesy of the History of Science Collections, University of Oklahoma Libraries.)

Archaeopteryx is classified as a bird. Like all birds, it must have been warm-blooded.

By the Cretaceous, birds had become much more like our modern creatures. They had developed extremely strong hollow skeletons, which made them much more effective flyers than their ancestors. An example is *Hesperornis*, a specimen of which was found well preserved in a fine-grained limestone in Kansas (Figure 20–21). Such early birds had teeth, a feature lost during the Cenozoic. This bird was similar to the modern loon or grebe, being highly specialized for swimming and diving.

WARM-BLOODEDNESS AND THE FOSSIL RECORD

An animal's metabolic rate—the rate at which internal chemical reactions break down food and convert it into energy—approximately doubles for every rise of 18° Fahrenheit (10° Celsius) rise in the animal's temperature. If the animal's temperature drops too low, too little energy is available to sustain life and the animal dies. If its temperature rises too high, life processes race and the animal dies of exhaustion.

COLD-BLOODEDNESS

A cold-blooded animal, like a snake or a lizard, is at the mercy of its environment. It must keep its internal temperature within a narrow range by basking in the sun or by seeking shade or water. With no way to make its own internal heat, it must not allow its temperature to fall too low during the night or the winter months. Not surprisingly, cold-blooded animals normally live in warm, temperate localities or in the tropics where there is little change in temperature and the days are of equal length year-round. Some devise elaborate cooling mechanisms like the sails on the Permian "sail-back" reptiles or the spinal plates on the Jurassic stegosaur. Others hibernate or estivate in order to survive seasonal temperature changes.

Larger size confers great advantages on many animals. The ratio of surface area to volume is greatest for small animals and babies; they lose very large amounts of heat from their skin and must eat large amounts of food to maintain their body temperature. A shrew, a mouse-sized insect-eating mammal, must eat its own body weight in food every single day. The enormous demands of flight on the small body of a hummingbird requires that it hibernate every single night to save energy. Human beings and elephants need eat only a small fraction of their body weight in food each day; in proportion to our mass, we lose little heat from our surface.

The tendency toward gigantism that we have seen in every lineage of reptiles reflects these simple energy equations. A greater mass of tissue tends to hold a constant temperature more easily; temperature fluctuations experienced by a large lizard are much smaller than those experienced by a small lizard. Lizards spend much of their life moving into and out of

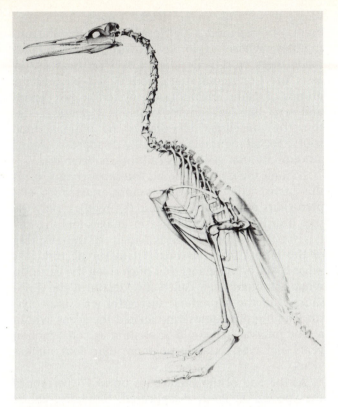

FIGURE 20–21

This is the skeleton of *Hesperonis regalis,* a Cretaceous wading bird adapted for eating fish along the shore. This specimen comes from a chalk layer in Kansas and has been fully restored in this artist's sketch. The skeletal features of this bird are quite similar to those of modern birds. (From O. C. Marsh, 1880, *Odontornithes,* courtesy of the History of Science Collections, University of Oklahoma Libraries.)

sunlight and basking on warm rocks in order to make heat loss through skin equal heat intake through the same skin.

Another disadvantage of cold-bloodedness is that cellular respiration captures only about half of the energy in the food that is eaten. The other half is released as waste heat; thus one-half of all that is eaten is wasted. On the other hand, because the sun provides much of the heat needed to maintain body temperature, cold-blooded animals require much less food than warm-blooded animals.

WARM-BLOODEDNESS

Warm-blooded animals like humans beings put the "waste" heat of metabolism to work by maintaining their body temperatures at about 40° Celsius above their surroundings. *The rate of metabolism and food in-* *take is two to ten times higher for warm-blooded animals than for cold-blooded creatures of the same weight.* Not only is the production of heat twice as efficient, but heat is also more easily retained by various kinds of insulation, including fur, fat, hair, and feathers. Warm-blooded animals have less difficulty maintaining their temperature in a cold climate; for example, the involuntary muscular activity we call shivering produces heat.

Only one climate can be easily lethal to a warm-blooded animal—an environment more than a few degrees *above* its normal body temperature. We humans are excellent furnaces but poor radiators. We can stand a great deal of cold, but very little heat— hence the universal rise in the use of air conditioning in climates that 40 years ago were singularly unattractive places to live.

We have mentioned several times that some reptilian groups might possibly have been warm-blooded. Let us look at the evidence for what is a distinctly unreptilian characteristic.

Warm-Blooded Dinosaurs?

How would we recognize the warm-bloodedness of an animal now preserved only as a fossil? Surprisingly, five different attributes can independently suggest warm-bloodedness in what is now only a pile of bones.

1. Warm-blooded animals require far more food per unit of their body weight. In carnivores this means the ratio of predators to prey is quite low, that is, it takes many prey to provide food for only a few predators.

2. Because much more blood must be circulated to maintain uniform body heat, the bones of warm-blooded animals have more abundant blood passages than do those of cold-blooded animals.

3. Warm-blooded animals can successfully live in colder climates. If we find dinosaur bones among the kinds of rocks or other fossils that indicate a cold climate, the dinosaurs must have been warm-blooded.

4. Warm-blooded animals tend to walk upright, but cold-blooded animals have their legs protruding from the sides of their bodies and have a sprawling posture.

5. All very large animals today are warm-blooded and have the four-chambered hearts necessary to produce the high blood pressures needed to

circulate blood over a very large body mass. What then can we say about the therapod and sauropod giants of the middle and late Mesozoic?

In still another related puzzle, paleontologists have long wondered why the cold-blooded reptiles were so dominant throughout the Mesozoic and the warm-blooded ancestors of the mammals occupied such an insignificant niche. Only after the Mesozoic reptiles were largely gone did the little mammal-like creatures become significant parts of the fauna. In the Cenozoic mammals expanded into the dominant animal group.

To one group of geologists the answer to all these puzzles is simple. From what can be reconstructed from predator:prey ratios, from the occurrence of dinosaur skeletons in areas of cold Mesozoic climate, from the nature of dinosaur footprints and legs, and from the examination of well-preserved dinosaur bones, they conclude that *the majority of Mesozoic dinosaurs were warm-blooded creatures.*

To other equally qualified geologists, most of the evidence in all but the ancestors to the birds indicates that dinosaurs were cold-blooded, like today's reptiles. The controversy continues, for dinosaur warm-bloodedness is an idea that contradicts a century-old belief that dinosaurs were reptiles, and therefore, by definition, cold-blooded.

The controversy illuminates the problem of attempting to classify transitional phenomena. What can we make of a birdlike creature like *Archaeopteryx* or its Early Cretaceous successor *Deinonychus*? *Archaeopteryx* is covered with feathers like a bird, but has teeth like a thecodont and feet like a theropod. Is it a bird, a reptile, or something in between? And what of the *therapsids*? They, too, are one-half mammal and one-half reptile, and *they* too may have been warm-blooded. Definitions cannot evade nature's complexity.

EXTINCTION AT THE CRETACEOUS-TERTIARY BOUNDARY

Second only to the extinction across the Permo-Triassic boundary at the close of the Paleozoic, the mass extinction across the Cretaceous-Tertiary boundary—often called the *K-T boundary extinction* from the geologic symbols for Cretaceous and Ter-

tiary—was a truly major dying. Among the invertebrates, the belemnites, or "fossil cigars," and the ammonites, whose explosive growth in the Jurassic made them worldwide index fossils for marine Jurassic and Cretaceous strata, both exited, as did large groups of clams, including big oysterlike pelecypods and coral-like reef-forming rudistids. Oceanic plankton, the unicellular life that is the ultimate food source for all marine life, virtually disappeared across a boundary that geologists call the *plankton line.*

Among the vertebrates, all the saurischians, ornithischians, icthyosaurs, plesiosaurs, mosasaurs, and pterosaurs disappeared. Among the reptiles, only the true lizards, the crocodilians, the snakes and the turtles survived. Only the birds remain as the offspring of the dinosaurs; your next Thanksgiving turkey is what is left of the giants that once ruled the earth. In what has often been called the "death of the dinosaurs," all the supremely successful terrestrial, marine, and aerial "dinosaurs" exited the scene within a time interval that could be as short as half a million years and was certainly no more than four million years.

At the end of the Cretaceous up to 75 percent of the species on earth disappeared. Similarly, at the Permo-Triassic boundary 96 percent of all Paleozoic species failed. What process or processes could fell life-forms across the land, sea, and air? What could remove the pterodactyl but leave true birds alone? What could fell the icthyosaur but leave the fish? What could fell *Tyrannosaurus* and then *Triceratops*, while leaving small mammals, including only one primate, our ancestor, alone?

CAUSES OF EXTINCTION

Biologists find it quite difficult to understand the causes of even modern extinctions; an abundance of data may still leave the causes uncertain. How then can we make sense of events millions of years ago for which most of the clues were destroyed? What we are left with are the patterns of dying, and these patterns appear to be broadly random. The supremely successful species disappeared just as quickly as those that were already struggling.

Scientists have suggested various possible causes of the Mesozoic extinctions. These include overwhelming disease, collapse of food chains, climatic changes, sea regressions, variations in solar output, supernovae explosions, widespread volcanism, and rise and fall of the sea level. Whatever the cause or causes, the effects extended across almost every environmental niche.

General Causes

Two worldwide changes can be recognized at the end of the Cretaceous. *Both of them are associated in time with mass extinctions throughout the Phanerozoic.*

1. **Decline in sea level**. Sea level declined throughout the Late Cretaceous as the seas withdrew from their maximum transgression across the land. Such a decline was associated with an *increasingly vigorous continental climate having temperature extremes and violent storms*, and with *regression of the seas* from the widespread shallow shelves, which contribute most of our marine fossils. Such declines signal decreases in the diversity of species and indicate severe stress among surviving species.

2. **Decline in worldwide temperature**. From the very, very mild warm temperatures that had prevailed through most of the Mesozoic, the world's temperatures in the Cenozoic dropped an average of 22° Fahrenheit (12° Celsius) over a span of 70 million years. This temperature drop eventually brought about the Late Cenozoic glacial ages.

These two events may have set the stage for massive extinction, but the gradual declines in sea level and temperature could hardly have wreaked such havoc alone. How would *gradually* declining sea level and temperature create an *abrupt* worldwide die-off in floating plankton? That sharp disappearance is reflected in the *plankton line*, a universal marker in the oceanic sediments deposited across the K-T boundary. Could something else have intervened?

K-T Impact Event

There seems little doubt from the geologic evidence that the earth was impacted at the K-T boundary by an asteroid many miles or kilometers in diameter. Such an impact would have struck the earth with an estimated energetic yield of millions of megatons of TNT and made the explosion of our most awesome nuclear weapons seem like firecrackers in comparison. It may have created huge firestorms that could have covered the earth in a long-lasting blanket of soot, a scenario similar to the events forecast to occur during a nuclear war.

The dust and ash created by the firestorms and the pulverization of the earth would have blocked out the sun and stopped all photosynthesis for a period of time lasting anywhere from several months to decades. This would have led to the immediate death of the photosynthetic plankton and a subsequent collapse of the oceanic food chain. Land plants would have been destroyed, and the surviving herbivorous dinosaurs would have died of starvation. This, in turn, would have led to the starvation of the carnivores who preyed upon them. Only some seeds would have survived, their long dormancy broken as sunlight slowly returned, and the earth began the process of repopulation of plants—somewhat like the repopulation still occurring on Mount St. Helens after the catastrophic eruption (see Chapter Three) there in 1980.

The evidence for such a meteoritic impact scenario lies in anomalously high contents of *iridium*, a rare metal allied to platinum and far more abundant in asteroids than in the earth's crust. Findings throughout the world of Late Cretaceous clays whose chemical analyses yield high values for iridium make it rather certain that an impact event occurred. The additional findings of shocked quartz grains—similar to those produced both by meteoritic impact and by explosive volcanism—and *microtektites*, small glassy spheroids created during an impact, strengthen the case for impact. Recent studies of K-T boundary sediments revealed layers of carbon-rich particles just above the iridium-rich layer; the carbon particles were analyzed by scanning electron microscopy and turned out to be soot—just what would be expected in the aftermath of firestorms. Other data seem to indicate that asteroidal impacts would have generated both heavy-metal and cyanide poisoning, and still other data show that a very large impact event would have put a huge hole in the upper ozone-rich layer that guards the earth from damage by ultraviolet rays. A truly large impact would have made the conditions for life difficult indeed.

CAUSE OF K-T EXTINCTION

In spite of the probable magnitude of the asteroidal impact, the evidence leading from probable meteoritic cause to certain biologic effect is riddled with inconsistencies. There is little doubt the earth was impacted at the K - T boundary, and it is reasonable to assume such an event would have had a profound effect on the delicate web of life on earth. But the evidence tying impact to the observed patterns of extinction is tenuous and subject to multiple interpretation.

Why should a massive impact event spare terrestrial plants, the small vertebrates like birds and mammals, and reptilian fauna like snakes, lizards, crocodiles, and turtles? The strange selectivity—indeed the utter randomness—of who died and who lived, has yet to be explained. The search for causes continues.

The Results of Mass Extinction

Late Cretaceous mass extinctions set the pattern for the Cenozoic world. Dominant life-forms were forced from the scene and ecological niches opened for life-forms that might otherwise never have gained their opportunity. Without the extinction of the terrestrial dinosaurs, the little mammals would never have had their chance.

Among the little mammals scurrying about at the end of the Cretaceous world was one single species of small primates, the only one of our particular subclassification to witness the great annihiliation. Had this one group not survived, the primates would have come to an end with the Cretaceous, and our lineage would never have evolved. It seems only random chance that this one line survived the great dying and continued into the Cenozoic world, where its distant progenitors would be upright human creatures who would learn to reconstruct the past. To use our brains to discover that our existence today is the improbable result of the random survival of a single species 660,000 centuries ago is both to recognize and to deny our own insignificance.

Perhaps the last word on a topic so full of import for humanity should come from poets rather than geologists. Consider, then, these lines from *Prometheus Unbound* by Percy Bysshe Shelley (1792–1822).

. . . *and they*
Yelled, gasped, and were abolished;
or some God Whose throne was in a Comet,
    passed, and cried—"
Be not!"—and like my words they were no more."

SUMMARY

1. The Mesozoic Era began 245 Ma with the breakup of Pangea, an occurrence marked in the East by the Palisadian event; that separation left part of Africa attached to form what is now Florida. In the West, subduction began, recognized by the creation of an island arc complex and the resultant Cassiar–Sonoman orogeny, which disturbed much of the Cordilleran area.

2. Subsequently, in Late Jurassic time, eastward subduction produced continental magmatic arcs and massive batholiths in what has been called the Nevadan orogeny. From these beginnings a great wave of deformation slowly spread eastward, engulfing the entire Cordilleran region in episodes of plutonism, volcanism, thrusting, and folding, which created massive wedges of conglomerates from the rising Cordilleran highlands. Usually called the Sevier orogeny, this event was caused by the westward-moving continent colliding with the seafloor. The accompanying subduction along a gently dipping Benioff zone deformed the entire Cordilleran mountain system.

3. The continental margin underwent rapid subsidence, with great amounts of coal and black shale being deposited. These coal and black shale beds record a large number of eustatic cycles producing repetitive sedimentation much like that affecting eastern North America in the Late Paleozoic. These materials were to be crushed eastward against the cratonic margin by the Laramide orogeny, which began in Late Cretaceous and continued into the Cenozoic, eventually creating the Rocky Mountains.

4. In an episode of accretion that was to last 150 million years, fragments of continental, oceanic, and mixed continental and ocean crust were docked with the western margin of North America from sites covering the whole of the Pacific Basin. In the trip across the Pacific, some microplates may have covered a distance equal to one-sixth the circumference of the earth. This zone of accreted slivers separated from one another by faults increased the size of Paleozoic North America by 25 percent.

5. Among the invertebrates, the ammonites became a dominant and varied member of the marine communities. Among the plants, the rise of the angiosperms at the end of the Mesozoic Era, along with the rise of pollinating insects, was the most important event.

6. Among the vertebrates, the rise of reptiles of thecodontal ancestry led to saurischians, ornithischians, icthyosaurs, plesiosaurs, mosasaurs, and pterosaurs. Informally termed dinosaurs, the saurischians and ornithischians became the ruling terrestrial life-forms and developed into the largest land animals ever to inhabit the earth. The pterosaurs were the first large group to adapt to an aerial life-style; birds evolved from birdlike dinosaurs and took wing in the Jurassic to fill the skies.

7. Although reptiles are by definition cold-blooded creatures, some of the dinosaurs and pterosaurs may have been warm-blooded, an adaptation that conferred immense advantages in environmental adaptation. The birds and the little mammals that developed in the Jurassic were certainly warm-blooded.

8. Extinction across the Cretaceous-Tertiary (K-T) boundary brought an end to all the dominant reptiles, except for snakes, turtles, crocodilians, and lizards. Other victims included most marine phytoplankton, the sturdiest species of clams, the belemnites, and the ammonites. The K-T extinction depleted the number of species by 50 to 75 percent, and it remains unexplained. The regression of the world's oceans, accompanied by a decided and prolonged cooling trend, may have set events in motion leading to the decline of many species. Then an asteroid impact during the K-T boundary time may have had profound and far-reaching effects on already weakened life-forms. The results of this second largest mass extinction opened the Cenozoic environmental niches to a host of random survivors, including a single primate, the mammalian ancestor to humankind.

KEY WORDS

angiosperm	Palisadian event
Cassiar–Sonoman orogeny	Sevier orogeny
Laramide orogeny	suspect terrane
microplate tectonics	suture
Nevadan orogeny	thecodonts
paleobotanist	therapsids

EXERCISES

1. Herbivorous dinosaurs were the prey for carnivorous dinosaurs. They responded to predator pressure by developing defensive weapons and by becoming rapid runners. How can these modes of defense be recognized in fossil skeletons?

2. Given that before the pterosaurs no creatures larger than insects could fly, why would certain dinosaurs develop wings before they could fly? *Hint:* See referenced article by Stephen Jay Gould.

3. What processes and products were common to the Palisadian event, the breakup of Late Proterozoic supercontinent, and the ongoing breakup of east Africa?

4. In what ways can birds be considered "feathered reptiles"?

5. Snakes apparently evolved from lizards. What lizardlike features remain in snakes?

6. In a general way, the western Mesozoic eugeocline was deformed, intruded by igneous batholiths, and accreted onto the continent. Later the miogeocline was deformed in an orogenic sequence that moved eastward from marginal sea toward the craton. What similarities are there in sequence between the deformation of the western and eastern geoclines? What is one very large difference?

7. Today, there are thick deposits of Jurassic salt in sediments beneath the Gulf of Mexico. What process could have led to closure of the Gulf Basin and the basin's prolonged evaporation? Is there a similarity (see Chapter Nineteen) to the Silurian Michigan Basin?

SUGGESTED READINGS

ARCHIBALD, J. DAVID, AND CLEMENS, WILLIAM A., 1982, Late Cretaceous extinctions, *American Scientist*, pp. 377–385.
▶ *Argues against rapid extinction and impact model; a thoughtful paper.*

BAKKER, ROBERT T., 1986, *The Dinosaur Heresies*, New York, William Morrow and Co., 481 pp.
▶ *Quoting a reviewer " . . . romping, stomping, warm-blooded science." A brilliant synthesis and a totally provocative book.*

COLBERT, EDWIN H., 1980, *Evolution of the Vertebrates*, New York, John Wiley Interscience, 510 pp.
▶ *The standard reference, written by a leading scholar who has devoted a long life to the study of the topic, especially dinosaurs.*

ELLIOTT, DAVID K., 1986, *The Dynamics of Extinction*, New York, John Wiley Interscience, 294 pp.
▶ *Excellent summary of knowledge of mass extinctions. Elegant book.*

GOULD, STEPHEN JAY, 1985, Not necessarily a wing, *Natural History*, October, pp. 12–25.
▶ *An elegant example of experimental confirmation of an intuitive idea, concerning warm-bloodedness and the origin of wings. Marvelous!*

HOWELL, DAVID G., (ed.), 1985, *Tectonostratigraphic Terranes of the Circum-Pacific Region*, Houston, Texas, Circum-Pacific Council for Energy and Mineral Resources, 581 pp.
▶ *Thorough technical review of microplate tectonics and suspect terranes.*

JONES, DAVID L., et al., 1982, The growth of western North America, *Scientific American*, November, pp. 70–84.

► *Clear statement of accretion of suspect terranes by one of the earliest investigators.*

MARAN, STEPHEN P., 1984, What struck Tunguska? *Natural History*, February. pp. 36–37.
► *Nice review of an event in 1908 that may shed light on the evidence of asteroidal impact—and near misses.*

NORSTOG, KNUT, 1987, Cycads and the origin of insect pollination, *American Scientist, v. 75, pp. 270–279.*
► *Excellent review of the reasons for the ascendancy of angiosperms.*

RAUP, DAVID M., 1986, *The Nemesis Affair*, New York, W. W. Norton & Co., 220 pp.
► *''A story of the death of dinosaurs and the ways of science.''*

RUSSELL, DALE A., 1982, The mass extinctions of the Late Mesozoic, *Scientific American*, January, pp. 58–65.
► *Clear statement of the asteroidal hypothesis by a vigorous researcher in the field.*

WILFORD, JOHN NOBLE, 1986, *The Riddle of the Dinosaur*, New York, Alfred A. Knopf, 304 pp.
► *Written for the general public, this book does a fine job of reviewing up-to-date dinosaur lore.*

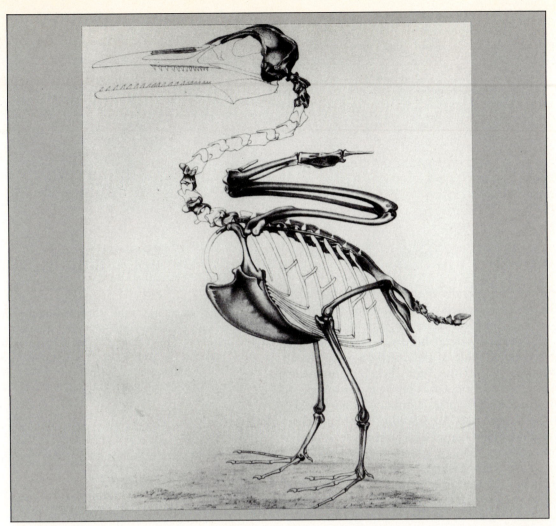

This figure illustrates the degree of restoration possible by skilled paleontologists when fewer than one-half the fossil bones have been preserved. Through an extremely detailed knowledge of comparative anatomy, the shape and location of the missing bones (shown in this figure as open and uncolored) can be filled in to yield a fairly complete restoration. (From O. C. Marsh, 1880, *Odornithes*, Memoirs of the Peabody Museum, Vol. 1, courtesy of the History of Science Collections, University of Oklahoma Libraries.)

CHAPTER *twenty-one*

The Cenozoic Era

The past is but the beginning of a beginning, and all that is and has been is but the twilight of the dawn.

H. G. WELLS (1866–1946),
THE DISCOVERY OF THE FUTURE (1901)

The great mystery is not that we should have been thrown down here at random between the profusion of matter and that of the stars; it is that from our very prison we should draw, from our own selves, images powerful enough to deny our nothingness.

ANDRÉ MALRAUX, 1901–1976,
LA CONDITION HUMAINE (1933)

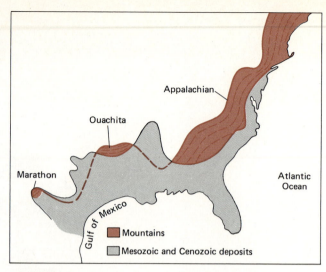

FIGURE 21–1

The Late Cretaceous transgression buried the folded rocks of three mountain ranges. The buried trends are shown as a dashed line. The geometric relation between the buried Ouachitas and the Appalachians is unknown. (Simplified from P. B. King and G. J. Edmonston, 1972, *Generalized Tectonic Map of North America*, U.S. Geological Survey.)

The Cenozoic Era, literally the age of "modern life," began 66 million years ago and continues. The span of time we call Cenozoic covers slightly more than 1 percent of the earth's total history. Evidence about this last second of time is preserved in a little-disrupted record. Within this brief flicker of time the North American continent was brought to its present state, the Rocky Mountains culminated their development, and mammals became the dominant life-form. Upright human beings appeared approximately 5 million years ago, leaving a fragmentary fossil record, and began their climb to ascendancy.

The rocks of the Cenozoic Era are subdivided into two periods, the Tertiary and Quaternary; the Quaternary, which began 1.6 Ma, is subdivided into Holocene and Pleistocene epochs, and the Tertiary, which lasted from 66 Ma to 1.6 Ma, is subdivided into Pliocene, Miocene, Oligocene, Eocene, and Paleocene epochs.

To many geologists the Cenozoic rocks and fossils are the most absorbing of all, because the evidence they present is relatively fresh and complete. For this reason the last 1 percent of geologic time is the most clearly understood interval. The culmination of the growth of the North American continent and its invasion by the earliest Native Americans in the Late Pleistocene have been clearly recorded.

CENOZOIC PHYSICAL EVENTS

The record of the events of the last 66 million years leaves us a wealth of detail about an extremely busy time. From that encyclopedia of truly ancient history, this text chooses the Cenozoic record in seven different major geographic areas to study.

1. Evolution of the Atlantic and Gulf coastal plains.
2. Development of the High Plains.
3. Origin of Rocky Mountain basins and lakes.
4. Evolution of the Colorado Plateau.
5. Development of the Basin and Range.
6. Origin of the Columbia Plateau and Snake River Plain.
7. Evolution of the Pacific Rim.

Each of these records in the seven areas demonstrates how geologists extract history from rocks and fossils—the evidence left behind by a restless earth. We will also briefly outline some of the changes worked by continental and alpine glaciation of North America, which began in the Early Pleistocene and continues today. This episode of multiple glacial advances and retreats has already lasted 1 million years and has greatly influenced both the scenery and the life-forms of North America.

THE COASTAL PLAINS

Throughout the Atlantic and Gulf coastal plains the Cenozoic history consists of discontinuous regression of the seas from the land (Figure 21–1), a continuation of the regression which had closed the Mesozoic Era. The climate remained generally warm and tropical, with tropical vegetation the norm in what is now the southeastern United States.

In Arkansas the weathering of aluminum-rich igneous rocks led to the formation of *bauxite,* an ore of aluminum. Bauxite is the residuum of a formerly clay-rich soil, stripped by intense weathering (see Chapter 4) of all its potassium, sodium, calcium, iron, magnesium, and silica. The residual soil is extremely rich in aluminum. Bauxite can form only in a tropical environment such as that in Cuba, where bauxite is being mined today.

As the Appalachians were worn down, the sediments deposited on the coastal plain became an inverted record of Appalachian history, that is, the youngest coastal plain sediments were eroded from the recently exposed core of the oldest Appalachian rocks. In Florida, which has no muddy rivers draining to the sea, the absence of eroded particles left the waters around it clear and open to a rich, diverse group of reef-building animals. Limestone deposition, which continues today, preserves the record of life in warm, sunlit, well-oxygenated water.

Within the Gulf of Mexico, Jurassic salt layers that formed after the initial rifting along our southern margin are now buried under as much as 24,000 feet (7300 meters) of younger rocks and recent sediments deposited by streams which drained the emergent North American continent. The salt layers, whose density was less than that of the sediment above them, began to rise in mid-Cenozoic time as plumes (Figure 6–9) to form *salt domes.* The rising salt pushed through the denser overlying sediment much like an igneous intrusion. The buoyant, low-density salt rises for the same reason that a wooden ball will not stay under water.

Oil and natural gas have accumulated in the flanking arched rocks where they are truncated or broken by impermeable salt. Many of the major oil fields of the Gulf Coast are found adjacent to such salt domes. *More than three-quarters of the world's supply of oil and natural gas comes from Cenozoic rocks.*

Throughout the coastal plains, the sands, silt, mud, and lime rest on a basement of Paleozoic and Mesozoic rock. Farther out at sea the same sedimentary accumulations rest on ocean floor basalts which are slowly subsiding under the great weight of sediment. As the sediments continue to accumulate on these passive shelves, they have all the characteristics of modern miogeoclines (see Chapter Seventeen).

The huge embayment of sediments, which may reach thicknesses in excess of 5 miles (8 kilometers), covers a particularly interesting Paleozoic relation. As far south as the Appalachian mountain trend can be traced (Figure 21–1) and as far east as the Ouachita–Marathon mountain trend can be traced through evi-

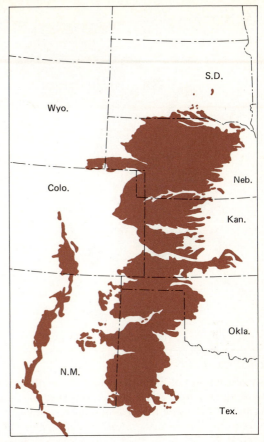

FIGURE 21–2

Surface distribution of Late Cenozoic rocks adjacent to the Rocky Mountains. These units of loosely consolidated sedimentary rocks form the High Plains and contain the major aquifer system (see Chapter Fifteen) that furnishes groundwater over this whole semiarid land. Severe depletion of this aquifer, especially in Texas, jeopardizes the lives and property of many people. The rocks are essentially a giant alluvial fan shed from the rising Rocky Mountains in Late Cenozoic time. (After P. B. King and H. M. Beikman, 1978, U.S. Geological Survey Professional Paper 904.)

dence recovered from the oil wells drilled in these two areas, the trends of these two ranges approach each other at right angles. The area of intersection remains a mystery, hidden under miles of Cenozoic sediment.

THE HIGH PLAINS

Stretching eastward from the Rocky Mountains is an immense composite alluvial fan of Cenozoic sedimentary rocks. In terms of association and position, these rocks are similar to the Queenston and Catskill rocks shed in Paleozoic times during the rise of the Appalachians. The coalescing composite fan forms a

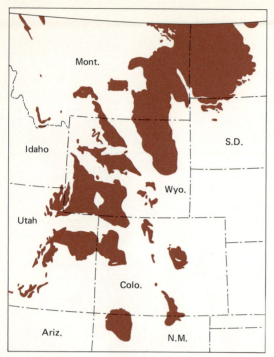

FIGURE 21–3

Surface distribution of Early Cenozoic rocks *within* the Rocky Mountain area. These are deposits shed into small basins among the rising Rocky Mountains, with each basin recording the history of nearby uplift. (After P. B. King and H. M. Beikman, 1978, U.S. Geological Survey Professional Paper 904.)

high, flat tableland to the east of the Rocky Mountains, which act as a major moisture barrier.

This part of the western United States is a region of flatlands with endless horizons and sparse population. Both irrigated and dry-land agricultures prevail; ranching and cattle grazing are other sometimes profitable activities on what are called *the High Plains*. The native vegetation is short grass, typical of a semiarid climate. To the east of the High Plains (Figure 21–2) are the Low Plains, which stretch from Kansas to West Virginia. This lower area, which receives more moisture, originally supported tall prairie grass but now supplies wheat and corn.

The High Plains were formed by essentially continuous sedimentation from the Rocky Mountains as they rose during the Middle and Late Tertiary Period. Their eastern margin is a sharp topographic break, created by eastward-draining streams cutting back into the soft underlying sedimentary rocks. Locally, the sharp escarpment is called the break in the "cap rock." The eastern margin or edge approximately parallels the hundredth meridian of west longitude;

the western margin is more gradual, as the High Plains grade into the flanks of the Rocky Mountains, which approximately parallel the one hundred and fifth meridian of west longitude (see Appendix D).

This gigantic alluvial fan, made up of a number of smaller ones, is an inverted record of the rise of nearby Rocky Mountains, since the oldest layers under the cap rock come from the erosion of the youngest layers at the top of the Rockies. The huge, coalescing alluvial fans are composed of silt, a minor amount of volcanic ash, and occasional layers of freshwater limestone, the product of transient lakes on the fans. The fans are a few hundred yards or meters in thickness and cover parts of seven states (Figure 21–2).

BASINS IN THE ROCKY MOUNTAINS

The rise of the Rocky Mountains in Early Tertiary was the culmination of the **Laramide orogeny**, a complex series of mountain-making events—the intrusion of batholiths, both block faulting and thrust faulting, minor volcanism, and regional metamorphism.

Another way in which we can recognize the uplift of mountains is by the sediments they shed into local closed basins. The rise of the Rocky Mountains was not the continuous rise of a single homogenous mountain range. Rather, parts of the marginal craton were either elevated or depressed. Between rising segments, numerous shallow basins received the sediment load flowing down from the nearby rising highlands.

The basin-fill deposits of the Rocky Mountain region (Figure 21–3) are widespread over the American West. Coal was deposited in many of these basins throughout the entire Tertiary Period, forming an important economic resource for the northern Great Plains and Rocky Mountains. Along with the coal came volumes of silt, freshwater limestone, volcanic ash, and well-preserved fossils of freshwater fish and air-breathing life.

Some of the basins were undoubtedly lakes for a time, and the abundance of animal and plant life in them left the lake sediments richly endowed with organic debris. These ancient lake sediments form what are now termed *oil shales* (see Chapter Sixteen), a potentially vital source of petroleum-based energy for the twenty-first century.

The soft sedimentary rocks that resulted from the compaction and cementation of these lake sediments are easily eroded into picturesque landforms and badland topography. Many of the splendid national

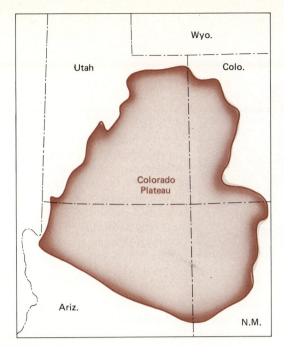

FIGURE 21–4
The Colorado Plateau is an area of stable cratonic rock overlain by generally flat-lying Phanerozoic sedimentary rock. The plateau has recently been uplifted many thousands of feet (several hundred meters) above its former mid-Cenozoic elevation.

parks of the west, such as Bryce Canyon, Badlands, and Cedar Breaks, present the beauty of these uplifted and eroded continental sedimentary rocks with their delicate hues of pink, orange, gold, yellow, and lavender.

The last major event in the long history of the development of the Rocky Mountains occurred about 40 million years ago. Older Tertiary rocks were tilted and beveled (cut across) to form a platform for younger Tertiary sediments throughout much of the Rocky Mountain region. The rocks trapped within Middle and Late Cenozoic basins clearly document this, recording a general drying trend as the rising Rocky Mountains cut off the flow of humid air from the west.

A general broad uplift began in the Late Cenozoic and still continues. The old basin-fill deposits have been raised half a mile (0.8 kilometer) or more above the elevation at which they were originally deposited. Coincident with this broad uplift, a large roughly circular mass of the craton west of the Rockies also began to rise. This area, known as *the Colorado Plateau*, has risen nearly a mile (1.6 kilometers) in the last 10 million years.

THE COLORADO PLATEAU

The Colorado Plateau is a roughly circular area centered about the Four Corners, a point shared in common by corners of the states of Arizona, Utah, Colorado, and New Mexico (Figure 21–4). The area remained undisturbed during the whole of the long string of orogenic events that so strikingly affected the rest of the West. Its Precambrian craton is covered with thin sequences of Phanerozoic sedimentary rocks in relatively undisturbed layers.

Training ground for generations of geologists, the Colorado Plateau displays geologic features with unmatched clarity. Throughout the plateau, the sedimentary layers are generally flat-lying but are occasionally interrupted by sharp folds (Figure 21–5) created when segments of the plateau moved upward along major high-angle faults. The plateau is marked by the development of large, broad basins and domes as well as sharp uplifts. Perhaps its best-known ornament is the incomparable Grand Canyon, but it is an area famous for its national parks and monuments and stunning geology.

Around the margin of the plateau, Tertiary and Quaternary volcanism (Figure 21–6) records the cracking and rifting of the crust associated with the intitiation of uplift. The plant fossils trapped in local basins also record the uplift, changing within 6 million years from those typical of semiarid scrub to those typical of high-elevation conifer forests. As the plateau rose, the rivers that flowed across the plateau gained renewed cutting power—power to cut through even the hardest rocks in the highest places. There is no better example of the power of a newly elevated river to carve deeply than the Grand Canyon, a 234-mile-long (376-kilometer) chasm across the southern edge of the plateau in Arizona. All over the plateau other rivers cut deeply incised streams, creating an area of high tablelands deeply cut by vigorous streams in a generally semiarid climate.

THE BASIN AND RANGE

Surrounding the Colorado Plateau on its northern, western, and southern margins is a region of complexly deformed sedimentary, metamorphic, and volcanic rock, broken along nearly vertical faults and moved up into mountain *ranges*, and down into intervening *basins*. The overall effect is as though every other key on a piano were depressed; the landscape is composed of uplifts separated by basins partially filled with sediment (Figure 21–7).

This lovely, barren landscape is composed of

FIGURE 21–5
A steep fold at Flaming Gorge on the north flank of the Uinta Mountains of northeastern Utah. During the Cenozoic uplift, the Green River cut a deep canyon directly through the fold, exposing its flexure to easy viewing. (Photograph courtesy of the U.S. Geological Survey.)

heavily eroded uplifts in what is a desert climate, and it contains splendid examples of Tertiary volcanism and complex structural geology. The vertical uplifts along high-angle faults have created block-shaped horsts (see Chapter Seven) in what is termed the **Basin and Range orogeny**, an episode of block faulting and volcanism which began in Late Oligocene (30 Ma) and continues. The intervening grabens may contain basin-fill sediments up to 3 miles

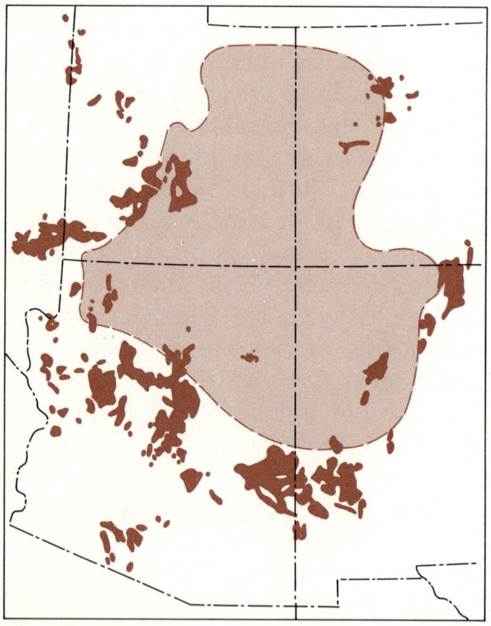

FIGURE 21–7
This Skylab photograph illustrates the essential character of the Basin and Range country. The Y-shaped object in the center is Lake Mead, which was produced when the Colorado River was dammed by Boulder Dam. Most of the land in the photograph is Nevada; here the boundaries between the sediment-filled basins and the uplifted mountain ranges are sharp. Running diagonally along the northeast (upper right) corner of this photograph is a very large, straight fault line which divides the Colorado Plateau on the right from the Basin and Range on the left. (Image courtesy of NASA.)

FIGURE 21–6
Surface distribution of volcanic rocks yielding radiometric age dates between 4 and 10 Ma. The boundaries of the Colorado Plateau have been superimposed to illustrate the striking relation between Colorado Plateau uplift margins and volcanic areas. (Adapted from P. B. King and H. M. Beikman, 1978, U.S. Geological Survey Professional Paper 904.)

(5 kilometers) thick; the sediments record continuing faulting and range uplift. Probably the best known of all the grabens is Death Valley, whose floor is the point of lowest elevation in North America (282 feet or 86 meters below sea level), from which one can see the top of Mount Whitney (elevation 14,494 feet or 4418 meters), the highest point in the 48 contiguous United States.

The block faulting of the Basin and Range resulted from mid-Tertiary thinning and extension of the continental crust. As the stretched crust broke along parallel fractures, blocks collapsed downward along numerous parallel high-angle fault planes. Among the older hypotheses to explain this anomalous event is the North American continent's overrunning of the East Pacific spreading center, which continues to spread beneath the Basin and Range.

In this model the Basin and Range structures, which extend from British Columbia south to Mexico, were produced when a ridge was subducted beneath an area of tension, extension, thinning, rifting, and volcanism. Another model presumes that back-arc spreading, similar to the process that created the Sea of Japan, is steadily thinning the continental crust, which will eventually rift to produce a new spreading ocean basin, perhaps analogous to the Red Sea. The causes for the formation of the Basin and Range remain an intriguing question, the subject of continued research into the processes beneath our feet.

THE COLUMBIA PLATEAU

During the Cenozoic Era one of the greatest known outpourings of continental volcanic rock occurred as a repetitive sequence of basaltic lava flows from numerous eruptive centers in the Pacific Northwest. The total extent of the various flows encompassed large parts of three states (Figure 21–8). The total volume of rock was four times that of the entire island of Hawaii.

The initiation of major basaltic volcanism is a common by-product of the early stages of continental rifting and separation, an event we have successively tracked through the Late Proterozoic-Early Paleozoic separation (see Chapter Eighteen) of a supercontinent, the Triassic separation of the eastern shoreline (see Chapter Twenty), and the Jurassic separation along the Gulf of Mexico. The Columbia Plateau seems to have resulted from a combination of rifting initiated at the same time as the Basin and Range orogeny and the development of a mantle plume, which is currently under Yellowstone (see Chapter Three, Figure 3–3). Perhaps as we look still

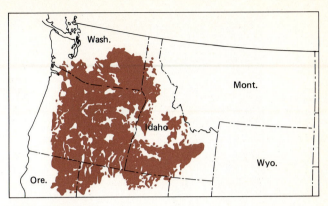

FIGURE 21–8

Distribution of Cenozoic volcanic rocks on the Columbia Plateau region of Idaho, Washington, and Oregon, the largest volcanic field in North America. Rifting and rotation beginning in the Miocene caused these rocks to erupt. (P. B. King and H. M. Beikman, 1978, U.S. Geological Survey Professional Paper 904.)

farther west we can find an answer to what processes were continuing to affect so strikingly the intermountain West.

THE PACIFIC RIM

The Cenozoic began with a continuation of marine deposition in a broadly Andean-type subductional regime as the eastern margin of the Pacific Ocean floor was being overridden by North America, creating a wide band of generally compressional landforms which stretch from British Columbia to Mexico and forming the Cordilleran region. Still farther east the craton was rising to create the Rocky Mountains, whose sediments created a broad apron beyond the mountains and filled numerous basins within the mountain trend.

By the Miocene, the overall picture along the Pacific Rim of North America was totally changed. The margin had shifted from one of compression to one of right-lateral (see Chapter Nine) shearing, with the initiation of the San Andreas and related fault systems, to be followed by clockwise rotation of the Cordilleran region and massive outbreaks of volcanism along rifted zones in the Columbia Plateau, the Basin and Range, and the margins of the Colorado Plateau. In Late Cenozoic, initiation of island arc processes in the far Pacific Northwest created the Cascade Range and developed a thermal plume under what is now Yellowstone (see Chapter Three). *The mid-Cenozoic was a time of profound change all over the American West*; we must look to the Pacific Ocean basin itself to understand the cause.

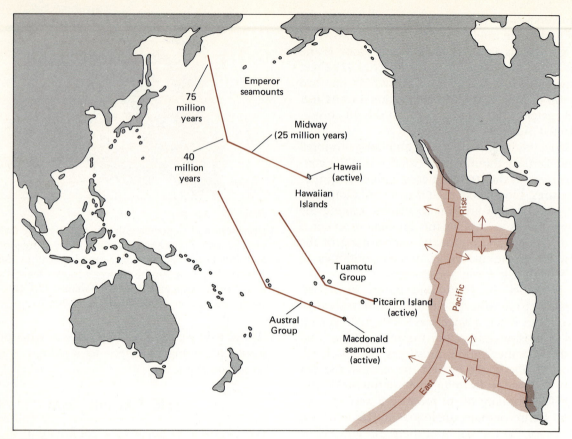

FIGURE 21–9

In 1838 James Dwight Dana, an American geologist, recognized from the comparative extent of erosion of each island in the Hawaiian chain that they were progressively older to the northwest. We can now explain his shrewd observation, since the Hawaiian, Tuamotu, and Austral island groups all have the same parallel trend. Islands in all three chains, including their still-older, now-submerged continuations, have radiometric ages that are steadily older the more northwesterly the island. The volcanic islands all appear to have formed as the Pacific Ocean plate moved northerly and then northwesterly across relatively stationary "hot spots," locations under the plate where mantle plumes are steadily driving magma through the plate in undersea eruptions. (Map much modified from Kevin C. Burke and J. Tuzo Wilson, Hot spots on the earth's surface, copyright © 1976, by Scientific American, Inc. All rights reserved. Adapted with permission.)

Motion of the Pacific Plate

Scattered across the earth's surface are about 40 isolated areas where basaltic volcanism appears to be continuous in a stationary spot. As we described the process in the Hawaiian Islands (see Chapter Three), the areas of constant volcanism have been called **hot spots** or **mantle plumes**. As the second name suggests, these narrow circular zones are interpreted as areas where hot material rises steadily from the upper mantle as convective plumes. The plumes elevate the local heat flow, extend, thin, and crack the crust, steadily injecting it with magma.

The hot spots *appear* to remain stationary in space throughout time. As an aside, we should note that some data suggest that the mantle plumes, and their associated volcanism, are also in motion; the issue is still a frontier of active research. When a plate overrides the mantle plume, volcanoes form as the molten material is vented onto the crust; a string of extinct volcanoes extends away from the hot spot *in a direction parallel to plate motion.*

Across the Pacific Ocean floor, stretching west-northwest across the ocean from Macdonald seamount, Pitcairn Island, and Hawaii (Figure 21–9) are three chains of volcanic islands and submerged volcanoes whose rocks yield progressively older dates northwestward. The abrupt "bend" in the line (Figure 21–9) of submerged volcanoes older than 40 million years suggests that the Early Cenozoic direction of Pacific plate motion was north-northwesterly, changing abruptly to west-northwesterly in mid-Cenozoic. Zones of seafloor sediments, rich in fossils of microscopic life that live only near the equator, support this interpretation. Progressively older deposits of these fossils, which must once have been near the equator, are found in progressively more northerly sites. The change from generally northerly to generally westerly motion of the Pacific plate *occurred at approximately the same time as the initial shift from compression to shearing along the Pacific Rim of North America.*

The collision between the westward-moving American plate and the northwestward-moving Pacific plate *transformed* the relative plate motion from *oblique consumption* of a plate which separated the North American and Pacific plates to oblique shear *along* the Pacific Rim. As more and more of the Pacific Rim came into contact with the Pacific plate, the San Andreas and other *transform faults* (see Chapter Ten, Figure 10–19) became longer and longer. A smaller and smaller percentage of the Pacific Rim consumed oceanic seafloor; more and more of it became a sliding boundary between the North American and Pacific plates.

The total horizontal shift along the San Andreas Fault since Miocene time is at least 125 miles (200 kilometers). Suspect terranes southeast of the San Andreas Fault record a northward shift during the Cenozoic that may total as much as 3500 miles (5600 kilometers), with minimal estimates in the range of several hundred miles or kilometers.

This mid-Cenozoic event, which transformed consumption of the sea floor to sliding the boundary between plates, affected in rather different ways the whole of the western United States and continues to affect the California Coast Ranges today. The Gulf of California began to open along a spreading axis coincident with the East Pacific Rise during the Pliocene (about 5 Ma); this spreading is an example of continuing continental fragmentation. Slippage along the San Andreas transform and other associated faults has created crustal stresses which have formed numerous polygonal Late Cenozoic basins and domes. They have been major sources of oil in southern and central California.

The Sierra Nevadas

The Sierra Nevada mountain range of southern California is part of a chain of granitic batholiths that

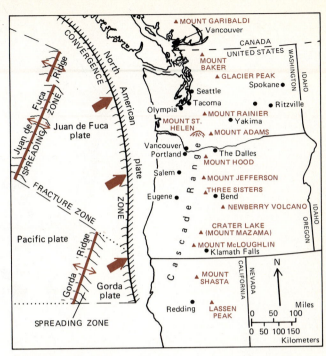

FIGURE 21–10

The volcanoes in the chain running south from Mount Garibaldi in British Columbia to Mount Lassen in northern California were formed through partial melting of the Gorda and Juan de Fuca plates as they were subducted beneath the North American continent. Farther to the south, the interaction of the western margin of the North American continent with an overridden spreading center caused oblique shearing. (Sketch modified from an outline map by the U.S. Geological Survey.)

stretch north to Idaho (see Figure 20–5) and into the coastal batholiths of British Columbia. Formed deep underground in the second half of the Mesozoic as a part of Nevadan orogenic events, the solid granitic batholiths were moved upward along a high-angle fault system that stretches along the east face of the Sierras for 450 miles (700 kilometers). The granites were eventually to move upward to their current elevations and be glaciated in the alpine glaciation that began in the Pliocene and reached full effect in the Pleistocene.

To the west of the Sierra Nevada range, the uplift was offset by the subsidence of the land surface to form the Sacramento and San Joaquin valleys. Within these depressions are nearly 6 miles (10 kilometers) of sediment derived from the rising Sierras and the Coast Ranges. These layers of sediment recorded continuing uplift of the Sierras amidst ongoing structural disturbances throughout the Middle and Late Cenozoic.

Continental ice sheet

FIGURE 21–11

The maximum extent of glacial advance in North America during the Pleistocene. The colored arrows give inferred directions of glacial motion. The black areas are inland lakes formed along the glacial margin in a cool temperate rainy climate, with a lowered potential for evaporation. Modified slightly from a map by the U.S. Geological Survey.)

The Cascades

The Cascade volcanoes are the products of partial melting of the Gorda and Juan de Fuca plates (Figure 21–10). This partial melting produced an island arc off the Pacific Rim; that arc system existed from the Triassic (240 Ma) to the Oligocene (30 Ma). Throughout this time, the western coast must have looked something like the west coast of South America, having volcanism above plutonism; turbidites and deep-water volcanic sediments were deposited along a rapidly subsiding foredeep. The Cascades are a classic magmatic arc, the only example of this type between the island arcs of the Aleutians to the far north and the arc volcanoes of central and southern Mexico to the far south.

The development of the arc throughout Late Cenozoic time was superimposed across older structures in what has been termed the **Cascadian orogeny**; this event was accompanied by mild folding, both high-pressure and regional metamorphism, and thrusting. The continuing subduction today of parts of the Pacific plate is paralleled by discontinuous eruptions from the Cascade Range volcanoes, including the spectacular eruption of Mount St. Helens in 1980 (see Chapter Three). The offshore trench is being buried under sediment from rivers draining a humid, recently glaciated area.

PLEISTOCENE GLACIATION

Within the last several million years, continental ice sheets have invaded parts of North America at least four times, dramatically altering the scenery of the United States and Canada. As described in Chapter Fourteen, continental glaciation spread southward to a limit approximately coincident with the Missouri and Ohio rivers in the Midwest and Long Island in the Northeast (Figure 21–11). The spread of the glacier from central Canada created a diverse array of landforms, also described in Chapter Fourteen. This chapter will concern itself with only two glacial effects.

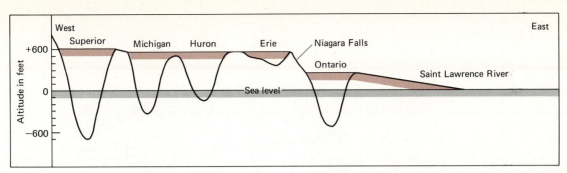

FIGURE 21–12

Diagrammatic profile of the Great Lakes from west to east. Notice that, except for Niagara Falls, the interconnected lakes form a large river system whose valley bottoms were carved well below sea level by past glacial erosion. (Adapted from Jack L. Hough, 1958, *Geology of the Great Lakes*, copyright © 1958, by University of Illinois Press. All rights used with permission.)

▶ The production and modification of lakes.

▶ The erosion of continental shelves.

The Great Lakes

Together, the Great Lakes make up the largest single body of fresh water in the world. They are also the world's largest inland river system. As Figure 21–12 suggests, the connected lakes are essentially a river system draining eastward into the Saint Lawrence River; the only interruption to a smooth longitudinal profile is the drop in elevation between shallow Lake Erie and deep Lake Ontario, a drop we know as Niagara Falls. The lip of the falls is a site of intense erosion; the lip moves westward at the rate of 1 or 2 yards or meters a year. In about 25,000 years the falls will reach Lake Erie and drain it completely.

The bottom of each lake, other than that of Erie, was cut well below modern sea level by Pleistocene glacial erosion of a river system which before the Pleistocene flowed south to the Gulf of Mexico. The event that changed the direction of flow occurred about 15,000 years ago, when the last of the glacial invasions from Canada began its retreat.

As the glacier retreated, lakes formed along its margins in the scoured river valleys, spilling south out of successively more northerly and *lower* outlets. When the ice had melted far enough back to reach the meltwater channel that is now the Saint Lawrence River, the lower elevation of the old channel made it the new drainage for the Great Lakes—and a much later major shipping seaway.

Lake Bonneville

As mentioned in Chapter Fourteen, the cool, rainy climates typical of glacial margins led to the widespread development of **pluvial lakes**, the lakes created by the ponding of precipitation in subglacial climates. Among these lakes were Agassiz, covering parts of Manitoba and the Dakotas; Missoula, covering parts of Alberta and Montana; and Bonneville (see Figure 14–33), covering parts of Idaho and Utah.

The remnant of glacial Lake Bonneville is the Great Salt Lake in north-central Utah. Lake Bonneville had its outlet in an area of southeastern Idaho known today as Red Rock Pass, through which it spilled into the Snake River. As the climate warmed at the beginning of the Holocene (10,000 years ago), the great inland lake shrank until it was lower than its outlet. Trapped, it slowly evaporated, leaving its salty remnant near what is now Salt Lake City, Utah.

The water level in glacial Lake Bonneville fluctuated many times throughout its existence as a function of climate; it still does, as wetter years in the late 1980s brought the lake to its highest level, which caused substantial property damage. Because of fluctuations in the former lake level, the shoreline moved up and down the sides of the Wasatch Range, ringing the lake margin with cyclic shorelines and deltas, which today are extremely well preserved (Figure 21–13). All the major cities of Utah are built on one or the other of these ancient lake levels which stretch along the margins of the Wasatch Range.

Lake Bonneville at its maximum extent was about the size of modern Lake Michigan. The filling of this great lake caused the continental crust to be depressed under the weight of its water. As evaporation followed climatic warming, the crust began slowly to rebound upward, a process that continues today. Precise measurements of the elevation of fossil shorelines of known radiometric age allow the rate and location of isostatic rebound to be accurately measured, thus providing another measure of the flow of the invisible asthenosphere deep beneath our feet.

U. S. GEOLOGICAL SURVEY   LAKE BONNEVILLE PL. XXVII

THE ANCIENT DELTAS OF LOGAN RIVER, AS SEEN FROM THE TEMPLE.

FIGURE 21–13

Rivers flowing out of the mountain into former Lake Bonneville formed deltas at the point of their entry into the lake. As the lake level changed, so did the level of each delta. What was left behind after evaporation of most of the lake is a series of essentially flat surfaces. The town of Logan, Utah is built on several of these surfaces, as are almost all other towns along the west margin of the Wasatch Range. This view, drawn in 1890, is looking east toward the Wasatch Range from Logan. (From Grove Karl Gilbert, 1890, *Lake Bonneville*, U.S. Geological Survey Monograph 1, Pl. 27.)

The Continental Shelf

Further evidence of the Ice Age glaciation of North America comes from examination of our continental shelf. During periods of major ice advance, when a larger percentage of the world's water is tied up in solid ice, sea level may fall by more than 300 feet (100 meters). Today we find fossils of continental animals in water over 100 yards (about 90 meters) deep, and anthropologists are able to recognize the remains of human dwellings—even camp fires—on what is now the flooded continental shelf.

As sea level fell worldwide, the erosive power of continental rivers was enhanced, not only because their base level (see Chapter Twelve) fell, but also because they were receiving more precipitation and losing less to evaporation in a generally cooler climate. Vigorous streams cut deep, wide canyons into the edge of the emergent continental shelf. Currents moving down those canyons carried huge volumes of coarse sediment out into the deep water covering the continental slope and rise. Sediment that would have been trapped on the shelf was washed right out to the deep sea. Today an impressive group of large submarine canyons score the continental shelf.

Both periods of major glaciation—that during the Proterozoic-Cambrian and that during the Permo-Triassic—were associated with similar periods of continental emergence, widespread shelf exposure, and vigorous erosion, as well as dramatic decreases in the diversity of life-forms. We might well ask how this last glacial period has affected life, which for the first time had included bipedal mammals of great complexity and potential. We human beings are very surely the first life-forms on earth who have acquired the power to destroy all life; we could easily extinguish a 4-billion-year-old experiment.

CENOZOIC LIFE

The Cenozoic is termed the "Age of Mammals" because this diverse group has dominated the land, sea, and air during all of Cenozoic time. Mammals are distinguished by a highly efficient four-chambered heart, a high metabolic rate, and precision temperature control, among many distinctive characteristics. After being dominated for over 100 million years by the Mesozoic reptiles, mammals emerged as highly adaptable life-forms.

Before we discuss a few examples of mammalian life, we first review the status of other plant and animal groups that survived the great die-off across the K-T (Cretaceous-Tertiary) boundary. The Cenozoic began with an impoverished fauna and flora whose diversity rapidly increased during its early generally mild climates. Diversity again increased when the climate turned cooler as the earth slipped into a glacial chill during the second half of the Cenozoic Era.

CENOZOIC PLANTS

The rise of the **angiosperms**, the "encased seed" plants, in Cretaceous time (125 Ma) rapidly placed the **gymnosperms**, or "naked seed" plants, in an inferior position. The gymnosperms include the familiar conifers and the less familiar cycads and gink-

FIGURE 21–14
This fossil leaf is from an Early Tertiary black shale and coal sequence in north-central New Mexico. It has the broad-leaved form typical of angiosperms but the poorly organized veinlet form that typifies more primitive angiosperms. (Photograph courtesy of the U.S. Geological Survey.)

goes, but they are wholly subordinate to angiosperms, which cover over 90 percent of the earth's vegetated surface and comprise all the species we eat.

The rise to dominance of the angiosperms was remarkably swift. Two facts about angiosperms stand out. Their broad leaves (Figure 21–14), in contrast to the needles typical of conifers, are able to catch far more sunlight, allowing them more efficient photosynthesis. Another element of their success is their reproductive strategy. Both the gymnosperms and angiosperms reproduce by sexual means, transferring the male component as *pollen* into the female *ovule*. Within the ovule, the plant embryo grows into a *seed*. Gymnosperms freely expose the ovules at the surface of the plant, but angiosperms generally enclose it in a fruit.

All gymnosperms and some angiosperms depend on wind to transfer the pollen from its source to the ovule. Because of the vagaries of wind transportation, gymnosperms must produce a million pollen grains for each ovule in order to ensure a measure of reproductive success. Such a profligate process is wasteful of the plant's energy stores; moreover, fertilization is random, with weak male microspores as likely to fertilize an ovule as vigorous ones. Angiosperms, in contrast, produce only about 6000 pollen grains for each ovule. They depend on pollinating insects, who eat most of the pollen, to carry the remainder to a distant ovule.

Once again we can watch life adjust to a problem, here the unreliability of wind as a fertilizing mechanism. By displaying its pollen as protein-packed food within the center of a nectar-rich colorful flower, the angiosperm entices an insect to move from flower to flower, inadvertently pollinating each ovule as it moves. But how to keep the insect from eating the ovule, which is as nutritious as the pollen? The angiosperm's solution is to enclose the ovule, keeping it out of harm's way. But how can the pollen fertilize the ovule if the ovule is enclosed? The answer was *pollen tubes*, which pierce the ovule and deliver the pollen directly to the egg. The egg then grows into a seed within the enlarging ovule, which we then call a fruit.

Insects deliver large numbers of pollen grains to each ovule. Only the single most vigorous pollen grain with the most rapidly growing pollen tube completes the fertilization of the ovule; less vigorous pollen grains reach an already fertilized ovule and are locked out. This selective process eliminates less vigorous genetic material while preserving the most vigorous material. *Among plants only the angiosperms have this highly effective method of increasing their vigor through natural selection of male vigor.* Thus they have placed gymnosperms in a subordinate position throughout the whole of the Cenozoic.

In an even more subordinate position are the horsetails, mosses, and psilopsids. These remnant populations, together with true ferns, cycads, ginkgoes, and conifers, occupy less than 10 percent of the earth's vegetated surface. The angiosperms, which include the cereal grains, grasses, vegetables, and all flowering plants, have had an explosive growth in diversity during the Cenozoic, paralleled by the rise of pollinating insects.

Only in the subarctic or polar high-latitude areas is Cenozoic plant and insect diversity severely diminished today. In the Antarctic, the *largest* insects are smaller than a human fingernail, and most are much smaller than that. Only the simple plants, such as photosynthetic bacteria and algae, survive as encrusting species under the ice in an area that is still very much in the grip of an ice age. The widespread glaciation we associate with the last 2 or 3 million years

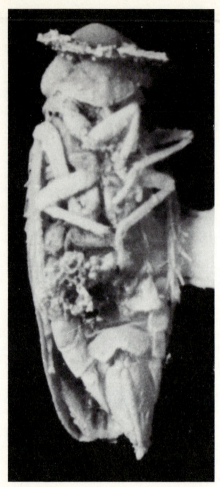

FIGURE 21–15

This photograph illustrates the mode by which fossilized insects were preserved. The creature is a Miocene ancestor to the grasshopper, preserved by lime saturation in the shores of a lake. (Photograph courtesy of the U.S. Geological Survey.)

in North America began in the Antarctic about 15 million years ago and continues, though there is some evidence that ice-free intervals occurred during the Pliocene.

INVERTEBRATES

Among unicellular life, the remnants of the *foraminifera* radiated and became an important part of marine planktonic communities. Their calcareous shells form a major part of the ocean floor ooze in some areas. Their siliceous cousins, the *radiolaria*, remain an important part of the marine world. The sponges maintain their solitary ways, as do two major groups of corals. The bryozoa also maintain themselves, but brachiopod populations and diversity are much diminished from their Mesozoic levels.

The shelled cephalopods are represented only by the simpler nautiloids like the *chambered nautilus*. Their belemnitic and ammonitic kindred did not survive the Mesozoic in the seas they once dominated. The clams and snails continue their steady course, as do the crinoids and echinoids.

Among the arthropods, the insects became supremely abundant. Insects (Figure 21–15) occupy every conceivable ecological niche on the continents; there are more species of insects than of all the other animal groups on the earth put together. In the sea the most common arthropods are the crustaceans, of whom the tasty lobster is probably the best known. The minute bean-shaped bivalved ostracods are another arthropod group that maintained their place through the Late Mesozoic annihilation.

VERTEBRATES OTHER THAN MAMMALS

Among fishes, the lungfish and lobe-finned fish remain minor components today; both the cartilaginous and bony-skeleton fish are dominant. The amphibians occupy a minor portion of the scene today—a half-water and half-land existence shared with some reptilian turtles, marine iguanas, and crocodiles. On land only the lizards, the tuatara, and the snakes—some of which have partially or wholly returned to the water—remain as remnants of the reptilians that dominated every environment of the Mesozoic Era.

The birds, which differentiated from the reptiles in the Jurassic (a Triassic fossil bird has been described; few paleontologists accept that description), have become an important part of life in the Cenozoic Era. It has been conservatively estimated that there are 100 billion birds alive at this moment. We know relatively little of the evolution of birds because they are rarely preserved as fossils. Their fragile, hollow skeletons are easily destroyed, and their carcasses fall on the surface to make easy prey for scavengers. The birds and their mammalian counterparts the bats are important aerial predators and scavengers. The birds continue as a curious mix, sharing the scales on their legs and their egg-laying traits with reptilian ancestors, but sharing their elaborate circulatory system and warm-bloodedness with mammals.

Surely the most spectacular fossil bird ever discovered is *Argentavis magnificens*, a Late Miocene (8 to 5 Ma) bird reminiscent in overall appearance (Figure 21–16) of a bald eagle or a California condor. Weighing as much as an average human (160 pounds or 72 kilograms), the bird stretched 11 feet (3.3 meters) from beak to tail and had a wingspan of 25 feet (7.6 meters). It was similar in size to a small

FIGURE 21–16
This scene includes a large teratorn bird which must have wheeled in the skies over the Pleistocene landscape. (Courtesy of the George C. Page Museum of La Brea Discoveries, a branch of the Los Angeles County Natural History Museum. Used with permission.)

single-engine private aircraft. Its dimensions made it the world's largest flying bird, capable of daily flights of over 200 miles (320 kilometers). The bird was the *teratorn*, a member of a group of very large birds that became extinct 10,000 years ago. The name literally translates as "wonder bird." They must have been the greatest of the Tertiary airborne predators.

MAMMALS

We observed that at the end of the Cretaceous all four-legged land animals with a body mass greater than 22 pounds (10 kilograms) became extinct, along with all marine and aerial reptiles. The absence of large reptiles left behind numerous ecological niches, ripe for the filling by the mammals. The mammals, which had been small and generalized members of the fauna since their evolution from the *therapsids* in the Triassic, became the stock from which the fauna of large four-legged animals (*tetrapods*) was rebuilt in the Tertiary.

Some general mammalian characteristics are

▶ Giving birth to fully developed live young.

▶ Warm-bloodedness, with hairy skin and fat layers for insulation.

▶ Movable lower jaw with strongly differentiated teeth.

▶ Large brain mass to body mass ratio.

▶ Development of mammary glands from sweat glands on the chest.

▶ Suckling of young with milk from breasts.

▶ Four-chambered heart separating oxygenated from deoxygenated blood.

▶ Breathing assisted by chest diaphragm.

▶ Palate separating breathing from eating functions.

Introduction

The most primitive living mammals are the *monotremes*, a group that includes only the duck-billed platypus and the spiny echidna, both confined to Australia and New Zealand. They lay eggs, yet secrete milk from nipples for their young. These "hybrids" are sometimes referred to as "living therapsids."

Another odd mammalian group are the *marsupials*, such as the opossum and the kangaroo, which give birth to very immature young that must finish their embryonic life attached to a nipple in their mother's pouch. The marsupial mammals had their greatest radiation within Australia and South America, two continents isolated from all others after their separation in the Cretaceous. The marsupials appear to have moved from South America to Australia via Antarctica before the two continents separated; no one knows why the *placental mammals*, which give birth to fully developed live young, did not follow the same route.

The Cenozoic record makes up only the last third of the evolutionary history of mammals, but it is the time of their worldwide radiation to replace the extinct larger reptilian tetrapods. By the start of the Tertiary, the continental landmasses were largely separate, a fact that made possible independent evolution of each fauna on the different continents. The isolation of Australia led to the development of wholly marsupial and monotreme mammals. The similar isolation for South American marsupials was ended when the two Americas rejoined in the Late Tertiary. This continental juncture was followed by an almost complete replacement of South American marsupial mammals with North American placental mammals.

To understand the evolution of faunas throughout the Cenozoic, we need to know that in North America in the first half of the Cenozoic the climate was warm and temperate. A dramatic cooling trend initiated at the end of the Eocene (36 Ma), bringing an 18°F (10°C) drop in average annual temperature and an increase in the yearly temperature range from winter to summer of about 9°F (5°C) to about 45°F (25°C). Post-Oligocene mammalian faunas had increasingly to adapt to a cooler climate with much greater temperature extremes.

The following will review only a few of the mammal groups whose radiation filled the earth's surface. Modern mammals include flying mammals, the bats, and aquatic mammals, such as the walrus, otters, whales, and dolphins. Both mammalian groups are poorly represented in the fossil record. Terrestrial mammals number many familiar creatures, including pigs, deer, horses, zebras, camels, opossums, moles, shrews, lemurs, monkeys, apes, chimpanzees, humans, elephants, armadillos, sloths, rabbits, cats, dogs, bears, mice, cattle, squirrels, beavers, porcupines, and many, many others.

North American continental deposits, primarily in the Far West, provide a rich record of the development of mammals. Geologists have been studying this record for a century or more. We will now briefly review some of the highlights of this record.

Insectivores

Insectivores are voracious insect-eating mammals. They are among the earliest true mammals; their tiny bones and teeth have been found among Paleocene sediments. The Early Paleocene insectivores were ancestral to the modern shrew, hedgehogs, and moles. They continue to survive by virtue of their small size, secretiveness, and nocturnal habits. The shrew is the tiniest of all insectivores; as mentioned in the last chapter, the shrew's entire life is a race to take in enough food to make up for the heat loss from its body surface. Shrews must locate and eat their own weight in insect food each day; it is little surprise that they are nervous, ill-tempered creatures.

Rodents

Rodents are herbivores with extremely large front incisor teeth which continue to grow throughout their lives. Rodents appeared first in the Paleocene and diversified in the Eocene. They are in some ways the most successful of all mammals, having invaded every land area on earth. They also outnumber all other mammals combined. The rodent with which we may unfortunately be most familiar is the rat. Rats steal our food, invade our homes, make us sick, and follow us wherever we go. Rodents are either **omnivores**, eating both plants and animals, or **herbivores**, eating only plants.

Other rodents include hamsters, guinea pigs, chipmunks and squirrels, porcupines, mice, rabbits, and beavers. Because their long, sharp incisors grow continuously, these teeth must be sharpened by gnawing on hard material. Farther back in the rodent's mouth, grinding cheek teeth are used for shredding their food. Among the oddest fossil rodents is *Epigaulus*, a burrowing rodent which looked like a prairie dog with small antlers, and *Casteroides*, a Pleistocene beaver which was about the size of a small brown bear.

Ungulates

Ungulates are the hoofed mammals, horses, deer, cattle, and swine. They are herbivores, with massive

FIGURE 21–17
These animals are *Phenacodus primaevus*, condylarths which appear to be the ancestors to all ungulates. (Courtesy of the American Museum of Natural History, used with permission.)

flat teeth modified for cropping and grinding large quantities of plant material, and digestive systems with specialized stomachs to convert bulky plant material into nourishment. To defend themselves, ungulates are equipped either with antlers, which they shed each year, or with permanent horns.

Many ungulates also have long, limber legs which allow them to outrun predators. The feet have been modified into horny hooves especially adapted for running over hard ground. Ungulates normally walk on the tips of their "toes," with the wrist and ankle far off the ground.

Plant-eating animals all share certain advantages and disadvantages. They are surrounded by food, and they can eat at their leisure because the food will not run away. The teeth of all herbivores include front incisors suitable for nipping vegetation and back molars modified for crushing and grinding hard material. The teeth must be tough and capable of surviving a great deal of abrasion from woody stalks and cellulose-rich brushy food. The disadvantage in being a herbivore is that the protein content of plants is generally lower than that of meat, so that more bulk must be eaten to gain the necessary nourishment.

Ungulates arose in Paleocene times, from ancestors which were probably insectivores, and evolved in two rather distinct stages. Primitive ungulates evolved during the warmer Paleocene and Eocene time from the *condylarths* (Figure 21–17), the root stock for all ungulates. From this stock two major groups followed.

▶ The *perissodactyls*, ungulates with an odd number of toes, including the ancestors to asses, horses, rhinoceroses, tapirs, and zebras.

▶ The *artiodactyls*, ungulates with an even number of toes, including pigs, the ancestors to hippopotamuses, camels, deer, bison, antelopes, sheep, goats, giraffes, and cattle.

The primitive ungulates went into a serious decline with the onset of cooler, more variable climates in the Early Oligocene (36 Ma); they were wholly replaced by modern ungulates in the Pliocene (3 Ma). We will briefly trace the history of a few examples of both the even- and odd-toed ungulates.

THE ODD-TOED UNGULATES OR PERISSODACTYLS

The perissodactyls evolved from the condylarths in Early Eocene (55 Ma) time, creating a wide-ranging primitive fauna. Among the best known of the early perissodactyls are the *titanotheres*, which evolved very rapidly into beasts that were up to 8 feet (2.4 meters) at the shoulder, much larger than the modern rhinoceros. By mid-Oligocene, the titanotheres became extinct, perhaps because their teeth were no match

FIGURE 21–18

Titanotherium was a rhinoceroslike creature of the Eocene. Like its modern offshoot, the titanothere must have been much at home in the open grasslands. (Photograph courtesy of the American Museum of Natural History. Used with permission.)

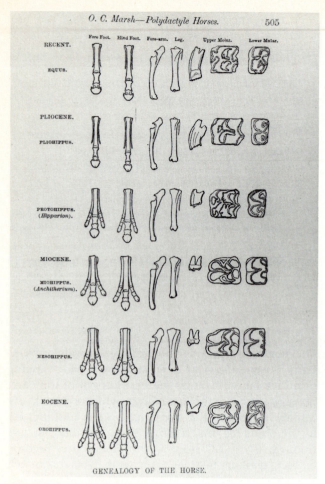

O. C. Marsh—Polydactyle Horses.   505

GENEALOGY OF THE HORSE.

FIGURE 21–19

This graph shows the overall evolution of horses' hooves, forelegs, hind legs, and teeth over the last 60 million years. We now know that this presentation omits some other modifications, but as a simplification it nicely describes the evolutionary development of critical skeletal components within a well-known group. (From O. C. Marsh, 1871, courtesy of the History of Science Collections, University of Oklahoma Libraries.)

FIGURE 21–20

This is *Mesohippus bairdi*, a Middle and Upper Oligocene horse which when fully grown was about the size of a Shetland pony. It was already well adapted to high-speed flight across hard surfaces. (Photograph courtesy of the American Museum of Natural History. Used with permission.)

for the abrasive grasses that became much more common in a cooler, temperate climate.

Left behind are bones of very large animals, like the *Titanotherium* (Figure 21–18), whose bones litter Eocene basin deposits all over the American West, including the John Day country of northeastern Oregon. Buried along with the titanotheres were fossil rhinos like *Baluchitherium*, which reached a height of 15 feet (4.5 meters) and was the largest land mammal that ever lived.

Also among the animals that shared the Eocene with titanotheres were the ancestors of the horse (*Equus*), whose evolution through all the Cenozoic is faithfully recorded in the Cenozoic sediments of

mountain basins throughout the Far West. The earliest horse, sometimes called the "dawn horse," was *Eohippus* or *Hydracotherium*, the most primitive known perissodactyl. *Hydracotherium* was a small creature, about the size of a collie dog. The animal had four toes on the front foot and three on the back.

The development of the horse took place on the North American continent and consisted of the following trends (Figure 21–19).

► Increase in size.
► Elongation of the legs.
► Reduction in the number of toes.
► Increase in brain complexity and size.
► Straightening and stiffening of the back.
► Changes in tooth structure and differentiation.

As grass became the dominant angiosperm in the cooling climate, one branch of the horse family took to the plains (Figure 21–20), where there was a premium on speed and agility in avoiding predators, and where the animals could use their ability to chew and digest increasingly tough grasses. Another branch of the horse family took to the Miocene woodlands, retaining more primitive teeth and a three-toed stance, but it became extinct in the Pliocene.

By the Pleistocene the modern horse (*Equus*) was

FIGURE 21–21

Protoceras was an Oligocene ancestor to the modern deer. A lightly built traguloid, it had rudimentary antlers and protected itself from predators by staying in open forests. (Photograph courtesy of the U.S. Geological Survey archives.)

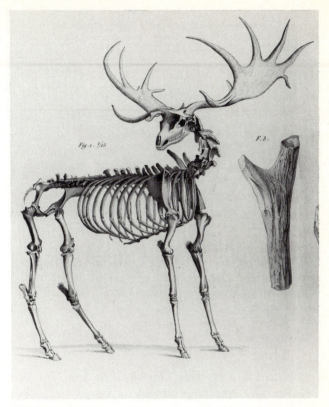

FIGURE 21–22

This engraving is of *Megaloceros*, often called the "Irish elk"; it is actually a rather extravagant deer. A common fossil in Pleistocene peat bogs, many complete skeletons have been located. (From G. Cuvier, 1836, *Recherches sur les Ossemens Fossiles*, courtesy of the History of Science Collections, University of Oklahoma Libraries.)

fully developed; during this epoch it migrated to other continents. In South America, about 3 Ma, when the two Americas joined, the horse was part of the migration that essentially replaced the marsupial fauna with placental fauna. The horse became extinct in North America about 10,000 years ago and had to be reintroduced from Europe by sixteenth-century European explorers. The cause of the extinction remains a mystery, since the horse had evolved here during progressively colder weather. There have been some suggestions that, like the woolly mammoth and other large mammals, the early American Indians hunted it to extinction.

THE EVEN-TOED UNGULATES OR ARTIODACTYLS

Like the perissodactyls, the *artiodactyls* are descendants of the condylarths. Modern artiodactyls include almost all hooved animals except the horse, tapir, and rhinoceros. The artiodactyls, like the perissodactyls, evolved in the Early Cenozoic into a primitive group, which became extinct. They were replaced by a modern group more able to cope with a cool, temperate climate.

In this early group were the *traguloids*, probable ancestors to the modern deer. An example is the Oligocene *Protoceras*, which had two horns above the eyes and two more on its nose (Figure 21–21). Like the horns of many horned animals of today, they may have been used more in ritual combat with other males and in sexual displays than in defense.

An extravagant example of the general tendency for larger and larger size among all ungulates is the Pleistocene deer *Megaloceros*, more commonly called the "Irish Elk" (Figure 21–22). Found by the hundreds in Ice Age peat bogs in Ireland, they are not true elk but rather deer with unusually large antlers, which might easily stretch 12 feet (3.5 meters) from tip to tip. Many complete skeletons are known; the antlers never fail to create wonder wherever they are displayed.

Carnivores

Carnivores are the meat-eating mammals. As a group they have the opposite problem of herbivores. Their food is in highly nutritious and rich in protein, containing up to one-quarter of all the food energy the prey has eaten—but the food runs away or hides or protects its life with antlers, horns, spikes, armored skins, unpleasant smells, stinging fluids, and the like. Carnivores must overcome these defenses by being even more fleet afoot and nimble, and by possessing vicious, sharp-pointed teeth and grasping

FIGURE 21–23

The creodonts were the earliest true carnivores; the example shown here is *Oxyaena lupina*, an agile, powerful predator of the Early Cenozoic. (Photograph courtesy of the American Museum of Natural History. Used with permission.)

FIGURE 21–24

Smilodon was a highly efficient killing machine whose greatly elongated incisors were used to stab prey to death. Powerfully built, the saber-toothed cat is a frequently encountered fossil in Pleistocene sedimentary rocks and in the famous La Brea tar pits in Los Angeles. (Photograph courtesy of the American Museum of Natural History. Used with permission.)

claws to kill their prey plus shearing teeth for later tearing and grinding the prey's bones and flesh. Many carnivores hunt in packs and have evolved a complex social order, much as early human beings had to do when they hunted very large, dangerous game.

Among the most primitive of the carnivores were the *creodonts*, like *Oxyaena* (Figure 21–23), a large and powerfully built Eocene carnivore that must have been one of the major predators of Early Cenozoic time. Many of these primitive carnivores became extinct at the end of the Eocene, but a few others survived up into the Pliocene Epoch before passing into oblivion.

The much more familiar group of modern carnivores includes bears, raccoons, minks, badgers, skunks, otters, dogs, and cats. Few are more familiar to us than the great cats, synonyms for stealth and sudden death. They are highly specialized for a life of killing and meat eating. They are very strong, muscular, and agile, with great grasping claws and long stabbing and cutting teeth. They have evolved from their civet ancestry since Oligocene times.

Few fossil cats command our attention with quite the dread of *Smilodon* (Figure 21–24), the Pleistocene saber-toothed tiger. This cat was as large as a modern lion, with two huge saberlike teeth within a widely hinged mouth. *Smilodon* was designed to hunt in stealth, jumping on an unsuspecting animal even larger than itself and stabbing it to death. It must

have been a terrifying Pleistocene predator, but as the large mammals that were dinner for *Smilodon* became extinct at the end of the Pleistocene, so did the saber-toothed tiger.

Proboscideans

As the name suggests, **proboscideans** are hooved herbivores distinctive for their possession of a much elongated nose or trunk. The first proboscidians originated in the Eocene as stout, heavy-footed ungulates and underwent evolution in two parallel lines. The general changes were

► Adaptation of teeth for chewing rough vegetable food.

► Formation of tusks from incisor teeth.

► Elongation of the nose and upper lip to form a trunk.

► General increase to a very large size.

► Development of large skulls and ears, with a shortened neck.

Among the early proboscideans were the dinotheres, represented by *Dinotherium* (Figure 21–25), a group which evolved separately from the elephants

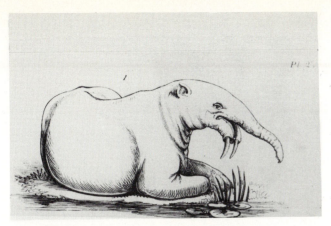

FIGURE 21–25

This reconstruction of an early proboscidean is *Dinotherium*, an animal of the Early Miocene whose evolutionary lineage is close to, but not in the same direct line, as the lineage of true elephants. When skeletons of this creature were first unearthed, their elephantlike outline created an immediate problem in classification. (From Reverend William Buckland, 1837, courtesy of the History of Science Collections, University of Oklahoma Libraries.)

FIGURE 21–26

Trilophodon was the ancestor both to the mastodon and to the modern elephant. With an elongated upper lip and nose, these animals were well equipped for low browsing. They were perhaps half the size of the modern elephant. (Photography courtesy of the American Museum of Natural History. Used with permission.)

and became extinct during the Pleistocene Epoch. The modern elephants evolved from the mastodon; the elephants are the last members of a group on its way to extinction. Down now to a total of only two species, this once-diverse group dominated the Plio-Pleistocene terrestrial world.

MASTODONS

The direct ancestor to the mastodon and to the elephant was *Trilophodon* (Figure 21–26; sometimes termed *Gomphotherium*), which originated in the Miocene. Its lower jaw was elongated, ending in two short tusks, and the trunk was already well developed. Another mode of feeding was employed by *Amebelodon*, whose lower tusks were modified into scoops. These ''shovel-tusked'' mastodons must have used their scoops to dig up aquatic plants from lake bottoms.

Among the best-known mastodons is *Mastodon americanus*, whose skeletons have been found all over North America. Although somewhat smaller than a modern elephant, it was even more powerfully built, with huge blunt feet, stout skeleton, and curving tusks (Figure 21–27). Mastodon fossils are found in deposits laid down through the Early Holocene (8000 years ago). The excellent preservation of some specimens allowed us to discover that they

were covered with dark rusty-colored hair and browsed primarily on brush and leaves. Their extinction was coincident with the spreading of human beings throughout North America. Humans may have hunted the mastodons to extinction.

Mammoth is the *name applied to all extinct elephants*. The mastodons can be distinguished by their teeth, which were generally a series of cones arranged in interlocking order (Figure 21–28). The elephants have wholly different teeth consisting of tall interlocking crenulated plates, which form massive flat surfaces for grinding (Figure 21–28). The grinding movement of their teeth when they eat roughage wears down the teeth and continually exposes a new tough surface.

MAMMOTHS AND ELEPHANTS

The Pleistocene was the age of mammoths; they wandered across every continent except the Antarctic, Australia, and South America. Like their kindred the mastodons, the mammoths walked among upright human beings, whom they must have terrified with their size, strength, and huge curving tusks (Figure 21–29). The largest of the mammoths was *Mammut imperator*, whose height at the shoulder stretched to 14 feet (4 meters). The imperial mammoths were joined by the woolly mammoths. Draw-

FIGURE 21–27

This model of *Mastodon americanus* reveals the robust, stout body plan of this powerfully built *Mammut*. Mastodons had a wholly different type of teeth from those of the true elephant. (Photograph courtesy of the U.S. Geological Survey archives.)

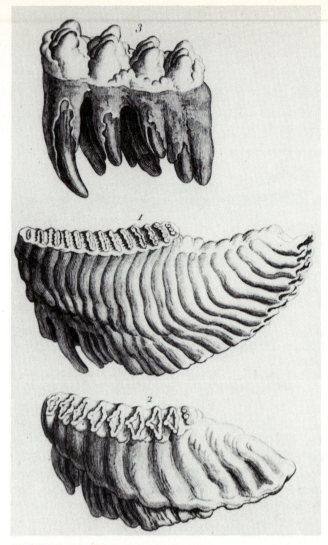

FIGURE 21–28

The upper teeth, with their rounded conical cusps, are those of a *mastodon*, which used them in browsing. The middle tooth with the crenulated top surface is that of a modern elephant; that below is from a direct ancestor to the elephant, called the *mammoth*. (From Bartholomew Faujas de Saint-Fond, 1809, *Essai di Geologie*, Vol. 1, courtesy of the History of Science Collections, University of Oklahoma Libraries.)

ings of them festoon the caves of our Ice Age ancestors who knew mammoths as fair game.

Still another of the giants is the Columbian mammoth, which had enormous spiral tusks that crossed at their tips. These animals disappeared from North America fewer than 8000 years ago. Mammoth remains have been found deep frozen in Siberia, so well preserved that the undigested food in the stomach can be easily identified.

Pleistocene Extinctions

What an adventure it would have been to have lived in North America 20,000 years ago! We would have known mastodons, mammoths, horses, giant beavers, saber-toothed tigers, ground sloths, hunting dogs, wild horses, large camels, wild pigs, and many other species now found only in central Africa. The causes of the decline of large mammals, leading to their almost total extinction in North America at the time of the human invasion 11,000 years ago, remain an enigma.

We noted earlier that the horse may have been hunted to extinction in North America. The disappearance of many other large mammals from North America coincident with the invasion of humankind suggests—but does not prove—a cause and effect relation. Since human beings are ruthless hunters, having run the bison to within a few hundred individu-

als of extinction, it has been proposed that human intervention caused the disappearance of the Ice Age mammalian fauna from North America.

Since human beings belong to the group of mammals called *primates*, and since we have driven over 400 species into extinction in our brief existence, perhaps we should close with a brief look at our own development as the earth's most successful primate.

FIGURE 21–29
These models of *Trilophodon* and *Mammut americanus* illustrate the difference in size between the mid-Cenozoic ancestor to elephants and the Pleistocene elephant or mammoth. (Photograph courtesy of the U.S. Geological Survey archives.)

Primates

Primates are highly complex creatures that have descended from insectivore ancestry. If we ask what are the odds the earth would host human beings, we should note that only *one* species of primate survived the latest Cretaceous mass extinction. Only one . . .

Modern primates include lemurs, tarsiers, monkeys, apes, and human beings. They acquired numerous adaptations for life in the trees.

► Long, flexible arms and legs, with grasping fingers and complex thumbs.
► Large brain size, which increased throughout the Cenozoic.
► Highly complex nervous systems, with correspondingly high levels of activity and alertness.
► Binocular vision with depth perception and color perception.
► Highly mobile joints, with mobility in several directions.

Each of these features conveys advantages to those whose only defense from danger is to swing away through the trees. As an example, binocular, color-sensitive, stereoscopic vision with highly developed depth perception is extremely important if a creature is to jump from limb to limb—or drive a car.

Primates can be divided into two groups, the *prosimians*, which include lemurs, tarsiers, and a few others, and the *anthropoids*, which include monkeys, apes, and **hominids**, a primate group belonging to the family *Hominidae*. Modern human beings, *Homo sapiens*, are the only living species of hominids.

The monkeys evolved first in the Early Oligocene

FIGURE 21–30
This painting depicts Cro-Magnon families crossing the Bering Strait from Asia to North America by means of a stretch of water that was frozen during a glacial advance. Having once reached North America, they found an enormous continent stocked with large mammals, and they proceeded to rapidly capitalize on this rich food resource. They may also have destroyed it. (Courtesy of the George C. Page Museum of La Brea Discoveries, a branch of the Los Angeles County Natural History Museum. Used with permission.)

(36 Ma), followed in the Late Oligocene (25 Ma) by the apes, which lack an external tail. Hominids developed in turn in the Pliocene (6 to 4 Ma).

Hominid ancestry appears to lie with a group of Miocene apes, the dryopithecines (of whom *Dryopithecus* is an example), which were direct ancestors to the modern chimpanzee with whom humans share the greatest genetic resemblance. The dryopithecine apes adapted to the cooling climate of the Late Tertiary. They walked the open woodlands where food was near the ground and enclosed in tough shells and skins. One of those dryopithecine apes was *Ramapithecus*, a Pliocene woodland ape similar to the probable ancestor of the first true hominid, an upright creature called *Australopithecus*. *The australopithecines were the intermediates between dryopithecine apes and primitive human beings.*

HOMINIDS

Bipedal hominids are creatures of the Pliocene. They split from the dryopithecine apes between 6 to 4 million years ago. The earliest upright hominid is *Australopithecus afarensis*, meaning the "southern ape of Africa." Partial remains of this creature have been found, as well as 3.75-million-year-old footprints at Laetoli, in Tanzania.

As climate continued to cool, open woodlands and savannas became more extensive and efficient upright walking became ever more important. By Early Pleistocene (2 to 1.6 Ma), australopithecines became extinct, replaced by a new group of hominids called *Homo habilis*, or "able man." Man is essentially a Pleistocene animal.

Among the differences between *Homo* and *Australopithecus* is brain size; the brain size of the most primitive *Homo* was nearly one and one-half times that of the australopithecine apes. By Middle Pleistocene, *Homo erectus*, "upright man," had evolved, with a still larger brain and slightly larger overall size. This classification includes the creatures once called "Java man" and "Peking man." By now our predecessors used tools, controlled fire, and cooked food. It is probable, based on studies of their skulls, that they were capable of an extremely primitive form of speech.

By 290,000 years ago, *Homo sapiens*, "wise man," arrived. He was announced by a still larger brain (two to three times that of the australopithecines), and the presence of skull structures that suggest individuals were fully capable of modern speech. All modern humans arose from Africa between 140,000 to 290,000 years ago. By 125,000 years ago, *Neanderthal man*, a variety of *Homo sapiens* named for the

Neander Valley near Dusseldorf, Germany, was fully developed. Neanderthals skillfully chipped beautiful stone tools and weapons, and they used fire. Neanderthals were a foot or more shorter than today's average humans, and they were more robust, probably as an adaptation to the glacial climates in which they lived.

The Neanderthals were the first people to bury their dead, which is one reason why so many well-preserved skeletons have been found. They often placed ornaments and special objects in the graves, a suggestion that they may have believed in an afterlife. About 30,000 years ago Neanderthals disappeared, replaced by *Cro-Magnon man*, a creature essentially indistinguishable from modern humans.

These human beings were highly skilled nomadic hunters who followed the reindeer and other quarry during the very last of the last glacial advance. Their tools are beautifully chipped stone, worked to a level indicating considerable artistic ability. They knew the glacial chill, and left the caves in which they paused filled with hundreds of figures of animals and other symbols of their imagination. They crossed the Bering Strait (Figure 21–30) to enter North America about 30,000 years before Columbus was to *rediscover* it.

They endured and created Indian cultures of great complexity and endurance. Our ancestors created agriculture and with it a primitive technology. Through the application of technologies human beings came to have the leisure time to pursue the *most distinctively human task—to ask questions and actively seek answers.*

Only human beings can wonder what makes human beings human. The answer is, of course, the very ability to ask that question. Only human beings could discover that the Ice Age is in both our past and our future. As we look back to a distant time, picturing in our imagination our Neanderthal kin burying a loved one and reaching for the mystery of life, we are watching ourselves.

SUMMARY

1. Coastal plain regression from the Cretaceous maximum transgression dominated the Cenozoic. The maximum regression during glacial periods produced severe erosion of the margins of the continental shelf, whose southern margin was penetrated by many salt intrusions. To the west, the culmination of the Laramide orogeny (Figure 21–31) formed the Rocky Mountains, which then shed asymmetrical coalescing alluvial fans

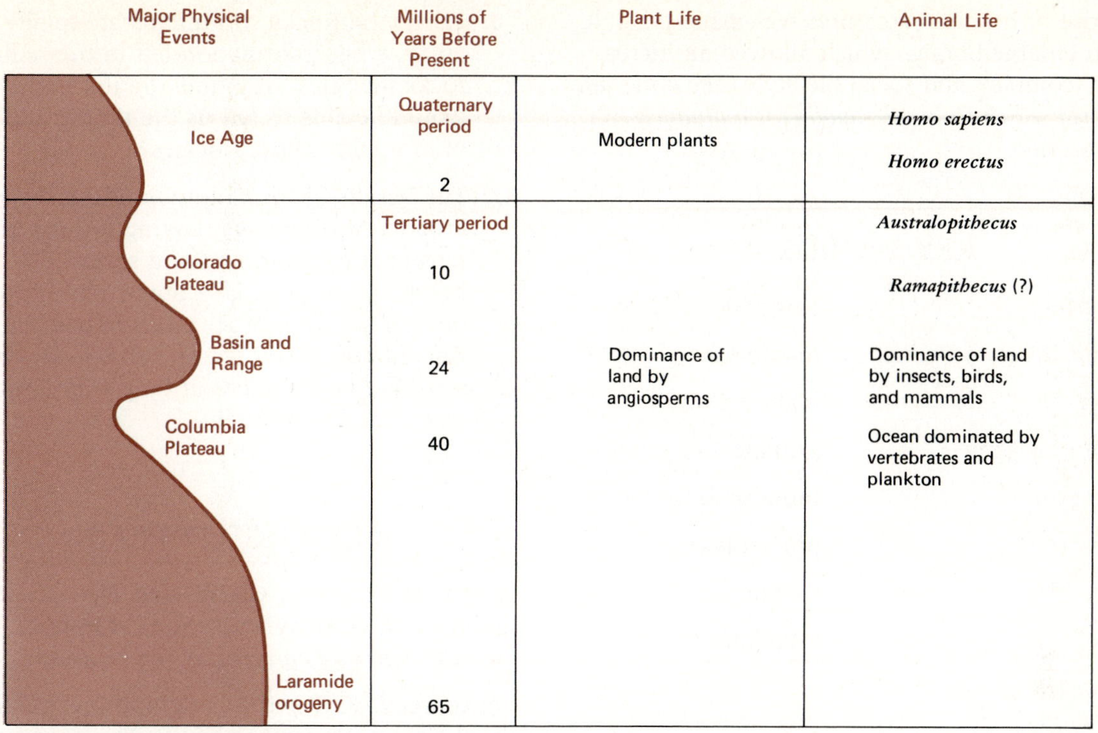

Major Physical Events	Millions of Years Before Present	Plant Life	Animal Life
Ice Age	Quaternary period	Modern plants	*Homo sapiens*
			Homo erectus
	2		
	Tertiary period		*Australopithecus*
Colorado Plateau	10		*Ramapithecus* (?)
Basin and Range	24	Dominance of land by angiosperms	Dominance of land by insects, birds, and mammals
Columbia Plateau	40		Ocean dominated by vertebrates and plankton
Laramide orogeny	65		

FIGURE 21–31

A summary of some of the major events during the Cenozoic Era.

to form the High Plains and deposited sediments in intramontane basins of many sizes and ages.

2. The overrunning of the Pacific plate by the advancing North American plate was coincident in time with the initiation of Basin and Range block faulting, crustal extension, and volcanism, and with the early development of the Cascade Range. The change from subduction to shearing along the west coast led to the initiation of the San Andreas Fault, which grew in length, accommodating the counterclockwise rotation of western North America. In Late Cenozoic the Sierra Nevada mountain range was formed by block faulting of Late Mesozoic granitic batholiths. The uplift of the Colorado Plateau was also a Late Cenozoic event.

3. The last physical event of the Cenozoic was repetitive continental and alpine glaciation, which modified scenery over large parts of middle- to high-latitude North America. The Great Lakes were formed by deep erosion and progressive reexposure of a preexisting river system in Late Pleistocene time.

4. Cenozoic flora is dominated by the angiosperms, whose reproductive strategy depends on mutually beneficial interactions with pollinating animals, chiefly insects. The angiosperms grow on 90 percent of all the earth's vegetated land surface.

5. Mammals evolved from therapsid ancestors, but they remained small and subdominant until the K-T boundary extinction opened the Cenozoic niches for large four-legged animals. The resulting radiation placed mammals in essentially every environmental niche, including insect eaters, gnawing omnivores and herbivores, hoofed herbivores, swift carnivores, tusked proboscidians, and tree-dwelling monkeys and land-dwelling apes, from whom hominid lineages diverged within the Pliocene.

6. The rise of hominid creatures was made possible by an enlarged brain, which allowed an increasingly complex and social life-style. *The active and reasoned search for understanding is a uniquely human trait.*

KEY WORDS

angiosperm	Laramide orogeny
Basin and Range orogeny	mammoth
carnivore	mantle plume
Cascadian orogeny	primate
gymnosperm	proboscideans
herbivore	pluvial lake
hominid	rodent
hot spot	ungulates
insectivore	

EXERCISES

1. Science has been termed "rationally organized experience." What does that mean? Is science more or less than this?

2. We obtain accurate radiometric ages from two different minerals in the same granite from the same Sierra Nevadan batholith and observe that there is 16 million years difference between the two dates. How can we explain this difference?

3. We observe that fossil pollen in lake sediment is of spruce and fir trees, an association not typical of modern pollen in the same lake, which is in a warm, low-elevation climate. Name two explanations for the change in pollen type recorded in past and present lake sediment.

4. The fossil hominid footprints at Laetoli allow scientists to say that the australopithecines were about 4 feet (1.3 meters) tall. How could this information have come from an exposure of numerous footprints?

5. Some Eocene elm leaves, rapidly buried under Eocene volcanic ash, were recently discovered. The leaves are almost unchanged and are still bright green. What information would you want to obtain from these leaves if they were available to you for research?

6. When the Rocky Mountains are totally eroded away, what information left in the sedimentary rocks that currently underlie the High Plains would be able to tell us the timing and direction of the uplift of the Rockies?

7. The basalts of the Columbia Plateau are of broadly Miocene age, having formed in a tensional environment and spread out as thin, highly continuous lava flows, which are piled one on top of another (see Chapter Three) over discontinuous soil zones. What evidence would lead us to conclude that they were forming in a tensional environment?

SUGGESTED READINGS

FLINT, R. F., 1971, Glacial and Pleistocene Geology, New York, John Wiley & Sons, 436 pp.
► *Basic reference on events of the ice ages.*

FLADMARK, K. R., 1986, Getting one's Berings, *Natural History*, November, pp. 8–16.
► *Thorough review of the invasion of North America by 11,000 years ago.*

GOULD, S. J., 1987, Empire of the apes, *Natural History*, May, pp. 17–26.
► *Splendid review of human ancestry, apes, outdoor science, prejudices, and so forth.*

GOULD, S. J., 1987, Life's little joke, *Natural History*, April, pp. 16–25.
► *The "bushiness" of evolution as applied to horses and other dwindling odd-toed ungulates.*

JARRET, R. D., AND MALDE, H. E., 1987, *Paleodischarge of the late Pleistocene Bonneville Flood, Snake River, Idaho, computed on new evidence*, Geological Society of America Bulletin, v. 99, pp. 127–134.
► *Fascinating account of the catastrophic emptying of Lake Bonneville; lovely example of how scientists work from evidence to theory.*

KING, PHILIP B., AND BEIKMAN, HELEN M., 1978, *The Cenozoic Rocks, a Discussion to Accompany the Geologic Map of the United States*, Washington, D.C., U. S. Government Printing Office, U. S. Geological Survey Professional Paper 904, 82 pp.
► *Excellent review of the evidence for major Cenozoic physical events in the different regions of the United States*

LEWIN, ROGER, 1982, *Thread of Life*, Washington, D.C., Smithsonian Institution, 256 pp.

▶ *Lucid writing on the total history of life with stunning illustrations.*

National Geographic Society, 1976, *Our Continent, A Natural History of North America*, Washington, D.C., National Geographic Society, 398 pp.
▶ *Elegantly illustrated, full of information, and a joy to read.*

PROTHERO, D. R., 1987, The rise and fall of the American rhino, *Natural History*, August, pp. 26–33.
▶ *Beautifully illustrated; an example of science as detective work.*

TAPPONIER, P., 1986, A tale of two continents, *Natural History*, November, pp. 56–65.
▶ *How the Eocene movement of India into Asia at 15 feet (4.6 meters) per century, formed the Himalayas. Great science, photos, and elevations.*

Early morning sunrise over the mountains of central Arizona.

CHAPTER *twenty-two*

Epilogue

Truth is that which changes most in time.

L. DURRELL

The story of the earth outlined in the chapters of this book tells us something of its nature, its history, its processes, its products, and its life. But what are the benefits that come from knowledge of the earth? It is a question I often ask myself and one that I am often asked by students in subtly different ways. What, overall, does the study of geology have to say to those who do not intend to follow it as a profession?

For most people the central value in a geology course is the appreciation of the immense variety of geologic processes and products, as well as the knowledge that the earth is always changing in occasionally catastrophic and more commonly very slow ways. Long after this book is laid aside, your understanding of geology will make a walk, a car ride, or an airplane trip a much more meaningful experience. Finding beautiful rocks and ancient fossils is certainly a part of the lifetime enjoyment that comes from an expanded understanding of the earth.

There are subtler values as well. We learn that humanity is a latecomer, preceded by 40 million centuries of ancestors. We may seem insignificant and unimportant—an ego-flattening experience produced by our own knowledge. Yet that same knowledge allows us to dominate our planet, alter our surroundings, and seize resources and wealth from the earth.

Over the course of our studies, we have learned that science is simply an extension of human curiosity augmented by rational thought—an insistence that time flows, that effect is the child of cause. We have also gained a recognition that science is an extremely human enterprise, constrained by the social view of the time, though radical new discoveries and the insights of a Copernicus, a Newton, a Hutton, a Darwin, or a Wegener may occasionally shatter these comfortable views. Such discoveries and insights are met with all the objectivity of a kangaroo court, but in time fruitful ideas prevail and become no longer heresy, but dogma.

Perhaps the greatest single lesson that geology and other sciences can offer is the recognition that nothing material is permanent. Life, planetary systems, chemical elements, and the earth all evolve. Even our perception of the passage of time changes. The "solar constant" of astronomy turns out not to be constant, and some geologists and cosmologists argue that the gravitational constant is also inconstant through time. Science can seek only to understand the material world, and the lesson from this world is that nothing ever stays the same.

Science cannot show that any material thing is permanent. What then is permanent?

The question haunts us all. It must have plagued anguished Neanderthal families as they laid their now-cold children in ancient graves. It continues as the fundamental question for us all. As we recognize that there are no gifts we can keep, perhaps our search for meaning in a transient world is all that is truly permanent.

We learn from this continuing search that life is not entirely bounded by the world of the senses and of reason. There are realities beyond these two that draw us toward what cannot be seen, touched, or proved.

Our human capacity to create images in the mind allows us to look backward and mentally watch the Appalachians rise, or root for a tiny speck of Archean life as it struggles to survive in a lifeless world. We discover that life succeeds only when it grows, when it struggles, when it has a purpose, and when it belongs.

We cannot live meaningful lives in isolation from the earth or from one another. We come from the earth and will return to it. We are both product and observer of billions of years of change. From this vantage point we all briefly share in the human experience.

APPENDIX A
IDENTIFICATION OF COMMON MINERALS

The determinative tables are *organized* to help you identify minerals through a simple process of elimination. FIRST, determine whether the luster is metallic/submetallic (reflects light like a metallic object) or nonmetallic (anything else).

If the mineral's luster is metallic/submetallic, proceed to the first part of the identification tables, which are for all minerals with a metallic to submetallic luster. If the luster is nonmetallic, proceed to the second part of the tables, which list all other minerals whose luster is nonmetallic. (If a mineral commonly exhibits either metallic or nonmetallic luster, it is shown in both parts of the tables.)

SECOND, determine the mineral's hardness on the Mohs scale. Within each table, minerals are arranged in groups of increasing hardness.

Once you have determined the luster and the hardness, you should have anywhere from one to eight minerals left from which to choose. Examination of other physical properties should help you to make the correct choice from those remaining.

THIRD, for some minerals you will have to determine the mineral's streak using a porcelain streak plate.

The identification tables in this appendix are organized as follows.

LUSTER: METALLIC OR
SUBMETALLIC
Hardness: less than 2½
Hardness: between 2½ and 5
Hardness: greater than 5

LUSTER:
Streak: definitely colored
Streak: colorless
Hardness: less than 2½
Hardness: between 2½ and 5
Hardness: between 5 and 7
Hardness: greater than 7

Note that within the nonmetallic group, minerals that yield a distinctively colored streak are separated into their own small group and are not subdivided by hardness: other properties allow rapid distinction among the members of this group.

For every mineral its specific gravity (G.), color,

streak (if any), chemical composition (see Appendix B), crystal system, and remarks about distinctive identifying properties are also given.

You will need a set of minerals of known hardness, a magnet, a glass plate, a copper penny, a common nail, dilute 10 percent hydrochloric acid, and a magnifying glass in order to complete all the tests used to identify minerals.

Mineral Identification Chart

Luster: Metallic or Submetallic
Hardness: Less Than 2½

Streak	Color	G.	H.	Remarks	Name, Composition, Crystal System
Black	Gray to iron black	2.3	1–1½	One direction perfect cleavage, greasy feel	Graphite, C, hexagonal
Black, slightly greenish	Blue black	4.7	1–1½	One direction perfect cleavage, greasy feel	Molybdenite, MoS_2, hexagonal
Gray black	Black to lead gray	7.6	2½	Perfect cubic cleavage; in cubic crystals, massive, granular	Galena, PbS, isometric
	Blue black to lead gray	4.5	2	One direction perfect cleavage, bladed with cross striations	Stibnite, Sb_2S_3, orthorhombic
Bright red	Red to vermilion	8.1	2–2½	One direction perfect cleavage. Usually granular massive	Cinnabar, HgS, hexagonal (R)[a]
Red brown	Red to vermilion	5.3	1 +	Earthy. Crystalline hermatite is harder and black	Hematite, Fe_2O_3, hexagonal (R)[a]

Hardness: Greater Than 2½, Less Than 5

Streak	Color	G.	H.	Remarks	Name, Composition, Crystal System
Black	Iron black	4.7	2½ +	Usually splintery or in radiating fibrous aggregates; sometimes softer	Pyrolusite, MnO_2 tetragonal
Gray black	Steel gray to black on exposure	5.7	2½–3	Usually compact massive; associated with other copper minerals	Chalcocite Cu_2S Orthorhombic
Black	Fresh surface, brownish bronze, purple tarnish	5.1	3	Usually massive, associated with other copper minerals	Bornite, Cu_5FeS_4, isometric
	Brass yellow	4.2	3½–4	Usually massive, may occur as tarnished crystals resembling tetrahedrons, associated with other copper minerals, darker than pryite	Chalcopyrite, $CuFeS_2$, tetragonal
Light to dark brown	Dark to yellowish brown	4.0	3½–4	Six directions perfect cleavage, tetrahedral crystals, streak lighter than specimen	Sphalerite, ZnS, isometric

[a](R) refers to the rhombohedral subdivision of the hexagonal system.
[b]Luster can be considered as vitreous unless specified otherwise.

Mineral Identification Chart (continued)

Streak	Color	G.	H.	Remarks	Name, Composition, Crystal System
Copper red	Copper red, tarnish black	8.9	2½–3	Malleable; usually in irregular grains	Copper, Cu, isometric
Silver white	Silver white, tarnish black	10.5	2½–3	Malleable; usually in irregular grains	Silver, Ag, isometric
Gold yellow	Gold yellow	15.0 to 19.3	2½–3	Malleable; usually as irregular grains, nuggets. density varies with silver content	Gold, Au, Isometric

Hardness: Greater Than 5

Streak	Color	G.	H.	Remarks	Name, Composition, Crystal System
Black	Pale brass yellow	5.0	6–6½	Often in striated cubes. Massive, granular; it's the most common sulfide mineral	Pyrite, FeS_2, isometric
	Black	5.2	6	Granular; strongly magnetic (attracted to magnets)	Magnetite, Fe_3O_4, isometric
Red brown	Dark brown red to black	5.0	5½–6½	Massive, radiating, sometimes micaeous (specular hematite). Earthy or metallic luster	Hematite, Fe_2O_3, hexagonal (R)[a]

Luster: Nonmetallic[b]
Streak: Defenitely Colored

Streak	Color	G.	H.	Remarks	Name, Composition, Crystal System
Yellow brown	Yellow to dark brown	3.3+	1–2	Dull luster, massive, earthy. Also called goethite	Limonite, hydrous iron oxides, orthorhombic (?)
Pale yellow	Pale yellow	2.1	1½–2½	Granular to earthy	Sulfur, S, orthorhombic
Yellow brown	Light to dark brown	4.0	3½–4	Six directions perfect cleavage; tetrahedral crystals, streak lighter than specimen. Luster is resinous	Sphalerite, ZnS, isometric
Brown	Light to dark brown	3.8	3½–4	Three directions (rhombohedral) cleavage, not at right angles, magnetic when heated in a candle flame	Siderite, $FeCO_3$, hexagonal (R)[a]
Red brown	Dark brown red to black	5.0	5½–6½	Massive, radiating, sometimes micaeous (specular hematite)	Hematite, Fe_2O_3, hexagonal (R)[a]
Light green	Bright green	4.0	3½–4	Radiating fibrous crystals; associated with azurite. Effervesces in cold hydrochloric acid	Malachite, $Cu_2CO_3(OH)_2$, monoclinic
Light blue	Intense azure blue	3.8	3½–4	Radiating fibrous, small crystals; associated with malachite. Effervesces in cold hydrochloric acid	Azurite, $Cu_3(CO_3)_2(OH)_2$, monoclinic

Mineral Identification Chart (continued)

Streak	Color	G.	H.	Remarks	Name, Composition, Crystal System
Very light blue	Light green to turquoise blue	2.2	2–4	Massive, compact; associated with oxidized copper minerals. Sticks to tongue	Chrysocolla, hydrous Cu silicate, amorphous

Luster: Nonmetallic[b]
Streak: Colorless
Hardness: Less Than 2½

Cleavage

Streak	Color	G.	H.	Remarks	Name, Composition, Crystal System
One direction, perfect cleavage	White, gray, apple green	2.7	1	Greasy feel. Often distinctly foliated or micaeous	Talc, $Mg_3(Si_4O_{10})(OH)_2$, monoclinic
	White	2.8	2–2½	In folited masses and scales. Cleavage flakes are elastic	Muscovite, $KAl_2(AlSi_3O_{10})(OH)_2$, monoclinic
	Dark brown, black	3.0	2½–3	Usually in irregular foliated masses. Cleavage flakes are elastic	Biotite, $K(Mg,Fe)_3(AlSi_3O_{10})(OH)_2$, monoclinic
	Green	2.8	2–2½	Usually in irregular foliated masses	Chlorite, complex Fe-Mg silicate, monoclinic
One direction, perfect cleavage, seldom seen	White, gray	2.6	2–2½	Compact, earthy; adheres to dry tongue	Kaolinite, $Al_4(Si_4O_{10})(OH)_8$, monoclinic
One direction perfect; two directions, good; fourth direction, fair	White	2.3	2	Occurs as crystals and compact cleavable masses; also fibrous	Gypsum, $CaSO_4 \cdot 2H_2O$, monoclinic
Fracture is uneven	Pale yellow	2.1	1½–2½	Granular to earthy; luster dull to resinous	Sulfur, S, orthorhombic
	Gray, yellow, brown	2.0 to 2.5	1–3	In rounded grains, often earthy and claylike. Often harder than 2½	Bauxite, hydroxides of Al, amorphous

Luster: Nonmetallic[b]
Streak: Colorless
Hardness: Greater Than 2½, Less Than 5

Streak	Color	G.	H.	Remarks	Name, Composition, Crystal System
Perfect cubic cleavage	Colorless, white, red blue	2.2	2½	Common salt. Soluble in water, salty taste. In granular masses or cubic crystals	Haltite, NaCl, isometric

Mineral Identification Chart (continued)

Cleavage	Color	G.	H.	Remarks	Name, Composition, Crystal System
Three directions, not at right angles (rhombohedral)	Colorless, white, blue, varied	2.7	3	Effervesces rapidly in cold hydrochloric acid.	Calcite, $CaCO_3$, hexagonal (R)[a]
	Colorless, white, pink	2.8	$3\frac{1}{2}$–4	Powdered mineral will slowly effervesce in cold acid. Usually harder than 3	Dolomite, $CaMg(CO_3)_2$, hexagonal (R)[a]
	Light to dark brown	3.8	$3\frac{1}{2}$–4	Becomes magnetic when heated in a candle flame	Siderite, $FeCO_3$, hexagonal (R)[a]
One direction	Colorless, yellow, pink bluish	3.5	8	Usually as prismatic crystals; found in pegmatites (coarse-grained granite)	Topaz, $Al_2SiO_4(OH,F)_2$, Orthorhombic
Four directions	Colorless, yellow, red, blue, black	3.5	10	Adamantine (brilliant) luster. Often as octahedral crystals	Diamond, C, isometric
	Pink, rose red	3.5	$3\frac{1}{2}$–$4\frac{1}{2}$	Characterized by rhombic cleavage and color	Rhodochrosite, $MnCO_3$, hexagonal (R)[a]
Three directions, two at right angles to third, unequal quality	Colorless, white, blue, yellow, red	4.5	3–$3\frac{1}{2}$	Aggregates of platy crystals. Characterized by high density for a nonmetallic mineral	Barite, $BaSO_4$, orthorhombic
Four directions, equal quality	Colorless, pink, green, purple, etc.	3.2	4	In cubic crystals, characterized by cleavage	Fluorite, CaF_2, isometric
Six directions, equal quality	Dark yellow brown	4.0	$3\frac{1}{2}$–4	Luster resinous. Usually in cleavable masses, sometimes as tetrahedrons	Sphalerite, ZnS, isometric
Cleavage not prominent	Olive to blackish green	2.2	2–5	Massive; fibrous in the asbestos variety. Frequently mottled green in the massive variety	Serpentine, $Mg_6Si_4O_{10}(OH)_8$, monoclinic
	Colorless, white, blue, varied	2.7	3	May be fibrous, fine-granular, banded as in Mexican onyx. Effervesces in cold acid	Calcite, $CaCO_3$, hexagonal (R)[a]
	Yellow, red, orange	6.8	3	Luster is adamantine. Usually in square tabular crystals. Color and high density are characteristic	Wulfenite, $PbMoO_4$, tetragonal

Mineral Identification Chart (continued)

Cleavage	Color	G.	H.	Remarks	Name, Composition, Crystal System

Luster: Nonmetallic[b]
Streak: Colorless
Hardness: Greater Than 5, Less Than 7

Cleavage	Color	G.	H.	Remarks	Name, Composition, Crystal System
One direction	Blue to gray blue	3.6	5–7	In bladed aggregates with cleavage parallel to length. H5 parallel to length, H7 across length. In metamorphic rocks	Kyanite, Al_2SiO_5, triclinic
	Hair brown to grayish	3.2	6–7	As long prismatic tiny crystals, may be in parallel groups. Looks like fine hair or slender crystals. In metamorphic rocks	Sillimanite, Al_2SiO_5, orthorhombic
	Yellowish blackish green	3.4	6–7	As short prismatic crystals striated parallel to length	Epidote, complex Fe-Mg silicate, monoclinic
Two directions, unequal quality	Colorless, white, gray	2.8	5–5½	Usually cleavable; massive to fibrous	Wollastonite, $CaSiO_3$, triclinic
Two directions, equal quality	Colorless, white, pink, gray	2.5	6	In cleavable masses or as irregular grains in rocks. *Cleavage directions are perpendicular*	Orthoclase, $KAlSi_3O_8$, monoclinic
	Colorless, white, gray, bluish	2.7	6	Same as for orthoclase. Also often exhibits a play of colors. Parallel twinning striations seen on one of two cleavages	Plagioclase, $NaAlSi_3O_8$ to $CaAl_2Si_2O_8$, triclinic
	Green, black	3.1 to 3.5	5–6	In stout prisms with rectangular cross sections. Cleavage angles of 87 and 93 degrees are characteristic	Pyroxene group, complex Ca,Mg silicates, monoclinic
	Green, black	3.0 to 3.3	5–6	Crystal are slender prisms. Characterized by cleavage angles of 56 and 124 degrees	Amphibole group, complex Ca, Mg, Fe, silicates, monoclinic
Cleavage not prominent	Yellow, blue, green, brown varied	3.2	5	Usually in hexagonal prisms; also massive, vitreous luster	Apatite, $Ca_5(PO_4)(F,Cl,OH)$, hexagonal
	Colorless, red, brown, gray, yellow, blue	2.0	5–6	Conchoidal fracture. Hardness and density less than quartz. Precious opal shows play of colors (irridescence)	Opal, $SiO_2 \cdot nH_2O$, amorphous
	Yellow, brown	3.5	5–5½	Luster is brilliant to resinous, as wedge-shaped crystals in rocks	Sphene, $CaTiSiO_5$, monoclinic

Mineral Identification Chart (continued)

Cleavage	Color	G.	H.	Remarks	Name, Composition, Crystal System
	Blue, green, bluish green	2.7	6	Usually appears as amorphous masses	Turquoise, $CuAl_5(PO_4)_4(OH)_8 \cdot 4H_2O$, triclinic
	Colorless, white, varied	2.7	7	Crystals usually show horizontally striated hexagonal prisms. Conchoidal fracture; striations parallel one another	Quartz, SiO_2, hexagonal (R)[a]
	Light brown, yellow, red, green	2.7	7	Luster is waxy to dull. May be banded or lining cavities	Chalcedony, SiO_2 cryptocrystalline quartz
	Olive to grayish green, brown green	3.3	6½–7	As rounded disseminated grains in rocks. May be massive granular	Olivine, $(Mg,Fe)_2SiO_4$, orthorhombic

Luster: Nonmetallic[b]
Streak: Colorless
Hardness: Greater than 7

Cleavage	Color	G.	H.	Remarks	Name, Composition, Crystal System
Cleavage not prominent	Green, red blue, pink, black	3.2	7–7½	In slender prismatic crystals with rounded triangular cross section. Found in pegmatites, usually black	Tourmaline, complex silicates, hexagonal (R)[a]
	Brown to brownish black	3.7	7–7½	In prismatic crystals; commonly cross shaped. Alteration on surface causes reduced hardness	Staurolite, $Fe_2Al_9O_6(SiO_4)_4(O,OH)_2$, orthorhombic
	Reddish brown to olive green	3.2	7½	Prismatic crystals with nearly square cross sections. May show central black cross (chiastolite)	Andalusite, Al_2SiO_5, orthorhombic
	Brown to red, green, black	3.5 to 4.3	6½–7½	Usually in crystals showing dodecahedrons, well formed. Common in metamorphic rocks.	Garnet, complex silicate, isometric
	Bluish green, yellow, pink colorless	2.8	7½–8	Commonly as hexagonal prisms; often found in pegmatites	Beryl, $Be_3Al_2Si_6O_{18}$, hexagonal
	Bluish gray, varied	4.0	9	Luster is vitreous. Crystals are rounded hexagonal prisms	Corundum, Al_2O_3, hexagonal (R)[a]

APPENDIX B
CHEMICAL PRINCIPLES IN GEOLOGY

Since this book assumes an acquaintance with some chemical principles, some fundamental chemical ideas are included here for your convenience in review and reference.

THE COMPOSITION OF MATTER

Elements are the fundamental chemical entity. They are composed of still smaller units and may link together to form still larger chemical units. *Atoms* are the smallest electrically neutral units that still have the chemical properties of the element.

An atom contains a nucleus; a nucleus consists of densely packed *protons*, which are particles of positive charge, and *neutrons,* which are particles with no electrical charge. The protons and neutrons comprise essentially all of the mass of the atom. The number of protons in an atom is called the *atomic number*.

The tightly packed nucleus contains essentially all of the mass, but most of the volume of an atom is composed of empty space in which tiny *electrons,* particles of negative electrical charge, orbit the nucleus in concentric shells. Since an atom is electrically neutral, the number of electrons and protons in an atom are the same, thereby providing a balanced overall charge. Thus the atomic number not only specifies the number of protons, but also the number of electrons in an atom.

The number of protons and electrons in an atom distinguishes one element from another. There are approximately 88 naturally occurring elements; each is distinguished from all the others by its atomic number.

Variations within one element arise because the number of protons is not always accompanied by the same number of neutrons. Variations of the same element, all having the same number of protons (and electrons) but different numbers of neutrons in their nucleus, are known as *isotopes* of that element. The *mass number* of an element is the number of neutrons plus the number of protons. Thus all isotopes have the same atomic number but different mass numbers.

In the same way, different isotopes of unlike elements have the same mass number, usually written as a raised subscript. N^{14} is an isotope of nitrogen, whose nucleus contains 7 protons and 7 neutrons; C^{14} is an isotope of carbon, whose nucleus contains 6 protons and 8 neutrons. The atomic number of nitrogen is 7, and the atomic number of carbon is 6, but isotopes of each may have the common mass number 14.

Both the particles in the nucleus and the electrons that orbit the nucleus are thought to be arranged in concentric shells. If all the available sites in a shell are filled, the resulting atom is quite stable. As the number of shells increases, the overall size of the atom increases.

If an atom, an electrically neutral particle, gains or loses electrons from its shells, it becomes an *ion* and possesses an electric charge. If the atom gains one or more electrons, it has an excess negative charge and is called a negatively charged ion, or *anion*. Oxygen is one example; it readily acquires two excess electrons and becomes an anion with an overall charge of -2. Silicon may readily lose 4 electrons and thus become a positively charged *cation* with an overall charge of $+4$ (having lost 4 electrons).

Molecules

If two oxygen anions (having a *combined* charge of -4) join with a single silicon cation (having a charge of $+4$), a new entity is formed that is *electrically neutral* and therefore stable. The resulting *molecule*, the fundamental chemical unit of compounds, is composed of 1 ion of silicon, contributing 4 positive charges, and $1 + 1$ ions of oxygen, contributing a combined charge of four negative charges.

Such a molecule is written in chemical shorthand as SiO_2, meaning that 1 ion of silicon is joined with 2 ions of oxygen. The resulting unit is a molecule of *silicon dioxide*, a *chemical compound* formed when 1 silicon cation and 2 oxygen anions share their electrons to create an electrically neutral unit.

A chemical compound is formed when two or more elements join together to form a new substance. A compound is distinguished from a mixture (for example, a mixture of iron filings and powdered sulfur) by being a single chemical unit that requires the expenditure of energy to reseparate into its components and whose components are locked together in *constant, fixed proportions*. Thus FeS_2 is iron sulfide, a chemical compound, but a mixture of powdered iron and powdered sulfur is just that, a mechanical mixture, easily separated with a magnet.

There are only 24 or more elements of interest in most geologic environments; the following table lists

Atomic Number	Element Name	Chemical Symbol
1	Hydrogen	H
6	Carbon	C
7	Nitrogen	N
8	Oxygen	O
9	Fluorine	F
11	Sodium	Na
12	Magnesium	Mg
13	Aluminum	Al
14	Silicon	Si
15	Phosphorous	P
16	Sulfur	S
17	Chlorine	Cl
19	Potassium	K
20	Calcium	Ca
22	Titanium	Ti
25	Manganese	Mn
26	Iron	Fe
28	Nickel	Ni
29	Copper	Cu
37	Rubidium	Rb
38	Strontium	Sr
79	Gold	Au
80	Mercury	Hg
82	Lead	Pb
86	Radon	Rn
88	Radium	Ra
92	Uranium	U

them along with their atomic numbers and shorthand symbols:

Radioactivity

Radioactivity is the result of spontaneous changes in the atomic nucleus, involving both the emission of detectable particles of many kinds and the emission of varieties of electromagnetic radiation, mainly heat.

One particle emitted is called the *alpha* (α) particle, which is identical with a helium nucleus, which consists of 2 protons and 2 neutrons (He^4).

Another particle emitted in addition to alpha particles may be a *beta* (β) particle. Beta particles include particles of the same mass as electrons but of opposite (positive) charge, called *positrons*, and electrons.

Emission of alpha or beta particles may also be accompanied by the emission of *gamma* rays (γ), a type of electromagnetic radiation that partially overlaps X rays in the total spectrum of electromagnetic radiation.

If an unstable element emits an alpha particle, its mass number decreases by 4 and its atomic number decreases by 2. Since the atomic number decreases, a new element is left behind—one whose atomic number is 2 less than its parent element. Thus radium (atomic number 88), becomes radon (atomic number 86) by emitting an alpha particle.

Emission of beta particles is a complex field; the particles emitted are presumably in the nucleus and include four types of particles. When the element gains an electron, a proton becomes a neutron; when an electron is lost, a neutron becomes a proton. Thus beta particle emission is characterized by no change in the mass number, but a change in the atomic number. For example, C^{14} becomes N^{14} by losing a beta particle, while K^{40} becomes Ar^{40} by gaining an electron or Ca^{40} by losing a beta particle. The isotopes of nitrogen and carbon share a common mass number (14) and the isotopes of calcium, argon, and potassium also share a common mass number (40).

Radioactivity is the result of spontaneous change and may be detected by a variety of chemical and photographic methods that allow recognition of emission of the particles and energy associated with radioactive decay.

APPENDIX C
ROCK IDENTIFICATION CHARTS

Chapters Three and Five both contain rock identification charts, and additional information and charts are presented here.

If you know absolutely nothing about a rock specimen given to you for identification, you must first decide whether it is of igneous, sedimentary, or metamorphic origin. For some rocks, even this broad distinction is remarkably difficult to make. Several hints of what to look for follow.

ROCK NONCRYSTALLINE: composed of fragments

1. If fragments are glassy or volcanic, rock is pyroclastic.
2. If fragments are plant fragments, rock is a low-grade coal.
3. If fragments are bits of shell and bone, rock is a clastic limestone.
4. If fragments are in regular layers and are cemented together, rock is sedimentary.

ROCK CRYSTALLINE: crystals are aligned and layered

1. Rock is most likely metamorphic, although flowage in cooling magma or lava may align some minerals in an igneous rock.

ROCK CRYSTALLINE: crystals not aligned

1. Rock may be igneous. Should contain minerals characteristic of igneous rocks.
2. Rock may be a nonfoliated metamorphic rock. If so, should be dense and composed of quartz crystals (quartzite), calcite crystals (marble), or metamorphic minerals.
3. Rock may be a chemical sedimentary rock. If so, should *not* be dense, and test should be made for gypsum, salt, dolomite, calcite (acid test) and similar minerals.

ROCK GLASSY OR POROUS

1. Probably a volcanic rock.
2. May be a poorly cemented breccia or a poorly cemented travertine or similar freshwater limestone. Test for calcite with dilute hydrochloric acid.

These tests outlined will not guarantee placement of any rock in the correct category, but they should help. Rock identification is an art as much as a science; the more rocks you can identify, the better you will become at it. Tables C-1 to C-3 for rock identification provide a systematic means to rock identification.

Table C-1 depends on your being able to identify the dominant feldspar type if the rock is plutonic (coarse grained) and on your being able to identify the dominant ferromagnesian minerals (the compounds of iron or magnesium that form dark-colored minerals). If the rock is fine grained and, hence, volcanic, unless there are feldspar phenocrysts, identification must be made on the basis of color, with darker rocks at the top of the chart. If the rock is glassy, obsidian is often a dark, volcanic glass; and pumice is usually a light-colored, frothy glass. Fragmental or pyroclastic rocks are classified by dominant grain size.

Table C-2 separates clastic rocks, which generally feel "gritty" to the touch and are well layered, from crystalline rocks. Clastic rocks are classified on the basis of dominant grain size, chemical rocks on the basis of composition.

Like igneous rocks, metamorphic rocks are classified based on both their composition and texture (Table C-3).

Classification of fine-grained rocks is always tenuous for any rock. Here the separation depends on being able to recognize slaty cleavage, or its absence.

FIGURE C–1
Identification of Common Igneous Rocks

Mineral Composition		Texture				
Dominant Feldspar	Dominant Ferromagnesian	Super Coarse 2 centimeters	Coarse Grained visible crystals	Fine Grained invisible crystals	Glassy	Fragmental[b]
Absent			Peridotite	Unknown		
Ca-plagioclase	Olivine Pyroxene		Gabbro	Basalt (scoria if frothy)		Agglomerate (large, rounded fragments)
Na-plagioclase			Diorite	Andesite		Breccia (large angular fragments)
Plagioclase = orthoclase	Hornblende		Quartz monzonite[a]	Dacite[a]		Tuff breccia (mixture of small and large fragments)
Orthoclase	Biotite	Pegmatite[a]	Granite[a]	Rhyolite[a]	Obsidian (massive) or Pumice (frothy)	Tuff (small fragments)

[a]Contain quartz.
[b]Rock composition may range from basalt to rhyolite.

FIGURE C–2
Identification of Sedimentary Rocks

Clastic or Fragmental Sedimentary Rocks		
Rock	Original Unconsolidated Debris	Average Diameter of Fragments
Conglomerate	Rounded pebbles of quartz, rock fragments, etc.	Larger than 2 millimeters (gravel)
Sandstone	Sand-sized fragments of quartz, feldspar, rock fragments	$\frac{1}{16}$-2 millimeters (sand) (pinhead to invisible)
Siltstone	Silt-sized fragments of quartz, feldspar, etc.	$\frac{1}{256}$-$\frac{1}{16}$ millimeter (silt) (invisible, but gritty)
Shale and mudstone	Clay minerals, very fine quartz. (Shale splits easily along the bedding. Mudstone breaks into angular blocks)	Smaller than $\frac{1}{256}$ millimeter (clay) (smooth to touch)

Chemical and Organic Sedimentary Rocks			
	Rock	Original Unconsolidated Material	Comments
Limestone	Sparry limestone	Interlocking crystals	Composed of the mineral calcite; "fizzes" with dilute HCl acid
Limestone	Coquina	Skeletal fragments	Composed of the mineral calcite; "fizzes" with dilute HCl acid
Limestone	Skeletal micrite	Lime mud plus skeletal fragments	Composed of the mineral calcite; "fizzes" with dilute HCl acid
Limestone	Micrite	Lime mud and silt	Composed of the mineral calcite; "fizzes" with dilute HCl acid
	Dolostone	Limestone or calcareous ooze altered to dolostone by solution	Composed of mineral dolomite; must be powdered to "fizz" in HCl
	Chert	Silicous parts of animals (sponges) and plants (diatoms). Replacement of organic tissue by solutions. Inorganic precipitation around submarine hot springs.	Cannot be scratched with knife
Evaporite	Gypsum	Gypsum	Scratched with fingernail
Evaporite	Salt	Halite	Salty taste

FIGURE C–3
Identification of Metamorphic Rocks

Common Minerals	Texture		
	Nonfoliated	Foliated or Schistose	Banded or Gneissose
Quartz	Quartzite		
Calcite, dolomite	Marble		
Mica, chlorite + quartz, feldspar		Fine-grained slate	
Mica, chlorite + quartz, feldspar	Coarse-grained Granofels	medium-grained phyllite	Coarse-grained gneiss
Mica, chlorite, garnet, amphibole + quartz, feldspar, pyroxene	Coarse-grained Granofels	Coarse-grained schist	Coarse-grained gneiss
Garnet, wollastonite pyroxene, calcite + amphibole, plagioclase	Coarse-grained skarn		
Amphibole, plagioclase	Coarse-grained amphibolite	Coarse-grained amphibolite	

APPENDIX D
TOPOGRAPHIC AND GEOLOGIC MAPS

Any map is a representation of some property of the earth. If the map displays the shape of the earth's surface and the location of specific features, it is a *topographic* map. The term topographic literally translates as a "place drawing." A *geologic* map is often overprinted onto a topographic map; it shows the configuration of the earth's surface and its surface features and the geologist's interpretation of the kinds and attitudes of rock, or other geologic features, within the mapped area.

TOPOGRAPHIC MAPS

All maps, printed on flat paper, inevitably involve some distortion of the earth's shape, which is essentially a sphere. Maps of small areas involve minimal distortion, while maps of entire continents, or of the whole world, involve more and more sophisticated methods of projecting the earth's curved surface onto a flat plane in order to minimize the distortion.

The amount of area that can be covered depends on the ratio of map distance (the distance between two points on the map) to true distance (the distance between two points on the earth). Most commonly the *scale* of a map is stated as a ratio of these two distances. For many topographic maps, common scales are 1:24,000, 1:62,500, and 1:250,000. The scale of 1:62,500 implies that one unit of anything (inches, centimeters, feet, pencil lengths) equals 62,500 of the same unit on the earth. By design, the scale of 1:62,500 is also approximately the scale of 1 inch on the map equal to 1 mile on the earth. Many maps also include a bar scale as another device for showing the relation between map distance and true distance.

Topographic maps are the only maps that attempt to show the configuration of the earth's surface, plus all drainage features, vegetation, buildings, highways, bridges, dams, and cities—anything constructed by human beings. On most topographic maps, five different colors are used to sort out the kinds of features displayed.

1. *Brown.* Landforms of the area are displayed by means of thin brown lines, some of which have elevations printed on them. These lines are called

contour lines; they connect all points on a map of equal elevation above sea level (Figure D–1). The contour interval (C.I.) is stated on the map and is the elevation difference between adjacent lines.

2. *Green.* An optional overprint that indicates areas of substantial vegetative cover.

3. *Blue.* All bodies of water, including lakes, rivers, streams, and glaciers. Intermittent streams, which flow only part of a year, are shown as alternate dashes and three dots. Perennial streams, which flow year around, are shown as solid blue lines.

4. *Red.* On more modern maps, highways, location information, and some metropolitan areas are shown in red and light red.

5. *Black.* All objects made by humans, including constructed objects and geographic names and boundaries, plus some information on position, including latitude and longitude information. Township-range information on older maps is in black, on more modern maps is in red.

Location of an object on a map is usually given in at least two complementary systems. The system of latitude and longitude are familiar systems, wherein objects north and south of the equator are described in terms of their *latitude* (degrees north or south of equator) and objects east and west of the prime meridian, running north-south through Greenwich, England, are described in terms of their longitude (degrees east or west of Greenwich).

The township-range system numbers *township* strips of 36 square miles (93 square kilometers) north and south of some arbitrary line and *ranges* east and west of some arbitrary line. The grid of township-range lines produces blocks of land containing approximately 36 square miles (93 square kilometers). Each block is subdivided and numbered into *sections* of approxiimately 1 square mile (2⁶/₁₀ square kilometers). Within any one section, a system of subdivsion by quarters is traditional.

As an example of how this system works, we could describe an area as located in the NE¼, NW¼, Sec. 20, T. 10 N., R. 5 W. Translating, the area is in the northeast quarter of the northwest quarter of section 20, 10 townships north and 5 ranges west of the intersection of the arbitrary baselines in the area.

GEOLOGIC MAPS

Geologic maps convey the geologist's interpretation of the geologic features of any one area. The map may be highly detailed and specialized in areas where rock is clearly exposed, or it may be quite

FIGURE D–1

Contour lines suggest the contours of the face and body of a former superintendent of the U.S. Geological Survey's Topographic Division. (Courtesy of the U.S. Geological Survey.)

generalized in areas of substantial regolith. The map may be of the regolith itself and may ignore the underlying bedrock.

Other kinds of maps describe the occurrence of groundwater, karst areas, economic ore deposits, gravel resources, landslide danger, quarrying ease, slope stability and form, or any of a host of other kinds of geologic information.

Most geologic maps, however, record the geologist's impression of the nature and attitude of the bedrock. Any geologic map, then, reflects the state of current geologic knowledge; any map is an interpretation—an interpretation that presumably is closer to reality as amount of exposed rock increases.

Geologic maps are made in many ways, but they all involve a geologist walking over the area, carefully noting the kinds of rock, soils, fossils, structures, attitudes, and stratigraphic relations among rocks. Such "on-the-ground" analysis is often supplemented by aerial photography and various types of imagery from airplanes and space satellites. Geologic maps of Mars are an example of what can be done without having actually placed a geologist there.

Additionally, the geologist plots all the observations on a field map, which is further refined by detailed study of the rocks in the laboratory and by detailed analysis of fossils, minerals, and structural

Clastics

 Glacial drift   Sandstone   Conglomerate   Siltstone   Shale

Chemical sedimentary

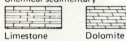 Limestone   Dolomite   Chert (in limestone)

Biochemical

 Fossiliferous limestone   Coal

Igneous

 Intrusive   Extrusive   Intrusive   Intrusive extrusive   Lava

Granitic — Basaltic

Metamorphic

 Slate   Schist   Gneiss

 Quartzite   Marble

FIGURE D–2
Some common symbols on geologic maps. Similar symbols are used in drawings throughout this book.

	Period	Epoch	Symbol	Color
Cenozoic era	Quaternary	Recent		Yellow
		Pleistocene	Q	Yellow–gray
	Tertiary	Pliocene	Tpl	
		Miocene	Tm	
		Oligocene	To	Yellow–brown
		Eocene	Te	
		Paleocene	Tp	
Mesozoic era	Cretaceous		K	Green
	Jurassic		J	Light green
	Triassic		T$_R$	Green–blue
Paleozoic era	Permian		P	Light blue
	Pennsylvanian		IP	Blue
	Mississippian		M	Blue
	Devonian		D	Blue–gray
	Silurian		S	Purple
	Ordovician		O	Red–purple
	Cambrian		E	Orange
	Precambrian era		PE	Brown, Gray–Brown

FIGURE D–3
Colors and symbols used to separate rock units of different ages on geologic maps. Although not used within the text, the names of the subdivisions of each era and the subdivisions of each of the periods within the Cenozoic era are given here to assist in interpretation of geologic maps.

information. Also, the geologist may measure and describe stratified rocks in many places in the area to be mapped and work out the correlation of these rocks. Most geologic maps have built into them many hours of laboratory work by many people and a lot of hiking and careful observing of the details of the earth's surface.

A synthesis of all the information available allows the geologist to draw the map using colors and symbols to describe both the age and rock types of the area (see Figure D–2 and D–3).

Typically, the map also contains an explanation area that illustrates what each symbol means and ar-

tificially "stacks" the geologic units in correct age sequence; the oldest unit is on the bottom of the stack, or column. Symbols are used in this geologic column to indicate the rock type characteristic of each unit. Some examples are given in Figure D–2; similar symbols are used throughout the drawings in this textbook.

Whatever the purpose of a map, it is finally an attempt to convey quickly a mass of information in a highly visual form. Its success depends on the skill of the geologist, the extent of exposure, the complexity of the geologic situation, and the combined skill of many artists, editors, and printers.

APPENDIX E
MEASUREMENT CONVERSION

For many years Americans have been used to stating measurements in the English system. As we join the rest of the world in using the metric system, a few words of explanation will be helpful.

The basic unit of *length* measurement in the metric system is the *meter,* a unit approximately equal to 1.1 yards. By using various **prefixes** *common to the entire metric system of measurement,* the basic unit can be expressed in terms of multiples of tens or tenths. A few of the common prefixes are listed.

$kilo$meter = 1000 meters

$centi$meter = $\dfrac{1}{100}$ of a meter

$milli$meter = $\dfrac{1}{1000}$ of a meter

$micro$meter = $\dfrac{1}{1000000}$ of a meter

In the same way, the basic unit of mass is the *gram,* a unit approximately equal to ½₈ of an ounce. One *kilo*gram is thus 1000 grams and is *very* approximately equal to 2 pounds.

Approximate conversion factors for various classes of measurement follow.

When you know:	You can find:	If you multiply by:
	Length	
inches	millimeters	25
feet	centimeters	30
yards	meters	0.9
miles	kilometers	1.6
millimeters	inches	0.04
centimeters	inches	0.4
meters	yards	1.1
kilometers	miles	0.6
	Area	
square inches	square centimeters	6.5
square feet	square meters	0.09
square yards	square meters	0.8
square miles	square kilometers	2.6
acre	square *yards*	4840
square centimeters	square inches	0.16
square meters	square yards	1.2
square kilometers	square miles	0.4
	Volume	
cubic inches	cubic centimeters	16.4
cubic feet	cubic meters	0.03
cubic yard	cubic meters	0.76
cubic mile	cubic kilometers	4
cubic centimeters	cubic inches	0.06
cubic meters	cubic feet	35.3
cubic kilometers	cubic miles	0.24
liters	gallons	0.25
	Mass	
ounces	grams	28
pounds	kilograms	0.45
tons (2000 pounds)	metric ton (1000 kilograms)	0.9
grams	ounces	0.035
kilograms	pounds	2.2
metric tons (1000 kilograms)	ton (short ton) (2000 pounds)	1.1
	Miscellaneous Conversions	
gram per cubic centimeter *(density)*	pounds per cubic foot	62.43
watt *(power)*	Btu per minute	0.057
gallons of *oil*	barrels	42
gallon *(volume)*	liters	3.78
miles per hour *(velocity)*	kilometers per hour	1.6
miles per hour	inches per second	17.6
kilometers per hour	centimeters per second	27.8
bar *(pressure)*	atmosphere (14.7 pounds per square inch)	0.987
kilometer of average rock	kilograms per square centimeter	265
kilogram per square centimeter	pounds per square inch	14.22
degrees Fahrenheit *(temperature)*	degrees Celsius	5/9 (after *subtracting* 32)
degrees Celsius	degrees Fahrenheit	9/5 (then *add* 32)

APPENDIX F
CLASSIFICATION OF LIFE

Biologic classification dates from the Swedish naturalist Carolus Linnaeus, who established a system for naming life in the eighteenth century. In this system, the categories of life are arranged in a hierarchy that expresses levels of biologic relationship. A *species* is a group of life forms so alike that they can interbreed. A *genus* (pl. *genera*) is a group of species so closely related that they share a common ancestry, but members of two separate genera cannot interbreed. A *family* is a group of related genera, while an *order* is a group of related families. A group of related orders is called a *class*, and a group of related classes is called a *phylum* (pl. *phyla*). A *kingdom* is a group of related phyla.

A human being is classified as follows:

KINGDOM: Animalia

PHYLUM: Chordata

CLASS: Mammalia

ORDER: Primates

FAMILY: Hominidae

GENUS: Homo

SPECIES: sapiens

The classification that follows is highly simplified and emphasizes the life forms that are described in this textbook. For a more complete classification of living forms, see any biology textbook.

KINGDOM PROKARYOTA: All prokaryotic life lacking a well-defined nucleus. Includes bacteria and blue-green algae

KINGDOM PROTISTA: Eukaryotic single-celled organisms, either solitary or colonial. Includes yellow-green and golden-brown algae, diatoms, coccoliths, amoeba, paramecium, etc.

KINGDOM PLANTAE: Includes all true plants

PHYLUM PSILOPHYTA: Extinct, leafless, root-less plants

PHYLUM LYCOPODOPHTA: Includes scale trees (Lycopsids) of the Paleozoic

PHYLUM EQUISETOPHYTA: Horsetails and scouring rushes and sphenopsids such as *Calamites* (Late Paleozoic)

PHYLUM POLYPODIOPHYTA: True ferns (Pteropsids)

PHYLUM PINOPHYTA: Gymnosperms ("naked seeds")

CLASS LYGINOPTERIODOPSIDA: Seed ferns, including *Glossopteris*

CLASS BENNETTITOPSIDA: Cycadeoids

CLASS CYCADOPSIDA: Cycads

CLASS GINKGOOPSIDA: Ginkgoes

CLASS PINOPSIDA: Conifers ("evergreens"); also *Cordaites*

PHYLUM MAGNOLIOPHYTA: All flowering plants ("angiosperms")

KINGDOM ANIMALIA: Includes all animal life

PHYLUM ARCHAEOCYATHA: Earliest Paleozoic spongelike forms of uncertain affinity

PHYLUM PORIFERA: Sponges

PHYLUM COELENTERATA: Includes all corals and jellyfish

PHYLUM PLATYHELMINTHES: Flatworms. Other worm varieties comprise separate phyla

PHYLUM BRYOZOA: Bryozoans ("moss animals")

PHYLUM BRACHIOPODA: Marine animals with two shell halves

PHYLUM ANNELIDA: Segmented worms

PHYLUM ARTHROPODA: Segmented animals with jointed appendages and chitonous external skeletons

SUBPHYLUM TRILOBITA: All trilobites of the Paleozoic

SUBPHYLUM CHELICERATA: Includes eurypterids, scorpions, spiders, ticks, and mites

SUBPHYLUM CRUSTACEA: Includes lobsters, barnacles, crabs, and ostracods

SUBPHYLUM LABIATA: Includes millipedes, centipedes, and all insects

PHYLUM MOLLUSCA: Soft-bodied, unsegmented animals, usually protected by a shell

CLASS SCAPHOPODA: Tusk shells or scaphopods

CLASS GASTROPODA: Snails

CLASS PELECYPODA: Clams and oysters

CLASS CEPHALOPODA: Marine animals with tentacles, well-developed eyes and central nervous systems, such as squids and octopuses

ORDER NAUTILOIDEA: Nautiloids, both straight and coiled, simple shells

ORDER AMMONOIDEA: More complexly ornamented shells

ORDER BELEMNOIDEA: Belemnites

PHYLUM ECHINODERMATA: Marine animals with spiny, calcareous plates, such as starfish and brittle stars

CLASS ECHINOIDEA: Sea urchins and sand dollars

CLASS BLASTOIDEA: Paleozoic blastoids

CLASS CRINOIDEA: Sea lilies and feather stars

PHYLUM PROTOCHORDATA: Includes graptolites

PHYLUM CHORDATA: Includes animals with central nervous system, including all vertebrates

PHYLUM UROCHORDATA: Sea quirts or tunicates

SUBPHYLUM VERTEBRATA: All animals with backbones

   CLASS AGNATHA: Primitive jawless fish ("ostracoderms")

   CLASS ACANTHODII: Primitive, spiny jawed fish

   CLASS PLACODERMII: Paleozoic jawed, armored fish

   CLASS CHONDRICHTHYES: Fish with cartilaginous skeletons, like the shark

   CLASS OSTEICHTHYES: Bony-skeleton fish

   SUBCLASS ACTINOPTERYGII: Ray-finned fish

   SUBCLASS SARCOPTERYGII: Lobe-finned fish, breathe air directly

      ORDER CROSSOPTERGII: Amphibian ancestors

      ORDER DIPNOI: Lungfish

   CLASS AMPHIBIA: Amphibians, including labbyrinthodonts

   CLASS REPTILIA: Reptiles, scaly skin, amniotic egg

   SUBCLASS ANAPSIDA: Mesosaurs, cotylosaurs, and turtles

   SUBCLASS SYNAPSIDA: Therapsids and mammallike reptiles

   SUBCLASS EURYAPSIDA: Plesiosaurs and icthyosaurs

   SUBCLASS DIAPSIDA: Lizards, snakes, tuatara, and all dinosaurs

CLASS AVES: True birds

CLASS MAMMALIA: Warm-blooded, hairy animals, young nursed at mammary glands

   SUBCLASS PROTOTHERIA: Egg-laying mammals, such as platypus

   SUBCLASS ALLOTHERIA: Multituberculates

   SUBCLASS EUTHERIA: Placental mammals. Young fetus develops in uterus with nutrition furnished thru placenta

      ORDER INSECTIVORA: Moles and shrews

      ORDER CHIROPTERA: Bats

      ORDER EDENTATA: Armadillos, tree sloths, glyptodonts, and ground sloths

      ORDER RODENTIA: Mice, rats, and beavers

      ORDER CETACEA: Whales

      ORDER CREODONTA: Carnivorous and extinct

      ORDER CARNIVORA: Modern meat eaters, including dogs and cats

      ORDER CONDYLARTHA: Ancestral ungulates

      ORDER AMBLYPODA: Primitive ungulates

      ORDER PERISSODACTYLA: Odd-toed hoofed mammals; horses, titanotheres, and chalicotheres

      ORDER ARTIODACTYLA: Even-toed hoofed mammals; oreodonts, cattle, and pigs

      ORDER PROBOSCIDEA: Elephants, mastodons, and mammoths

      ORDER PRIMATES: Monkeys, apes, and humans

GLOSSARY

aa Surface of a lava flow which is very rough, having jagged, tilted blocks and slabs of congealed lava.

Acadian orogeny A Middle Paleozoic phase of deformation in the northern Appalachians.

acid rain Precipitation whose acidity has been increased above pristine levels.

accordant stream junction A junction of two or more streams at a common elevation.

active margin A plate margin which is undergoing continuing deformation.

aggradation A buildup of the surface of the earth by deposition of materials.

A horizon The soil horizon nearest the earth's surface; commonly the leached zone.

Alleghany orogeny A Late Paleozoic deformation in the central and southern Appalachians.

alluvium All loose clay, silt, sand, gravel, and other detrital material deposited by streams.

amniotic egg One which surrounds the embryo with fluid; typical of birds and reptiles.

amphibole Any of a common group of complex aluminosilicate minerals which contain calcium, sodium, manganese, magnesium, or iron ions and have a crystal structure with well-developed prismatic cleavage planes intersecting at 56° and 124°.

andesite A volcanic rock of intermediate composition and color found only on continents. Common minerals are intermediate plagioclases, pyroxenes, biotite, and amphiboles.

angiosperm Broadly, the flowering plants, with highly specialized flowers and seeds.

angle of repose The maximum slope at which a heap of loose material is stable and will not collapse or slide.

angular unconformity A surface cut by past erosion across layers whose strata dip at an angle different from that of the layers above the erosional surface.

anthracite A hard, black lustrous coal containing a large amount of carbon and a small amount of volatiles.

anticline A fold in which the limbs dip away from its center; commonly older rocks are exposed in the central region.

aquifer A permeable geologic stratum of rock, gravel, sand, or silt capable of yielding groundwater to wells and springs. An aquifer below the water table contains groundwater.

Archean Eon The interval of time from oldest known rocks on earth to 2500 Ma.

arête A ridge crest, often between cirques, sharpened by alpine glaciation.

artesian well A well whose water is naturally pressurized; the water rises above the aquifer containing it.

aseismic creep Slow, steady movement along a fault surface without the release of significant elastic energy as earthquakes.

ash Fine particles of mineral and glassy material erupted from a volcanic vent. When consolidated, ash forms *tuff*.

asthenosphere Semifluid zone within the earth's mantle at the base of the rigid lithosphere; the asthenosphere allows isostatic vertical motion of the lithosphere and horizontal plate motion.

atom The smallest unit of a chemical element, consisting of a central *nucleus* surrounded by orbital electrons. It is held together by the electromagnetic force. Briefly, the smallest particle of an element that enters into the composition of a molecule.

atomic number The number of positive charges in the nucleus of an atom.

atomic radius The distance, commonly expressed in angstrom units (10^{-8} centimeter), from the center of the nucleus to the outermost electron. The atomic radius changes with the type of bonding.

atomic weight The average weight of atoms of an element compared to the average weight of oxygen atoms = 16.0000; atomic weight expresses the relative mass of an atom compared to the mass of an oxygen atom as a standard.

aulacogen A trough on a continent that is one part of a complex trilateral fault system formed above a mantle spreading center; a fault-bounded cratonic rift.

autotrophic Literally, "self-feeding." An autotroph is an organism which can synthesize organic nutrients directly from simple inorganic substances such as carbon dioxide and nitrogen compounds. Broadly, photosynthetic and chemosynthetic plants and prokaryotes are autotrophic; animals are not.

axial plane (fold) An imaginary plane which symmetrically bisects the crest or trough of a simple fold.

axiom A self-evident truth; a statement universally accepted as true.

barchan dune A dune which when seen from above is crescent-shaped and when seen in profile from the ground has a gentle, convex slope facing the wind; the two arms of the crescent point downwind. Its leeward side is steep and forms at the angle of repose.

barchanoid ridge dune A dune which when seen from above consists of a series of barchan dunes closely connected into a sinuous ridge.

basalt A common dark-colored volcanic rock composed of pyroxene and calcic plagioclase, often with olivine.

base level The elevation where the potential energy of a rock or mineral particle is zero. For streams, *temporary* base levels include lakes and tributary junctions; the *ultimate* base level for streams is sea level. Base level is often defined as the elevation below which streams cannot erode. Both glacial erosion and submarine landsliding, however, occur below base level for streams.

basement Broadly, igneous and/or metamorphic rocks overlain unconformably by younger sedimentary rocks; the basement rocks may be of any age.

Basin and Range orogeny A Tertiary deformation in the western United States with block faulting on broadly north - south trends, accompanied by volcanism and major crustal extension.

batholith An exposure of plutonic rock greater than 40 square miles (100 square kilometers) in area.

bedding In sedimentary rock, the naturally formed planes dividing rock units of different appearance. In undisturbed sedimentary rocks, bedding planes are commonly horizontal.

bed load The total mass of rock or mineral particles carried by a stream as fragments rolled along its valley floor.

bedrock Solid rock, especially that exposed under unconsolidated rock.

B horizon The soil layer beneath the A horizon; commonly the horizon into which materials leached from the A horizon are deposited.

Benioff zone A zone of earthquake foci at plate margins dipping from a shallow depth to depths as great as 700 kilometers (or beyond?); these zones of foci define the geometry of the subduction zone.

biofacies An assemblage of fossilized life-forms.

block A large (greater than 256 millimeters) angular volcanic fragment.

blue-green algae The common name for *cyanobacteria*, each of which is a single prokaryotic cell.

body waves Waves of seismic energy which travel outward from an earthquake focus and pass through the earth's interior.

bomb A mass of lava exploded from a volcanic vent and immediately solidified in the air into a rounded, irregular mass.

Bowen reaction series The sequence of crystallization events giving the order in which different minerals crystallize from a cooling magma. As minerals crystallize, they may separate from the magma, thereby steadily changing the chemical composition of the remaining fluid. This process yields minerals whose composition is unlike that of the parent magma. Or plagioclase minerals may incrementally change their overall composition in continuous crystallization; the dark-colored minerals rich in iron and magnesium may react with the melt to recrystallize as other minerals in discontinuous crystallization, as orthoclase and quartz form as leftovers. Then the resulting mass of minerals has the overall composition of its parent magma.

breaking strength The greatest stress which a substance may withstand before it loses cohesion and the ability to support further stress.

breccia A coherent rock composed of coarse angular fragments.

brittle strain Failure by instantaneous loss of cohesion and strength along an extension fracture.

BTU British thermal unit—the heat required to raise 1 pound of water from 60° to 61° Fahrenheit at a constant pressure of one standard atmosphere. One BTU is approximately equivalent to the heat produced by one wooden match.

bulk modulus The ratio of compressive stress applied to a substance to the change in its volume; a measure of the compressibility of a substance or its resistance to a change in volume.

burial metamorphism High-pressure metamorphism at low temperature.

butte A small, isolated, flat-topped hill with steep sides.

calcite A common mineral composed of calcium carbonate crystallized in hexagonal form.

caldera A more or less circular volcanic depression whose diameter is far larger than that of any associated vents. Calderas may be formed through erosion, explosion, collapse, or any combination of the three.

Caleodonian orogeny An Early Paleozoic deformation in northwestern Europe, Ireland, and Scotland.

caliche A generally dense crust of soluble alkalic material precipitated from evaporating water in desert soils; sometimes termed "hardpan."

Cambrian Period From 570 to 505 Ma; the oldest period of the Paleozoic Era.

capacity Of a stream, the total load it can move; measured in mass per hour.

carbonation Replacement of minerals by carbonates during chemical weathering.

carnivore An animal which eats only meat.

Cascadian orogeny A Tertiary deformation affecting the far northwestern United States and Canada.

Cassiar - Sonoman orogeny A Late Permian to Early Triassic deformation which stretched from Nevada to British Columbia, causing a substantial westward expansion of the North American continent.

cataclastic metamorphism Localized change in rock texture through granulation, shattering, rotation, and flattening of preexisting material.

catastrophism The belief that sudden, violent, short-lived events formed all landscapes a few thousand years ago. Catastrophists may also believe that all change in species came about through divine intervention after worldwide catastrophes.

catenary profile The ∪ - shaped cross section of a valley modified by valley glaciation. Holding a limp string aloft by both ends produces the same profile.

cave A natural hollow chamber formed underground, usually found in carbonate rocks upslope from a river or in areas exhibiting karst topography. Caves may or may not open to the earth's surface.

cement Natural mineral material chemically precipitated in the spaces between grains in a clastic rock; cement forms solid rocks from loose materials.

cementation The precipitation of binding material around grains of minerals, followed by consolidation into rock.

Cenozoic Era The last 66 million years; mammals dominate this time.

chemical compound A substance whose molecules consist of unlike atoms in fixed proportions. A compound cannot be separated into its elemental components by simple physical means.

chemical sedimentary rocks Rocks whose crystalline components are formed largely through precipitation, evaporation, or other chemical change.

chemical weathering A weathering process whereby the chemical composition of a mineral is altered, commonly in the presence of water.

chemosynthetic Created by a chemical process; often said of life-forms that derive their nutrition by inorganic chemical processes such as methanogenic bacteria.

C horizon Commonly the lowest soil horizon adjacent to bedrock, consisting of slightly altered bedrock.

chronologic time Time's passage measured in fixed intervals by a device working at a constant rate; time as kept with a watch.

cinder cone A roughly conical hill composed of loose cinders and other volcanic debris accumulated around a vent.

cirque In the highest part of mountains, a bowl-like, steep-walled hollow formed by intense glacial erosion of the headwaters of a stream. Cirques commonly form the blunt end of a valley.

clastic sedimentary rock One composed of mixed cemented fragments of preexisting rocks, minerals, or volcanic debris.

clay A collective name for a group of generally platy, extremely finely crystalline hydrous aluminosilicate minerals. Any soft, plastic, extremely fine-grained earthy sediment.

claystone A sedimentary rock composed largely of clay minerals and having little stratification.

cleavage Of minerals, the tendency to split along preexisting planes of weakness. A planar surface created by striking a mineral along a natural plane or planes of weakness is a *cleavage plane*. Metamorphism or unbalanced stresses during folding or faulting may also induce cleavage. Types of rock cleavage include *slaty cleavage* and *schistosity*, among others.

coal A combustible rock composed of highly modified organic matter and volatiles.

cohesion The measurable resistance of a material to shear; broadly, the resistance to coming apart.

color A phenomena produced in our eyes as certain wavelengths of incident white light are absorbed by a substance.

compaction The process of diminishing the volume and increasing the density of material; the process by which soil and sediment lose pore space as the weight of overlying material increases.

competency Of a stream, the maximum-size particle that it can carry; a large, slowly moving stream has low competency but high capacity.

composite cone A large volcanic cone composed of intermixed masses or alternating layers of ash and lava flows; commonly a stratovolcano is formed from the eruption of magma of intermediate composition.

compositional banding Layering in rocks caused by changes in mineral content.

compressional stress Stress created when forces act in the same plane in opposite directions, causing loss of volume and shortening in the direction of applied force.

conduction Transmission of heat or cold through a solid; temperature change in a medium which does not require movement of the medium.

Conduction of heat in a skillet eventually makes the handle hot.

conglomerate A sedimentary rock composed of rounded gravel plus cement.

conservational margin A plate margin along which surface area is neither created nor destroyed.

contact metamorphism Adjacent to igneous intrusions, the baking or other altering of rock in the contact zone as its temperature is elevated.

continental crust The rocks, characterized by generally higher silica content and lower specific gravity than mantle material, which form the continents and continental shelves. Both the lower and the lateral boundaries of continental crust are marked by transition from the broadly granitic continent to broadly basaltic mantle or seafloor.

continental drift A theory that continents move by plowing through oceanic crust.

continental glacier An extensive permanent ice sheet unconfined by valley walls and covering a large part of a continent.

continental shelf The part of the continental crust which surrounds a continent and is covered by seawater. The continental shelf ends seaward in the continental slope.

continuous reaction series The sequence of reactions by which early-formed plagioclase feldspar crystals continuously react with a cooling magma, thereby maintaining chemical equilibrium with the surrounding magma. These interactions continually change the feldspar's overall composition but do not cause abrupt changes in its crystal structure.

conventional well One whose water level does not reach the earth's surface; a pump (or a bucket!) must be used to gain water from a conventional well. Compare with *artesian well*.

convergent margin An active plate margin produced as two plates contact one another.

correlation In geology, two rock units being of the same age.

core The central zone of the earth, beginning at a depth of 1800 miles (2900 kilometers) and consisting of high-density iron-rich metallic material. The outer shell of the core has the physical properties of a liquid nickel–iron alloy.

covalent bond A bond in which atoms share electrons. Each atom of a bound pair contributes one electron to form a pair of shared electrons.

crater Bowl-shaped, generally steep-sided depression in a planetary surface created by impact or volcanism.

craton Ancient continental crust commonly composed of igneous and/or metamorphic rocks; see *basement*. Cratons are usually formed by repetitive welding or accretion of fragments of continental crust rifted from numerous geographic sources. The boundary or suture zones of these fragments may or may not become zones of seismicity. Cratons are generally less seismically active than the plate margins that terminate them.

creep An imperceptibly slow downhill movement of earth materials or regolith. The term is also applied to the slow commonly aseismic slippage of fragments of crust along fault planes.

Cretaceous Period From 144 to 66 Ma; the youngest period of the Mesozoic Era.

crevasse An open, nearly vertical fissure in the upper part of a glacier.

cross-bedding A condition in which lenticular sedimentary layers are inclined and transverse to the main stratification; sets of inclined parallel layerings are frequently and abruptly terminated by other sets.

crosscutting relation, principle of A unit that crosscuts or intrudes another is younger than the penetrated unit.

cross-stratification Cuspate, interfingering layering in sedimentary rock. See *cross-bedding*.

crust The outermost solid layers of the earth consisting of crystalline rock and extending downward to the Mohorovičić discontinuity. The crust is *not* the same as the *lithosphere*; the crust is the outermost part of the lithosphere above the M-discontinuity.

crystal form The common shape exhibited by an *uncleaved* mineral formed in an environment that allows crystal faces to form. Molecules within a crystal are internally arranged in a regularly repeating pattern, which may be expressed as planar crystal faces.

crystallization The natural collection of atoms or molecules into highly symmetrical, tightly bound, three-dimensional networks which repeat themselves in space.

cyanobacteria Commonly called blue-green algae; members of this diverse group consist of single prokaryotic cells.

dacite A finely crystalline volcanic rock composed of quartz, amphibole, and intermediate plagioclase. It is intermediate in composition between rhyolite and andesite.

debris fall The free downward fall of loose earth and unconsolidated rock fragments from a vertical face.

debris flow Rapid downslope movement of coarse, dense, muddy unconsolidated soil and rock fragments. See *earthflow*.

debris slide Rapid downslope sliding of comparatively dry, unconsolidated soil and rock fragments without backward rotation of the mass. See *landslide*.

deductive reasoning Reasoning from a known principle to its logical conclusion, from an accepted generality to a specific instance. An inference in which the conclusion about particulars follows from general or universal premises.

deflation The sweeping erosive action of wind on unconsolidated material, removing finer material in suspension and coarser material by saltation. The coarsest material is left behind as *deflation armor*.

deflation armor Surface residual coarse material after deflation of finer material. Synonyms are desert pavement and lag gravel.

delta Alluvial deposit at base level. Usually alluvial fans are triangular as viewed from above and form at the mouth of river, stream, or tidal inlet.

dendritic pattern As seen from above, the incised stream pattern taken by irregularly branched tributaries joining successively larger branched tributaries, which join the main stream; all stream junctions are accordant. This fractal pattern obeys universal laws of fluid and energy distribution and so is analogous to the branching pattern of a tree limb, our blood vessels, our bronchial tubes, our nervous system, lightning, fault plane terminations, and so forth.

density Mass of a substance per unit of volume, as in grams per cubic centimeter.

desert To geographers, an area receiving less than 10 inches (25 centimeters) of precipitation in an average year; to geologists, an area having internal drainage.

desertification The creation of dessicated, barren land conditions through natural changes in climate or through agricultural and population mismanagement which causes deserts to enlarge.

desert pavement See *deflation armor*.

desert varnish Dark yellowish-brown to purplish-black film of clay and manganese/iron oxides coating long-exposed rocks in deserts.

detrital sedimentary rocks Those composed of fragments cemented together.

Devonian Period From 408 to 360 Ma; middle period of Paleozoic Era.

D horizon Lowermost soil layer consisting of unmodified bedrock.

diffusion Slow migration of fluid or vapor through a slightly permeable solid.

dike A tabular body of igneous rock which has been injected while molten into a fissure and which cuts across layering of surrounding rock; any tabular intrusive into massive unlayered rock.

dilatancy Deformation in which masses of granular material expand in volume through rearrangement of grains.

dilatancy - diffusion Increase in volume accompanied by an intake of fluid by diffusion.

diluvialist One who believes that all rocks, or in a more restricted sense all unconsolidated rocks, were deposited by the worldwide flood as recorded in Genesis 7 and 8.

discharge Volume of flow of a fluid per unit of time as in cubic meters per second. Discharge is the product of stream velocity, depth, and width.

disconformity A planar to irregular surface of erosion separating layers of sedimentary rock that are broadly parallel to the disconformable surface and are parallel to one another. The erosional surface may or may not exhibit any relief.

discontinuous reaction series Sequence of reactions in which early-formed ferromagnesian minerals successively react with the surrounding, cooling magma to recrystallize as different minerals, whose composition is in chemical and thermal equilibrium with the cooling magma.

dissolved load The part of the total stream load that is carried in solution.

distributary Sluggish, low-gradient meandering stream on delta surface which distributes water out onto that surface.

divergent margin Plate boundary which is filled in with new lithosphere as the plate rifts and the two fragments move apart. The Red Sea is a modern example.

dolomite Common mineral composed of calcium magnesium carbonate. This term is sometimes applied to a sedimentary rock that is rich in the mineral dolomite. See *dolostone*.

dolostone A limestone whose carbonate fraction contains more than 50 percent dolomite. In some usage, dolostone is a synonym for dolomite.

drainage basin Area drained of surface runoff by a stream or series of streams. The boundaries of a drainage basin consist of a series of stream divides.

drainage divide The rim of a drainage basin; the area of highest elevation separating flow directions for runoff.

drift All unconsolidated rock material picked up, transported, and then deposited by a glacier. If unstratified, it is *till*; if layered, it is *stratified drift*.

dripstone Rock, commonly limestone, formed by dripping water in a cave.

drumlin Streamlined hill made up of drift whose long axis parallels the flow direction of the glacier under which it formed by deposition from the sole of the glacier.

ductile strain Partial loss of cohesion and strength following an extensive period of support of unbalanced stress. Failure occurs by repetitive sliding along multiple shear planes that intersect one another at acute angles.

dune A hill, ridge, or mound of sand piled up by the wind.

dynamothermal metamorphism Progressive changes in solid rock through simultaneous application of high temperature and pressure. These changes form a dense rock whose minerals are stable at moderate temperatures.

earthflow Downslope movement of soil and weathered rock along a basal and marginal shear surface. The addition of water forms a *mudflow* in fine material or a *debris flow* in coarser material.

earthquake Sudden trembling of rock masses caused by abrupt release of stored elastic strain energy.

Ediacaran fauna Assemblage of fossil remains of flat, marine animals unique to latest Proterozoic time.

elastic Capable of sustaining deformation without permanent loss of size or shape. In an elastic substance, observed strain is directly proportional to applied stress; the material regains its original shape after stress is released.

elastic rebound Movement back along a fault plane in rock when stored elastic strain is released by rupture.

elastic strain Fully recoverable deformation; the deformation which disappears when stress is removed.

electron A light elementary particle (lepton) having one unit of negative charge. It is a constituent of all atoms and has 1/1836 the mass of a proton.

element A substance whose atoms have the same atomic number.

empirical Relying on or based on experience and observation.

end moraine Hill of till marking limit of glacial advance; synonym *terminal moraine*.

energy The capacity to do work and overcome resistance.

eolian Pertaining to the action or effect of wind. Examples: eolian landform, eolian deposit.

Eon Largest subdivision of geologic time. Example: Proterozoic Eon.

epeirogenic Relating to epeirogeny. Example: epeirogenic basin.

epeiorogeny Slow uplift or depression of a very broad area. Compare to *orogeny*.

epicenter Point on the earth's surface vertically above the earthquake focus.

Era Subdivision of geologic time that includes two or more periods.

erg (1) Unit of energy in metric system for the amount of work done when a force of 1 dyne is applied through a distance of 1 centimeter. (2) Extensive desert region covered by shifting sand.

erode To abrade, to dissolve, to quarry, and to remove surface material by the action of gravity, water, ice, and wind; to shape and wear away the land surface.

erosion A collective term for the processes that wear away and remove surface material.

erosional unloading Rock expansion through release of stored elastic strain as overlying rock is removed. This process may cause rock bursts in mines or quarries and contribute to surface exfoliation of crystalline rocks.

erratic A boulder carried far from its source by a glacier.

esker A low, narrow, sinuous ridge of irregularly stratified drift. Eskers are formed by a stream flowing under a glacier.

eugeocline The seaward part of a passive continental margin that is commonly marked by volcanism. Older synonym: *eugeosyncline*.

eukaryote Cell whose genetic material (chromosomes) is in a nucleus. Cell type observed in all life except *cyanophytes* (blue-green algae) and bacteria.

eustatic Pertaining to worldwide changes in sea level, to the rise and fall of seas on the continents.

exfoliation The breaking off of concentric slabs, shells, sheets, or plates of rock from its surface exposure.

facies Overall distinguishing aspect of a geologic unit that separates it from other units having similar characteristics. There are three major categories of facies: (1) *lithofacies*, (2) *biofacies*, and (3) *stratigraphic facies*.

falls A mass-wasting category; dislodged particles freely dropping through the air.

fault A fracture in the earth's crust along which the adjacent rock surfaces are displaced in relation to each other. Along the failure plane there is discontinuous motion that may be felt as earthquakes.

faulting Movement producing relative displacement of rock masses along a faulted surface.

fault plane Roughly planar surface(s) along which adjacent rock surfaces have been displaced in relation to each other.

faunal assemblage, principle of An assemblage of specific fossils is unique to a single interval of time.

faunal succession, principle of Life-forms change in a definite and recognizable order through geologic time; once a species becomes extinct, it never recurs.

field reversal Flip-flop in direction along the force lines in the earth's magnetic field between normal polarity and reversed polarity. Human beings would experience field reversal as a magnetic compass needle that once pointed north changing to point south.

fissility Paper-thin prominent layering in clay-rich rocks.

flexural-slip fold One produced by buckling or flexing of layers accompanied by the slip of one layer over another. Bedding planes are the active surfaces of slip.

floodplain A relatively level valley floor composed of alluvium. Floodplains border the streams that form them during overflow or flooding.

flows Mass-wasting category; all semifluid movements of unconsolidated soil and rock fragments.

fluvialist One who believes that unconsolidated materials, especially *drift*, was formed by past streams.

focus Of earthquakes, the point of first motion, where seismic energy waves originate. Plural: foci; synonym: hypocenter. Contrast with *epicenter*.

fold axis Line connecting the points of maximum curvature on a folded surface; the line which when moved parallel to itself generates the folded surface.

foliated Composed of or separable into layers; parallel arrangement of planar minerals in a rock. Most commonly applied to metamorphic rocks, the term has been used to describe flow textures in igneous rocks and glaciers, coarse compositional banding in some crystalline rocks, and parallel structural features in any rock.

footwall Block of rock lying vertically beneath the fault plane.

force The agency which accelerates the speed or changes the direction of a mass or places a mass at rest in motion. The magnitude of the force is equal to the product of the mass and acceleration.

foredeep Long, narrow oceanic trench formed at subducting plate margin.

fossil Preserved evidence of past life.

fossil fuel Any combustible hydrocarbon concentrated from past life-forms; includes coal, natural gas, oil, and other petroleum products.

fractional crystallization A natural cooling process which leads to successive change in the chemical composition of magma as each mineral that crystallizes at progressively lower temperatures leaves the magma chamber so that they no longer interact with the magma. This process creates a succession of minerals; the chemical composition of each set of minerals produced is unlike that of the parent magma.

fracture An irregular surface of breakage across a mineral; mechanical failure along *other than* cleavage planes.

frost wedging Shattering of rock by growth of ice crystals in cracks.

gabbro Any of a group of dark-colored plutonic rocks composed of pyroxene and plagioclase; many gabbros also contain olivine.

geocline A linear site of deposition on a passive margin. Older synonym: *geosyncline*.

geologic time scale The relative age of geologic periods and their absolute time intervals. The time scale orders planetary events during the last 45 million centuries.

geology Study of the origin, materials, processes, landforms, history, resources, hazards, and ancient life of planet Earth and other planets.

geosyncline See *geocline*.

geothermal energy Energy derived from heat released from the earth's interior.

glacial erosion Scouring and shaping of earth's surface and transporting of soil and rock material by flowing ice.

glacial polish Smooth rock surface created by glacial erosion.

glacier A permanent flowing mass of snow and ice moving slowly down a slope or valley or spreading outward on a land surface.

gneissosity Coarse compositional banding of alternating rock or mineral layers formed by high-grade regional metamorphism.

Gondwanaland The hypothetical Permian supercontinent of the Southern Hemisphere; the southern part of Pangea.

Gondwanaland strata Similar sequences of Carboniferous to Jurassic rocks found in all continents that were once part of Gondwana.

graben A depressed elongate block of the earth's crust bounded on at least two sides by normal faults; its scenic expression is a rift valley.

gradient Of a stream, loss in elevation per unit of horizontal distance, as in feet per mile.

granite A very hard, visibly crystalline plutonic rock containing at least 10 percent quartz, with potassium feldspars dominant.

granular disintegration Grain-by-grain loss of rock coherence through weathering.

ground moraine Till deposited beneath a glacier and back from its edge or end, forming a gently rolling land surface.

ground motion Displacement of the earth's surface during earthquakes; ground motion is measured by seismographs.

Gutenberg discontinuity The zone of abrupt velocity change of seismic waves at the mantle - core boundary; *P*-wave velocity declines at this discontinuity while *S* waves disappear.

gymnosperm Plant whose seeds are not enclosed in an ovary.

half-life Time required for half of the atoms of a radioactive substance to decay into other isotopes or elements.

hanging valley Tributary valley whose mouth is above the floor of the main glaciated valley. The discordance in elevation is due to the greater erosive capacity of the main glacier.

hanging wall Fault block lying vertically above the fault plane.

hardness Resistance to scratching or indentation of a mineral or other material, as measured on a relative scale.

hardpan See *caliche*.

harmonic tremor Pulsating ground motion caused by alternate expansion and partial collapse of feeder fracture systems beneath a volcano.

headwaters The area where a stream begins; the source of a stream.

herbivore Animal which eats only plants.

hominid Any of a family of bipedal primate mammals that includes human beings, their immediate ancestors, and creatures of related genera.

homogeneous crystallization Continuous crystallization of magma to rock having the composition of the original magma.

horizon Of soil, a distinctive layer more than several inches or centimeters thick; the vertical sequence of horizons forms the soil.

hot spot A persistent surface location from which there is discontinuous magma eruption over periods of millions of years. Hot spots were once believed to remain fixed in absolute location; now there is some evidence to suggest that they may slowly move.

hundred-year flood A flood whose probability of occurrence is 1 percent per year.

Hutton, James Eighteenth-century Scottish geologist who advocated uniformity of process throughout vast geologic time on an earth whose cyclic processes were driven by plutonism for the benefit of humankind.

hydration Of minerals, the addition of water into the internal lattice.

hydrologic cycle Continuous circulation of water from the oceans, lakes, and rivers, through the atmosphere, to the land, and back to the ocean.

hydrolysis Of minerals, the chemical interaction of one mineral with water to create new minerals.

hypocenter Synonym of earthquake *focus*; point of first motion of earthquake.

hypothesis A proposition tentatively assumed in order to draw out its consequences for testing; less well tested than a *theory*.

Ice Age Latest of the epochs of glacial advance; equivalent to Pleistocene Epoch of last 1.6 million years.

igneous rock One that solidified from magma; igneous rocks include plutonic and volcanic rocks.

inclusion, principle of Included fragments are older than surrounding rock.

index fossil A widespread life-form that rapidly evolved; it dates the strata in which it is found.

induction Reasoning from the individual to the universal; deriving fruitful explanations from examination of individual facts.

insectivore Animal which eats insects.

interglacial Time periods between glacial advances.

intermittent stream One which flows for only part of a year; occasionally or seasonally the bed seepage and evaporation exceed the water supply and the stream bed is dry. Streams flowing only after storms may be termed ephemeral.

invertebrate Any animal without a backbone.

ion An atom or a molecule which by loss or gain of one or more electrons has acquired an overall electrical charge.

ionic bond Bond type in which outer shell of electrons is completed by transfer of electrons. In ionic bonding one or more electrons are transferred completely from one atom to another, thus converting the neutral atoms into electrically charged ions.

isochron Of radiometric age dating, a straight line containing all points of the same age.

isostasy, principle of The elevation of crust in buoyant equilibrium is proportional to its density and thickness. This equilibrium is maintained by the yielding flow of asthenospheric material under long-endured gravitational stress.

isotope One of several species of a single element that have different atomic weights.

joint Naturally occurring rock fracture which does not show any displacement of one side of the fracture in relation to the other side. Displacement is perpendicular to the fracture, not parallel to it. Joint surfaces are commonly subparallel to one another and form joint sets.

joint sets A group of subparallel joints in a geologic formation; regional pattern produced by groups of parallel joints.

Jurassic Period From 208 to 144 Ma; the middle period of the Mesozoic Era.

karst An irregular, rolling topography formed on carbonate or gypsum rock having sinkholes, caves, and underground drainage. Some karstic areas have no surface streams, all water flow occurring in subsurface streams along enlarged joints and cavern systems.

kerogen Brownish organic material in sedimentary rocks, especially in shales; kerogen can be distilled into petroleum products.

kettle Steep-walled depression in drift; kettles form when a buried block of stagnant ice melts during glacial retreat.

kettle lake Water-filled kettle.

kinetic energy The energy which a body possesses because of its motion.

komatiite Basalt containing large amounts of magnesium-rich minerals; it is typical of Archean Eon.

laccolith Concordant mass of igneous rock intruded between sedimentary layers and raising a domelike bridge in the overlying strata. The intrusion commonly has a flat base, is lenslike in cross section, and is circular in plan view.

lag gravel See *deflation armor*.

Laramide orogeny Period of deformation in the eastern Rocky Mountains whose phases ranged from Cretaceous to Paleocene.

lahar Mudflow of pyroclastic material on the flank of a volcano. Lahars may occur during eruption or much later.

landscape Series of associated landforms that can be seen in a single view.

landslide Rapid downslope movement of relatively dry unconsolidated soil and rock fragments, usually sliding on a basal plane.

lateral moraine Low ridge of drift formed along the side margin of a glacier.

lava (1) Molten extrusive material which reaches the earth's surface through volcanic vents and fissures. (2) Rock which solidified from molten extrusive material.

lava dome A broad, steep-sided volcano shaped like a flattened dome, formed by the extrusion of very viscous lava.

law (principle) Formal statement of invariant phenomena under specified conditions; uncontested natural truth.

Le Châtelier's principle A system at equilibrium if disturbed will shift so as to regain equilibrium.

left-lateral fault One along which the side opposite to the observer appears to have been moved horizontally and to the left. Compare with *right-lateral fault*.

lignite "Brown coal," intermediate between peat and subbituminous coal.

limb Of a fold, one of its sides.

limestone A sedimentary rock containing more that 50 percent calcium carbonate minerals.

liquefaction The temporary transformation of water-rich sediment or soil into a fluid mass with high pore pressure as cohesion and structure are lost.

lithification Conversion of loose sediments into hard rock.

lithofacies A subdivision of sedimentary deposits distinguished from others by an identifiable character.

lithosphere The part of the earth lying above the asthenosphere and having substantial rigidity. The lithosphere is *not* the same as the *crust*; rather, the lithosphere is composed of both oceanic and continental crust plus a small part of the uppermost mantle.

load Of a stream, the total quantity of material moved by it.

loess Unconsolidated deposit of homogeneous windblown silt, commonly buff to dark brown in color, deflated from glacial valleys and outwash plains. Loess maintains a stable vertical slope, owing to the cohesion of its extremely angular grains which are electrically attracted for one another.

longitudinal dune Long, narrow, low sand ridge oriented parallel to the direction of prevailing wind; occurs in semiarid and arid areas.

longitudinal profile Concave upward profile formed by a line that connects all elevations from the headwaters to the mouth of the stream. Sometimes called long profile or profile of equilibrium.

luster Of a mineral, the quality and overall appearance of its surface as it reflects light.

Lyell, Sir Charles Nineteenth-century English geologist who limited the concept of uniformity by insisting that both rates and processes had remained invariant through all time.

magma Naturally occurring molten mantle or crystal material. It is composed of a mixture of dissolved gases, a viscous fluid commonly enriched in silicon and aluminum, and crystals that are usually various metallic aluminosilicates and quartz. Other compositions are known, including magma whose composition approaches that of limestone!

mammoth Extinct member of the elephant family.

mantle The part of the earth above the core and below the crust.

mantle plume See *hot spot.*

mass extinction Extinction of numerous species within a short interval of geologic time.

mass wasting Dislodgement and downslope movement of unconsolidated soil and rock fragments driven solely by gravity.

matrix Finer-grained material filling the spaces around larger particles in a sedimentary rock.

matter What occupies space and makes up bodies that are perceptible to the senses.

mechanical weathering Disaggregation of rock by physical forces into unchanged smaller fragments.

Mercalli Intensity Scale A twelve-point scale used for measuring the extent of earthquake damage as reported by eyewitnesses. An earthquake detectable only by seismographs is designated I, an earthquake creating total devastation XII.

Mesozic Era From 245 to 66 Ma; the interval of time between the Paleozoic and Mesozoic eras. ''The Age of Reptiles.''

metallic bond The type of chemical bond in all metals. A sea of highly mobile valence electrons free to move throughout the crystal lattices holds the atoms together.

metamorphic processes Those changing solid rocks at temperature and/or pressure levels intermediate between sedimentary and igneous processes.

metamorphic rock Rock changed in the solid state by mineralogical, structural, and chemical processes in response to elevated temperature and/ or pressure and/or shearing stress.

metamorphism Mineralogical and textural adjustment of solid rock to conditions of elevated temperature, shear stress, pressure, or any combination of the three agents.

metamorphism, principle of Rock that is metamorphosed predates any agent that metamorphosed it.

metazoa The multicellular animals which make up the major portion of the animal kingdom; cells are differentiated into tissues and organs.

methanogen Bacterium whose energy is derived from respiration of methane; requires completely anaerobic conditions for growth.

mica Silicate mineral group having intermediate hardness and perfect basal cleavage, which allows the mineral to be readily separated into very thin, somewhat flexible sheets.

microplate tectonics Interaction of small plates sutured within a larger plate.

mineral Naturally formed crystalline chemical compound having physical properties and chemical formulae that are invariant or vary within narrow limits.

mineralogy Science which studies the occurrence, structure, form, and chemistry of minerals.

miogeocline Linear depositional area adjacent to a coastline on a passive margin. Older synonym: *miogeosyncline.*

Mississippian Period From 360 to 320 Ma; in North America the first part of the Carboniferous Period.

modulus Number expressing the degree to which a property is possessed by a substance. Examples: bulk modulus, shear modulus.

Moho Informal or common name for Mohorovičić discontinuity.

Mohorovičić discontinuity Briefly, the boundary between crust and mantle. The boundary, a planar surface ranging from 3 to 5 miles (5 to 8 kilometers) beneath the ocean basin floor to 20 to 60 miles (32 to 96 kilometers) beneath the continental surface, is recognized by an abrupt increase in seismic velocity of body waves, thought to result from an abrupt shift in chemical composition and mineral density between the crust and the mantle.

Mohs' hardness scale A scale measuring *relative* resistance of minerals to being scratched. The scale ranges from 1, the value given to the softest mineral, talc, to 10 for the hardest, diamond.

molding, principle of Material formed around a mold is younger than the object molded.

molecule The simplest, smallest structural unit of an element or a compound having all its physical and chemical properties.

mud cracks Dessication cracks formed through tensile stress of shrinkage on the surface of drying mud; mud cracks are polygonal in plan view and wedge-shaped in cross section.

mudflow Rapid downslope movement of water-soaked, fine-grained soil and rock fragments (regolith).

mudstone Bulky, extremely fine-grained sedimentary rock lacking fissility and rich in clay minerals.

natural gas Gaseous hydrocarbons generated within the earth; methane is commonly the dominant gas.

natural levee An elongate embankment made up of sand and silt deposited along both stream banks, formed during the sudden velocity drop when rising waters in a flooding stream suddenly spill out of their normal channel onto the much wider floodplain. See *discharge, floodplain.*

neocrystallization Formation of new crystalline material during metamorphism.

neptunist One who believes that all crustal rocks consist of material deposited or crystallized from water.

neutron An uncharged composite particle within the atomic nuclei of all elements except ordinary hydrogen. Its mass is almost identical to that of the proton. Like the proton, we now know that the neutron is not truly elementary, but consists of combinations of still smaller particles called quarks, each of which carries a fractional unit of electric charge.

Nevadan orogeny Deformation in the western North American Cordillera between Early Jurassic and Early Cretaceous time.

nonconformity An eroded surface in igneous or metamorphic rocks that is overlain by sedimentary rocks.

nonfoliated Having a homogeneous crystalline rock texture.

normal fault Fault in which the hanging wall has moved downward relative to the footwall.

nucleus The tiny (10^{-14} meter), dense, positively charged central core of an atom. The nucleus is composed of neutrons and protons held together by the strong force and contains almost all the atomic mass.

nueé ardénte Swiftly flowing turbulent and eruptive cloud of volcanic ash intermixed with hot, often incandescent gases.

oceanic crust Basaltic crust underlying the ocean basins and formed at oceanic ridge zones.

oceanic ridge Continuous mountain range on the ocean floor at the edge of diverging plate margins. The ridge exhibits high heat flow, frequent shallow seismicity, a large central rift, and abundant underwater volcanism.

oil Naturally formed complex liquid hydrocarbon, often associated with natural gas.

oil shale Finely layered drab-colored shale which contains *kerogens.*

olivine Common magnesium–iron silicate mineral.

ophiolite suite Group of dark, sodium-rich, low-silica igneous rocks and low-grade metamorphic rocks in association with gabbros and radiolarian cherts.

Ordovician Period From 505 to 438 Ma; the second oldest period in the Paleozoic Era.

original continuity Principle which asserts that identical strata separated across a canyon were continuous before canyon cutting.

original horizontality Principle which asserts that sedimentary deposits are laid down in originally horizontal blankets of sediment, so that bedding or stratification mark a former horizontal surface.

orogenic belt A linear zone of folded, faulted, metamorphosed, plutonized, and abnormally thick sedimentary rocks. Orogenic belts are formed by deformation at interacting plate margins.

orogeny A discontinuous deformational event which forms the internal structures of mountains over periods of millions of years. The processes occurring during an orogeny include folding, faulting, metamorphism, plutonism, volcanism, crustal extension, marginal subduction, formation of foredeeps and island arcs, back-arc spreading, crustal depression or uplift, creation of seismic zones, and other energetic processes.

orthoclase feldspar A common potassium aluminum silicate mineral.

Ouachita orogeny Late Paleozoic deformation on the southern margin of North America.

outwash plain Broad, gently sloping till sheet deposited beyond the glacial margin by meltwater streams.

overturned fold One whose limb is tilted beyond the perpendicular.

oxidation Of minerals, chemical weathering through interaction with oxygen with minerals.

pahoehoe Cooled hard lava flow having an undulating, often billowy surface. Pahoehoe is the Hawaiian word for ropy-surfaced lava.

paleomagnetics Study of the intensity and direction of the earth's past magnetic field as recorded in susceptible rocks.

paleontologist A geologist who studies the origin, ecology, evolution, and distribution of past life through fossil remains.

Paleozoic Era From 570 to 245 Ma; the oldest era containing abundant multicellular life with skeletons.

Palisadian event Triassic rifting in eastern North America.

Pangea Proposed Late Paleozoic supercontinent whose Mesozoic separation sent all the parts of modern continents to their present positions after plate tectonic movement.

partial melting Melting of the components with lower melting points from a complex solid.

passive fold A fold formed by differential offset of layering by distributive flow across it.

passive margin Aseismic continental margin where deposition is the dominant process.

pedalfer Soil group typically formed in a humid environment; pedalfers are rich in aluminum and iron.

pediment A broad, gently sloping bedrock surface of transportation cut back into the front of a much steeper mountain slope in a desert region. The pediment is usually covered with a thin veneer of alluvial gravel and sand.

pedocal Soil group typically formed in semiarid to arid environment; pedocals are rich in calcium and aluminum and usually contain a calcium carbonate *hardpan*.

Pennsylvanian Period From 320 to 286 Ma; the younger subdivision of the Carboniferous Period.

perched water table An elevated water table of perched water, which is groundwater that is unconfined and separated from underlying groundwater by an impermeable zone or layer.

percolation Slow movement of water through a permeable material.

perennial stream Stream which flows year around. Synonym: *permanent stream*.

Period An interval of geologic time that can be subdivided into epochs and combined into eras.

periodic table of elements Elements arranged in a matrix based on increasing atomic numbers. A pattern of periodic changes in physical and chemical properties emerges.

permeability The percentage of *interconnected* pore spaces in a rock.

permeable Having interconnected pores or opening that permit liquid or gases to pass through.

Permian Period From 286 to 245 Ma; the youngest period of the Paleozoic Era.

petroleum See *oil*.

phenocryst A large, conspicuous crystal in a fine-grained crystalline rock.

pH scale Measure of the acidity or alkalinity of a system. The scale ranges from 1, extremely acid, to 14, extremely alkaline, with 7 being neutrality.

physical properties Fixed characteristics of a substance that are related to its composition. Examples: hardness, color.

plagioclase feldspar A common group of calcium and sodium aluminum silicates.

plate In plate tectonics, a lithospheric fragment bounded by seismic zones in map view and the asthenosphere in cross section.

plate tectonics A theory proposing that the earth's lithosphere is subdivided into semirigid plates in constant relative motion. These plates float more or less independently on a viscous layer (the asthenosphere) in the mantle. The plate margins grind against one another and are the seat of most deformation, earthquakes, and volcanic activity. Plate tectonics is the process by which the earth cools itself by convection within the mantle (and outer core?).

playa The flat-floored bottom of an enclosed desert basin which occasionally becomes a shallow lake. Playas are often encrusted with highly soluble, generally whitish, alkalic minerals produced by repetitive evaporation of the shallow playa lake.

Pleistocene Epoch The last 1.6 million years less the last ten thousand years; equivalent to the Ice Age.

plutonic rock Coarsely crystalline igneous rock formed by slow underground cooling of magma.

plutonism General term for all phenomena associated with formation and emplacement of magma.

plutonist One who believes that the earth formed by the solidification of a molten mass, and that its continuing supply of internal heat drives uplift and orogenic processes. The theory of plutonism was proposed by James Hutton as an alternative to neptunism.

pluvial lake A lake formed in warmer climates during a period of exceptionally heavy rainfall as a glacier advance, especially during the Pleistocene.

polar wandering Path of the loci of the earth's magnetic poles as determined from rocks of increasing age. This plot of the fixed position of the pole records the relative motion of continents through time with respect to that fixed point.

porosity The percentage of pore space in a rock; the ratio of the volume of pore space to the volume of the rock.

porous Having numerous internal open spaces.

porphyritic Pertaining to an igneous rock having distinctly coarse crystals (phenocrysts) enclosed in a fine-grained matrix.

postdict To predict what past events or processes have occurred.

potential energy The energy which a piece of matter has because of its position.

Precambrian From consolidation of the earth's crust approximately four billion ears ago to 570 Ma; the interval of time prior to the Cambrian Period.

precipitation The discharge of water in any form from the sky.

pressure solution Selective dissolution of areas under pressure.

primate The most highly developed order of mammals. The creatures have a generalized limb structure and dentition, advanced finger mobility, stereoscopic binocular vision, and an enlarged, differentiated brain.

principle An explanatory statement accepted with the highest level of confidence based on its extensive testing; current statement of truth.

Proboscidean A group of herbivorous animals which have a nose elongated down into a trunk, incisors enlarged to become tusks, and pillarlike legs.

profile of equilibrium See *longitudinal profile*.

prokaryote A cell which does not have a distinct nucleus. The area containing dioxyribonucleic acid does not have a limiting membrane. Only cyanobacteria and bacteria are prokaryotic. For contrast, see *eukaryote*.

Proterozoic Eon From 2500 to 570 Ma; the younger part of the Precambrian.

proton The positively charged composite particle found in all atomic nuclei. Its diameter is about 10^{-15} meter.

P **wave** Seismic rarefaction and compression wave.

pyroclastic Composed of particles of rock, mineral, and glass fragmented through explosive volcanic action. Pyroclastic translates as "fire-broken." If the fragments are welded together as they fall out of a *nuee ardente* eruption, they form a unique volcanic rock termed a *welded tuff* or *ignimbrite*.

pyroxene Group of common complex metallic silicate minerals having moderate hardness and cleavage planes whose intersection is at angles of 87° and 93°.

quartz Mineral composed of silicon dioxide; the most abundant and widespread of all minerals. Its Mohs' hardness of 7 makes it the hardest common mineral; when growth conditions allow, it forms lustrous six-sided prisms which range in color from water clear to jet black.

Quaternary Period The last 1.6 million years of earth history.

quicksand A noncohesive mixture of sand saturated with water, yielding easily to pressure and readily engulfing objects on its surface.

radioactive decay See *radioactivity*.

radioactivity The property possessed by some elements of spontaneously emitting alpha, beta, and gamma rays by the disintegration of the atomic nuclei.

radiometric dating Calculating an age in years based on the ratio of the abundance of a long-lived radioactive element to the abundance of its decay product.

recharge Addition of water to the zone of saturation.

recrystallization The formation of new crystals having the same chemical composition as the original rock.

recumbent fold An overturned fold whose axial plane is nearly horizontal.

reflection In seismology, the return of some seismic wave energy from a discontinuity surface on which it is incident. The returning energy is reflected back through the medium it originally traversed and is recorded by a *seismograph*.

refraction In seismology, the deflection of a ray path of seismic energy as the ray path passes into an earth layer having different physical properties that cause its velocity to speed up or slow down. As the speed of passage varies, the ray path traced by the passage of seismic energy is bent or refracted from its initial orientation, producing curving ray paths through the earth's interior.

regional metamorphism See *dynamothermal metamorphism*.

regolith The blanket of loose soil and unconsolidated rocky debris of any thickness that overlies bedrock and forms the earth's surface.

regression The retreat of the sea from the land over thousands or millions of years.

rejuvenation Of a stream, stimulation to renewed erosion owing to a decline in base level and/or an uplift of the land surface.

relict bedding Preserved sedimentary bedding in metamorphic rocks.

relief The difference in elevation between the highest and lowest points on the earth's surface in a landscape.

reservoir rock A permeable rock capable of storing fluid; often denotes a rock containing oil or gas in appreciable quantities.

reverse fault One in which the hanging wall has moved up relative to the footwall.

rhyolite A high-silica volcanic rock equivalent in composition to granite.

Richter Magnitude Scale A measure of the amplitude of ground motion during an earthquake. Each increase in number value corresponds to a tenfold increase in ground motion and very, very roughly to a thirtyfold increase in released energy. A magnitude 2 earthquake can barely be felt; a magnitude 9 earthquake is the greatest ever to have occurred.

rift A linear valley created by crustal extension—by the collapse of a fractured block between two faults caused by tensile stress. Rift valleys often exhibit high heat flow, abundant microseismic activity, and abundant basaltic volcanic activity.

right-lateral fault One along which the side opposite to the observer appears to have moved horizontally and to the right. See *left-lateral fault*.

ripple marks An undulating surface pattern on a bedding plane formed when oscillating wave energy or current flow disturbed the soft sediment.

rock A naturally formed substance having some coherence and composed of minerals and other crystalline, fragmental, glassy, and organic materials.

rock cycle The repetitive sequence of energetic events that convert one major rock type to another through time.

rockfall A free fall of newly detached rock debris from a vertical face.

rock-forming minerals The minerals which frequently occur in significant amounts in common rocks.

rock glacier A flowing mixture of snow, ice, and rock debris. Its surface is made up of boulders and fine material cemented by ice.

rock slide Rapid downslope turbulent motion of newly detached bedrock segments from a weakened surface.

rodent Small mammal which has a pair of large, continuously growing incisors in each jaw.

roof rock Impervious rock which provides a seal over a reservoir rock.

saltation Transport of sediment in which particles are moved forward in a series of intermittent leaps as they bounce along a surface; a common mode of *deflation*.

sandstone Sedimentary rock composed dominantly of sand-sized grains.

schistosity Strongly parallel layering formed during total recrystallization of a parent rock into a moderately coarse-grained metamorphic rock.

science Organized, tested knowledge of the natural world.

seafloor spreading Formation of new seafloor by shouldering aside and intrusion of existing seafloor at oceanic ridges, causing the seafloor to move perpendicular to the ridge axes.

sediment Unconsolidated fragmented material which comes from weathering of rock and is carried and deposited by wind, water, and ice.

sedimentary rock A rock made up of naturally consolidated sediment or crystalline material deposited in layers in water. Only sedimentary rocks may contain *fossils*.

seiche Standing wave oscillations created in water trapped in an enclosed basin through the disturbance of the surrounding basin by an earthquake.

seif dune See *longitudinal dune*.

seismic moment A measure of the total ground motion produced by an earthquake along the length of the plane of failure.

seismogram The record of ground motion.

seismograph The device which records ground motion as a seismogram.

seismology The branch of geology which studies earthquakes.

sequence Of cratonic sedimentary rocks, a packet of those rocks, bounded by major continent-wide unconformities.

sequential time The time referred to in designating when a specific event happened in comparison to another event. Also termed *relative time*. Examples: tomorrow, yesterday, next winter, last century.

Sevier orogeny Mesozoic deformation along the eastern edge of the Great Basin.

shadow zone Surface area of the earth, between 103° and 143° from the epicenter of an earthquake, which receives no seismic waves because of refraction of the *P* waves and absorption of the *S* waves by the outer core.

shale Fissile sedimentary rock composed largely of clay minerals.

shear joint Joint formed by surfaces which intersect at acute angles and along which there has been movement in response to shearing stresses.

shear modulus Measure of a material's rigidity or resistance to change in shape.

shear stress Opposing stresses acting *along* a plane. More formally, opposing stresses acting parallel to a plane that is also pressured by other opposing forces acting perpendicular to it.

shield See *craton*.

shield volcano A broad, low dome having very gentle slopes and formed from low-silica less viscous lava.

shock metamorphism Deformation produced by hypervelocity impact. The complete and permanent shattering and shocking of the impact site produced by transient, hypervelocity, high-pressure waves.

silicates In geology, the most common group of minerals; composed of silicon and oxygen or silicon, aluminum, and oxygen.

sill A tabular igneous concordant intrusion into layered rock; the boundaries of the intrusive body are parallel with the planes of the enclosing rock.

siltstone Sedimentary rock composed dominantly of cemented silt-sized sediment.

Silurian Period From 438 to 408 Ma; post-Ordovician and pre-Devonian time.

similarity, principle of Items of identical appearance have a common origin.

sinkhole A roughly circular surface depression in a region of *karst* topography, produced by a collapse of a cave roof. Sinkholes whose closed bottoms are below the water table form sinkhole lakes or sinks.

slaty cleavage A finely laminated structure unique to slate; slaty cleavage allows relatively easy splitting of slate into thin slabs much used in the billiard table and construction industries. The cleavage presumably took place during expulsion of water from highly pressured muddy sediment. Other genetic processes have also been proposed.

slide Mass movement of earth, rock, or snow along a distinct flat basal plane of failure.

slip face Steeply sloping surface on the lee side of a dune caused by sand standing at its angle of repose.

slump Downward and outward rotational movement of a mass of poorly consolidated surface material along a concave upward spoon-shaped failure plane.

snow line The elevation above which snow remains year around.

soil The unconsolidated and usually weathered surface material of the earth made up of mineral and organic matter and capable of sustaining plant roots; mature soils have distinct separations into vertical layering or *horizons*.

soil profile A vertical section of a soil showing its parent material and horizons.

solifluction The mass movement of soggy soil down a slope, caused by a seasonal thaw of upper soil above an impermeable permafrost horizon.

sorting (1) The degree to which all fragments in a sedimentary rock are of similar diameter. (2) The processes in the sedimentary environment which sort fragments of similar size, shape, and weight into separate deposits.

source rock The sedimentary rock whose included organic content was later transformed into oil.

specific gravity The ratio of the density of a substance to the density of pure water at 4°C.

spreading center A divergent plate margin. See *ocean ridge*.

spring A flow of groundwater from an area where the water table meets the land surface.

stable interior Continental area of old seismically stable igneous and metamorphic rock overlain by sedimentary rock.

star dune A dune form which when seen from above has several small sand ridges radiating from a common center.

Steno, Nicolaus Recognized similarity of interfacial angles on minerals; proposed laws of similarity and molding.

strain Deformation produced by stress.

stratification The common rock structure produced by depositing sediment in sheetlike layers.

stratigraphic facies Rock units distinguished by form, the nature of their boundaries, and their mutual relations.

stratovolcano See *composite cone*.

stratum A single tabular or sheetlike mass of sedimentary rock of one kind formed by natural causes and made up usually of a series of layers lying between beds of other kinds. A stratum is bounded by bedding planes. Plural: strata.

streak The color of a powdered mineral obtained by rubbing the mineral on a white unglazed porcelain plate.

stream erosion The incision of the land surface and the transportation of load by streams.

strength Resistance of material to loss of cohesion or shape.

stress The magnitude of a force acting across a unit area. Examples: pounds per square inch, dynes per square centimeter.

strike-slip fault One whose direction of movement has been horizontal along the fault plane.

stromatolite Characteristic domal matlike calcareous deposit formed by cyanobacteria which both trap sediment and produce calcareous laminae.

subduction zone A sloping area along collisional margins where continental crust overrides and consumes oceanic crust.

succession of facies Observation that facies succeed one another in vertical order which parallels the horizontal succession of facies.

superposition, principle of In an undisturbed sequence of sedimentary rocks, the older rocks are on the bottom.

surface faulting Faulting which offsets the earth's surface.

surface wave A seismic wave whose travel is confined to the earth's crust.

suspended load The load carried in suspension owing to turbulence of a stream.

S wave Seismic wave which exposes material to elastic shear strain as it passes.

symmetry Of a crystalline substance, the repeating pattern of crystal faces that indicates the solid geometry of its ordered internal arrangement.

syncline A downfold whose limbs dip toward each other and intersect in a trough.

Taconian orogeny Late Ordovician to Early Silurian deformation in the northern Appalachians.

talus Coarse, angular rock fragments derived by rockfall from a steep cliff and accumulated in an apron at its base.

tarn A small lake occupying a cirque.

tar sand Sand deposit whose pores contain heavy residual oils and asphalt, as the lighter portions have already escaped.

technology The practical or industrial arts; the processes that attempt to control or limit the natural world.

tensional stress Force per unit area that tends to cause separation of the material across the plane on which it acts.

terminal moraine A low ridge of drift marking the limit of glacial advance.

Tertiary Period From 66 to 1.6 Ma; the oldest of the two periods in the Cenozoic Era.

thecodonts Reptiles which evolved during the Triassic Period and ran on their hind legs. Thecodonts were ancestors to the dinosaurs.

theory A generally accepted explanation of sensed data; theories have a higher level of certainty than hypotheses and a lesser level of certainty than a *law or principle.*

therapsids Mammal-like reptiles which evolved in mid-Permian time and had complex teeth and legs positioned underneath their body. Therapsids were ancestors to the mammals.

thermal metamorphism Rock change owing to the elevated temperature adjacent to an igneous intrusion.

thixotropy Property of some colloidal substances which abruptly liquefy when shaken. Examples: house paint, quicksand.

thrust fault Reverse fault whose fault plane angles down less than 30° from the horizontal; formed in environments of strong horizontal compressive stress.

time A system of reckoning duration or interval from a recurring reference point.

transform fault A strike-slip fault normal to and connecting oceanic ridge axes offset by the fault.

transgression The advance of the seas across the land over thousands or millions of years.

transverse dune A sand deposit elongated normal to the direction of prevailing winds with a gentle windward slope and a steep leeward slope.

trap Any structural or stratigraphic change that limits fluid migration and causes the accumulation of fluids, like oil, having commercial value.

Triassic Period From 245 to 208 Ma; the oldest of three Periods in the Mesozoic Era.

trough lake A linear lake occupying a glacially eroded stream valley floor.

truncated spur Projection into a preglacial valley that was beveled as the glacier passed and widened, deepened, and straightened the valley.

tsunami A seismic sea wave having low amplitude, long period, high speed propagation, and long wavelength in the open ocean. Tsunamis are caused by a submarine earthquake which abruptly distorts the shape of the ocean basin. When tsunamis approach shallow coastal areas, they may pile up to great heights and cause substantial damage.

turbidite A sedimentary rock consisting of material that was carried down the continental slope onto the deeper ocean floor by a sudden turbid flow kicked off by an earthquake. Turbidity flows have been observed to travel downslope at speeds approaching that of race cars.

ultimate base level Elevation below which erosion cannot occur. For a stream this is sea level; for a turbidity flow, ultimate base level is the deepest ocean floor.

unconformity A buried erosional surface that creates a gap in an otherwise continuous rock record of geologic time. The three types of unconformity are *angular unconformity, disconformity, or nonconformity.*

ungulates A large group of herbivorous mammals which possess hooves. Even-toed ungulates include all cattle, sheep, goats, pigs, and so on; odd-toed ungulates include horses, tapirs, and rhinos.

uniformitarianism Principle asserting that past geologic events can be explained by processes operating today; earth processes and natural laws have remained invariant throughout geologic time. According to modern understanding, rates and intensities of geologic processes have not necessarily remained invariant through geologic time, though the relation between process and product has remained invariant.

uniformity, principle of Method by which geologists attempt to reconstruct the past in terms of modern processes; the principle is often stated as "The present is the key to the past."

unit cell The simplest polyhedron which forms a crystal lattice by regular repetition in space.

upright fold A fold whose axial plane is essentially vertical.

valley glacier One confined by stream valley walls. Typically valley glaciers form above *snow line* in mountainous areas. Synonym: *alpine glacier*.

valley train Elongate body of alluvium deposited beyond the margin of a glacier by meltwater streams.

Van der Waals bond The weakest form of atomic bonding, caused by the electrostatic attraction of each atomic nucleus for electrons of adjacent nuclei, diminished by the repulsion of adjacent electrons.

velocity The rate of travel per unit of time, as in miles per hour or meters per second.

vent Channelway or opening within a volcano from which volcanic materials are extruded.

vertebrate Animal having a backbone or spinal column.

viscosity The measure of resistance to flow; numerically, viscosity is the ratio of applied shear stress to observed rate of flow.

viscous strain Strain whose rate of flow is directly proportional to the applied shear stress.

volatiles Substances which are easily vaporized.

volcanic neck A column of igneous rock formed when lava congealed in the conduit of a volcano.

The neck records the shape of the plumbing system that fed the vent and may later be left standing alone above the adjacent country when erosion removes adjacent rocks.

volcanic rock Finely crystalline to glassy rock formed by the crystallization of lava on or near the earth's surface.

volcanism Processes by which magma is extruded onto the earth's surface.

volcano A broadly conical hill created by the eruption of lava and pyroclastic material from its vent(s).

water table The surface separating the zones of aeration and saturation; the upper limit of the portion of the ground wholly saturated with water.

wave In seismology, a repetitive disturbance to elastic material created by the passage of seismic energy. Waves have a measurable amplitude, wavelength, period or frequency, and velocity.

wave velocity The speed of passage of the wave front past a given point.

weathering Mechanical disintegration and chemical decomposition of earthy and rocky materials through exposure to the elements and to life-forms. Weathering brings rock into equilibrium with the earth's surface environment without transportation of the weathered material.

weathering zone The part of the earth's surface affected by weathering.

Wilson cycle Cycle of spreading (opening) and consumption (closure) of ocean basins as continental margins change from being passive depositional margins to being active consumptive margins.

wind erosion Carving of the earth's surface and transportation of debris by wind. Synonym: *deflation*.

xenolith An accidental fragment included in an igneous rock. Xenolith fragments are torn away from rocks surrounding both intrusive and extrusive centers.

yield stress Stress level above which strain becomes permanent.

zonal soil Soil which contains detectable horizons.

zone of aeration The part of an aquifer whose pores are not saturated with water.

zone of accumuluation Soil horizon(s) which accumulate material leached from overlying horizons; for many soils the B horizon is the zone of accumulation.

zone of saturation The part of an aquifer whose pores are saturated with water; if the aquifer is *permeable* and pressurized, *percolation* ensues.

INDEX